D1498837

Methods in Microbiology
Volume 26

Series Advisors

Gordon Dougan Department of Biochemistry, Wolfson Laboratories, Imperial College of Science, Technology and Medicine, London, UK

Graham J Boulnois Zeneca Pharmaceuticals, Mereside, Alderley Park, Macclesfield, Cheshire, UK

Jim Prosser Department of Molecular and Cell Biology, Marischal College, University of Aberdeen, Aberdeen, UK

Ian R Booth Department of Molecular and Cell Biology, Marischal College, University of Aberdeen, Aberdeen, UK

David A Hodgson Department of Biological Sciences, University of Warwick, Coventry, UK

David H Boxer Department of Biochemistry, Medical Sciences Institute, The University, Dundee, UK

Methods in Microbiology

Volume 26
Yeast Gene Analysis

Edited by

Alistair JP Brown

Department of Molecular and Cell Biology
Institute of Medical Sciences
University of Aberdeen
Aberdeen, UK

and

Mick Tuite

Department of Biosciences
University of Kent at Canterbury
Canterbury
Kent, UK

ACADEMIC PRESS
San Diego London Boston
New York Sydney Tokyo Toronto

Copyright © 1998 by ACADEMIC PRESS, except Chapter 7, pp. 101–118, which is a US Government work in the public domain and is not subject to copyright.

Academic Press
525 B Street, Suite 1900, San Diego, California 92101-4495, USA
http://www.apnet.com

Academic Press Limited
24–28 Oval Road, London NW1 7DX, UK
http://www.hbuk.co.uk/ap/

A catalogue record for this book is available from the British Library

ISBN 0-12-521526-6 hardback
ISBN 0-12-136655-3 paperback
ISSN 0580-9517

Typeset by Phoenix Photosetting, Lordswood, Chatham, Kent
Printed in Great Britain by WBC Book Manufacturers Ltd, Bridgend, Mid Glamorgan
98 99 00 01 02 03 WB 9 8 7 6 5 4 3 2 1

Contents

Contributors

F Baganz Department of Biomolecular Sciences, UMIST, Sackville Street, PO Box 88, Manchester M60 1QD, UK

D Bailey Incyte Pharmaceuticals, 3174 Porter Drive, Palo Alto, California 94304, USA

M Bolotin-Fukuhara Laboratoire de Génétique Moléculaire, Institut de Génétique et Microbiologie (URA 2225 du CNRS), Université Paris-Sud, 91405 Orsay Cedex, France

A Brachat Institut für Angewandte Mikrobiologie, Biozentrum, Universität Basel, Klingelbergstrasse 70, CH-4056 Basel, Switzerland

SCC Brekelmans Fungal Cell Wall Group, Institute for Molecular Cell Biology, University of Amsterdam, Kruislaan 318, NL-1098 SM Amsterdam, The Netherlands

AJP Brown Department of Molecular and Cell Biology, Institute of Medical Sciences, University of Aberdeen, Foresterhill, Aberdeen AB25 2ZD, UK

LHP Caro Fungal Cell Wall Group, Institute for Molecular Cell Biology, University of Amsterdam, Kruislaan 318, NL-1098 SM Amsterdam, The Netherlands

MJ Curcio Wadsworth Center for Laboratories and Research, New York State Department of Health, PO Box 22022, Albany, New York 12201-2002, USA

M-C Daugeron Département de Biochimie Médicale, Centre Médical Universitaire, 1 rue Michel Servet, CH-1211 Genève 4, Switzerland

J de la Cruz Département de Biochimie Médicale, Centre Médical Universitaire, 1 rue Michel Servet, CH-1211 Genève 4, Switzerland

K-D Entian Institute for Microbiology, J-W Goethe-Universität, Marie-Curie-Strasse 9, D-60439 Frankfurt, Germany

C Friddle Department of Genetics, Stanford University, Stanford, California 94305, USA

D Frishman Munich Information Center for Protein Sequences, Max-Planck-Institute for Biochemistry, 82152 Martinsried, Germany

LM Furness Central Research, Molecular Sciences, Pfizer, Sandwich, Kent, UK

DJ Garfinkel ABL-Basic Research Program, Movable Genetic Elements Section, Gene Regulation and Chromosome Biology Laboratory, PO Box B, Building 539, Frederick, Maryland 21702-1201, USA

RD Gietz Department of Human Genetics, University of Manitoba, 770 Bannatyne Avenue, Winnipeg, Manitoba, Canada R3E 0W3

IM Hagan School of Biological Sciences, 2.205 Stopford Building, University of Manchester, Oxford Road, Manchester M13 9PT, UK

K Heumann Munich Information Center for Protein Sequences, Max-Planck-Institute for Biochemistry, 82152 Martinsried, Germany

JMX Hughes Posttranscriptional Control Group, Department of Biomolecular Sciences, UMIST, PO Box 88, Manchester M60 1QD, UK

JC Kapteyn Fungal Cell Wall Group, Institute for Molecular Cell Biology, University of Amsterdam, Kruislaan 318, NL-1098 SM Amsterdam, The Netherlands

FM Klis Fungal Cell Wall Group, Institute for Molecular Cell Biology, University of Amsterdam, BioCentrum Amsterdam, Kruislaan 318, NL-1098 SM Amsterdam, The Netherlands

N Koloteva Posttranscriptional Control Group, Department of Biomolecular Sciences, University of Manchester Institute of Science and Technology (UMIST), PO Box 88, Manchester M60 1QD, UK

P Kötter Institute for Microbiology, J-W Goethe-Universität, Marie-Curie-Strasse 9, D-60439 Frankfurt/M, Germany

P Linder Département de Biochimie Médicale, Centre Médicale Universitaire, 1 rue Michel Servet, CH-1211 Genève 4, Switzerland

EJ Louis Yeast Genetics, Institute of Molecular Medicine, John Radcliffe Hospital, Oxford OX3 9DS, UK

PT Magee Department of Genetics and Cell Biology, University of Minnesota, St Paul, Minnesota 55108, USA

WH Mager Department of Biochemistry and Molecular Biology, IMBW, Biocentrum Amsterdam, Vrije Universiteit, NL-1081 HV Amsterdam, The Netherlands

H Martín Departamento de Microbiologia II, Facultad de Farmacia, Universidad Complutense, 28040 Madrid, Spain

K Maurer Department of Biochemistry and Molecular Biology, IMBW, Biocentrum Amsterdam, Vrije Universiteit, NL-1081 HV Amsterdam, The Netherlands

JEG McCarthy Posttranscriptional Control Group, Department of Biomolecular Sciences, UMIST, PO Box 88, Manchester M60 1QD, UK

H-W Mewes Munich Information Center for Protein Sequences, Max-Planck-Institute for Biochemistry, 82152 Martinsried, Germany

M Molina Departamento de Microbiologia II, Facultad de Farmacia, Universidad Complutense, 28040 Madrid, Spain

RC Montijn Fungal Cell Wall Group, Institute for Molecular Cell Biology, University of Amsterdam, Kruislaan 318, NL-1098 SM Amsterdam, The Netherlands

PA Moore Human Genome Sciences, Department of Molecular Biology, 9410 Key West Avenue, Rockville, Maryland 20850, USA

C Nombela Departamento de Microbiología II, Facultad de Farmacia, Universidad Complutense, 28040 Madrid, Spain

SG Oliver Department of Biomolecular Sciences, UMIST, Sackville Street, PO Box 88, Manchester M60 1QD, UK

P Philippsen Institut für Angewandte Mikrobiologie, Biozentrum, Universität Basel, Klingelbergstrasse 70, CH-4056 Basel, Switzerland

PW Piper Department of Biochemistry and Molecular Biology, University College London, London, UK

RJ Planta Department of Biochemistry and Molecular Biology, Vrije Universiteit, De Boelelaan 1083, NL-1081 HV Amsterdam, The Netherlands

P Pokorni Mikrobiologisches Institut, ETH Zürich, CH-8029 Zürich, Switzerland

AFJ Ram Fungal Cell Wall Group, Institute for Molecular Cell Biology, University of Amsterdam, Kruislaan 318, NL-1098 SM Amsterdam, The Netherlands

C Rebischung Institut für Angewandte Mikrobiologie, Biozentrum, Universität Basel, Klingelbergstrasse 70, CH-4056 Basel, Switzerland

GS Roeder Department of Biology, Yale University, PO Box 208103, New Haven, Connecticut 06520, USA

J Rosamond School of Biological Sciences, 2.205 Stopford Building, University of Manchester, Oxford Road, Manchester M13 9PT, UK

P Ross-Macdonald Department of Biology, Yale University, PO Box 208103, New Haven, Connecticut 06520, USA

M Sánchez Departamento de Microbiologia II, Facultad de Farmacia, Universidad Complutense, 28040 Madrid, Spain

A Sheehan Department of Biology, Yale University, PO Box 208103, New Haven, Connecticut 06520, USA

V Smith Department of Molecular Biology, Genentech Inc., 460 Point San Bruno Boulevard, South San Francisco, California 94080, USA

M Snyder Department of Biology, Yale University, PO Box 208103, New Haven, Connecticut 06520, USA

I Stansfield   Department of Molecular and Cell Biology, Institute of Medical Sciences, University of Aberdeen, Foresterhill, Aberdeen AB25 2ZD, UK

MJR Stark   Department of Biochemistry, University of Dundee, Dundee DD1 4HN, UK

S Steiner   Max-Planck-Institut für Züchtungsforschung, D-50829, Köln-Vogelsang, Germany

CJ Stirling   School of Biological Sciences, 2.205 Stopford Building, University of Manchester, Oxford Road, Manchester M13 9PT, UK

S te Heesen   Mikrobiologisches Institut, ETH Zürich, CH-8029, Switzerland

B Teusink   EC Slater Institute, BioCentrum, University of Amsterdam, Plantage Muidergracht 12, NL-1018 TV, Amsterdam, The Netherlands

MF Tuite   Department of Biosciences, University of Kent, Canterbury, Kent CT2 7NJ, UK

MAA Van Berkel   Fungal Cell Wall Group, Institute for Molecular Cell Biology, University of Amsterdam, Kruislaan 318, NL-1098 SM Amsterdam, The Netherlands

H Van den Ende   Fungal Cell Wall Group, Institute for Molecular Cell Biology, University of Amsterdam, Kruislaan 318, NL-1098 SM Amsterdam, The Netherlands

JH Vossen   Fungal Cell Wall Group, Institute for Molecular Cell Biology, University of Amsterdam, Kruislaan 318, NL-1098 SM Amsterdam, The Netherlands

A Wach   Institut für Angewandte Mikrobiologie, Biozentrum, Universität Basel, Klingelbergstrasse 70, CH-4056 Basel, Switzerland

HV Westerhoff   Department of Microbial Physiology, BioCentrum, Free University, De Boelelaan 1087, NL-1081 HV Amsterdam, The Netherlands

RA Woods   Department of Biology, University of Winnipeg, 515 Portage Avenue, Winnipeg, Manitoba, Canada R3B 2E9

A Zollner   Munich Information Center for Protein Sequences, Max-Planck-Institute for Biochemistry, 82152 Martinsried, Germany

Preface

The completion of the DNA sequence of the genome of the yeast *Saccharomyces cerevisiae* in the Spring of 1996 heralded the completion of a major international collaboration (Goffeau *et al.*, 1997). Although a major milestone in modern biological research, it is simply the first step in the long road to the complete understanding of the genetic make-up of this simple eukaryotic cell. The challenge we now face is to assign a biological function to the products of each of the 6000+ genes identified by computer analysis of the nucleotide sequence of the 16 yeast chromosomes. With the experimental analysis of gene function in *S. cerevisiae* having been on-going for some three decades, such an analysis will not start from a zero knowledge base. As of August 1997, a large number of *S. cerevisiae* genes have been identified as encoding a protein or RNA of known function (Table 1). This means that for some 35% of the genes (i.e. 2000+) we have no clues to their function. Yet even when the function of all the genes has been assigned, we must then consider how these gene products are integrated into the biology of this unicellular eukaryote. The task facing us is therefore a daunting one, but not beyond the realms of possibility. In the light of this major challenge, we realized that it would be timely to produce a volume that reviewed the approaches available to us for the analysis of gene function in *S. cerevisiae*.

The ability to transform *S. cerevisiae* with plasmid DNA (or fragments thereof), first described 20 years ago (Beggs, 1978; Hinnen *et al.*, 1978), undoubtedly revolutionized how we can best meet this challenge, allowing researchers to make precise and controlled changes to the yeast genome – from single base pair "mutations" to large-scale chromosomal deletions. Yet it is important not to focus simply on recombinant DNA technology *per se*. Rather, the introduction of a mutation or deletion within the target sequence is just the first step in gene analysis. For about one in six of the genes analysed by gene disruption so far, the outcome has been simple to interpret – the cells die. This has told us unequivocally that the encoded protein is essential to the yeast cell but it has provided little insight into its

Table 1. What is known about the identity of the open reading frames (ORFs) in the *Saccharomyces cerevisiae* genome

What is known about the ORF?	No. of ORFs	%
Known protein	2611	41
Strong similarity to known protein	346	6
Similarity to known protein	733	12
Similar to unknown protein	653	10
No similarity	1544	25
Questionable ORF	398	6

Data taken from Goffeau *et al.* (1997).

biological function. Similarly, disruption of the other genes has told us that these genes encode a protein or RNA that is not essential for cell viability, leaving the researcher to identify a phenotype associated with the loss of that gene product. In many cases no distinctive phenotype can be found.

Bioinformatics can of course guide us in the elucidation of gene function either by identifying homologs of known function from other species, or by locating amino acid motifs diagnostic of proteins of a certain class. Ultimately, however, we must tackle several important questions if we are to address properly the question "what is the function of the product of my chosen gene?" Although numerous questions may be asked, the following are some of the most important:

- Is the gene transcribed and its mRNA translated and, if so, do the levels of expression of the gene change specifically in response to environmental changes?
- Does the encoded protein have any sequence features that might help in the diagnosis of its function?
- Where does the gene product locate to in the cell?
- Does the protein interact physically and/or functionally with any other proteins or RNA in the yeast cell and, if so, which ones?
- What is the phenotype of a cell unable to produce the gene product in question or of a cell that makes an aberrant gene product?

This volume is designed to help a researcher identify the best experimental approaches to these questions by providing a user-friendly overview of the strategies available for the analysis of genes in *S. cerevisiae*. In so doing, we have recognized that not everyone studying genes in yeast will necessarily be an experienced yeast researcher. Indeed, we expect many researchers picking up this volume to be looking at yeast for the first time. For example, we hope that this volume will be useful to researchers whose interests are in gene function in animal or plant cells, but who wish to take the opportunities provided by yeast molecular genetics to explore the function of the yeast homolog of their favorite gene or to dissect their favorite gene in yeast.

It has not been our intention to produce a recipe book *per se*, but rather to include comprehensive yet critical overviews of the protocols that can be found in the published literature. Where new techniques or strategies are proposed that have not been previously published in a "How to" format, then more detailed protocols have been included. For a comprehensive compendium of basic techniques in yeast genetics and molecular biology, the reader is referred to one of several current "yeast methods" volumes (Guthrie and Fink, 1991; Johnston, 1994; Evans, 1996; Adams *et al.*, 1997).

This volume focuses almost exclusively on one particular species of yeast (and remember there are over 500 different yeast species). The reason for this should be obvious because *S. cerevisiae* is the only yeast whose full genome sequence is known, and this yeast is the subject of major international projects to unravel genome function. This will not, however, be the only yeast genome to be sequenced and it is likely that by the end of 1998 the genome sequences of both the fission yeast, *Schizosaccharomyces pombe*, and the human pathogen, *Candida albicans*, will be released. At such time new volumes will need to be produced because both of these yeast species present their own unique set of technical challenges when it comes to studying gene function.

In achieving our aim we have been very fortunate to attract as contributors many of the leading international experts in the field of yeast gene analysis. Many of the authors are also actively involved in EUROFAN, the EC-funded network established to study the function of novel genes revealed by yeast genome sequencing, or in related international "functional analysis" projects. We are grateful to all the authors for their help in assembling this volume and for the speed with which they produced their excellent manuscripts. Finally we must express our gratitude to Tessa Picknett and Duncan Fatz at Academic Press for their efforts in keeping this project on the rails and (almost) on schedule.

August 1997
Mick F Tuite
Alistair JP Brown

References

Adams, A., Gottschling, D. and Kaiser, C. (1997). *Methods in Yeast Genetics: A Laboratory Course Manual*. Cold Spring Harbor Laboratory Press, New York.

Beggs, J. D. (1978). Transformation of yeast by a replicating hybrid plasmid. *Nature* **275**, 104–109.

Evans, I. H. (ed.) (1996). *Methods in Molecular Biology, Volume 53: Yeast Protocols*. Humana Press, New Jersey.

Goffeau, A., Aert, R., Agostini-Carbone, M. L., Ahmed, A., Aigle, M., Alberghina, L., Albermann, K. *et al.* (1997). The yeast genome directory. *Nature* **387** (Suppl.), 5–105.

Guthrie, C. and Fink, G. R. (eds) (1991). *Guide to Yeast Genetics and Molecular Biology (Methods in Enzymology, vol. 194)*. Academic Press, London/San Diego.

Hinnen, A., Hicks, J. B. and Fink, G. R. (1978). Transformation of yeast. *Proc. Natl. Acad. Sci. USA* **75**, 1929–1933.

Johnston, J. R. (ed.) (1994). *Yeast Molecular Genetics: A Practical Approach*. IRL Press, Oxford.

1 Introduction to Functional Analysis of the Yeast Genome

Stephen G. Oliver
Department of Biomolecular Sciences, UMIST, Manchester, UK

CONTENTS

Introduction
Genome organization and redundancy
Informatic analysis: usefulness and limitations
Generation of specific deletion mutants
Quantitative phenotypic analysis
The transcriptome
The proteome
Comparative functional genomics and human medicine

List of Abbreviations

CHR	Cluster homology region
LTR	Long terminal repeat
MCA	Metabolic control analysis
MIPS	Martinsried Institute for Protein Sequences
ORF	Open reading frame
PCR	Polymerase chain reaction
SAGE	Serial analysis of gene expression
snRNAs	Small nuclear RNAs

I. INTRODUCTION

In April 1996, the complete genome sequence of the brewers' and bakers' yeast *Saccharomyces cerevisiae* was deposited in the public data libraries (Goffeau *et al.*, 1996, 1997). This was an important event, not just because it was the first complete eukaryotic genome sequence, but also because it was the first total sequence for an important model organism for which there is a large constituency of researchers ready and able to exploit the sequence data. Thus, we are moving into the post-sequencing phase of

METHODS IN MICROBIOLOGY, VOLUME 26
ISBN 0-12-521526-6
Copyright © 1998 Academic Press Ltd
All rights of reproduction in any form reserved

genomic research in which attention will shift to the determination of the function of the novel genes revealed by the sequences and to understanding how the overall organization of these genomes affects their evolution and operation. This chapter will provide an overview on how these problems are being tackled for *S. cerevisiae* and will concentrate on the approaches being adopted to elucidate the functions specified by the 6000 or so protein-encoding genes that the sequence reveals (Oliver, 1996). The remainder of this volume will deal with specific techniques that can be applied in such a functional analysis.

◆◆◆◆◆◆ II. GENOME ORGANIZATION AND REDUNDANCY

The complete yeast genome sequence contained several surprises. The first was the number of genes that it contains. At the start of the sequencing project, there were about 1000 markers on the yeast genetic map (Mortimer *et al.*, 1992), but the complete sequence reveals approximately 6000 protein-encoding genes, 120 ribosomal RNA genes in a large tandem array on chromosome XII, 40 genes encoding small nuclear RNAs (snRNAs) scattered throughout the 16 chromosomes, and 273 tRNA genes (belonging to 43 families) that are also widely distributed (Goffeau *et al.*, 1996). The reason for the lack of precision concerning the number of protein-encoding genes is not that gene identification is a problem in yeast, as it is for the large eukaryotic genomes (Xu *et al.*, 1996). Introns are rare in *S. cerevisiae* nuclear genes (Dujon, 1996) and the transcript analyses published so far (Yoshikawa and Isono, 1990; Capieaux *et al.*, 1991; Tanaka and Isono, 1993; Richard *et al.*, 1997) show a good agreement between the mRNA molecules detected and the open reading frames (ORFs) predicted by the sequence. Rather, the problem comes with the small ORFs. From the outset of the sequencing project, a somewhat arbitrary cut-off of 300 bp has been used to define ORFs (Oliver *et al.*, 1992). While this has the merit of excluding spurious ORFs (there is a probability of < 0.2% of such an ORF occurring by chance), there are several known genes whose protein products are less than 100 amino acids long, e.g. the ribosomal protein genes *RPL46*, *RPL47A*, *RPL47B*, *RPS25*, *RPS27A*, *RPS27B*, *RPS33A* and *RPS33B* (Planta *et al.*, 1996). Special software packages have now been developed to identify small ORFs that are likely to represent real genes (Barry *et al.*, 1996; Termier and Kalogeropoulos, 1996; Andrade *et al.*, 1997) and to identify ORFs that may have been artificially terminated owing to sequencing errors. Careful transcript analyses or the use of gene fusion approaches (see below) will be required to confirm experimentally the predictions made using these programs, both for genes encoding very small proteins and in order to distinguish sequencing errors from genuine incidences of pseudogenes.

The other major surprise, after the sheer number of genes revealed by the sequence, was the large proportion of these genes that are apparently redundant because identical, or extremely similar, copies could be found elsewhere in the genome (Mewes *et al.*, 1997; Wolfe and Shields, 1997a). Readers may view this redundancy for themselves at either of two Web sites (Wolfe and Shields, 1997b; Heumann *et al.*, 1997). They will find that all yeast chromosomes share blocks of homology that can be recognized either at the nucleotide sequence level or at the level of the amino-acid sequence homology displayed by the predicted protein products of the ORFs. These blocks, or cluster homology regions (CHRs; Goffeau *et al.*, 1996), are found both at the chromosome ends and at internal sites within chromosome arms. It had long been known that yeast contained several multigene families whose members are predominantly associated with chromosome ends (e.g. the *SUC, MAL* and *MEL* genes, which are involved, respectively, in the fermentation of sucrose, maltose and melibiose; Carlson *et al.*, 1985; Charron *et al.*, 1989; Naumov *et al.*, 1995). In the case of chromosome I (which is the smallest), there is good reason to believe that redundant sequences have accumulated at the ends simply to increase its size and ensure correct disjunction in meiotic nuclear division (Bussey *et al.*, 1995). Similar arguments may be made for some of the other very small chromosomes (Oliver, 1995). In other cases (e.g. chromosomes III; Oliver *et al.*, 1992, and XI; Dujon *et al.*, 1994), the presence, at internal sites on chromosome arms, of repeat sequences that are normally associated with telomeres indicates that the chromosomes may have grown over evolutionary time by recombination events involving their ends. Finally, several smaller CHRs are associated with recombination events between yeast transposons or their LTRs, either by the excision of a section of a chromosome and its reintegration at a new site (Melnick and Sherman, 1993) or by unequal crossing-over events caused by the misalignment of Ty elements at different positions within a chromosome arm (Wicksteed *et al.*, 1994).

Whatever the means whereby this redundancy was generated, understanding its true nature, and developing experimental strategies with which to deal with it, will be central to any systematic attempts to elucidate gene function on a genome-wide scale. One approach, which Dr. Ed Louis (Inst. Molecular Medicine, Oxford) and I are adopting, is to try to construct a "minimalist" yeast genome in which every gene is an essential one. Our definition of essentiality is (necessarily) an operational one; nevertheless, this exercise should define the basic set of functions required by a eukaryotic cell and enable the contribution of non-essential yeast genes (as well as genes from other organisms) to be assessed by adding them back to this minimalist set. The intermediates in this process will be a library of strains containing large deletions either within (Fairhead *et al.*, 1996) or at the ends of chromosomes. This library will represent an important resource for the systematic study of gene interactions through the identification of synthetic phenotypes.

◆◆◆◆◆◆ III. INFORMATIC ANALYSIS: USEFULNESS AND LIMITATIONS

The first, and very important, step in the elucidation of the function of a novel gene is to compare the amino-acid sequence of its predicted protein product with those of other protein sequences in the public data libraries to see whether it is similar to a protein of known function that has previously been characterized in another organism. Powerful as this approach is, its results need to be treated with caution and intelligence (Oliver, 1996) because informatic analysis is not, on its own, sufficient to determine function.

First, functionality may be assigned in biochemical terms while giving no clear indication of the biological role of the novel protein. For instance, recognizing that an ORF encodes a protein kinase or phosphatase tells you nothing about the metabolic or developmental pathway in whose regulation such an enzyme may be involved. Second, the assignment of function in the organism where the gene or protein was originally discovered may have been mistaken or, at least, superficial. For instance, yeast chromosome III contains an ORF (YCL017c) showing greater than 40% amino-acid sequence identity to the nifS proteins of nitrogen-fixing bacteria (Oliver et al., 1992). S. cerevisiae does not fix N_2, yet the nifS homolog is an essential gene. Similar genes have now been found in several other bacteria (Mehta and Christen, 1993; Sun and Setlow, 1993; Leong-Morgenthaler et al., 1994), none of which fix N_2, and experimental and informatic analyses (Ouzounis and Sander, 1993; Leong-Morgenthaler et al., 1994) suggest that they encode a class of transaminases that use pyridoxal phosphate as a co-factor. In the nitrogen-fixing bacteria, the nifS gene product is now known to be responsible for the insertion of the sulphur into the Fe-S centre of the nitrogenase, again using pyridoxal phosphate as a co-factor (Zheng et al., 1993). In yeast, on the other hand, YCL017c has been demonstrated to play a role in pre-tRNA splicing (Kolman and Soll, 1993) and a mitochondrially related phenotype has also been recorded (B. Corner, personal communication).

These examples demonstrate that "wet" experiments will be necessary in order to elucidate the functions of the novel genes discovered by systematic sequencing. Nevertheless, informatic analysis remains an efficient route for assessing not "what's in a genome" but "what we know is in this genome" (Mewes et al., 1997; Chapter 3). While admitting that function assignment via sequence homology is a process fraught with danger, particularly as far as the physiological or developmental context in which the predicted proteins carry out their functions, it is possible to construct a provisional functional catalog for yeast's protein-encoding genes (Mewes et al., 1997). The MIPS catalog delineates 11 categories describing the cellular function of the individual ORFs (e.g. metabolism, energy, transcription, etc.). These categories are subdivided into further subcategories. Of the 5800 or so protein-encoding genes, only 22% are genetically and biochemically well characterized and specify products of known function. For approximately 20% of the remaining genes, the experimental data regarding their roles are so sparse as to allow only an estimation of their functions in vivo. Another 20% of the yeast proteins show similarities to known proteins and the remaining 38% show either similarities to other uncharacterized proteins or show no similarities at all. There is thus much experimental work to do in elucidating the function of novel yeast genes. Fortunately, as will become evident from the rest of the contributions to this volume, we have both the information and the genetic tools required to carry out analyses of these novel genes on an individual basis.

◆◆◆◆◆◆ IV. GENERATION OF SPECIFIC DELETION MUTANTS

The "minimalist" genome represents a top-down strategy to the analysis of gene function in *S. cerevisiae*. However, the small size of the yeast genome (at 12 Mb, it is less than four times the size of that of *Escherichia coli*), and the facility with which it may be manipulated using recombinant DNA techniques, means that it is perfectly feasible to undertake the systematic deletion of each and every one of yeast's 6000 or so ORFs. An efficient approach to gene deletion in yeast has been developed using a polymerase chain reaction (PCR)-mediated strategy (Baudin *et al.*, 1993) which exploits the precision and efficiency of the organism's mitotic recombination system. An important development of this approach was made by Wach and Philippsen (Wach *et al.*, 1994; see Chapter 5). In their method, a gene replacement cassette containing a gene (*kanMX*) which confers geneticin resistance on *S. cerevisiae* is tailed, via a PCR reaction, with sequences homologous to those flanking the target ORF in a yeast chromosome. Geneticin-resistant cells are selected following transformation with this PCR product and 80% or more of the transformants are found to have incorporated *kanMX* in place of the target ORF. The efficiency and accuracy of this replacement event is due to the fact that *kanMX* consists of a drug-resistance determinant from a bacterial transposon which is expressed in yeast by the use of promoter and terminator sequences from the filamentous ascomycete fungus, *Ashbya gossypii*. Thus the only regions of homology to the yeast genome that the replacement cassette contains are those sequences, complementary to the flanks of the target ORF, that the experimenter has designed and used as primers in the PCR reaction. Longer flanks may be generated using a nested PCR reaction (Amberg *et al.*, 1995; Wach, 1996) and this improves the site-specificity of the replacement event and also enables gene replacement to be carried out in almost any *S. cerevisiae* strain, including industrial yeasts, because *kanMX* is a positively selectable marker. Improved vectors have now been produced that permit reuse of the marker through *in vivo* excision (Fairhead *et al.*, 1996; Guldener *et al.*, 1996) or the facile cloning of the gene replaced (Mallet and Jacquet, 1996).

A development that will have important consequences for the worldwide effort to generate a complete library of 6000 yeast strains, each with a specific ORF deleted from its genome, is a method that allows each deletant to be given a specific oligonucleotide tag (a so-called "molecular bar-code"; Shoemaker *et al.*, 1996). This will not only permit the efficient management of the deletants within culture collections but also offers the prospect of the efficient determination of the proportions of different mutants in a mixed population by employing hybridization-array technology (Schena *et al.*, 1996). This has relevance to the quantitative analysis of phenotype.

◆◆◆◆◆◆ V. QUANTITATIVE PHENOTYPIC ANALYSIS

There are two main reasons for taking a quantitative approach to the analysis of the phenotypes produced by specific gene deletions. The first is that, as the different systematic genome sequencing projects progress, there is a growing set of genes that have homologs in a range of organisms but in none of these organisms is their function understood. It is possible that these genes have been missed because molecular geneticists usually design their experiments so as to provide qualitative answers to the questions posed. It may be that we are now presented with a group of genes whose functions will only be revealed if we undertake a quantitative analysis of phenotype. The second reason to expect some genes to have a quantitative impact on phenotype is the genetic redundancy revealed by the yeast genome sequence. If a particular gene is a member of a paralogous set of identical (or nearly identical) genes then, provided that they are all regulated in a similar manner and their products are targeted to the same cellular location, the contribution which any individual member of the set makes to phenotype will, necessarily, be some fraction of the whole.

I have previously suggested (Oliver, 1996) that "top-down" metabolic control analysis (MCA; Brand, 1996) provides a conceptual and mathematical framework for the quantitative analysis of gene function. At the highest level of this "top-down" analysis, quantitative data can most usefully be taken in two main types of experiment. In the first, the effect of deleting or overexpressing a particular gene on the growth rate (or fitness) of the organism is measured. Differences in growth rate are most easily revealed in competition experiments because these are very sensitive to small effects (Danhash *et al.*, 1991). Competition experiments using mutant strains generated by either Ty insertions (Smith *et al.*, 1995, 1996) or by the PCR-mediated gene replacement protocol described above (Baganz *et al.*, 1997) have been performed using various forms of quantitative PCR analysis to determine the proportions of the different competing strains in the yeast cell population. The analysis of yeast chromosome V by the "genetic footprinting" approach of Smith *et al.* (1996) succeeded in revealing a phenotype for approximately 60% of the chromosome's 268 predicted genes. Genes of known function were included in the analysis and acted as important controls. However, it was striking that, even for these known genes, novel (or, at least, unanticipated) phenotypes were often found. Such results suggest that quantitative data obtained by systematic approaches to phenotypic analysis will be important in achieving an holistic view of the cell in what is often termed "integrative biology".

The second type of data required for the MCA approach is the measurement of the change in the relative concentrations of metabolites as a result of the deletion or overexpression of a gene. There are some 563 low-molecular-weight intermediates on the metabolic map that are relevant to yeast metabolism (P. R. Butler, personal communication). Thus the number of metabolites is an order of magnitude lower than the number of genes and that fact alone should prompt a consideration of the usefulness of following metabolites rather than gene products. It should also be possible to take advantage of the fact that the functions of 40–50% of the genes are already known. We have attempted to combine these two ideas in an analytical strategy which we term FANCY (Functional ANalysis by Co-responses in Yeast) (Chapter 17). Theoretical work by Rohwer *et al.* (1996) has shown that some enzymes can be grouped into so-called "monofunctional units". The useful property of such units is that perturbations in any of the enzymes within them will always produce exactly the same co-responses outside the unit, irrespective of which enzyme in the unit was perturbed, and irrespective of the magnitude of that perturbation. Thus, across the metabolic map, there will be units of enzymes (of many different sizes) that will produce identical co-responses outside that unit. We can use this property for functional analysis. When two deletion mutants have the same co-responses, they will affect the same monofunctional unit. Knowing the unit in which one of the genes causes an effect, we can infer that of the other and so locate the origin of the changes in metabolites. Such an approach requires a fast and reliable way of sampling the concentration of as many metabolites as possible to produce a kind of "metabolic snapshot" of each of the deletants (Chapter 17). We are developing a two-stage strategy to obtaining such data. In the first (or cladistic) phase, deletants of novel genes are grouped with genes of known function by comparing their infrared spectra (Naumann *et al.*, 1996; Goodacre *et al.*, 1996) when grown under a range of conditions. This information then decides the type of "metabolic snapshot" to be taken in the second (or analytical) phase, using tandem mass spectrometry (Rashed *et al.*, 1995).

◆◆◆◆◆◆ VI. THE TRANSCRIPTOME

A powerful way to elucidate the function of novel genes uncovered by systematic genome sequencing projects is to determine the physiological conditions, or developmental stage, where those genes are expressed and relate their expression patterns to those of genes whose function is well known. The availability of complete genome sequences presents opportunities for the efficient study of gene expression at the RNA level on a genome-wide scale. For instance, any yeast ORF is uniquely identified by a sequence of just 12 nucleotides. This makes yeast an ideal subject for the analysis of gene expression using SAGE technology (Velculescu *et al.*, 1995). This has been undertaken by Velculescu *et al.* (1996) who have coined the term "transcriptome" to convey the complete set of expressed genes, and their levels of expression, under a given set of conditions. In their initial studies, they were able to detect a total of 4665 gene transcripts using this approach (approximately 75% of the total predicted by the sequence) with most genes belonging to the low expression class. Alternative approaches to transcriptome analysis include differential display (Liang and Pardee, 1992) and hybridization-array technologies (Schena *et al.*, 1996), with the latter offering the best hope for high-throughput analysis of the transcriptome under a wide range of physiological and developmental conditions. However, ultimately, the different approaches will compete not only in terms of their relative efficiency but also on which provides the most accurate quantitation of transcript levels. What is certain is that classical Northern approaches are likely to disappear, with even the correlation between transcript and ORFs being undertaken using overlapping panels of oligonucleotides in hybridization arrays.

As with the FANCY approach, it will be important to use transcript analysis to group genes into functionally related groups that will, hopefully, contain some members of well-defined function. A useful route is via the analysis of strains deleted for single genes specifying transcription factors. Ninety transcription factors have so far been identified in yeast (Svetlov and Cooper, 1995), but the complete genome sequence suggests that this is but a fraction of the total. Thus deletion of genes specifying known transcription factors can identify novel genes under their control and provide important clues as to their function (Dang *et al.*, 1994; Chapter 11). The complementary approach is to delete a gene specifying a novel transcription factor and get some idea of the domain of biological activity under its control from the functions of the known genes whose transcription it regulates.

◆◆◆◆◆◆ VII. THE PROTEOME

The protein-level correlate of the transcriptome is the proteome – the complete set of proteins found in a cell under a given set of conditions (Wilkins *et al.*, 1996). The yeast proteome is being defined by two-dimensional gel analysis (Sagliocco *et al.*, 1996; Shevchenko *et al.*, 1996), using mass spectrometry as the method by which to identify the proteins contained within the spots on the gel (Ferrara *et al.*, 1994; Wilm *et al.*, 1996). Whilst this approach will now proceed apace, it is important (as usual) to take maximum advantage of a complete genome sequence. Because it is only necessary to identify a protein fragment whose mass is unique to, and therefore diagnostic of, a particular yeast protein, rather than the total set of fragments produced from each protein, it should be possible to analyse mixtures of proteins. Among other things, this will allow the identification of the components of protein complexes and thus provide an important biochemical correlate to *in vivo* data on protein–protein interactions provided by the yeast two-hybrid approach. That technique has already been used to provide a complete protein linkage map of 'phage T7 (Bartel *et al.*, 1996) and similar analyses for whole organisms, including yeast, can be expected.

◆◆◆◆◆◆ VIII. COMPARATIVE FUNCTIONAL GENOMICS AND HUMAN MEDICINE

If yeast can be established as a good system with which to pursue the functional analysis of novel genes, then it should be possible to exploit *S. cerevisiae* in the functional analysis of the genomes of other organisms, including humans (Chapter 22). Of 170 human disease genes identified by cloning for function, and entered in the OMIM database, 72 find homologs in yeast (as defined by a BLAST analysis with a cut-off of e^{-10}); 28 of these yeast genes were discovered by systematic genome sequencing (Foury, 1997). Correspondingly, of 80 human disease genes identified by positional cloning, some 33 find homologs in yeast including 12 genes identified by the genome sequencing project. This may suggest that about a third to a half of genes relevant to human disease may have close relatives in the yeast genome. It might be possible to increase this proportion if we can extend the analysis from a comparison of protein sequences to a functional mapping of the genes from the human, and other eukaryotic, genomes onto that of yeast.

There are already several examples of the functional complementation of lesions in the yeast genome with cDNA clones from humans (Schild *et al.*, 1990; Plon *et al.*, 1993; Thon *et al.*, 1993; Glerum and Tzagaloff, 1994). Moreover, it has recently proved possible to clone cDNA copies of genes from the human malarial parasite *Plasmodium falciparum* in a similar manner (Volkman *et al.*, 1990; Wooden *et al.*, 1997). This is a particularly hopeful sign for the generality of this approach because, with a G+C content of approximately 20%, the *P. falciparum* genome is very disparate from that of yeast. Thus it should be possible to express coding sequences from a range of eukaryotic human pathogens in yeast and establish fast and effective screens for new chemotherapeutic agents. Such an approach, of course, is not confined to screening for anti-infectious agents but, via the expression of heterologous G-protein-coupled receptors and all or part of a human or mammalian signal transduction pathway (King *et al.*, 1990), it is possible to use yeast to screen for drugs effective against systemic human diseases. Several of the technical difficulties that have beset such screening systems seem now to have been resolved by the use of a chimeric yeast/mammalian Gα protein to couple the signal from a rat somatostatin subtype 2 receptor (SSTR2) located in the yeast cell membrane to a pheromone-responsive *HIS3* reporter gene, whose expression enabled the host yeast cells to grow in the absence of histidine (Price *et al.*, 1995). Whilst this impressive achievement probably heralds an increasing use of smart yeast screens for human medicines, it is still probably true to say that the major contribution that the functional genomics of *S. cerevisiae* will make to human medicine will come from the vastly increased understanding of eukaryotic biology that it will provide.

Acknowledgements

Work on yeast genome analysis in my laboratory has been supported by the European Commission (in both the Yeast Genome Sequencing and

EUROFAN Networks), the BBSRC, the Wellcome Trust, and by Pfizer Central Research, Applied Biosystems, Amersham International, and Zeneca.

References

Amberg, D. C., Botstein, D. and Beasley, E. M. (1995). Precise gene disruption in *Saccharomyces cerevisiae* by double fusion polymerase chain reaction. *Yeast* **11**, 1275–1280.

Andrade, M. A., Daruvar, A., Casari, G., Schneider, R., Termier, M. and Sander, C. (1997). Characterization of new proteins found by analysis of short open reading-frames from the full yeast genome. *Yeast* **13**, 1363–1374.

Baganz, F., Hayes, A., Gardner, D. C. J. and Oliver, S. G. (1997). Suitability of replacement markers for functional analysis studies in *Saccharomyces cerevisiae*. *Yeast* **13**, 1563–1573.

Barry, C., Fichant, G., Kalogeropoulos, A. and Quentin, Y. (1996). A computer filtering method to drive out tiny genes from the yeast genome. *Yeast* **12**, 1163–1178.

Bartel, P. L., Roecklein, J. A., SenGupta, D. and Fields, S. (1996). A protein linkage map of *Escherichia coli* bacteriophage T7. *Nature Genetics* **12**, 72–77.

Baudin, A., Ozier-Kalogeropoulos, O., Denouel, A., Lacroute, F. and Cullin, C. (1993). A simple and efficient method for direct gene deletion in *Saccharomyces cerevisiae*. *Nucleic Acids Res.* **21**, 3329–3330.

Brand, M. D. (1996) Top down metabolic control analysis. *J. Theor. Biol.* **182**, 351–360.

Bussey, H., Kaback, D. B., Zhong, W. W., Vo, D. T., Clark, M. W., Fortin, N., Hall, J. *et al.* (1995) The nucleotide sequence of chromosome I from *Saccharomyces cerevisiae*. *Proc. Natl. Acad. Sci. USA* **92**, 3809–3813.

Capieaux, E., Ulaszewski, S., Bazi, E. and Goffeau, A. (1991) Physical, transcriptional and genetic mapping of a 24 kb region between the *PMR1* and *ATE1* loci of chromosome VII from *Saccharomyces cerevisiae*. *Yeast* **7**, 275–280.

Carlson, M., Celenza, J. L. and Eng, F. J. (1985) Evolution of the dispersed suc gene family of *Saccharomyces* by rearrangements of chromosome telomeres. *Mol. Cell. Biol.* **5**, 410–419.

Charron, M. J., Read, E., Haut, S. R. and Michels, C. A. (1989) Molecular evolution of the telomere-associated *MAL* loci of *Saccharomyces*. *Genetics* **122**, 307–316.

Dang, V. D., Valens, M., Bolotin-Fukuhara, M. and Daignan-Fornier, B. (1994) A genetic screen to isolate genes regulated by the yeast CCAAT-box binding protein HAP2p. *Yeast* **10**, 1273–1283.

Danhash, N., Gardner, D. C. J. and Oliver, S. G. (1991) Heritable damage to yeast caused by transformation. *Bio/Technology* **9**, 179–182.

Dujon, B. (1996) The yeast genome project – what did we learn? *Trends Genet.* **12**, 263–270.

Dujon, B. *et al.* (1994) Complete DNA sequence of yeast chromosome XI. *Nature* **369**, 371–378.

Fairhead, C., Llorente, B., Denis, F., Soler, M. and Dujon, B. (1996) New vectors for combinatorial deletions in yeast chromosomes and for gap-repair cloning using split-marker recombination. *Yeast* **12**, 1439–1458.

Ferrara, P., Rosenfeld, J., Guillemot, J. C. and Capdeville, J. (1994) Internal peptide sequence of protein digested in-gel after one- or two-dimensional gel electrophoresis. *Tech. Protein Chem.* **IV**, 379–387.

Foury, F. (1997) Human genetic diseases: a cross-talk between man and yeast. *Gene* **195**, 1–10.

Glerum, D. M. and Tzagaloff, A. (1994) Isolation of a human cDNA for heme A: farnesyltransferase by functional complementation of a yeast *cox10* mutant. *Proc. Natl. Acad. Sci. USA* **91**, 8452–8456.

Goffeau, A., Barrell, B. G., Bussey, H., Davis, R. W., Dujon, B., Feldmann, H., Galibert, F. *et al.* (1996) Life with 6000 genes. *Science* **274**, 546–567.

Goffeau, A. *et al.* (1997) The yeast genome directory. *Nature* **387** (Suppl.): 5–105.

Goodacre, R., Timmins, E. M., Rooney, P. J., Rowland, J. J. and Kell, D. B. (1996) Rapid identification of *Streptococcus* and *Enterococcus* species using diffuse reflectance-absorbency Fourier-transform infrared-spectroscopy and artificial neural networks. *FEMS Microbiol. Letts.* **140**, 233–239.

Guldener, U., Heck, S., Fiedler, T., Beinhauer, J. and Hegemann, J. H. (1996) A new efficient gene disruption cassette for repeated use in budding yeast. *Nucleic Acids Res.* **24**, 2519–2524.

Heumann, K., Harris, C. and Mewes, H. W. (1997) Genome browser *http://speedy.mips.biochem.mpg.de/mips/programs/GENOME*

King, K., Dohlman, H. G., Thorner, J., Caron, M. G. and Lefkowitz, R. J. (1990) Control of yeast mating signal transduction by a mammalian β-2-adrenergic receptor and GSα subunit. *Science* **250**, 121–123.

Kolman, C. and Soll, D. (1993) *SPL1-1*, a *Saccharomyces cerevisiae* mutation affecting transfer-RNA splicing. *J. Bact.* **175**, 1433–1422.

Leong-Morgenthaler, P., Oliver, S.G., Hottinger, H. and Soll, D. (1994) A *Lactobacillus nif*S-like gene suppresses an *Escherichia coli* transaminase B mutation. *Biochimie* **76**, 45–49.

Liang, P. and Pardee, A. B. (1992) Differential display of eukaryotic messenger RNA by means of the polymerase chain reaction. *Science* **257**, 967–971.

Mallet, L. and Jacquet, M. (1996) Intergenic flip flop, a method for systematic gene disruption and cloning in yeast. *Yeast* **12**, 1351–1358.

Mehta, P. and Christen, P. (1993) Homology of pyridoxal-5′-phosphate-dependent aminotransferases with the cobC (cobalamin synthesis), *nif*S (nitrogen fixation), pabC (p-aminobenzoate synthesis and malY (abolishing endogenous induction of the maltose system) gene products. *Eur. J. Biochem.* **211**, 373–376.

Melnick, L. and Sherman, F. (1993) The gene clusters *ARC* and *COR* on chromosome V and chromosome X, respectively, of *Saccharomyces cerevisiae* share a common ancestry. *J. Mol. Biol.* **233**, 372–388.

Mewes, H. W., Albermann, K., Bähr, M., Frishman, D., Gleissner, A., Hani, J., Heumann, K. *et al.* (1997) Overview of the yeast genome. *Nature* **387** (Suppl.): 7–73.

Mortimer, R. K., Contopoulou, C. R. and King, J. S. (1992) Genetic and physical maps of *Saccharomyces cerevisiae*, 11. *Yeast* **8**, 817–902.

Naumann, D., Schultz, C. and Helm, D. (1996) What can infrared spectroscopy tell us about the structure and composition of intact bacterial cells? In: *Infrared Spectroscopy of Biomolecules* (H. H. Mantsch and D. Chapman, eds), pp. 279–310. New York: Wiley-Liss.

Naumov, G. I., Naumova, E. S. and Louis, E. J. (1995) Genetic mapping of the α-galactosidase *MEL* gene family on right and left telomeres of *Saccharomyces cerevisiae*. *Yeast* **11**, 481–483.

Oliver, S. G. (1995) Size is important, but … *Nature Genet.* **10**, 253–254.

Oliver, S. G. (1996) From DNA sequence to biological function. *Nature* *379*, 597–600.

Oliver, S. G. *et al.* (1992) The complete DNA sequence of yeast chromosome III. *Nature* **357**, 38–46.

Ouzounis, C. and Sander, C. (1993) Homology of the nifS family of proteins to a new class of pyridoxal phosphate-dependent enzymes. *FEBS Lett.* **322**, 159–164.

Planta, R. J., Goncalves, P. M. and Mager, W. H. (1996) Global regulators of ribosome biosynthesis in yeast. *Biochem. Cell Biol.* **73**, 825–834.

Plon, S. E., Leppig, K. A., Do, H-N. and Groudine, M. (1993) Cloning of the human homolog of the *CDC34* cell cycle gene by complementation in yeast. *Proc. Natl. Acad. Sci. USA* **90**, 10484–10488.

Price, L. A., Kajkowski, E. M., Hadcock, J. R., Ozenberger, B. A. and Pausch, M. H. (1995) Functional coupling of a mammalian somatostatin receptor to the yeast pheromone response pathway. *Mol. Cell. Biol.* **15**, 6188–6195.

Rashed, M. S., Oznand, P. T., Bucknall, M. P. and Little, D. (1995) Diagnosis of inborn errors of metabolism from bloodspots by acylcarnitines and amino acids profiling using automated tandem mass spectrometry. *Pediatr. Res.* **38**, 324–331.

Richard, G.-F., Fairhead, C. and Dujon, B. (1997) Complete transcriptional map of yeast chromosome XI in different life conditions. *J. Mol. Biol.* **268**, 303–321.

Rohwer, J. M., Schuster, S. and Westerhoff, H. V. (1996) How to recognize monofunctional units in a metabolic system. *J. Theor. Biol.* **179**, 213–228.

Sagliocco, F., Guillemot, J., Monribot, C., Capdevielle, M., Perrot, M., Ferran, E., Ferrara, P. *et al.* (1996) Identification of proteins of the yeast protein map using genetically manipulated strains and peptide-mass fingerprinting. *Yeast* **12**, 1519–1534.

Schena, M., Shalon, D., Heller, R., Chai, A., Brown, P. O. and Davis, R. W. (1996) Parallel human genome analysis – microarray-based expression monitoring of 1000 genes. *Proc. Natl. Acad. Sci. USA* **93**, 10614–10619.

Schild, D., Brake, A. J., Kiefer, M. C., Young, D. and Barr, P. C. (1990) Cloning of three human multifunctional *de novo* purine biosynthetic genes by functional complementation of yeast mutations. *Proc. Natl. Acad. Sci. USA* **87**, 2916–2920.

Shevchenko, A., Jensen, O. N., Podtelejnikov, A. V., Sagliocco, F., Wilm, M., Vorm, O., Mortensen, P. *et al.* (1996) Linking genome and proteome by mass spectrometry – large-scale identification of yeast proteins from 2-dimensional gels. *Proc. Natl. Acad. Sci. USA* **93**, 14440–14445.

Shoemaker, D. D., Lashkari, D. A., Morris, D., Mittmann, M. and Davis, R. W. (1996) Quantitative phenotypic analysis of yeast deletion mutants using a highly parallel molecular bar-coding strategy. *Nature Genetics* **14**, 450–456.

Smith, V., Botstein, D. and Brown, P. O. (1995) Genetic footprinting – a genomic strategy for determining a gene's function given its sequence. *Proc. Natl. Acad. Sci. USA* **92**, 6479–6483.

Smith, V., Chou, K. N., Lashkari, D., Botstein, D. and Brown, P. O. (1996) Functional analysis of the genes of yeast chromosome V by genetic footprinting. *Science* **274**, 2069–2074.

Sun, D. and Setlow, P. (1993) Cloning and nucleotide sequence of the *Bacillus subtilis nad*B gene and a *nif*S-like gene both of which are essential for NAD biosynthesis. *J. Bacteriol.* **175**, 1423–1432.

Svetlov, V. V. and Cooper, T. G. (1995) Compilation and characteristics of dedicated transcription factors in *Saccharomyces cerevisiae*. *Yeast* **11**, 1439–1485.

Tanaka, S. and Isono, K. (1993) Correlation between observed transcripts and sequenced ORFs of chromosome III of *Saccharomyces cerevisiae*. *Nucleic Acids Res.* **21**, 1149–1153.

Termier, M. and Kalogeropoulos, A. (1996) Discrimination of fortuitous and biologically constrained open reading frames in DNA sequences of *Saccharomyces cerevisiae*. *Yeast* **12**, 369–384.

Thon, V. J., Khalil, M. and Cannon, J. F. (1993) Isolation of human glycogen branching enzyme cDNAs by screening complementation in yeast. *J. Biol. Chem.* **268**, 7509–7513.

Velculescu, V. E., Zhang, L., Vogelstein, B. and Kinzler, K. W. (1995) Serial analysis of gene expression. *Science* **270**, 484–487.

Velculescu, V. E., Zhang, L., Vogelstein, J., Basrai, M. A., Bassett, D. E. Jr, Hieter, P., Vogelstein, B. *et al.* (1996) Characterization of the yeast transcriptome. *Cell* **88**, 243–251.

Volkman, S. K., Cowman, A. F. and Wirth, D. F. (1990) Functional complementation of the *ste6* gene of *Saccharomyces cerevisiae* with the *pfmdr1* gene of *Plasmodium falciparum*. *Proc. Natl. Acad. Sci. USA* **92**, 8921–8925.

Wach, A. (1996) PCR-synthesis of marker cassettes with long flanking homology regions for gene disruptions in *S. cerevisiae*. *Yeast* **12**, 259–265.

Wach, A., Brachat, A., Pohlmann, R. and Philippsen, P. (1994) New heterologous modules for classical or PCR-based gene disruptions in *Saccharomyces cerevisiae*. *Yeast* **10**, 1793–1808.

Wicksteed, B. L., Collins, I., Dershowitz, A., Stateva, L. I., Green, R. P., Oliver, S. G., Brown, A. J. P. *et al.* (1994) A physical comparison of chromosome III in six strains of *Saccharomyces cerevisiae*. *Yeast* **10**, 39–47.

Wilkins, M. R., Pasquali, C., Appel, R. D., Ou, K., Golaz, O., Sanchez, J. C., Yan, J. X., *et al.* (1996) From proteins to proteomes – large-scale protein identification by 2-dimensional electrophoresis and amino-acid analysis. *Bio/Technol.* **14**, 61–65.

Wilm, M., Shevchenko, S., Houthaeve, T., Brei, D., Schweigerer, L., Fotsis, T. and Mann, M. (1996) Femtomole sequencing of proteins from polyacrylamide gels by nano-electrospray mass spectrometry. *Nature* **379**, 466–469.

Wolfe, K. H. and Shields, D. C. (1997a) Yeast gene duplications. *http://acer.gen.tcd.ie/~khwolfe/yeast/nova/index.html*

Wolfe, K. H. and Shields, D. C. (1997b) Molecular evidence for an ancient duplication of the entire yeast genome. *Nature* **387**, 708–713.

Wooden, J. M., Hartwell, L. H., Vazquez, B. and Sibley, C. H. (1997) Analysis in yeast of antimalarial drugs that target the dihydrofolate reductase of *Plasmodium falciparum*. *Mol. Biochem. Parasitol.* **85**, 25–40.

Xu, Y., Mural, R. J., Einstein, J. R., Shah, M. B. and Uberbacher, E. C. (1996) Grail – a multiagent neural-network system for gene identification. *Proc. IEEE* **84**, 1544–1552.

Yoshikawa, A. and Isono, K. (1990) Chromosome III of *Saccharomyces cerevisiae* – an ordered clone bank, a detailed restriction map, and analysis of transcripts suggest the presence of 160 genes. *Yeast* **6**, 383–401.

Zheng L., White R. H., Cash V. L., Jack R. F. and Dean D. R. (1993) Cysteine desulfurase activity indicates a role for NifS I metallocluster biosynthesis. *Proc. Natl. Acad. Sci. USA* **90**, 2754–2758.

2 Whole Chromosome Analysis

Edward J. Louis

Yeast Genetics, Institute of Molecular Medicine, John Radcliffe Hospital, Oxford, UK

◆◆

CONTENTS

List of Abbreviations

ARS	Autonomously replicating sequence (origin of replication)
ATCC	American Type Culture Collection
CEN	Centromere
CHEF	Clamped homogeneous electric field
DSB	Double-strand break
FIGE	Field inversion gel electrophoresis
FISH	Fluorescent in situ hybridization
IF	Immunofluorescence
ORF	Open reading frame
PFGE	Pulsed-field gel electrophoresis
TEL	Telomere
YAC	Yeast artificial chromosome

◆◆◆◆◆◆ I. INTRODUCTION

A. General Properties of Whole Chromosomes

Chromosomes, as genetic elements in and of themselves, are of interest for many reasons. The analysis of their function can be different but complementary to the functional analysis of open reading frames (ORFs). In addition to the issues concerning fidelity of transmission and the functions of the known elements (centromeres (CENs), autonomously replicating sequences (ARSs) and telomeres (TELs)), there are several other interesting properties. There are hot and cold regions of recombination, as well as hot spots of transposable element accessibility. There are domains of transcriptional "silencing" and higher-order chromatin structure. Other issues are domains of recombinational interaction, timing of replication and locations of ARSs, new functional chromosome elements, chromosome size dependence on stability and recombination, GC-rich and GC-poor domains and their distribution, nuclear architecture and the localization of specific chromosomal elements and domains in the nucleus, stable versus dynamically variable chromosomal regions, and chromosome evolution. These issues are not mutually exclusive and there may be a great deal of overlap between various properties and functions.

Some of the questions of interest about whole chromosomes are as follows. Are there any functional elements other than CENs, ARSs and TELs? Yeast artificial chromosomes (YACs) function with just these three elements, yet they are not as stable as native chromosomes. Perhaps there is something else that enhances the stability of native chromosomes. What are the elements that govern accessibility of regions, either for transposable elements, recombination, general transcriptional repression or mating type switching, and is there any relationship among these? What is the relationship between sequence elements and their localization within the nucleus, such as telomere clustering and peripheral location, for example? Is there a relationship between these architectural arrangements of chromosomal domains and the ability of different regions to interact or their accessibility? In order to address these questions, techniques designed to deal with chromosomes as physical and genetic entities are needed.

B. Physical Tools

Whole chromosomes can be studied in a variety of ways both physically and genetically. The advent of pulsed-field gel electrophoresis (PFGE) technology has made it possible to study whole genomes physically in ways not possible before. For example, the small size of yeast chromosomes and the lack of visible condensation of the chromosomes preclude standard karyotype analysis. Now, electrophoretic karyotypes allow a physical analysis of chromosomes for species comparisons, and genetic aberrations such as aneuploidies, translocations or truncations. Assessment of chromosomal events at different life stages of yeast such as *HO* endonuclease cutting at the *MAT* locus during mating type switching or meiotic double-strand break (DSB) formation is also now possible.

Advances are being made with immunofluorescence (IF) and fluorescent in situ hybridization (FISH) that are allowing analysis of the physical location of chromosomal domains and the proteins that interact with them within the nucleus. This architectural understanding will be an important part of understanding general cell biology. Questions concerning the relationships of nuclear localization and function of various chromosomal elements and domains such as TELs, CENs and ARSs can now be addressed.

C . Genetic Tools

The genetic analysis of whole chromosomes is aided by having the entire sequence of *Saccharomyces cerevisiae*. The bioinformatics of the yeast genome (see Chapter 3) has been useful in determining some of the properties described above and can be used to help define potential chromosomal elements and domains of interest through comparative anatomy of chromosomes within and between *Saccharomyces* species. Whole chromosomes can be transferred between strains and species, which allows for testing of general haplo-sufficiency (complementation and compatibility of genotypes) as well as other general properties such as timing and location of replication, efficiency of the centromere and recombinational interactions. Virtually any alteration (insertion of genetic and physical markers, small-scale mutations, large-scale changes such as terminal truncations, deletions and translocations) can be made at will with the current technologies available (see Chapters 5 and 6). This allows for the determination and testing of function of *cis*-acting components of the properties described above.

The combination of genetic and physical tools makes whole chromosome analysis and the correlation of the above-mentioned properties a feasible endeavour. The methods described in this chapter can be applied to artificial and foreign chromosomes as well as to the native chromosomes of *S. cerevisiae*. This chapter will give examples of techniques that anyone can use, as well as some that require specialized equipment. Included are PFGE techniques for electrophoretic karyotyping for species identification and chromosome evolution, for aneuploid detection and other chromosome abnormalities, and for physical mapping of meiotic DSBs. Other techniques include genetic marking of difficult regions for physical and genetic mapping, whole chromosome transfers for functional and compatibility studies, and the analysis of chromosomal alterations for the search of new functional elements. The more difficult techniques included are the analysis of sublocalization of chromosomal elements within the nucleus using FISH and IF.

◆◆◆◆◆◆ II. PULSED-FIELD GEL ANALYSIS

A. Basic Principles and Techniques

The advent of technology that allows for separation of large DNA molecules over a broad range of sizes has greatly enhanced our ability to analyse whole chromosomes. PFGE techniques are varied, but all result in the electrophoretic separation of large molecules in an efficient and effective way. DNA molecules with length larger than the pore size in the agarose gel travel end on or looped over such that any molecule over a certain length will move at the same rate governed by the relative size of the end at the moving front. PFGE technology involves alternating the direction of the electric field (anywhere from 90° to complete inversion of 180°) requiring the large DNA molecules to reorient themselves relative to the electric field. The time to reorient depends on the length of the DNA, hence the ability to separate large molecules. The chromosomes therefore "snake" along or fold back over themselves in PFGE. For a theoretical discussion and experimental tests of the theories of large DNA molecular movement see Southern and Elder (1995).

Two types of pulsed-field gels are commonly in use. One is the clamped homogeneous electric field (CHEF) technique, which is useful for separating linear molecules in the range of 20 kb to several megabases and generally uses a 120° change in direction. The other is field inversion gel electrophoresis (FIGE) technology, using a complete 180° reverse of direction. This is generally used for a few kilobases up to 200 kb and fills the gap between standard agarose gels and CHEF gels. These types of gels can be complementary and are often used together for physical analysis (see Louis and Borts, 1995, for an example).

There are several methods for preparing genomic DNA in agarose plugs for PFGE (see, for example, Monaco, 1995). In general, the methods require digestion of the cell wall with any one of a number of enzyme preparations available commercially, followed by treatment with proteases to yield whole chromosomal "naked" DNA molecules without any associated chromatin. There are also quick methods available that generally work but are not always reliable. For efficient preparation of genomic DNA from a large number of different strains (tens to hundreds at a time) simultaneously, the procedure outlined in Protocol 1 is recommended. It was used to efficiently screen 384 primary transformants for the marked

---

Protocol I. Small-scale DNA preparation for large numbers of PFGE samples

1. 1 ml overnight YEPD cultures of strains (including YACs) pelleted in a 96-well, 2 ml plate (cells can be cultured in the plate).
2. Wash once with 0.5 ml 50 mM EDTA and repellet.
3. Resuspend in 200 µl 50 mM EDTA, add 100 µl SCE-zymolyase–β-mercaptoethanol solution.
4. Add 0.5 ml 1% LM agarose in 0.125 M EDTA quickly, mix well.
5. Pipette into plug formers (100 µl volume) or spot on parafilm.
6. When set, place in a 1.5 ml Eppendorf tube or in a clean 96-well 2 ml plate and cover with 0.5 ml overlay solution.
7. Incubate for at least 4 h at 37°C.
8. Remove overlay and cover with 0.5 ml PK-RNase solution.
9. Incubate overnight at 37°C.
10. Remove overlay and cover with 0.45 M EDTA, 0.1 M Tris for storage at 4°C.

Solutions
- SCE-zymolyase–β-mercaptoethanol solution: 3 mg/ml zymolyase 20T, 5% β-mercaptoethanol, 1 M sorbitol, 0.1 M sodium citrate, 0.05 M EDTA (pH 5.6–7.0).
- Overlay solution: 0.45 M Tris (pH 8.0), 0.05 M EDTA, 5% β-mercaptoethanol.
- PK-RNase solution: 0.4 M EDTA, 1% Sarkosyl, 1 mg/ml proteinase K, 0.1 mg/ml RNase A

chromosome ends (see section IV.A). The main differences between this and other protocols is the use of RNAse A, a 37°C incubation temperature for all steps, and the small scale. DNA prepared in this way has been successfully stored and used for several years and can be easily digested with a variety of restriction enzymes for further analysis. These DNA samples embedded in low-melt agarose can be used in any gel electrophoresis system.

B. Assessment of Chromosome Size Variation

Chromosome size, or length variation is a common property of most fungi as reviewed by Zolan (1995) and *S. cerevisiae* strains are highly polymorphic in chromosome sizes, which is evident on CHEF gels (Figure 1, lanes 1–7). Other related species also exhibit chromosome size polymorphisms as can be seen in the isolates of *S. paradoxus* shown in lanes 8–10 and *S. bayanus* in lanes 11–13 of Figure 1. These size polymorphisms are due to several factors, including the presence of Ty elements and subtelomeric Y' elements as reviewed in Zolan (1995). They can also be due to insertion/deletion polymorphisms of non-essential regions that may occur between Ty elements such as is the case for chromosome III in some strains analysed by Wicksteed *et al.* (1994). Chromosome size changes can be observed in the lab, either as a result of transformation or recombination (see section II.E), but also in adapting cultures over long periods of time as was seen by Adams *et al.* (1992).

C. Species Identification and Chromosome Evolution

In the *Saccharomyces* sensu stricto taxon there are several species that are difficult to distinguish by classical physiological taxonomic means. Their species designations come from the biological species definition in that hybrids are sterile but the parental species are fertile. The advent of electrophoretic karyotypes has enhanced our ability to distinguish some of these species as well as to assess their relatedness, as demonstrated by Naumov *et al.* (1992). In Figure 1, the electrophoretic karyotypes of several isolates of three of these species is shown. It is clear that the third species, *S. bayanus* (in lanes 11–13), has a karyotype that is quite distinct from that of *S. cerevisiae* and *S. paradoxus* (the other species represented on the CHEF). There are now six species in the *Saccharomyces* sensu stricto group that are all closely related and electrophoretic karyotyping can be used to distinguish three of them. The other three all have similar karyotypes, as demonstrated by Naumov *et al.* (1995a,b). When collecting yeast from the wild, this technique greatly speeds up the classification of isolates.

It is quite clear that all *Saccharomyces* sensu stricto species have essentially the same number of chromosomes and DNA content although some rearrangements are visible. Hybridization with unique sequence probes has confirmed the general maintenance of synteny in these species. There are, however, some obvious differences in chromosome sizes and some obvious chromosomal translocations. Probing with the *HO* gene for the

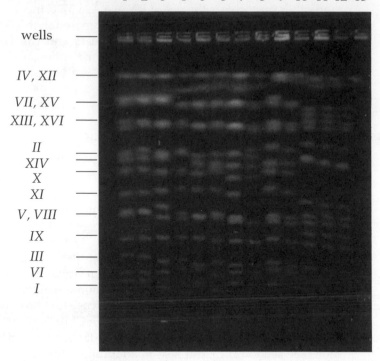

Figure 1. CHEF gel analysis of several strains from three species of *Saccharomyces*. The first seven lanes are different strains of *S. cerevisiae*. Lanes 8–10 are *S. paradoxus* and lanes 11–13 are *S. bayanus*. These were separated on a 1% agarose gel (0.045 M Tris-borate and 0.001 M EDTA (1/2× TBE)) for 15 h with a constant switch time of 60 s followed by 9 h with a constant switch time of 90 s and 6 V/cm. The switch angle was 120° and the temperature was 14°. Naumov *et al.* (1992, 1995a,b) among others have shown that these and several other newly recognized *Saccharomyces* sensu stricto species that were previously all classified as *S. cerevisiae* are distinguishable by the biological species definition. It is quite clear that *S. bayanus* is distinguishable from the other two by karyotype. Chromosome size polymorphisms are quite evident, especially for *S. cerevisiae*. A reciprocal translocation in the first *S. paradoxus* strain can be seen in lane 8 in which chromosomes VIII and XI are missing and there is a new band near chromosome II and a doublet at chromosome IX.

chromosome IV homologs shows variation in size in some of the species (Naumov *et al.*, 1995b). Southern analysis with unique probes can be used to determine the nature of the rearrangements. Ryu *et al.* (1996) have shown that *S. bayanus* (the furthest relative of *S. cerevisiae* in the group) has two apparent reciprocal translocations relative to the *S. cerevisiae* genome resulting in rearrangements of four of the chromosomes.

D. Aneuploid Detection

Genetic analysis of aneuploids is complicated by the fact that in general they yield poor viability upon crossing. Genetic detection of aneuploidy can be undertaken with the appropriate markers but can also be detected and analysed physically by electrophoretic karyotyping. In many situations there will not be any genetic markers for aneuploid detection, leaving physical methods as the only option. CHEF gels allow for visual detection of abnormal chromosomal contents by relative intensity of stained bands or more reliably by multiple bands for a chromosome when size polymorphisms exist. In an experiment with a hybrid cross between *S. cerevisiae* and *S. paradoxus*, Hunter *et al*. (1996) demonstrated that the rare viable meiotic progeny were highly aneuploid using CHEF gels. Figure 2 shows several examples of these aneuploids. Chromosome I (lane 6) and chromosome II (lane 13) disomes are quite obvious owing to the difference in size of the chromosome in the two parental genomes. Chromosome IX and X disomes, on the other hand, are detected by relative intensity of the ethidium bromide staining (lanes 4, 5, 6, 10 and 18). In cases where aneuploids are expected at high frequencies, electrophoretic karyotyping can be used for primary detection whereas in lower-frequency cases it can be used for confirmation of suspected aneuploids.

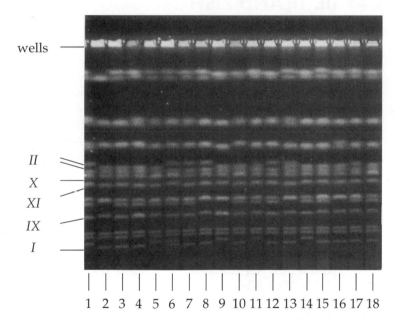

Figure 2. CHEF gel analysis of aneuploids. In a study of a hybrid cross between *S. cerevisiae* and *S. paradoxus*, Hunter *et al*. (1996) found that among the rare viable meiotic spores, there was a great deal of aneuploidy, which they detected using CHEF gels. The parental strains are in lanes 1 and 2 and it is clear that there is a difference in size between some of the homologous chromosomes, such as chromosome I (the lowest band on the gel). These size differences make it easy to identify I disomes (lane 6) as they have both parental bands. For chromosomes in which the sizes are nearly identical (such as IX, the fourth from the bottom), disomes are detectable by a doubly intense band, as in lanes 4 and 9. These were run as in Figure 1.

E. Detection of Translocations, Deletions and Size Changes

Chromosomal abnormalities that cause changes in size are readily detectable on CHEF gels and can be used to assay for translocations and deletions/truncations. In Figure 1 there is an example of a reciprocal translocation in the first *S. paradoxus* strain relative to the others (lane 8). The chromosome bands for VIII and XI are missing and a new band is seen near chromosome II and a doubly intense band at chromosome IX. Recombination-induced translocations can be detected in this way. Examples of this can be seen in the work of Goldman and Lichten (1996) in which they assessed the ability of ectopically located sequences to recombine with each other. Transformation sometimes results in coincident changes in chromosome size, which can also be detected by CHEF as seen by Louis and Borts (1995). Deletions and terminal truncations can also be detected using CHEFs and this is routinely used when making nested truncations of YACs as well as in native chromosomes (see section V.C).

◆◆◆◆◆◆ III. IF AND FISH

A. Immunofluorescence

Various techniques are used to determine cellular localization of proteins which include immunofluorescence when antibodies are available and tagging with another protein such as LacZ (for which antibodies are available) or green fluorescent protein (which fluoresces under appropriate stimulation) as described in Chapters 10 and 12. These techniques are useful for determining cytoplasmic vs. nuclear localization but subnuclear localization is more difficult because of the small sizes involved. When proteins are of high enough abundance and have specific subnuclear associations (at least for some stages) it is possible to visualize their sublocalization. For example, the protein Rap1 is known to be associated with many aspects of telomere biology in yeast and Klein *et al.* (1992) demonstrated that much of the Rap1 in a cell can be detected at a few foci (5–8), near the nuclear periphery, using anti-Rap1 antibodies. These foci are clusters of telomeres with which Rap1 is associated. Other telomere-associated proteins (Sir3 and Sir4) have been analysed in a similar fashion and shown to colocalize with the Rap1 foci by Gotta *et al.* (1996). Similar approaches are possible for proteins associated with other chromosomal elements such as CENs and for proteins known to be associated with chromosomes at specific stages, such as synaptonemal complexes during meiosis. This technique should be applicable to any protein that is abundant enough and has a strong affinity for a subnuclear location or chromosomal domain.

B. Fluorescent in situ Hybridization

Fluorescent in situ hybridization is difficult in yeast because of the small size of the nucleus and the technical difficulties in getting probes into the nucleus without destroying the nuclear structure. Generally large probes are necessary for detecting a signal. Using large regions (65–70 kb) as probes, Weiner and Kleckner (1994) were able to measure the distance between homologs through meiosis to assess pairing. Similarly Guacci *et al.* (1994) looked at chromosome condensation and sister chromatid pairing through the mitotic cell cycle using a variety of probes at different distances along a chromosome. Guacci *et al.* (1994) have improved the FISH technology such that smaller probes (only 4 kb) yield reasonable signals; however, this is at the expense of the ability to use IF to look at associated proteins.

C. Combined IF and FISH

One problem with these techniques is being able directly to show colocalization with the DNA elements because the methods for preparing cells for FISH are generally not amenable to maintaining the protein associations necessary for IF and the methods for preparing cells for IF do not generally allow the probes for FISH access to the DNA. Gotta *et al.* (1996) have overcome some of the technical difficulties in combining IF and FISH techniques and have successfully demonstrated colocalization of three telomere-associated proteins (Rap1, Sir3 and Sir4) with the telomeric Y' repeated sequence. The proteins are abundant and strongly associated with telomeres. The Y' probe, although on the small side (5.2 kb), is highly repetitive such that the telomere clustering amplifies the signals. The next technical difficulty will be in differentially labeling probes such that the simultaneous localization of different sequence elements can be assessed.

This combined approach not only confirms the DNA–protein associations by colocalization but provides a powerful tool for analysing in detail genes and conditions that are known to have effects on various properties of chromosomes. For telomeres and the subtelomeric region, it is known from the sequence data that there are ends that share large tracts of homologies. Genetic data indicate that there are groups of ends that recombine with each other more often than with other ends (Louis *et al.* 1994). Along with the physical clustering seen at a three-dimensional architectural level described by Gotta *et al.* (1996), it should be possible to address how correlated these three types of association are. Furthermore, questions of whether each type of clustering (sequence, genetic and physical) is governed by primary sequence, location or something else can now be addressed.

◆◆◆◆◆◆ IV. MARKED ENDS FOR PHYSICAL AND GENETIC MAPPING

A. Marking Chromosome Ends Uniquely

The lack of genetic markers and unique sequences at the ends of chromosomes has precluded genetic and physical analysis of chromosome ends.

In order to mark and clone all of the telomeres for the yeast genome sequencing project, Louis and Borts (1995) developed a strategy for uniquely marking each chromosome end with a vector sequence and a genetic marker, as seen in Figure 3. This construct was integrated into the TG_{1-3} telomeric sequences via homologous recombination. The frequency of integration was greatly enhanced by the use of DNAse I in the presence of $MnCl_2$, which in the right concentration causes a single random DSB in the plasmid. This method can be used in any situation where there are no convenient restriction sites. The marked ends are not only useful for obtaining the adjacent sequences by marker rescue into *E. coli*, or by long-range PCR, but can also be used in physical and genetic studies. The ends of foreign chromosomes or YACs can also be marked for rescue of the ends in this manner. A complete set of 32 strains, each with a different telomere marked with *URA3* and vector, is available from the American Type Culture Collection (ATCC) (accession numbers 90972–90999 and 96000–96004). Other difficult regions of the genome (such regions of a repetitive nature or where no useful markers exist) can be marked in a similar fashion using a different target sequence and standard transformation techniques.

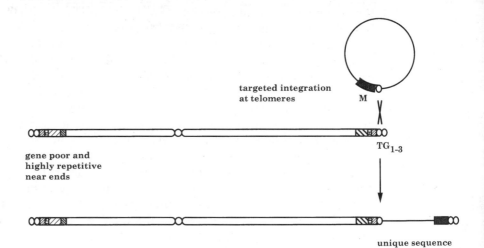

Figure 3. Marking each chromosome end with a unique sequence (vector) and genetic marker (M). The chromosome ends of yeast are composed of a mosaic of repeats that are highly variable and polymorphic, making it difficult for physical and genetic mapping in the region. Louis and Borts (1995) developed a technique for efficiently integrating a *URA3* based vector into each telomere of one strain. As no convenient restriction site existed to target the vector to the TG_{1-3} sequences, the plasmid was randomly cut with DNAse I in the presence of $MNCl_2$, which generated a fraction of molecules with single DSBs in the telomeric TG_{1-3} sequence, thus targeting the plasmid integration. These marked ends were useful in completing the physical map of the yeast genome as well as in cloning each end for the sequencing project. They are also useful for genetic mapping (see section IV.D) and physical mapping of meiotic DSBs (see section IV.C and Figure 4). Any genetic marker can be used and the method can be applied to artificial as well as foreign chromosomes.

B. Physical Mapping of Chromosome Ends

Having a unique sequence at the ends (or in any region filled with repeated sequences) allows for physical mapping of sequences and structures near the ends. Restriction maps of the ends were completed (Louis and Borts, 1995), which could not be accomplished before because of the repeated nature of the subtelomeric sequences (Riles *et al.*, 1993). More useful, though, is that this marker provides a unique sequence which can be used to assess various properties of one end without the complications of shared homologies with other ends. For example, the study of replication very close to the ends of chromosomes is complicated by the repetitive nature but having a unique insertion allows that sequence to be used for probing. In the next section there is an example of the usefulness of chromosomes marked at the end in the study of initiation of meiotic recombination.

C. Physical Mapping of Meiotic DSBs at the Whole Chromosome Level

One of the recent advances in the understanding of meiosis in yeast is the detection of DSBs early in the process of recombination. The levels of DSBs are correlated with levels of recombination and are non-random in distribution. Fine structure mapping of the DSBs by Liu *et al.* (1995) and de Massy *et al.* (1995) reveals much about the local distribution. The study of the larger-scale whole-chromosome distribution of these DSBs was pioneered by Zenvirth *et al.* (1992). For YACs it is possible to measure the DSBs from the end of the molecule but for native chromosomes this is difficult for the reasons outlined above. The meiotic DSBs are normally transient and are processed in a way that makes them hard to detect. By using the unique mutation *rad50s*, which is not radiosensitive yet still is recombination defective (Cao *et al.*, 1990), it is possible to study the initiating events of meiotic recombination as the DSBs that form early accumulate and do not get processed. Until now only YACs with unique markers at the telomere could be monitored from the end, though DSBs have been measured from near the ends (Klein *et al.*, 1996). By using the end-labelled chromosomes we can look at the pattern of meiotic DSBs from any end. Figure 4 shows the breaks from nine different ends. Each chromosome has a different pattern of DSBs but some generalizations can be made by comparing them. First, there are very few breaks in the last 40–50 kb of any end measured and this is likely to hold for all ends. Second, there are large internal areas that appear to be deficient in breaks yet we know that there is genetic recombination in many of these areas. We can now use these marked ends and PFGE to address issues about the genes and sequences involved in the distribution of meiotic DSBs at a whole chromosome level and whether there is any correlation with other properties of chromosomes.

D. Genetic Mapping Near Chromosome Ends

With a complete set of marked chromosome ends it is possible to genetically map telomeric loci. This can sometimes be a more rapid method for

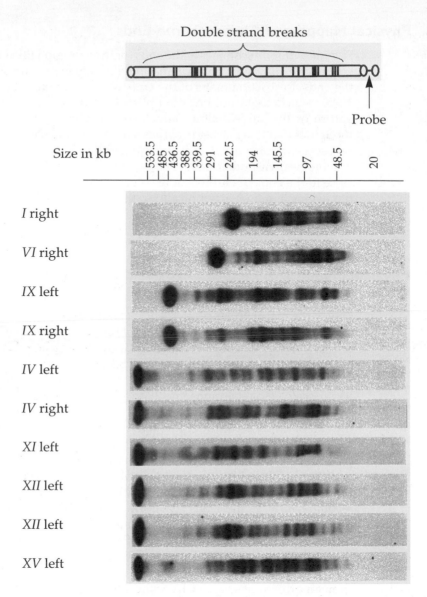

Figure 4. Meiotic DSB mapping along whole chromosomes using uniquely marked ends. DSBs are thought to be an early signal of meiotic recombination as well as the initiating event. Numerous people have studied meiotic DSBs at the local level, finely mapping them to small regions if not the base pair. At the whole chromosome level, the study of meiotic DSBs has been pioneered by Zenvirth *et al.* (1992). A problem with these studies is the lack of a unique marker at the end of a chromosome so that previously probes were to more internal unique regions. Using the marked ends as described and the *rad50s* mutation (Cao *et al.*, 1990) it is possible to visualize the meiotic DSBs from the ends. The figure is a Southern analysis of meiotic time course DNA sampled at 20 h for several different strains marked at different ends in a *rad50s* background. The fragments were separated on a CHEF gel using 3–48 s switching time over 24 h with standard 6 V/cm and 120° angle and 14°C in 1% agarose 1/2 × TBE. The probe was pGEM3Zf(–), which is the vector integrated along with the *URA3* marker. There are clear differences between ends and large regions, particularly near the telomeres, where there are few DSBs.

determining left or right end rather than going through other molecular means which are complicated by the high degree of relatedness and variation found at ends in different strains. These end-marked strains have been used to map each member of the large subtelomeric multigene family *MEL1-10* by Naumov *et al.* (1995c). These strains can also be used to address some of the issues concerning rates of homologous recombination near the ends of chromosomes.

◆◆◆◆◆◆ V. CHROMOSOME MANIPULATIONS

A. Whole Chromosome Transfer

Using a karyogamy-deficient mutation, *kar1*, it is possible to transfer a chromosome from one cell to another during cytoduction. The use of this process to study chromosomes of other species has been pioneered by Nilsson-Tillgren *et al.* (1980, 1981, 1986). They and Chambers *et al.* (1996) have used this method to study various properties of foreign chromosomes in a *S. cerevisiae* background which would be difficult in the donor strain (*S. carlsbergensis* does not sporulate well or yield many viable progeny) or in the hybrid cross (meiotic progeny are inviable or aneuploid). By selecting for the presence of the new chromosome and for the recipient strain with the possible selection against the other donor chromosomes, it is possible to get disomic strains containing one extra copy of the selected chromosome. Using either segregation or selection against the host copy of the chromosome, it is possible to get to the euploid state with the host chromosome substituted by the donor chromosome.

Haplo-sufficiency can be tested as well as segregation and recombinational properties. In addition, the functions of the CEN, ARS and TEL elements can be assessed in the *S. cerevisiae* background. Nilsson-Tillgren *et al.* (1981, 1986) and Chambers *et al.* (1996) used these whole chromosome transfers to study the segregation and recombination properties of these chromosomes, which was not possible in the original strains or full hybrids.

B. Directed Chromosomal Rearrangements

Dresser *et al.* (1994) wanted to test the effects of a paracentric inversion on meiotic pairing, recombination and segregation in yeast to compare to a similar situation previously studied in mice. They created a specific inversion using PCR to clone and generate the desired junctions. By co-transforming in these junctions they could select for the inversion. This method could be used to create any inversion or translocation that was desired. This is a more direct approach than using recombination between heteroallelic markers to generate specific rearrangements.

C. Chromosomal Deletions in Search of Functional Elements

In mating type switching there is a preference by *MATa* cells to chose the *HMLα* cassette and by the *MATα* cell to chose the *HMRa* donor cassette. This donor preference is clearly due to both *cis*- and *trans*-acting factors. In order to determine what the *cis*-acting elements are, Wu and Haber (1996) performed a detailed deletion analysis (both large internal deletions and terminal truncations) of chromosome III and found a 700 bp element on the left arm that "opens" the whole left arm in *MATa* cells specifically. This "opening" causes the preference of the choice of *HML*, while in *MATα* cells the whole left arm is "closed" not only in terms of donor preference, but in terms of the ability to find an ectopic recombinational partner. In constructing the test strains they used a terminal truncation vector with a target sequence, a selectable marker and a telomere "seed". They also created internal deletions by cloning small sequences flanking the desired deletion on each side of a selectable marker. The terminal truncation approach is the most efficient for deleting out to the end of a chromosome. Their internal deletion approach can be made more efficient using the PCR-based strategies outlined for ORF disruptions in Chapter 5. This was a clear demonstration of the powerful combination of using the sequence information as well as current transformation technology to discover and locate a new functional chromosome element.

◆◆◆◆◆◆ VI. COMPARATIVE ANATOMY OF RELATED GENOMES

We now have powerful tools to analyse various properties of chromosomes. One of the underlying problems in studying non-ORF elements and properties is that not all of the elements and properties have been identified. This is where bioinformatics can help. In sequence comparisons of non-coding regions it is possible to identify shared motifs that may have functional significance. It will then be up to the experimenter to determine their function if any. In addition it is not always clear what phenotypes to test for. The problems of assaying for function are potentially more complicated than those of the ORF analysis outlined in Chapter 1. The sequence of the yeast genome can clearly guide a logical plan of ORF disruptions for functional analysis but cannot necessarily provide such a plan with other chromosomal elements.

The problems associated with non-ORF functional analysis can be illustrated in the following example. The sequence comparison of all of the ends of the chromosomes reveals a small shared element called core X that resides amidst highly variable sequences. All 32 ends have an ARS and 31 out of 32 have an associated Abf1 binding site. This conservation clearly hints at function, but does not provide a clue as to what the function is. Deletion of all of these elements from a pair of homologs has a negligible effect on chromosome stability (F. E. Pryde and E. J. Louis, unpublished) so they do not have a major role in fidelity of chromosomal transmission. If there is a function to these elements, other phenotypes must be tested. Comparison of sequences of close species will help determine what sequences are conserved and therefore more likely to be important for such an analysis. The conservation of the core X could be accidental; however, when Zou *et al.* (1995) cloned the integration sites of Ty5 elements (which unlike the other Ty elements preferentially integrates near the ARSs of core X elements and *HML* and *HMR*) from *S. paradoxus*, they found conservation of the core X ARS and Abf1 site with no obvious flanking sequence conservation. Clearly there is something important about this element that has yet to be discovered. This comparison of close species genomes will help in determining which potential sequences are likely to be functional as they should have diverged in the close species if there is no selection for maintenance.

This kind of comparison is also useful in studying ORFs as it can lead to determination of the important promoter elements as well as the status of questionable ORFs. Just as the components of the core X are conserved across the species, the various binding sites for regulatory proteins will likely be conserved. Similarly, accessory sequence elements, such as those found with the ARS consensus in functioning origins, may be more easily determined by comparative genome anatomy. Any ORF that is questionable from the *S. cerevisiae* genome project, either by size or codon usage or lack of any homologies, will exist in the related species if it is a real part of being a *Saccharomyces* yeast. There is a great deal of value that can come from sequencing some close relatives of *S. cerevisiae*. Such a pilot project with a random library from several of the *Saccharomyces* sensu stricto species, which can be directly mapped onto the *S. cerevisiae* sequence, is underway at Washington University, St Louis under Mark Johnston and this will be another important tool for studying the genome of our favourite *Saccharomyces*.

Acknowledgements

I would like to thank Rhona H. Borts and Fiona E. Pryde for comments, Neil Hunter for use of unpublished CHEF gel pictures and the rest of the yeast genetics lab for comments and support. This work is supported by the Wellcome Trust and by the EU yeast genome projects.

References

Adams, J., Puskas-Rozsa, S., Simlar, J. and Wilke, C. M. (1992). Adaptation and major chromosome changes in populations of *Saccharomyces cerevisiae*. *Curr. Genet.* **22**, 13–19.

Cao, L., Alani, E. and Kleckner, N. (1990). A pathway for generation and processing of double-strand breaks during meiotic recombination in *S. cerevisiae*. *Cell* **15**, 1089–1101.

Chambers, S. R., Hunter, N., Louis, E. J. and Borts, R. H. (1996). The mismatch repair system reduces meiotic homeologous recombination and stimulates recombination dependent chromosome loss. *Mol. Cell. Biol.* **16**, 6110–6120.

de Massy, B., Rocco, V. and Nicolas, A. (1995). The nucleotide mapping of DNA double-strand breaks at the *CYS3* initiation site of meiotic recombination in *Saccharomyces cerevisiae*. *EMBO J.* **14**, 4589–4598.

Dresser, M. E., Ewing, D. J., Harwell, S. N., Coody, D. and Conrad, M. N. (1994). Nonhomologous synapsis and reduced crossing over in a heterozygous paracentric inversion in *Saccharomyces cerevisiae*. *Genetics* **138**, 633–647.

Goldman, A. S. and Lichten, M. (1996). The efficiency of meiotic recombination between dispersed sequences in *Saccharomyces cerevisiae* depends upon their chromosomal location. *Genetics* **144**, 43–55.

Gotta, M., Laroche, T., Formenton, A., Maillet, L., Scherthan, H. and Gasser, S. M. (1996). The clustering of telomeres and colocalisation with Rap1, Sir3 and Sir4 proteins in wild-type *Saccharomyces cerevisiae*. *J. Cell Biol.* **134**, 1349–1363.

Guacci, V., Hogan, E. and Koshland, D. (1994). Chromosome condensation and sister chromatid pairing in budding yeast. *J. Cell Biol.* **125**, 517–530.

Hunter, N., Chambers, S. R., Louis E. J. and Borts, R. H. (1996). Mismatch repair is involved in meiotic sterility of a yeast hybrid. *EMBO J.* **15**, 1726–1733.

Klein, F., Laroche, T., Cardenas, M. E., Hofmann, J. F., Schweizer, D. and Gasser, S. M. (1992) Localisation of Rap1 and topoisomerase II in nuclei and meiotic chromosomes of yeast. *J. Cell Biol.* **117**, 935–948.

Klein, S., Zenvirth, D., Sherman, A., Ried, K., Rappold, G. and Simchen, G. (1996). Double-strand breaks on YACs during yeast meiosis may reflect meiotic recombination in the human genome. *Nat. Genet.* **13**, 481–484.

Liu, J., Wu, T. C. and Lichten, M. (1995). The location and structure of double-strand DNA breaks induced during yeast meiosis: evidence for a covalently linked DNA–protein intermediate. *EMBO J.* **14**, 4599–4608.

Louis, E. J. and Borts, R. H. (1995). A complete set of marked telomeres in *Saccharomyces cerevisiae* for physical mapping and cloning. *Genetics* **139**, 125–136.

Louis, E. J., Naumova, E. S., Lee, A., Naumov, G. I. and Haber, J. E. (1994). The chromosome end in yeast: Its mosaic nature and influence on recombinational dynamics. *Genetics* **136**, 789–802.

Monaco, A. P. (1995). *Pulsed-Field Gel Electrophoresis: A Practical Approach*. IRL Press, Oxford.

Naumov G. I., Naumova, E. S., Lantto, R. A., Louis, E. J. and Korhola, M. (1992). Genetic homology between *Saccharomyces cerevisiae* with its sibling species *S. paradoxus* and *S. bayanus*: electrophoretic karyotypes. *Yeast* **8**, 599–612.

Naumov, G. I., Naumova, E. S., Hagler, A. N., Mondonca-Hagler, L. C. and Louis, E. J. (1995a). A new genetically isolated population of the *Saccharomyces* sensu stricto complex from Brazil. *Antonie van Leeuwenhoek* **67**, 351–355.

Naumov, G. I., Naumova, E. S. and Louis, E. J. (1995b). Two new genetically isolated populations of the *Saccharomyces* sensu stricto complex from Japan. *J. Gen. App. Microbiol.* **41**, 499–505.

Naumov, G. I., Naumova, E. S. and Louis, E. J. (1995c). Genetic mapping of the a-galactosidase *MEL* gene family to right and left telomeres of *Saccharomyces cerevisiae*. *Yeast* **11**, 481–483.

Nilsson-Tillgren, T., Peterson, J. G. L., Holmberg, S. and Kielland-Brandt, M. C. (1980). Transfer of chromosome III during *kar*-mediated cytoduction in yeast. *Carlsberg Res. Commun.* **45**, 113–117.

Nilsson-Tillgren, T., Gjermansen, C., Kielland-Brandt, M. C., Peterson, J. G. L. and Holmberg, S. (1981). Genetic differences between *Saccharomyces carlsbergensis* and *Saccharomyces cerevisiae*: analysis of chromosome III by single chromosome transfer. *Carlsberg Res. Commun.* **46**, 65–71.

Nilsson-Tillgren, T., Gjermansen, C., Holmberg, S., Peterson, J. G. L. and Kielland-

Brandt, M. C. (1986). Analysis of chromosome V and the *ILV1* gene from *Saccharomyces carlsbergensis*. *Carlsberg Res. Commun.* **51**, 309–326.

Riles, L., Dutchik, J. E., Baktha, A., McCauley, B. K., Thayer, E. C., Leckie, M. P., Braden, V. V. *et al.* (1993). Physical maps of the six smallest chromosomes of *Saccharomyces cerevisiae* at a resolution of 2.6 kilobase pairs. *Genetics* **134**, 81–150.

Ryu, S. -L., Murooka, Y. and Kaneko, Y. (1996). Genomic reorganisation between two sibling yeast species, *Saccharomyces bayanus* and *Saccharomyces cerevisiae*. *Yeast* **12**, 757–764.

Southern, E. M. and Elder, J. K. (1995). In *Pulsed-Field Gel Electrophoresis: A Practical Approach* (A. P. Monaco, ed.), pp 1–19. IRL Press, Oxford.

Weiner, B. M. and Kleckner, N. (1994). Chromosome pairing via multiple interstitial interactions before and during meiosis in yeast. *Cell* **77**, 977–991.

Wicksteed, B. L., Collins, I., Dershowitz, A., Stateva, L. I., Green, R. P., Oliver, S. G., Brown, A. J. *et al.* (1994). A physical comparison of chromosome III in six strains of *Saccharomyces cerevisiae*. *Yeast* **10**, 39–57.

Wu, X. and Haber, J. E. (1996). A 700 bp *cis*-acting region controls mating-type dependent recombination along the entire left arm of yeast chromosome III. *Cell* **87**, 277–285.

Zenvirth, D., Arbel, T., Sherman, A., Goldway, M., Klein, S. and Simchen, G. (1992). Multiple sites for double-strand breaks in whole meiotic chromosomes of *Saccharomyces cerevisiae*. *EMBO J.* **11**, 3441–3447.

Zolan, M. E. (1995). Chromosome-length polymorphism in fungi. *Microbiol. Rev.* **59**, 686–698.

Zou, S., Wright, D. A. and Voytas, D. F. (1995). The *Saccharomyces* Ty5 retrotransposon family is associated with origins of DNA replication at the telomeres and the silent mating locus *HMR*. *Proc. Natl Acad. Sci. USA* **92**, 920–924.

3 The Bioinformatics of the Yeast Genome

Hans-Werner Mewes, Dmitrij Frishman, Alfred Zollner and Klaus Heumann

Munich Information Center for Protein Sequences (MIPS), Max-Planck-Institute for Biochemistry, Martinsried, Germany

◆◆◆

CONTENTS

List of Abbreviations

EU	European Commission
GSG	Genome similarity graph
LTR	Long terminal repeats
MIPS	Martinsried Institute for Protein Sequences
NMR	Nuclear magnetic resonance
ORF	Open reading frame
PCR	Polymerase chain reaction
PDB	Protein database
SGD	Saccharomyces Genome Database
3D	Three dimensional
TCA	Tricarboxylic acid
WWW	World Wide Web
YPD	Yeast Protein Database

◆◆◆◆◆◆ I. INTRODUCTION

The sequence of the complete yeast genome was released on 24 April, 1996 as the result of an international, worldwide collaborative effort. This event marked the final success of an unprecedented experiment in molecular biology: a distributed, data-oriented approach to sequence the 16 chromosomes of a small, but very well studied eukaryotic organism.

In this chapter, we describe organizational aspects of a distributed sequencing network, the principles of primary sequence data evaluation and the implementation of data resources accessible to the biological community. Emphasis is given to methods developed during the sequencing of the yeast genome such as annotation and classification of open reading frames (ORFs), structural characterization of yeast proteins, and visualization of the yeast genome data including evolutionary redundancy.

◆◆◆◆◆◆ I. ORGANIZATION OF THE PROJECT

The project was launched by an initiative of A. Goffeau (1989) and the European Commission (EU) to sequence chromosome III in a pilot study. Based on libraries donated by M. Olson and C. Newlan, regions of about 11 kb were distributed to more than 30 European laboratories which published the 315 kb sequence in 1992 (Oliver *et al.*, 1992). Average sequencing throughput increased from 100 kb per year in 1991 to more than 3 Mb in 1995.

After the feasibility of the approach was shown with the release of the chromosome III sequence, the remaining chromosomes were distributed among the EU sequencing consortium, and laboratories in Canada, England, the USA and Japan. The European yeast sequencing project followed a collaborative, decentralized approach where chromosome-specific "DNA-coordinators", the participating laboratories, and an informatics coordination center (Martinsried Institute for Protein Sequences, MIPS) actively cooperated. While the DNA-coordinators were responsible for selecting and distributing clones from an ordered library based on precise physical maps, the laboratories performed the experimental sequencing work. MIPS served as the informatics coordinator and was responsible for data processing and chromosome assembly of about 60% of the genome. Sequence analysis was performed as a three step procedure: (1) preparation and distribution of ordered clone libraries by the DNA-coordinators; (2) experimental double-strand sequence determination and verification of the restriction map by the laboratories followed by data submission to MIPS; and (3) contig assembly, quality control, sequence analysis, and annotation at MIPS.

The bioinformatics of the yeast genome involves several technical procedures within the scope of applied informatics such as data processing, evaluation, and handling. On the other hand, the analysis and annotation of the data must be performed in the context of molecular biology and genetics. The close collaboration of computer scientists and experienced molecular biologists is a necessity. First of all, the quality of the underlying sequence must be ensured. Secondly, an in depth-analysis must be performed on the entity of the data. Thirdly, the data must be rendered accessible to the scientific community. In the following, we will give an overview of the individual steps of the project and discuss its relevance with respect to systematic genome analysis.

◆◆◆◆◆◆ II. DATA QUALITY

The quality of a nucleic acid sequence can be defined in terms of identity to the underlying natural biomolecules. Database entries of the same genetic object may display discrepancies for a variety of reasons. The individual sequences can reflect correctly the investigated molecule, but minor differences as a result of strain variations or mutagenic events in cloning procedure may occur. On the other hand, single reads of amplified DNA sequences display a high error rate, up to 1–2%. Fragment assembly of single reads may not correct for these ambiguities owing to problems in the data evaluation. In addition, incorrect assembly of fragments may cause rearrangements in the assembled sequence (Waterman, 1995). The independent verification by the informatics coordinator has certainly improved the quality of the *Saccharomyces cerevisiae* genomic sequence. For instance, to ensure the quality of the sequence, all submitted data have been inspected for consistency with the physical map, for frameshift anomalies, and for consistency among overlapping sequences submitted by different laboratories. Homology searches were carried out at the nucleotide level in order to detect possible vector contaminations, genetic elements in non-coding regions such as LTRs, and regulatory sequences such as splicing signals, promoters, and enhancers.

It is important to realize the consequences of sequencing errors for the interpretation of the final data. Gene density in *S. cerevisiae* is approximately 1 ORF per 2 kb. An error rate of 5 per 10 000 bp is equivalent to 1 error per gene, i.e. on average every gene is incorrect. Unfortunately, a significant proportion of errors are frameshifts, not base differences. Intensive evaluation of error rates has been performed in the EU sequencing project. A variety of methods has been applied in case of chromosome XI (B. Dujon and H. W. Mewes, unpublished results) and more than 5% of the genome has been resequenced by G. Valle (personal communication). As a result, the error rate of the early work was clearly higher, but dropped from more than 3 to lower than 2 errors per 10 000 bp. Approximately 2000 errors are estimated to be present in the current version of the yeast genome sequence. The comparison to earlier independent publications in the database with an average length of 2.5 kb per entry reveals a much higher error rate (>10 per 10 000 bp). Obviously, progress in sequencing technology as well as independent quality control has resulted in a very significant improvement in accuracy.

The systematic sequencing effort has the goal to unravel the complete genomic information. However, genomes may contain regions with low complexity or extended repeats causing technical problems in the sequence determination. In the case of the yeast genome, some 120 repeats of ribosomal RNA genes located in a tandem array on chromosome XII have not been determined, leaving the complete analyzed sequence at 12.1 Mb.

III. PRIMARY DATA ANALYSIS

A primary goal for sequence annotation in yeast is the identification and characterization of the different genetic elements. Table 1 summarizes the genetic elements in *S. cerevisiae*. Almost 70% of the genome is covered with coding elements (ORFs) that are first translated electronically into hypothetical proteins and then scrutinized by different methods for detailed homology and structure prediction analysis. The selection criterion for the identification of ORFs is a sequence not containing any stop signals of more than 300 bases, equivalent to 100 amino acids. However, smaller ORFs were scanned for homologies to other, known proteins. Approximately 6% of the ORFs identified remain questionable (Termier and Kalogeropoulos, 1996); these are close to the 100 amino acid threshold, have a low codon adaptation index (Sharp and Lloyd, 1993) and do not show significant similarity to any known proteins. Therefore the total number of expressed genes in *S. cerevisiae* is close to 6000, constituting the proteome of yeast (Goffeau *et al.*, 1996).

Individual chromosomes have been subject to detailed sequence data analysis (Bork *et al.*, 1992; Koonin *et al.*, 1994; Ouzounis *et al.*, 1995). An

Table I. Number of genetic elements in *S. cerevisiae*. (a) ORFs; (b) RNA genes and DNA elements

(a)	
extracted ORFs > 100 aa	6279
ORFs <100 aa (identified by similarity)	87
ORFs containing introns	220
5′-UTR introns	15
questionable ORFs	384
Retrotransposons:	
TY1	66
TY2	26
TY3	4
TY4	6
TY5	2
(b)	
tRNA	277
intron-containing tRNA	58
rRNA (5S, 5.8S, 18S, 25S)	100–200
snRNA	1
SRP-associated RNA (SRC1)	1
RNase P RNA	1
Telomerase template RNA	368
TY-LTRs	104
solo-LTRs	103
remnant LTRs	161

aa, amino acids

ever increasing fraction of characterized genes has been reported (Casari *et al.*, 1995). Most of the assignments have been based on the powerful paradigm that sequence homology reflects similarity in structure and function (Figure 1a). Despite the strong evidence for a common ancestor within a protein family, the functional classification is not equivalent to the understanding of the cellular function of a protein. For example, the

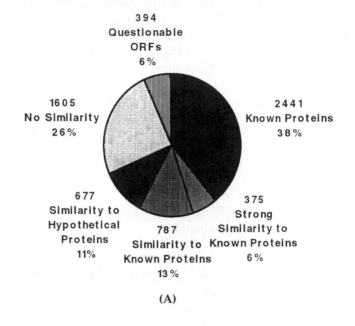

(A)

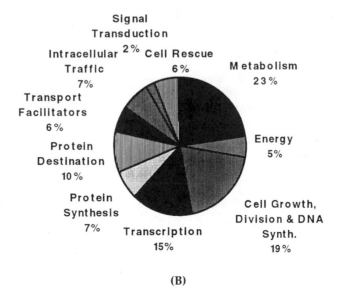

(B)

Figure 1. (A) Classification of *S. cerevisiae* ORFs according to known functions and homology to protein sequences in the databases. (B) Classification of characterized proteins according to the functional catalog (FunCat).

transporters of the inner mitochondrial membrane are undoubtedly related to each other, but display different substrate specificity and translocation properties. Therefore, the identification of a relationship to a well-characterized protein is not always sufficient for the proper assignment of its cellular role.

The aim of the genome sequence analysis is an overall functional classification of yeast proteins. Much information on the function of yeast proteins has been collected experimentally through decades of research on yeast genetics and cell biology. To classify the yeast ORFs, we assign these to the category "known proteins". A second group displays clear homology to well-characterized proteins in other organisms. In addition, homologies to other uncharacterized proteins are frequently found as a result of the systematic analysis of other genomes. According to the data shown in Figure 1, only half of the yeast proteins can be classified as well

Table 2. Functional classification of identified yeast proteins (FunCat)

	Number of ORFs found
Metabolism	1105
Amino acid metabolism	186
Nitrogen and sulfur metabolism	64
Nucleotide metabolism	132
Phosphate metabolism	33
Carbohydrate metabolism	390
Phospholipid/glycolipid/sterol/sphingolipid/ fatty acid metabolism	145
Biosynthesis of vitamins, co-factors and prosthetic groups	72
Ionic homeostasis	83
Energy	229
Glycolysis	30
Gluconeogenesis	13
Pentose phosphate pathway	9
TCA pathway	22
Respiration	67
Fermentation	34
Energy reserves	34
Others	20
Cell growth/cell division/DNA synthesis	914
Cell growth	64
Budding/cell polarity/filament	135
Pheromone response/mating type	135
Sporulation/germination	72
Meiosis	74
DNA synthesis	75
Recombination/DNA repair	66
Cell cycle/cell cycle control	265

Cytokinesis	21
Others	7
Transcription	641
rRNA synthesis	35
rRNA processing	28
tRNA synthesis	22
tRNA processing	23
tRNA modification	10
mRNA synthesis	334
mRNA processing	83
RNA transport	24
Others	62
Protein destination	456
Folding and stabilization of proteins	44
Targeting/sorting/translocation of proteins	97
Modification of proteins	124
Assembly of protein complexes	75
Proteolysis	112
Others	4
Transport facilitators	304
Ion channels	7
Ion transporters	60
Sugar/carbohydrate transporters	41
Amino acid transporters	25
Allantoin/allantoate transporters	6
Lipid transporters	15
Purine/pyrimidine transporters	11
Transport ATPases	36
ABC transporters	26
Drug transporters	18
Others	58
Signal transduction	112
Pheromone response pathway	26
Morphogenic pathway	24
Osmosensing pathway	13
Nutritional response pathways	18
Cell rescue	289
Stress response	125
DNA repair	62
Detoxification	84
Cell death and aging	11
Degradation of exogenous polynucleotides	1
Others	6

characterized. If the questionable ORFs are withdrawn from the total number, this fraction increases and will soon reach 60%. The application of sensitive methods to sequence data using weak, local relations (Bork and Gibson, 1996) is extremely helpful for the assignment of putative functions. However, these findings must be complemented by experimental procedures such as phenotypic analysis of gene disruptions, overexpression analysis and a wide range of other specialized methods. Nevertheless, roughly one-third of yeast proteins remain uncharacterized, a situation that is similar to the genomes of *Mycoplasma genitalium*, *Haemophilus influenzae* and *Synechocystis* sp. In the archebacterial genome of *Methanococcus jannaschii* (Bult *et al.*, 1996), two-thirds of the proteins have unknown functions.

Sequence data analysis based on sequence similarity and evolutionary relationships organizes proteins into families (Barker *et al.*, 1996), homology domains, blocks (Henikoff and Henikoff, 1996), and motifs (Bairoch, 1992). These properties allow information generation by induction, but do not necessarily reflect biochemical equivalence. For example, members of the dehydrogenase class or the membrane-bound transporter class may be evolutionarily unrelated. The functional catalog serves as a framework to relate the yeast genome data to cellular biochemistry and physiology. A classification scheme developed by Riley (1993) has been applied to the genomes of *M. genitalium* (Fraser *et al.*, 1995) and *H. influenzae* (Fleischmann *et al.*, 1995). We have adopted and extended this scheme, resulting in a functional catalog of the yeast proteins. The catalog contains seven major and 173 minor classes (Mewes *et al.*, 1997) (Table 2). The distribution of the characterized yeast proteins is shown in Figure 1b.

◆◆◆◆◆◆ IV. STRUCTURAL CHARACTERIZATION OF YEAST PROTEINS

The biochemical pathways of yeast involve a rich set of functionally and structurally different proteins. However, three-dimensional (3D) structures are known for only 2% of them, as determined by X-ray crystallography or nuclear magnetic resonance (NMR) spectroscopy. As seen in Table 3, among the structurally characterized yeast proteins are physiologically important proteins, such as enolase, TATA-box binding protein, Matα2 homeodomain protein, etc. If one takes into account the yeast ORFs reliably related to sequences from other organisms with known folds, the amount of structural information rises to 11% (Figure 2). For the remaining yeast proteins no experimental 3D data are available, thus making structure prediction the only, albeit imperfect, source of information.

We attempted a possibly complete structural characterization of the yeast proteins using a combination of sequence comparison and prediction techniques. The goal of our work was not only to provide cross-references to the known 3D structures related to yeast, but also to annotate sequences with secondary structures obtained through pairwise alignment with the closest sequence with known structure or through prediction.

For each yeast ORF a FASTA (Lipman and Pearson, 1985) search was performed against the full sequence database representing a non-redundant combination of the PIR-International (George *et al.*, 1996) and TREMBL (Bairoch and Apweiler, 1996) sequence sets. Sequences significantly related to the query sequence were extracted from the database using the SRS retrieval system (Etzold and Argos, 1993) and aligned using the ClustalW software package (Thompson *et al.*, 1994). Another FASTA search was made against the sequences with known 3D structures

Table 3. A representative set of the *S. cerevisiae* proteins with known 3D structures. Atomic resolution is given only for X-ray structures

PDB code	Resolution	Name
1aky	1.630	Adenylate kinase
1apl	2.700	Matα2 homeodomain
1asz	3.000	Aspartyl tRNA synthetase
1bgw	2.700	Topoisomerase
1csm	2.200	Chorismate mutase
1d66	2.700	Transcription regulator GAL4
1ebh	1.900	Enolase
1fcb	2.400	Flavocytochrome
1gcb	2.200	Bleomycin hydrolase DNA-binding protease (GAL6)
1gky	2.000	Guanylate kinase
1ncs	–	Transcriptional factor SWI5
1plq	2.300	Proliferating cell nuclear antigen
1pvd	2.300	Pyruvate decarboxylase
1pxt	2.800	Peroxisomal 3-ketoacyl-CoA thiolase
1pyc	–	DNA-binding domain CYP1 (HAP1)
1pyi	3.200	Pyrimidine pathway regulator 1
1pyp	3.000	Inorganic pyrophosphatase
1qpg	2.400	3-phosphoglycerate kinase
1sdy	2.500	Cu, Zn superoxide dismutase
1tkb	2.300	Transketolase
1ukz	1.900	Uridylate kinase
1yea	1.900	Cytochrome C (isoform 2)
1ypi	1.900	Triose phosphate isomerase
1yrn	2.500	Mating-type protein A-1
1ysa	2.900	GCN4 basic region leucine zipper
1ysc	2.800	Serine carboxypeptidase
1ytb	1.800	TATA-box binding protein
2cyp	1.700	Cytochrome C peroxidase (ferrocytochrome C)
2pcc	2.300	Cytochrome C peroxidase (CCP)
2uce	2.700	Ubiquitin conjugating enzyme
2yhx	2.100	Yeast hexokinase
3pgm	2.800	Phosphoglycerate mutase

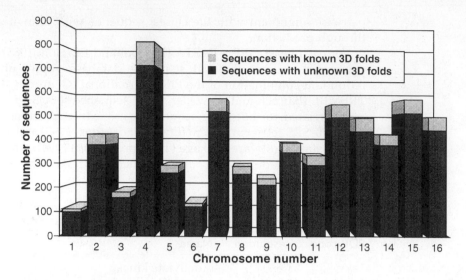

Figure 2. Number of *S. cerevisiae* sequences with known and unknown 3D structures.

extracted from the latest release of the PDB database (Bernstein *et al.*, 1977), and for significantly related hits, the secondary structure assignment from known atomic coordinates used, as produced by the program STRIDE (Frishman and Argos, 1995).

Secondary structure and transmembrane region prediction was effected using the programs Predator (Frishman and Argos, 1996a,b) and TMAP (Persson and Argos, 1994), respectively. A sequence without an identifiable 3D fold was considered an integral membrane protein if it was predicted to have at least one transmembrane region. Otherwise it was considered to be a globular protein with unknown 3D structure and the secondary structure prediction was used. Every protein with known or predicted secondary structure was additionally assigned to one of the five structural classes (all-α, all-β, α/β, irregular, and membrane). Figure 3 shows the distribution of known and predicted yeast protein folds over the five structural classes. Every third yeast protein has at least one hydrophobic region (Goffeau *et al.*, 1993). The actual number of membrane-associated proteins is probably lower (e.g. 18% if one considers proteins with only two or more transmembrane regions). Non-membrane proteins have a predominantly α/β folding arrangement. There is also a noticeable fraction of proteins with an irregular structure, most of which are probably non-globular.

Structural characteristics on every yeast ORF can be accessed through the MIPS WWW server[1]. This information is continuously and automatically updated as new sequence and structure information becomes available. Other sources of 3D structural information related to the yeast genome are the GeneQuiz resource[2] at EMBL (Heidelberg) (Scharf *et al.*,

[1] http://www.mips.biochem.mpg.de
[2] http://www.embl-heidelberg.de/~genequiz/yeast.html

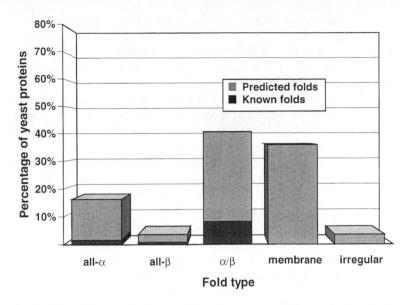

Figure 3. Distribution of *S. cerevisiae* proteins over five structural classes. Structural classes for globular proteins were defined on the basis of their secondary structural content according to the rules of Nakashima *et al.* (1986). All-α: more than 15% α-helix, less than 10% β-sheet; all-β: less than 15% α-helix, more than 10% β-sheet; α/β: more than 15% α-helix, more than 10% β-sheet; irregular: less than 15% α-helix, less than 10% β-sheet; or more than 20% of low-complexity sequence (Wooton, 1994). Integral membrane proteins are required to have at least one predicted transmembrane region.

1994) and the Sach3D facility[3] provided by the Saccharomyces Genome Database (SGD) at Stanford. Both sites contain exhaustive information on the corresponding protein database (PDB) entries and offer numerous options for visualizing the structures. In addition, GeneQuiz includes homology models for many yeast proteins generated by the WhatIf software (Vriend, 1990) while Sach3D includes a secondary structure prediction capability based on the program DSC by King and Sternberg (1996).

◆◆◆◆◆◆ V. VISUALIZATION OF THE GLOBAL PROPERTIES OF THE YEAST GENOME

With the complete genomic sequence at hand, available methods for comparative genome analysis and visualization are needed. Global properties of the genome such as gene density, gene orientation, gene duplication, GC content and location of transposable elements can be investigated. This type of information cannot be deduced from gene-directed sequencing work. Analysis of individual chromosomes has been done and published (e.g. Dujon *et al.*, 1994). Also, homologous, redundant regions between yeast chromosomes have been described (Lalo *et al.*, 1993). These

[3] http://genome-www.stanford.edu/Sacch3D

efforts could not provide a systematic, global view of the *S. cerevisiae* genome.

Although smaller in size and complexity, the set of sequenced bacterial genomes is increasing rapidly. Their protein content ranges from 470 (*M. genitalium*) to 3168 (*Synechocystis* sp). As for the yeast genome, inter- and intragenomic comparisons are of great interest to investigate evolutionary relations within and among genomes. These questions require not only the analysis of single chromosomes or genomes, but ask for exhaustive procedures for cross comparison.

Systematic sequencing efforts are paralleled by attempts to sequence closely related genomes, saving time and costs to detect organism-specific genes. For the elucidation of cell cycle events, particularly in the context of gene disruption and functional analysis, genome-wide expression and phenotypic analysis is likely to produce data on a large scale (Davis, 1996; Smith *et al.*, 1996). Obviously, the numerical representation of such data does not allow for an easy interpretation. In addition, most of the current methods do not allow for an intuitive parametrization because visual inspection of the results is not possible.

To fulfill these requirements, data analysis must be efficient, automated and flexible. While raw data and their related information (annotation) are complex data structures *per se*, the first step is the database design and its implementation by a suitable technology. In a next step, the data must be indexed to allow for rapid query processing. Finally, data and query interfaces have to be crafted. The visualization of results to allow the interpretation of complex genome properties becomes an essential feature of a successful comprehensive genome database resource. Certain types of queries can be anticipated and precomputed on the entire data set. For example, the search for protein families and their location within the genome is a standard query that should not require special knowledge or time-consuming computations. For a large number of similar questions, the exhaustive all-against-all comparison of the database is a prerequisite. The global view on genome redundancy allows users to focus their scope of interest on particular regions.

Several bioinformatic groups have addressed this topic. The first successful software suite for genome analysis, ACeDB, was presented by Durbin and Mieg (1991, unpublished). Recent work presented visual browsing applied to the *Drosophila* genome[4] and a top-down strategy for whole genome analysis by visualization (Heumann *et al.*, 1996). The latter approach was applied to the yeast sequence database on CD-ROM, which includes a local implementation of the genome browser[5]. Advanced WWW applications rely on the object-oriented programming language JAVA that allows one to generate platform-independent applications for the Internet. However, independent from the technology of user-interfaces and wide area network access, the power of any visualization technique depends on the underlying database structure and the known relationships of the data to be visualized: for example, (i) sequence

---

[4] http://fruitfly.berkeley.edu
[5] The Yeast Sequences CD-ROM is available from the authors.

similarity relationships between gene products identified on different chromosomes or in different organisms; (ii) functional classification of gene products; and (iii) relationships to other data resources such as protein or EST-database homologies.

In addition, the needs for data presentation may differ widely. For example, a global presentation of retrotransposons in yeast requires another resolution of genome presentation than the inspection of a collinear gene cluster duplication. Different granularity in the presentation of query results asks for appropriate methods to access and display data. The very same data, for example the detailed description of an ORF, may be accessed from different data access points.

Our data access and visualization tool for the yeast genome (Figure 4) relies on: (i) graphical browsing (right), (ii) browsing of tables (top left), and (iii) full text search (bottom left). Figure 4 shows an example of how the asparagine synthetase YGR124w may be identified from different access points. The genome browser display identifies YGR124w as a member of a gene cluster duplicated between chromosomes VII and XVI. Scanning the table of ORFs from chromosome VII shows a high value for the codon adaptation index (Sharp and Lloyd, 1993). The full text retrieval of the query string "ASN" returns the homologous ORF located on chromosome XVI, gene name ASN1, which is represented by a link in the genome browser display.

This example indicates only a few possible approaches to identify a data item of interest and to link to related annotation. Ongoing work in functional analysis projects[6] will generate a wealth of information on specific properties of yeast proteins and broaden the scope of access points and data relations. For example, it will be possible to search for lethal genes or a combination of gene disruptions leading to well-defined phenotypes.

A very important feature of the visualization tools is the integration of information from different sources into a genome-specific database. The genomic sequence is not a data item *per se*, but functions as a topological backbone to interconnect very different types of information that reflect current biological knowledge. The access and visualization tool integrates information into a complex genomic database. As a consequence, the homology principle can be used as a powerful relation between sequences, which provides a path to navigate through the genome and links to other data sources.

Our method to visualize genomic sequence data necessitates this navigational power. An all-against-all comparison of the genomic sequence data ensures the full exploration of all relationships above variable thresholds that may be restricted to certain genetic elements (e.g. DNA, proteins, TY-elements, etc.). The comparison itself is done by local alignments of blocks of the DNA sequence of fixed size. For each block the six-frame translation into protein sequences is also generated allowing for the concurrent comparison of DNA and amino acid sequences. No prior interpretation of the genomic sequence data by extraction of ORFs is

---

[6] For example, EUROFAN by the European Commission, and The Yeast Functional Analysis Program in Germany.

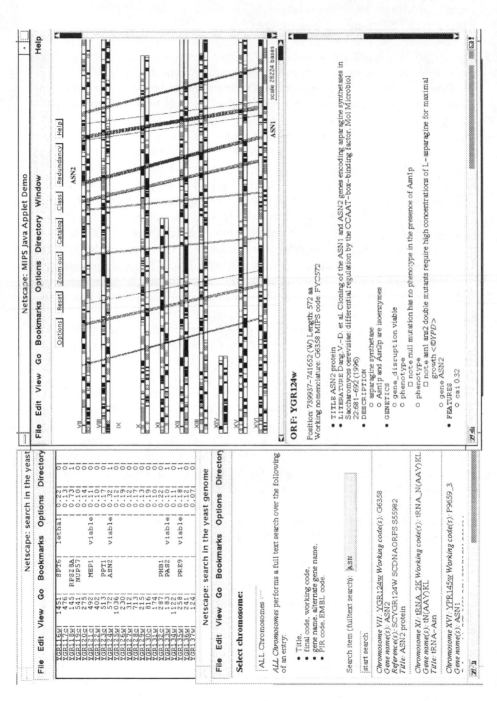

Figure 4. Genome browser display. Upper left: Section of the summary table of chromosome VII. Lower left: Full text search form and query result showing *ASN* related genes. Upper right: Detailed display of the largest collinear gene cluster between chromosomes VII and XVI. Horizontal frames represent a set of neighboring genes that assemble to the individual chromosomes. The location of *ASN2* is marked on chromosome VII. Interchromosomal connecting lines represent sequence similarity relationships between blocks of 500 nucleotides. Lower right: Display of the annotation related to gene *ASN2* which can be identified from access points as presented in the other windows.

necessary. This is an important prerequisite for the analysis of genomes whenever splice sites are not well defined. As a result this analysis can be applied to higher eukaryotes.

The all-against-all comparison of the yeast genome is computationally challenging. Classical sequence alignment programs must scan the complete genome for each of the n blocks to be compared (where n is proportional to the size of the genome). This problem can be solved by the application of index data structures such as suffix or position trees for rapid and selective access to similar blocks in the genome. The all-against-all matching of the preprocessed database will then scale linear instead of quadratic compute time. We have applied the Hash Position Tree as a dynamic, persistent index data structure applicable for large-scale sequence database searching and genomic pattern matching (Mewes and Heumann, 1995).

The result of the exhaustive comparison is a set of similarity relationships, the genome similarity graph (GSG). The GSG allows for easy investigation of multi-homology relationships, because its graph properties relate to multi-copy elements in yeast. A prominent example is the identification of clustered gene duplications by well-defined graph properties. The GSG displays the location of homologous gene clusters on different chromosomes (Figure 4).

As a result, the systematic analysis of the complete yeast genome has revealed that at least 30% of the genome has arisen through clustered duplications. The evolutionary development of the yeast genome is based on the duplication of regions within single blocks up to 100 kb.

◆◆◆◆◆◆ VI. THE YEAST GENOME ON THE WEB

The complete yeast genome is presented on the (World Wide Web) WWW by several groups that have either independently or collaboratively developed different approaches to organize, present, and access yeast genome data. The WWW as a medium demonstrates the rapid progress in the development of genome databases and related interfaces[7]. A more detailed summary is available electronically[8]. The following descriptions are restricted to the major Web sites that frequently update relevant information.

The MIPS WWW resource[9] includes a large set of detailed annotations on yeast proteins and provides comprehensive access to several query and graphical interfaces to browse the genome (see above) as well as the

---

[7] Some genome sites with browsing functionality are:
- Cyanobase; http://kazusa.or.jp/cyano/index.html
- SUN; http://www.sunlabs.com
- Flybase; http://fruitfly.berkeley.edu

[8] http://www.sciencemag.org/science/feature/data/genomebase.shl

[9] http://speedy.mips.biochem.mpg.de/mips/yeast

functional classification of the characterized yeast proteins. Annotations include information from the literature, input from yeast scientists, and database sources such as the YPD. In addition, secondary structures for proteins with significant homologies to PDB entries are available as well as predictions for globular proteins (see above). Yeast protein classification by protein families and functional categories are also available. MIPS has developed a comprehensive version of the Yeast Sequence Database on a CD-ROM that allows one to overcome restrictions of network data transfer by local access to the data[10].

The SGD at Stanford[11] provides access to knowledge associated with yeast genes for both yeast and non-yeast researchers. It administers a *S. cerevisiae* Gene Name Registry and is interconnected to other major resources as YPD, GenBank, Medline, MIPS, SWISS-PROT, and the Kyoto Encyclopedia of Genes. In addition BLAST and FASTA searches against the yeast genome sequence are provided and polymerase chain reaction (PCR) primer design is available.

The YPD by Proteome, Inc.[12] contains protein reports with detailed descriptions of characterized proteins derived from the literature. In addition to the calculations of sequence-derived properties such as molecular weight, isoelectric point, codon bias, predicted transmembrane domains, and predicted sequence motifs, proteins are classified by subcellular localization, posttranslational modifications, precursor cleavages, and functional categories. YPD includes links to other major data resources, e.g. SGD, MIPS, XREFdb, and Entrez.

The group of B. Barrell at the Sanger Centre at Hinxton[13] provides information on *S. cerevisiae*, particularly on the chromosomes IV, IX, XIII, and XVI that were entirely or partially sequenced at the Sanger Centre. In addition, they provide information on the fission yeast *Schizosaccharomyces pombe* and, like other yeast resources, they also provide links to major databases as well as sequence analysis tools.

XREFdb at the National Library of Medicine[14] is a database to cross-correlate human disease states to genes in model organisms. The data in XREFdb can be accessed by submission of protein sequence queries or queries against selected mouse and human map regions.

The GeneQuiz[15] effort of the Sander group at the EMBL in Heidelberg is based on a software system that relies on sequence similarity searches within or across species and that applies advanced bioinformatics methods. In addition to the information on protein function, GeneQuiz reports 3D structures modeled by similarity. Results are routinely improved and extended by database searches.

[10] The CD-ROM is available from the authors
[11] http://genome-www.standord.edu
[12] http://www.proteome.com
[13] http://www.sanger.ac.uk
[14] http://www.ncbi.nlm.nih.gov/XREFdb
[15] http://www.embl-heidelberg.de/~genequiz

VII. SUMMARY AND PERSPECTIVES

We have described our approach to assemble, annotate, and analyze genomic sequence data. These genomic sequence data represent a coordinated system for the identification of all relevant cellular entities in a closed form. Despite the uncertainties caused by sequencing errors, no additional information is required to detect and characterize all relevant functional items for the life cycle of yeast. The functional characterization of a large part of the genome is still missing. Nevertheless, the impact of having the complete sequence at hand for the detailed analysis of the organization and function of yeast genomes is already visible. A complex analysis of phenotypes by systematic genetic footprinting of 268 ORFs has been published by Smith *et al.* (1996). Major functional analysis programs to perform systematic gene disruptions for all yeast genes have been launched in the USA and Europe. The complex data generated by a large variety of experimental procedures are the next challenge for the bioinformatics of the yeast genome.

Our experience led us to realize that to compile and maintain genome data and to represent heterogeneous data in a consistent form is a complex process. Genome analysis must combine informatics technology with experimental molecular, structural, and cellular biology to keep up with the highly dynamic process of assimilating biological knowledge at the molecular level. Current examples of yeast databases reflect that endeavor in a competitive but collaborative effort.

Acknowledgements

The authors acknowledge the work of all DNA-coordinators and laboratories that contributed to the sequencing project and thank A. Goffeau for his ever-encouraging, ever-lasting gentle support. K. Albermann, J. Hani, and H. Tettelin have compiled the functional catalog, and M. Baehr, A. Gleissner, and A. Maierl have developed the software of the MIPS WWW interface. This work was supported by grants from the European Commission (BIOCT2-930172 and BIOCT4-960110) and the Bundesministerium f. Bildung und Forschung (FKZ 0310577).

References

Bairoch, A. (1992). PROSITE: a dictionary of sites and patterns in proteins. *Nucleic Acids Res.* **20** (Suppl.), 2013–2018.

Bairoch, A. and Apweiler, R. (1996). The SWISS-PROT protein sequence data bank and its new supplement, TREMBL. *Nucleic Acids Res.* **24**, 21–25.

Barker, W. C., Pfeiffer, F. and George, D. G. (1996). Superfamily classification in PIR-International Protein Sequence Database. *Methods Enzymol.* **266**, 59–71.

Bernstein, F. C., Koetzle, T. F., Williams, G. J., Meyer, E. F., Brice, M. D., Rodgers, J. R., Kennard, O. *et al.* (1977). The protein data bank: a computer-based archival file for macromolecular structures. *J. Mol. Biol.* **112**, 535–542.

Bork, P. and Gibson, T. J. (1996). Applying motif and profile searches. *Methods Enzymol.* **266**, 162–184.

Bork, P., Ouzounis, C., Sander, C., Scharf, M., Schneider, R. and Sonnhammer E. (1992). Comprehensive sequence analysis of the 182 predicted open reading frames of yeast chromosome III. *Prot. Sci.* **1**, 1677–1690.

Bult, C. J., White, O., Olsen, G. J., Zhou, L., Fleischmann, R. D., Sutton, G. G., Blake, J. A. *et al.* (1996). Complete genome sequence of the methanogenic archaeon, *Methanococcus jannaschii. Science* **273**, 1058–1073.

Casari, G., Andrade, M. A., Bork, P., Boyle, J., Daruvar, A., Ouzounis, C., Schneider, R. *et al.* (1995). Challenging times for bioinformatics [letter]. *Nature* **376**, 647–648.

Davis, R. (1996). Whole genome functional analysis of yeast. *Microb. Compar. Genomics* **1**, 202–203.

Dujon, B., Alexandraki, D., Andre, B., Ansorge, W., Baladron, V., Ballesta, J. P., Banrevi, A. *et al.* (1994). Complete DNA sequence of yeast chromosome XI. *Nature* **369**, 371–378.

Etzold, T. and Argos, P. (1993). SRS – an indexing and retrieval tool for flat file data libraries. *Comput. Appl. Biosci.* **9**, 49–57.

Fleischmann, R. D., Adams, M. D., White, O., Clayton, R. A., Kirkness, E. F., Kerlavage, A. R. *et al.* (1995). Whole-genome random sequencing and assembly of *Haemophilus influenzae Rd. Science* **269**, 496–512.

Fraser, C. M., Gocayne, J. D., White, O., Adams, M. D., Clayton, R. A., Fleischmann, R. D., Bult, C. J. *et al.* (1995). The minimal gene complement of *Mycoplasma genitalium. Science* **270**, 397–403.

Frishman, D. and Argos, P. (1995). Knowledge-based secondary structure assignment. *Proteins: Struct. Funct. Genet.* **23**, 566–579.

Frishman, D. and Argos, P. (1996a). Incorporation of long-range interactions in secondary structure prediction from amino acid sequences. *Prot. Eng.* **9**, 133–142.

Frishman, D. and Argos, P. (1996b). Seventy-five percent accuracy in protein secondary structure prediction. *Proteins* **27**, 329–335.

George, D. G., Barker, W. C., Mewes, H. W., Pfeiffer, F. and Tsugita, A. (1996). The PIR-International Protein Sequence Database. *Nucleic Acid Res.* **24**, 17–20.

Goffeau, A. (1989) (ed) *Sequencing the Yeast Genome – A Detailed Assessment.* European Commission, DG XII.

Goffeau, A., Slonimski, P., Nakai, K. and Risler, J. L. (1993). How many yeast genes code for membrane-spanning proteins. *Yeast* **9**, 691–702.

Goffeau, A., Barrell, B. G., Bussey, H., Davis, R. W., Dujon, B., Feldmann, H., Galibert, F. (1996). Life with 6000 genes. *Science* **274**, 546–567.

Henikoff, J. G. and Henikoff, S. (1996). Blocks database and its applications. *Methods Enzymol.* **266**, 88–105.

Heumann, K., Harris C. and Mewes, H. W. (1996). A top-down approach to whole genome visualization. In *Proceedings of the International Conference on Intelligent Systems in Molecular Biology*, pp. 98–108. European Commission, Brussels.

King, R. D. and Sternberg, M. J. E. (1996). Identification and application of the concepts important for accurate and reliable protein secondary structure prediction. *Prot. Sci.* **5**, 2298–2310.

Klein, P., Kanehisa, M. and DeLisi, C. (1985). The detection and classification of membrane-spanning proteins. *Biochim. Biophys. Acta* **815**, 468–476.

Koonin, E.V., Bork, P. and Sander, C. (1994). Yeast chromosome III: new gene functions. *EMBO J.* **13**, 493–503.

Lalo, D., Stettler, S., Mariotte, S., Slonimski, P. P. and Thuriaux, C. R. (1993). Two yeast chromosomes are related by a fossil duplication of their centromeric regions. *Acad. Sci. Paris* **316**, 367–373.

Lipman, D. J. and Pearson, W. R. (1985). Rapid and sensitive protein similarity searches. *Science* **227**, 1435–1441.

Mewes, H. W. and Heumann K. (1995). Genome analysis: pattern search in biological macromolecules. In *Combinatorial Pattern Matching* (Z. Galil and E. Ukkonnen, eds), pp. 261–285. Springer, Berlin.

Mewes, H. W., Albermann K., Baehr M., Frishman, D., Gleissner A., Hani, J., Heumann K. *et al.* (1997). Overview of the yeast genome. *Nature* **387**, 7–65.

Nakashima, H., Nishikawa, K. and Ooi, T. (1986). The folding type of a protein is relevant to the amino acid composition. *J. Biochem.* **99**, 153–162.

Oliver, S. G., van der Aart, Q. J., Agostoni-Carbone, M. L., Aigle, M., Alberghina, L., Alexandraki, D., Antoine, G. *et al.* (1992). The complete DNA sequence of yeast chromosome III. *Nature* **357**, 38–46.

Ouzounis, C., Bork, P., Casari, G. and Sander, C. (1995). New protein functions in yeast chromosome VIII. *Prot. Sci.* **4**, 2424–2428.

Persson, B. and Argos, P. (1994). Prediction of transmembrane segments in proteins utilising multiple sequence alignments. *J. Mol. Biol.* **237**, 182–192.

Riley, M. (1993). Functions of the gene products of *Escherichia coli*. *Microbiol. Rev.* **57D**, 862–952.

Scharf, M., Schneider, R., Casari, G., Bork, P., Valencia, A., Ouzounis, C. and Sander, C. (1994). GeneQuiz: a workbench for sequence analysis. In *Proceedings of the Second International Conference on Intelligent Systems for Molecular Biology* (R. Altman, D. Brutlag, P. Karp, R. Lathrop and D. Searls, eds) pp. 348–353. AAAI Press, Menlo Park, California.

Sharp, P. M. and Lloyd, A. T. (1993). Regional base composition variation along yeast chromosome III: evolution of chromosome primary structure. *Nucleic Acids Res.* **21**, 179–183.

Smith, V., Chou, K. N., Lashkari, D., Botstein, D. and Brown, P. O. (1996). Functional analysis of the genes of yeast chromosome V by genetic footprinting. *Science* **274**, 2069–2074.

Termier, M. and Kalogeropoulos, A. (1996). Discrimination between fortuitous and biologically constrained open reading frames in DNA sequences of *Saccharomyces cerevisiae*. *Yeast* **12**, 369–384.

Thompson, J. D., Higgins, D. G. and Gibson, T. J. (1994). CLUSTAL W: improving the sensitivity of progressive multiple sequence alignment through sequence weighting, position-specific gap penalties and weight matrix choice. *Nucleic Acids Res.* **22**, 4673–4680.

Vriend, G. (1990). WHATIF: a molecular modeling and drug design program. *J. Mol. Graph.* **8**, 52–56.

Waterman M. S. (1995). *Introduction to Computational Biology*, pp. 135–160. Chapman & Hall, London.

Wooton, J. C. (1994). Sequences with "unusual" amino acid composition. *Curr. Opin. Struct. Biol.* **4**, 413–421.

4 Transformation of Yeast by the Lithium Acetate/Single-Stranded Carrier DNA/PEG Method

R. Daniel Gietz[1] and Robin A. Woods[2]

[1]Department of Human Genetics, University of Manitoba and [2]Department of Biology, University of Winnipeg, Winnipeg, Manitoba, Canada

◆◆

CONTENTS

List of Abbreviations

DMSO	Dimethyl sulfoxide
LiAc	Lithium acetate
PEG	Polyethylene glycol, molecular weight 3350
SC-His	Synthetic complete medium minus histidine
SC-Leu	Synthetic complete medium minus leucine
SC-Trp	Synthetic complete medium minus tryptophan
SC-Ura	Synthetic complete medium minus uracil
SS-DNA	Single-stranded carrier DNA
TE	10 mM Tris-HCl, 1mM Na_2 EDTA buffer, pH 8.0
UAS	Upstream activating sequence
YPD or YEPD	Yeast extract, peptone, dextrose medium (*see Appendix II: Table 1*)
YPAD	YPD supplemented with adenine

METHODS IN MICROBIOLOGY, VOLUME 26
ISBN 0–12–521526–6

I. INTRODUCTION

The transformation of intact yeast cells after treatment with a lithium salt was first reported by Ito *et al.* (1983). Their protocol included treatment of the cells with lithium chloride, the addition of plasmid DNA and polyethylene glycol (PEG), followed by a heat shock at 42°C. They obtained 400 transformants per microgram of plasmid DNA. Although this protocol was less efficient than contemporary procedures for the transformation of yeast spheroplasts (Broach *et al.*, 1979; Struhl *et al.*, 1979), it was simpler and did not require that cells be plated in regeneration agar. Several workers reported modifications of this basic protocol that increased the number of transformants. The most significant observation was that sonicated carrier DNA increased transformation efficiency to 10 000–20 000 per microgram of plasmid DNA (Gietz and Sugino, 1988; Stearns *et al.*, 1990). Baker (1991) reported the transformation of cells taken directly from plates and also found that carrier DNA improved the recovery of transformants. The addition of dimethyl sulfoxide (DMSO) prior to the heat shock was found by Hill *et al.* (1991) to increase the yield of transformants to 4×10^4. However, these workers reported that double-stranded carrier DNA did not enhance efficiency.

Although these transformation frequencies are adequate for work with yeast genes, they are low if one wishes to clone specific sequences from higher eukaryotes. In 1989 Schiestl and Gietz reported an improved procedure, now referred to as the lithium acetate (LiAc)/single-stranded carrier DNA (SS-DNA)/PEG protocol, that resulted in up to 100 000 transformants per microgram of plasmid DNA. Yeast cells were grown overnight in yeast extract, peptone, dextrose medium supplemented with adenine (YPAD), diluted into fresh YPAD and regrown for two divisions, harvested, incubated in 10 mM Tris-HCl, 1 mM Na$_2$ EDTA buffer, pH 7.5 (TE)/LiAc for 60 min at 30°C, and then with SS-DNA and plasmid in TE/LiAc for 30 min. PEG was then added and after a further 30 min the cells were heat shocked for 15 min, washed in TE and plated onto selective medium. The most significant increases in transformation efficiency were (a) a 3-fold increase due to regrowth for two divisions, (b) an 8-fold increase due to heat shock for 15 min, and (c) a 1000-fold increase using SS-DNA.

Subsequent modifications to the LiAc/SS-DNA/PEG protocol that increase transformation efficiency include: (a) reduction of duration of the exposure to LiAc (Gietz *et al.*, 1992); (b) omission of TE from the TE/LiAc and suspension of transformed cells in water rather than TE (Gietz and Woods, 1994); (c) optimization of plasmid, carrier DNA and cell concentrations (Gietz *et al.*, 1995); and (d) growth of cultures overnight on YPAD agar medium followed by a second incubation for two divisions in YPAD liquid medium (Schiestl *et al.*, 1993). The preparation of some of the reagents required for the protocol has also been simplified: (a) the carrier DNA does not need to be sonicated and the concentration can be reduced to 2 mg ml^{-1}, which makes it easier to dispense accurately; and (b) the LiAc and PEG solutions can be sterilized by autoclaving (Gietz *et al.*,

1997). Most recently we have modified the protocol for use in microtiter plates and found that transformation efficiency is enhanced by regrowth in double-strength YPAD.

◆◆◆◆◆◆ II. HIGH EFFICIENCY TRANSFORMATION AND GENETIC SCREENS

High efficiency transformation of yeast is used most frequently for applications such as the two-hybrid screen for protein–protein interactions (Fields and Song, 1989; Bartel and Fields, 1995), for one-hybrid screens that detect protein–DNA interactions (Li and Herskowitz, 1993; Wang and Reed, 1993), for inverted one-hybrid screens to identify DNA elements that bind a transcription factor (Liu *et al.*, 1993) and for screens that detect protein–protein interactions dependent on phosphorylation (Osborne *et al.*, 1995; Keegan and Cooper, 1996; Gietz *et al.*, 1997).

All these systems use two or more plasmids. The first plasmid contains either a known cDNA encoding a protein of interest fused to a GAL4 or lexA DNA binding domain, or a reporter gene linked to an upstream activating sequence (UAS)-less (basal) promoter containing a defined promoter element to be studied. The second plasmid usually contains a complex library of either cDNA or genomic DNA fragments. These two plasmids are introduced into yeast by sequential or co-transformation and one can then select or screen for the activation of a reporter gene. The libraries used in these screens can have a complexity of over 2×10^6 independent clones. To achieve a high probability of recovering a clone of interest at least 10 to 20 million transformants are necessary. The transformation protocols given below will generate the numbers of transformants needed to screen any complex library.

◆◆◆◆◆◆ III. YEAST TRANSFORMATION

The most efficient application of the two-hybrid system involves the sequential transformation of the bait and library plasmids into the two-hybrid yeast strain (see Chapter 15). The initial transformation of the strain with the bait plasmid requires only a small number of transformants and can be done using the Rapid Transformation Protocol (section III. B). Before carrying out a library screen it is important to investigate the relationship between plasmid concentration and the recovery of transformants for your host yeast strain using the High Efficiency Transformation Protocol (section III.C). Gietz and Schiestl (1991) found that 30–40% of transformation-competent cells take up more than one plasmid molecule. Thus increasing the plasmid concentration increases the yield of transformants but also increases the proportion of yeast transformants that contain more than one plasmid. A procedure for maximizing the number of transformants is set out in Gietz and Woods (1994). When optimal parameters for transformation have been defined you are ready to carry out a library screen. The strain carrying the bait plasmid is maintained on selective medium and then transformed with the cDNA library plasmid using the Library Screening Transformation Protocol (section III.D).

If the fusion protein produced by the bait plasmid is detrimental to yeast, introduction of the bait plasmid by the Rapid Transformation Protocol may only result in a small number of slow-growing transformants. The preparation of transformation-competent cells from this type of transformant is difficult and time-consuming. In this situation it is best to co-transform the bait and cDNA library plasmids into the host strain using the Library Screening Transformation Protocol. Co-transformation is less efficient than sequential transformation. However, careful attention to the amounts of each plasmid, which can be tested using the High Efficiency Transformation Protocol, will usually result in the desired number of transformants.

We have also included protocols for: (a) the preparation of frozen competent yeast cells (section IV), (b) the transformation of oligonucleotides directly into yeast (section V), and (c) the transformation of multiple samples using 96-well microtiter plates (section VI). These protocols have specific applications for research with yeast.

A. Solutions and Media

The solutions listed in Table 1 are required for all of the LiAc/SS-DNA/PEG transformation protocols. We routinely culture host strains lacking plasmids on YPAD medium; this is YPD medium supplemented with adenine hemisulfate at 100 mg/liter^{-1} (Sherman, 1991). Commercial YEPD broth and YEPD agar media (Becton Dickinson Microbiology Systems, Cockeysville, Maryland 21030, USA) are more convenient than making your own. Double-strength YEPD plus adenine (2×YPAD) is supplemented with adenine hemisulfate at 100 mg liter^{-1}. Synthetic complete (SC) drop-out media are prepared by adding a complete set of supplements, minus one or more components (SC-His, SC-Leu, SC-Trp, SC-Ura, SC-Trp-Leu, etc.), to select for specific plasmids (Rose, 1987; Sherman, 1991).

Table I. Solutions

- LiAc 100 mM: Make 200 ml by adding 20 ml of 1.0 M LiAc to 180 ml of sterile water
- LiAc 1.0 M: Make 50 ml and sterilize by autoclaving.
- PEG, molecular weight 3350, 50% w/v (PEG 3350 FW; Sigma): Dissolve 50 g PEG in 30 ml distilled/deionized water, make the volume up to 100 ml and sterilize by autoclaving. Store in a securely capped bottle
- Carrier DNA (Salmon sperm DNA; Sigma D1626), 2 mg ml^{-1}: Dissolve 200 mg in 100 ml sterile TE buffer on a stir plate overnight in a cold room. Dispense 90 ml into sterile 15 ml screw-capped plastic centrifuge tubes and the remainder as 1.0 ml samples into 1.5 ml microfuge tubes. Store at −20°C

B. Rapid Transformation Protocol

We have found that most yeast strains give $2–4 \times 10^3$ transformants per microgram of plasmid DNA when transformed according to Protocol 1. This should be sufficient for the transformation of the bait or reporter plasmid into a host strain. The plates should be incubated at 30°C until colonies can be isolated, typically 2–4 days.

Protocol 1. Rapid Transformation Protocol

1. Inoculate strain(s) to be transformed onto a YPAD plate and grow overnight at 30°C.
2. Suspend a loop full of cells for each transformation reaction (approximately 10–20 µl of cells) in 1 ml of sterile water in a 1.5 ml microfuge tube.
3. Centrifuge at top speed for 30 s and remove the supernatant.
4. Resuspend the cells in 1 ml of 100 mM LiAc and incubate at 30°C for 10–15 min.
5. Prepare a microfuge tube of carrier DNA (2 mg ml⁻¹) by denaturing in a boiling water bath for 5 min and chilling immediately in an ice/water bath.
6. The transformation mix contains the following:

PEG	240 µl
LiAc 1 M	36 µl
Carrier DNA	52 µl
Plasmid DNA + water	32 µl
Total volume	360 µl

   The reagents can either be added directly to the cell pellet in descending order (above) in step 8 and then vortexed, or mixed up in bulk for several transformations.
7. Centrifuge the cell suspension at top speed for 30 s and remove the LiAc.
8. Add 360 µl of transformation mix to each cell pellet and resuspend the cells by vortexing vigorously.
9. Incubate at 30°C for 30 min.
10. Heat shock at 42°C for 20 min.
11. Centrifuge at top speed for 30 s and remove the transformation mix.
12. Pipette 400 µl of sterile water or SC drop-out medium onto the pellet(s) and leave at room temperature for 5 min.
13. Resuspend the cells by vortexing thoroughly.
14. Pipette 200 µl samples onto each of two selective medium plates (SC drop-out medium).
15. Subculture transformants after 2–4 days, incubation at 30°C.

C. High Efficiency Transformation Protocol

The High Efficiency Transformation Protocol typically yields more than 2×10^6 transformants per microgram of plasmid per 10^8 cells (Schiestl *et al.*, 1993). If the strain(s) to be tested do not contain a plasmid they can be grown overnight either on YPAD agar plates or in a shaking incubator in 10 ml of $2 \times$ YPAD broth. Strains carrying a plasmid, a bait or reporter construct, should be incubated overnight in 20 ml of SC drop-out medium. Protocol 2 gives sufficient cells for 10 transformations.

The transformation efficiency (transformants per microgram of plasmid DNA per 10^8 cells) can be calculated as follows:

$$\text{Colonies per plate} \times 10^3 \text{ (dilution factor}^a) \times 10 \text{ (plasmid factor}^b)$$

[a] This accounts for the sequential 10^{-1} and 10^{-2} dilutions in Step 12, Protocol 2.
[b] In Step 10 of Protocol 2 we use 100 ng plasmid DNA per transformation.

An efficiently transformed strain should give 200–400 colonies per plate for a transformation efficiency between 2×10^6 and 4×10^6 µg^{-1} per 10^8 cells. This will yield 200 000–400 000 total transformants per reaction. If the yield of transformants obtained with a particular strain is low it may be possible to increase it by optimizing the carrier DNA concentration and duration of heat shock (Gietz and Woods, 1994). If highly efficient transformation is needed for a screen, it is best to use a yeast strain with good transformation characteristics (Gietz *et al.*, 1997). Examples of strains with excellent transformation characteristics include AB1380, Y190, CTY10-5D (Gietz *et al.*, 1995) and PJ69-4A (James *et al.*, 1996).

D. Library Screening Transformation Protocol

A typical two-hybrid screen requires $1-2 \times 10^7$ transformants. This necessitates a scaled-up version of the High Efficiency Transformation Protocol to allow use of at least 50 µg of the cDNA plasmid library. Simply adding more plasmid to a limited number of cells reduces transformation efficiency (Gietz *et al.*, 1995) and increases the proportion of transformants containing multiple plasmids. This complicates subsequent analysis. Protocol 3 contains the procedure for a 60-fold scale-up. In our hands this results in sufficient transformants for most screens. The strain carrying the plasmid must be incubated overnight in the appropriate SC drop-out medium to ensure plasmid retention. It can then be grown in $2 \times$ YPAD for two divisions with minimal loss of plasmid. If the bait plasmid being used is detrimental to yeast growth and it is decided to co-transform the bait and cDNA plasmids, the host yeast strain can be grown overnight in YPAD and inoculated into 250 ml of $2 \times$ YPAD to give 5×10^6 cells ml^{-1} (Protocol 3, Step 4).

The number of positives from any particular screen may range from 0 to over 1000. If enough transformants are obtained it should be possible to isolate the desired clone from the library.

Protocol 2. High Efficiency Transformation Protocol

1. Inoculate strain(s) to be transformed into liquid medium or onto agar medium and grow overnight at 30°C.
2. (a) Plate cultures: transfer about a 50 μl volume of cells into 50 ml of 2 × YPAD (YPAD or SC; see section III. C) broth in a 250 ml culture flask warmed to 30°C. This should give about 5×10^6 cells ml^{-1}. (b) Liquid cultures: determine the cell titer by measuring the optical density of 10 μl diluted into 1 ml (10^{-2} dilution) or counting the cells using a hemocytometer. For most strains an OD$_{600}$ of 0.1 is equivalent to about 1×10^6 cells ml^{-1}. Pipette 2.5×10^8 cells into 50 ml of 2 × YPAD as in (a) above.
3. Incubate at 30°C with vigorous aeration for 3–4 h. During this time most yeast strains will undergo at least two divisions. Determine OD$_{600}$ to ensure the correct cell concentration has been achieved. The total yield should be about 1×10^9 cells.
4. Harvest the cells by centrifugation and wash them in 10 ml of sterile water.
5. Transfer samples of 1×10^8 cells to microfuge tubes, one for each transformation, and centrifuge for 30 s.
6. Remove the supernatant and resuspend the cells in 1 ml of 100 mM LiAc.
7. Incubate the suspension at 30°C for 10 min.
8. Prepare a microfuge tube of carrier DNA (2 mg ml^{-1}) by denaturing in a boiling water bath for 5 min and chilling immediately in an ice/water bath.
9. Centrifuge the cell suspension at top speed for 30 s and remove the LiAc.
10. The transformation mix contains the following:

PEG	240 μl
LiAc 1 M	36 μl
Carrier DNA	52 μl
Plasmid DNA + water	32 μl
Total volume	360 μl

11. Add transformation mix to each cell pellet and resuspend the cells by vortexing vigorously. The reagents can either be added directly to the cell pellet in descending order and then vortexed or premixed in bulk for several transformations.
12. Incubate at 30°C for 30 min.
13. Heat shock at 42°C for 20 min.
14. Centrifuge at top speed for 30 s and remove the transformation mix.
15. Pipette 1.0 ml of sterile water or SC drop-out medium onto the pellet(s) and incubate at room temperature for 5 min. Resuspend the cells by vortexing vigorously.
16. Dilute 10 μl into 1.0 ml of water and plate 100 μl samples of each transformation onto SC drop-out medium.
17. Incubate at 30°C for 2–4 days and count the numbers of transformants per plate.

Protocol 3. Library Transformation Protocol

1. Inoculate the yeast strain containing the bait plasmid into 100 ml of SC drop-out medium in a 2000 ml flask and incubate at 30°C overnight at 200 rpm.
2. Determine the cell titre and transfer the volume containing 1.5×10^9 cells into sterile centrifuge tubes.
3. Centrifuge the cells at 3000g for 5 min and remove the supernatant.
4. Resuspend the cell pellets in 50 ml of prewarmed (30°C) $2 \times$ YPAD and add to 250 ml of medium in a sterile 4 litre culture flask. This will give a titer of 5×10^6 cells ml^{-1}.
5. Incubate at 30°C and 200 rpm for 4 h or until the titer reaches 2×10^7 cells ml^{-1}.
6. Harvest the cells by centrifugation in 6×50 ml centrifuge tubes at 3000g for 5 min.
7. Wash each pellet in 10 ml sterile water and transfer the suspensions to one tube.
8. Centrifuge and discard the supernatant.
9. Resuspend the cells in 25 ml of sterile 100 mM LiAc and incubate at 30°C for 10 min.
10. Prepare 4 microfuge tubes of carrier DNA (2 mg ml^{-1}) by denaturing in a boiling water bath for 5 min and chilling immediately in an ice/water bath.
11. Pellet the cells and remove the LiAc supernatant.
12. Add the transformation mix to the cell pellet in descending order.

1 PEG	14.40 ml
2 LiAc 1.0 M	2.16 ml
3 SS-DNA (2 mg ml^{-1})	3.12 ml
4 Plasmid DNA (50–100 µg)	1.92 ml
Total volume	21.60 ml

    Vortex vigorously until the cells are completely resuspended.
13. Incubate at 30°C for 30 min.
14. Heat shock at 42°C for 30 min. Mix by inversion for 15 s every 5 min.
15. Centrifuge at 3000g for 5 min and gently resuspend the cells in 20 ml sterile water.
16. Plate 200–400 µl samples of the transformed cells onto 50–100 large (150 × 15 mm) plates of SC drop-out medium selecting for the phenotype required.
17. To determine the overall transformation frequency plate 1 and 5 µl samples into 100 µl puddles of sterile water on normal plates of SC drop-out medium to detect the presence of the plasmid.
18. Incubate the large plates at 30°C for 3–5 days in loosely sealed plastic bags to prevent them drying out.
19. Subculture and test putative positives.

◆◆◆◆◆◆ IV. FREEZING AND STORAGE OF TRANSFORMATION-COMPETENT CELLS

Procedures for the transformation of *Escherichia coli* (e.g. Chung and Miller, 1993) allow the storage of frozen competent cells. Samples can be removed from storage, thawed and transformed at a moment's notice. The LiAc/SS-DNA/PEG protocols described here require forethought. The strain has to be grown at least overnight before transformation if reasonable yields are required. Dohmen *et al.* (1991) have described a protocol (Protocol 4) for making competent cells that can be frozen, stored, thawed and transformed. We have used this protocol and obtained transformation efficiencies of 1×10^4 transformants μg^{-1} with frozen samples of the strain PJ69-4a (James *et al.*, 1996). It is likely that a yeast strain that transforms well with the High Efficiency Transformation Protocol will give good transformation frequencies with frozen competent cells.

Protocol 4. Frozen competent cells

1. Inoculate strain into 10 ml YPAD and grow to an OD_{600} of 0.6–1.0 (about 1×10^7 cells ml^{-1}).
2. Centrifuge and wash cells in 5 ml of 1.0 M sorbitol, 10 mM Bicine-NaOH (pH 8.35), 3% ethylene glycol, 5% DMSO.
3. Centrifuge and resuspend cells in 200 µl of the same solution.
4. Freeze in dry ice/acetone and store at –70°C.
5. Remove from freezer and add plasmid DNA (100 ng to 5 µg) plus 40 µg SS-DNA in a total volume of 25 µl.
6. Thaw by brief incubation at 37°C until melting begins and then mix by hand until completely thawed.
7. Add 1.4 ml of 40% PEG, 0.2 M Bicine-NaOH and mix by inversion.
8. Incubate at 30°C for 60 min.
9. Centrifuge, wash, and resuspend the cells in 0.15 M NaCl, 10 mM Bicine-NaOH (pH 8.35) and plate onto selective medium.

◆◆◆◆◆◆ V. *IN VIVO* MUTAGENESIS BY TRANSFORMATION WITH SYNTHETIC OLIGONUCLEOTIDES

Yamamoto *et al.* (1992) described a procedure for *in vivo* mutagenesis of *CYC1*, coding for iso-1-cytochrome *c*, by LiAc/PEG transformation with specific oligonucleotides. *CYC1* transformants with very low levels of activity were identified against a background of *cyc1* after incubation for 4 days on YEPD. They found that the yield of transformants was dependent on oligonucleotide concentration and that efficiency was maximal (50 transformants per microgram), when 100 µg oligonucleotide was used to transform 1.3×10^8 cells. A recent paper by Linske-O'Connell *et al.*

(1995) indicates that the recovery of transformants is optimized when 1×10^6 cells are treated with 100 µg of oligonucleotide. Yamamoto *et al.* (1992) reported that low concentrations of SS-DNA enhanced yields but that high concentrations were inhibitory. This observation is consistent with the model that we have proposed for the differential binding of double-stranded plasmid DNA and SS-DNA to reactive sites in the yeast cell wall (Gietz *et al.*, 1995).

Site-specific mutagenesis is a powerful technique in studies of the relationship between DNA sequence and protein function. We think that transformation with gene-specific oligonucleotides will be useful for such studies in yeast.

◆◆◆◆◆◆ VI. MICROTITER PLATE TRANSFORMATION

Although we have simplified the High Efficiency Transformation Protocol it is still quite laborious to carry out more than 10 to 20 transformations at any one time. Tests of a number of strains or meiotic segregants for transformation efficiency, or the response of a particular strain to various levels of plasmid, carrier DNA or LiAc, can be done in round-bottomed 96-well microtiter plates. For example, we have used this protocol to isolate and characterize transformation-deficient mutants of yeast and to demonstrate that the mutant phenotypes segregated 2:2 (Woods and Gietz, unpublished results).

Before transformation, cells are incubated in microtiter plates in $2 \times$ YPAD for 4 h in a rotary shaker incubator (New Brunswick G-24) at 400 rpm (Protocol 5). The lids must be loose to ensure adequate aeration. This requires that the bottom of the plates be attached to the platform of an incubator shaker. The "Jitterbug" incubator (Mandel Scientific Co., Guelph, Ontario, Canada) has a rotary motion of 1 mm and speeds adjustable from 575 to 1500 rpm and is made specifically for incubating microtiter plates.

The Microtiter Plate Protocol (Protocol 5) can be used in two ways: (a) multiple transformations of a single yeast strain, or (b) transformation of many different strains such as a mutant screen or meiotic segregants. The medium ($2 \times$ YPAD) and solutions can be dispensed into the wells in a sterile fashion with a multichannel pipettor. Multiple transformations of a single strain can be done by inoculating $2 \times$ YPAD medium with the strain and dispensing samples into the wells with a pipettor. Alternatively, many different strains can be patched onto a large plate of YPAD that has been "imprinted" with a sterile 96-prong replicator by placing it gently onto the medium and pressing lightly to leave an impression.

The replicator is sterilized by flaming in a bunsen burner after dipping the prongs in 95% ethanol. We find it best to position the replicator prongs-up in a beaker and to lower the YPAD plate onto the prongs until all the patches make contact. The surface of the plate is then gently scraped with the prongs to ensure that a sample of cells from each patch adheres to the appropriate prong. It is important to scrape the cells from

Protocol 5. Microtiter Plate Transformation Protocol

1. (a) Subculture strain(s) onto YPAD and incubate overnight at 30°C, or
   (b) patch yeast strains to 96-well grid on a YPAD plate and incubate overnight at 30°C.
   Warm 2 × YPAD and microtiter plates at 30°C overnight.
2. (a) Inoculate sample of culture into 2 × YPAD and adjust cell titer to 1.5×10^7 ml^{-1}. Pipette 150 µl into wells. This will result in 2.25×10^6 cells per well;
   (b) Pipette 150 µl of 2 × YPAD into each well. Use the replicator to transfer samples of cells from the master plate to the microtiter plate wells.
3. Incubate the microtiter plates with the lids on at 400 rpm and 30°C. Start the shaker at low speed and increase to 400 rpm gradually. After incubation for 4 h there should be about 9×10^6 cells per well.
4. Weigh the microtiter plates and prepare a "balance plate" if necessary.
5. Centrifuge the plates at 3500 rpm for 10 min using a microtiter plate rotor in a bench-top centrifuge.
6. Remove the medium by aspiration with a yellow (200 µl) micropipette tip attached to a vacuum line. Be careful not to touch the cell pellet with the tip. Alternatively shake the medium out of the wells into a sink. This takes practice but is much faster than aspiration!
7. Pipette 150 µl water into wells and resuspend the cells at 400 rpm for 1–2 min.
8. Centrifuge as before.
9. Remove the water by aspiration or shaking.
10. Pipette 150 µl 100 mM LiAc into wells and resuspend cells on the shaker for 5 min at 400 rpm. Turn the shaker off and continue incubation for 10 min.
11. Centrifuge the microtiter plate as before.
12. Remove the LiAc by aspiration or shaking.
13. Boil carrier DNA (2 mg ml^{-1}) for 5 min and chill in ice/water.
14. Prepare transformation mix minus PEG (TM-PEG). The volumes below are for a single well; multiply up and allow some extra! Keep the TM-PEG in ice/water:

LiAc 1.0 M	15.0 µl
Carrier DNA (2 mg ml^{-1})	20.0 µl
Plasmid DNA + water	15.0 µl
Total volume	50.0 µl

15. Pipette 50 µl TM-PEG into each well.
16. Incubate the plate on the shaker at 400 rpm for 2 min to resuspend the cell pellets.
17. Pipette 100 µl PEG 3350 (50% w/v) into each well.

18. Incubate the plate at 500 rpm for 5 min, and then without shaking for 25 min.
19. Place the microtiter plate in a ZipLoc™ bag and float it in a 42°C water bath for 50 min.
20. Centrifuge the microtiter plate as before.
21. Remove the transformation mix by aspiration.
22. Pipette 100 µl water into the wells.
23. Resuspend the cells on the shaker for 5–10 min.
24. (a) Pipette 10 µl samples into 1 ml water and plate 100 µl samples onto SC drop-out medium; or
    (b) Sample the wells with a sterile replicator and print onto SC drop-out medium.
25. Incubate at 30°C for 2–4 days and count the numbers of colonies.

the plate gently. Avoid breaking the agar surface as fragments of agar in the transformation mix reduce efficiency. The plate is lifted off and the prongs inspected. Smaller than normal samples can be remedied by transfer of cells from the YPAD plate to specific prongs using a sterile loop or toothpick. The replicator is then taken from the beaker and the prongs lowered into 150 µl samples of 2 × YPAD in the wells of a 96-well microtiter plate and agitated to wash the cells into the medium. This transfers about 4×10^6 cells to each well. The plate is then processed according to steps 5 to 24 of the Microtiter Plate Transformation Protocol. The wells can be sampled with a sterile replicator or a multichannel pipette. The volume of fluid adhering to the prongs of the replicator ranges between 5 and 10 µl. The prongs of the replicator are lifted from the wells of the microtiter plate and the replicator positioned prongs-up in a suitable-sized beaker. A plate of SC drop-out medium is then inverted and lowered onto the prongs of the replicator to transfer a sample onto the agar surface. This sampling procedure can be repeated up to five times to transfer a greater volume of cell suspension onto the selective medium. It is best if the spots are allowed to dry between serial transfers.

VII. CONCLUDING REMARKS

The high efficiency LiAc/SS-DNA/PEG transformation procedure used to induce the yeast *Saccharomyces cerevisiae* to take up exogenous DNA has aided in the development and utilization of sophisticated technologies, such as the one, two and inverted one-hybrid systems. Also, the procedure is important for the efficient construction of libraries of null mutants for the functional analysis of novel yeast genes (Oliver, 1996; Shoemaker *et al.*, 1996). These and other molecular genetic techniques have increased the usefulness of this premier eukaryotic model organism so that it can now be considered a laboratory work horse which will find its way into many research programs. It is likely that if a large number of transformants is needed for any application, one of the protocols we have presented will be suitable.

Acknowledgements

This work was supported, in part, by a grant from the MRC of Canada to R. D. Gietz and from the University of Winnipeg to R. A. Woods. We should also like to acknowledge Ms Sheryl Stoyka, a University of Winnipeg student, who contributed to the development of the microtiter plate transformation assay.

References

Baker, R. (1991). Rapid colony transformation of *Saccharomyces cerevisiae*. *Nucleic Acids Res.* **19**, 1945.

Bartel, P. L. and Fields, S. (1995). Analysing protein–protein interactions using the two-hybrid system. *Methods Enzymol.* **254**, 241–263.

Broach, J. R., Strathern, J. N. and Hicks, J. B. (1979). Transformation in yeast: development of a hybrid cloning vector and isolation of the *CAN1* gene. *Gene* **8**, 121–133.

Chung, C. T. and Miller, R. H. (1993). Preparation and storage of competent *Escherichia coli* cells. *Methods Enzymol.* **218**, 621–627.

Dohmen, R.J., Strasser, A. W. M., Hohner, C. B. and Hollenberg, C. P. (1991). An efficient transformation procedure enabling long-term storage of competent cells of various yeast genera. *Yeast* **7**, 691–692.

Fields, S. and Song, O. (1989). A novel genetic system to detect protein–protein interactions. *Nature* **340**, 245–246.

Gietz, R. D. and Schiestl, R. H. (1991). Applications of high efficiency lithium acetate transformation of intact yeast cells using single-stranded carrier DNA. *Yeast* **7**, 253–263.

Gietz, R. D. and Sugino, A. (1988). New yeast–*E. coli* shuttle vectors constructed with *in vitro* mutagenized yeast genes lacking six-base pair restriction sites. *Gene* **74**, 515–522.

Gietz, R. D. and Woods, R. A. (1994). High efficiency transformation of yeast with lithium acetate. In *Molecular Genetics of Yeast: A Practical Approach* (J. R. Johnston, ed.), pp 121–134. Oxford University Press, Oxford.

Gietz, D., St. Jean, A., Woods, R. A. and Schiestl, R. H. (1992). Improved method for high efficiency transformation of intact yeast cells. *Nucleic Acids Res.* **20**, 1425.

Gietz, R. D., Schiestl, R. H., Willems, A. R. and Woods, R. A. (1995). Studies on the transformation of intact yeast cells by the LiAc/SS-DNA/PEG procedure. *Yeast* **11**, 355–360.

Gietz, R. D. Triggs-Raine, B., Robbins, A., Graham, K. C. and Woods, R. A. (1997). Identification of proteins that interact with a protein of interest: applications of the yeast two-hybrid system. *Mol. Cell. Biochem.*, **172**, 67–79.

Hill, J., Donald, K. A. G. and Griffiths, D. E. (1991). DMSO-enhanced whole yeast cell transformation. *Nucleic Acids Res.* **19**, 5791.

Ito, H., Fukuda, Y., Murata, K. and Kimura, A. (1983). Transformation of intact yeast cells treated with alkali cations. *J. Bacteriol.* **153**, 163–168.

James, P., Halladay, J. and Craig, E. A. (1996) Genomic libraries and a host strain designed for highly efficient two-hybrid selection in yeast. *Genetics* **144**, 1425–1436.

Keegan, K. and Cooper, J. A. (1996). Use of the two-hybrid system to detect the association of the protein-tyrosine-phosphatase, SHPTP2, with another SH2-containing protein, Grb7. *Oncogene* **12**, 1537–1544.

Li, J. L. and Herskowitz, I. (1993). Isolation of ORC6, a component of the yeast origin recognition complex by a one-hybrid system. *Science* **262**, 1870–1874.

Linske-O'Connell, L. I, Sherman, F. and McLendon, G. (1995). Stabilizing amino acid replacements at position 52 in yeast iso-1-cytochrome *c: in vivo* and *in vitro* effects. *Biochemistry* **34**, 7094–7102.

Liu, J., Wilson, T. E., Milbrandt, J. and Johnston, M. (1993). Identifying DNA-binding sites and analyzing DNA binding domains using a yeast selection system. *Methods: A Companion to Methods in Methodology* **5**, 126–137.

Oliver, S. G. (1996). From DNA sequence to biological function. *Nature* **379**, 597–600.

Osborne M. A., Dalton, S. and Kochan, J. P. (1995). The yeast tri-brid system – genetic detection of *trans*-phosphorylated ITAM-SH2-interactions. *Biotechnology* **13**, 1474–1478.

Rose, M. D. (1987). Isolation of genes by complementation in yeast. In *Methods in Enzymology* (S. L. Berger and A. R. Kimmel, eds), Vol. 152, pp. 481–504. Academic Press, San Diego.

Schiestl, R. H. and Gietz, R. D. (1989). High efficiency transformation of intact yeast cells using single-stranded nucleic acids as carrier. *Curr. Genet.* **16**, 339–346.

Schiestl, R. H., Manivasakam, P., Woods, R. A. and Gietz, R. D. (1993). Introducing DNA into yeast by transformation. *Methods: A Companion to Methods in Enzymology* **5**, 79–85.

Sherman, F. (1991). Getting started with yeast. In *Methods in Enzymology* (C. Guthrie and G. R. Fink, eds), Vol. 194, *Guide to Yeast Genetics and Molecular Biology*, pp. 3–21. Academic Press, San Diego.

Shoemaker, D. D., Lashkari, D. A., Morris, D., Mittmann, M. and Davis, R. (1996). Quantitative phenotypic analysis of yeast deletion mutants using a highly parallel molecular bar-coding strategy. *Nature Genetics* **14**, 450–456.

Stearns, T., Ma, H. and Botstein, D. (1990). Manipulating yeast genome using plasmid vectors. *Methods Enzymol.* **185**, 280–297.

Struhl, K., Stinchcomb, D. T., Scherer, S. and Davis, R. W. (1979). High-frequency transformation of yeast: autonomous replication of hybrid DNA molecules. *Proc. Natl. Acad. Sci. USA* **76**, 1035–1039.

Wang, M. M. and Reed, R. R. (1993). Molecular cloning of the olfactory neuronal transcription factor Olf-1 by genetic selection in yeast. *Nature* **364**, 121–126.

Yamamoto, T., Moerschell, R. P., Wakem, L. P., Ferguson, D. and Sherman, F. (1992). Parameters affecting the frequencies of transformation and co-transformation with synthetic oligonucleotides in yeast. *Yeast* **8**, 935–948.

5 PCR-Based Gene Targeting in *Saccharomyces cerevisiae*

Achim Wach[1], Arndt Brachat[1], Corinne Rebischung[1], Sabine Steiner[1], Karine Pokorni[2], Stephan te Heesen[2] and Peter Philippsen[1]
[1]*Institut für Angewandte Mikrobiologie, Biozentrum, Universität Basel, Basel, Switzerland, and*
[2]*Mikrobiologisches Institut, ETH Zürich, Zürich, Switzerland.*

◆◆◆

CONTENTS

List of Abbreviations

GFP	Green fluorescent protein
LFH	Long flanking homology
LiAc	Lithium acetate
ORF	Open reading frame
PCR	Polymerase chain reaction
PEG	Polyethylene glycol
SFH	Short flanking homology
SS-DNA	Single-stranded carrier DNA
YPD	Yeast peptone glucose medium (*see Appendix II: Table I*)

◆◆◆◆◆◆ I. INTRODUCTION

Transformations with polymerase chain reaction (PCR)-generated DNA fragments for one-step gene targeting in *Saccharomyces cerevisiae* were first described by McElver and Weber (1992) and Baudin *et al.* (1993). Several groups have published modifications of this method (Wach *et al.*, 1994, 1997; Amberg *et al.*, 1995; Eberhardt and Hohmann, 1995; Horton, 1995; Laengle-Ronault and Jacobs, 1995; Lorenz *et al.*, 1995; Manivasakam *et al.*, 1995; Fairhead *et al.*, 1996; Güldener *et al.*, 1996; Maftahi *et al.*, 1996; Mallet and Jacquet, 1996; Wach, 1996) and PCR-based

METHODS IN MICROBIOLOGY, VOLUME 26
ISBN 0–12–521526–6

gene targeting has now become a standard technique in many yeast laboratories.

The purpose of this chapter is to guide newcomers through the different steps of this very efficient method and to discuss possible applications for basic and advanced gene analyses. In the first part we will describe our experience with the selection marker *kanMX*, a hybrid gene expressing a bacterial aminoglycoside phosphotransferase under control of a strong fungal promotor (Wach *et al.*, 1994; Steiner *et al.*, 1995). The introduction of this marker led to two essential improvements. First, gene targeting was no longer dependent on the presence of auxotrophic markers in the host strain and, second, high backgrounds of false-positive transformants often obtained in PCR-targeting using *S. cerevisiae* genes as selectable markers were virtually eliminated because *kanMX* lacks homology to yeast DNA. In the second part of this chapter we will summarize the features of several new modules for PCR-targeting: a heterologous *HIS3* marker called *HIS3MX6*, two green fluorescent protein (GFP) reporter modules with *HIS3MX* or *kanMX* as selection marker which are useful for generating carboxy-terminal GFP fusions, as well as a *kanMX-GAL1* promotor module for making promotor exchanges, and a *kanMX-GAL1-GFP* reporter module for generating amino-terminal fusions.

◆◆◆◆◆◆ II. ONE-STEP GENE DELETIONS USING *kanMX4*

A. The Plasmid pFA6a-*kanMX4*

pFA6a-*kanMX4* belongs to a series of marker/reporter plasmids designed for classical or PCR-based gene targeting in fungi by homologous recombination (Wach *et al.*, 1994; Steiner *et al.*, 1995; Wach, 1996; Wach *et al.*, 1997). Figure 1 shows a restriction map of pFA6a-*kanMX4* and a partial sequence of its multiple cloning site. The 1.44 kb *kanMX4* marker is a hybrid of the open reading frame (ORF) of the kanamycin resistance gene *kan^r* from transposon Tn903 (Oka *et al.*, 1981) flanked by promotor and terminator sequences of the strongly expressed *TEF* gene of *Ashbya gossypii* (Steiner and Philippsen, 1994).

S. cerevisiae as well as other yeasts and certain filamentous fungi acquire resistance to the drug geneticin (G418) when transformed with a *kanMX* module. This selection system was first described by Jimenez and Davies (1980) and was further developed by Hadfield *et al.* (1990). Working with *kanMX* has the additional advantage that it renders *Escherichia coli* resistant to kanamycin.

B. Gene Targeting Using SFH-PCR Amplified *kanMX4*

Targeted alterations of the *S. cerevisiae* genome with *kanMX* bearing PCR-generated short flanking homology (SFH-PCR) has been applied successfully in many laboratories. For example, we have replaced 20 different

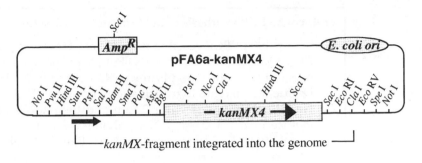

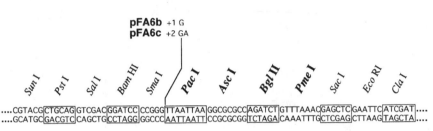

Figure 1. Restriction map and sequence of the multiple cloning site of pFA6a-kanMX4. The 2.5 kb cloning vector pFA6a is a derivative of pSP72 (Promega Corp., Madison, WI) with a newly designed multiple cloning site (Wach *et al.*, 1994). The 1.44 kb heterologous *kanMX4* selection module has been cloned into the *Pme*I site of this vector. The black arrow between *Nco*I and *Sca*I represents the bacterial *kan*ᵣ ORF inserted between fungal transcription control sequences. The complete sequence can be retrieved from EMBL Data Base Acc. No. AJ002680. *kanMX4* differs from the original *kanMX0* module in plasmid pAG224 (Steiner *et al.*, 1995) by a deletion of 460 bp downstream of the *kan*ᵣ coding sequence. This shortening is advantageous in PCR experiments. Grey arrows mark the recommended plasmid specific sequences in the chimeric PCR primers (see also Figure 2). The highlighted restriction sites can be used for addition of other modules (see below, and Wach *et al.*, 1994, 1997).

ORFs with *kanMX4* and found that only six out of 350 tested geneticin-resistant transformants carried incorrectly targeted *kanMX4*. The principle for construction of SFH-PCR fragments aiming at deletions of ORFs is shown in Figure 2 and the features of the two chimeric PCR primers S1 and S2 are described in the legend to this figure. The recommended PCR steps resulting in 1–5 µg of *kanMX4* modules flanked by 40–45 bp of target guide sequences are outlined in Protocol 1.

The 3′ ends of the chimeric primers consist of 18 to 19 nucleotides designed to bind to the *kanMX*-adjacent multiple cloning site rather than to the *kanMX* sequence (see Figure 1). This has two advantages. First, the same set of oligonucleotides can be used for PCR-targeting if a second heterologous marker is needed (*e.g. HIS3MX6*) for a two-step construction of homozygous deletants (see below). Second, after cloning these ORF replacement cassettes, convenient restriction sites are available for marker exchange or for reporter integration. If the same linker sequences are selected by each laboratory and if the PCR fragments are cloned, preferably in a way outlined at the bottom of Figure 6, the

Protocol 1. PCR synthesis of SFH-disruption cassettes

1. Set up 100 µl PCR reaction mix [10 mM Tris-HCl (pH 8.3), 50 mM KCl, 2.5 mM MgCl$_2$, 0.2 mM dNTP, 1 µM primers, 0.2–0.5 µg (≈1 nM) plasmid template (e.g. pFA6-kanMX4 cut with NotI)] and vortex. Divide the mix into two 50 µl aliquots and run PCR.

2. Heat PCR mix to 94°C for 2 min and then add 1 µl DNA polymerase mix (2.5 U *Taq* DNA polymerase and 0.5 U DNA polymerase with 3′–5′ exonuclease activity, like *Vent* DNA polymerase, to decrease the error rate during DNA polymerization and to prevent non-specific 3′ nucleotide overhangs).

3. Run PCR:
   120 s, 94°C (initial denaturation)
   (hot start)
   30 s, 94°C (denaturation)   ⎫ this sequence
   30 s, 54°C (annealing)   ⎬ is repeated
   90 s, 72°C (elongation)   ⎭ 20 times
   120 s, 72°C (final elongation)
   Samples are then cooled to 2°C

4. Add 5 µl of 3 M sodium acetate followed by 120 µl of ethanol, vortex and centrifuge at 12 000 rpm for 15 min.
   Optional: Samples can be treated with phenol/chloroform prior to precipitation

5. Wash pellet with 70% ethanol, air-dry for 10 min, and resuspend pellet in 10 µl of TE.

6. Analyse 1 µl by agarose gel electrophoresis.

7. Use 5 µl (usually 1–5 µg of DNA) to transform *Saccharomyces cerevisiae*.

whole yeast community will benefit in the near future by having access to several thousand isostructural ORF replacement cassettes. This goal has already been achieved for nearly 1000 novel ORFs investigated by EUROFAN, the European Yeast Functional Analysis Network.

PCR products synthesized with S1–S2 primers are directly used for transformation of *S. cerevisiae* cells (see Protocol 2 for experimental procedure). An example showing the size heterogeneity of colonies on a 2-day-old transformation plate and the result of restreaking putative transformants on a second selection plate is shown in Figure 3. Generally, transformants that form large colonies on the original transformation plate have the *kanMX* module integrated into their genome.

C. Verification of Correct ORF Replacements

As a rule, both novel joints that are formed upon integration of the marker must be verified. This can be done either by Southern blot analysis or much faster, and just as reliably, by PCR experiments. The PCR method

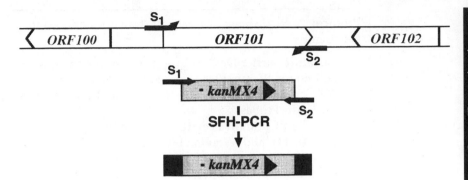

Figure 2. Sequences of SFH-PCR primers for complete ORF deletions. The chimeric PCR primers have the following features. Primer S1 consists of 40–45 nucleotides of the coding strand upstream of the second codon followed by 18 nucleotides <u>CGTACGCTGCAGGTCGAC</u>-3′ (*Sun*I, *Pst*I, *Sal*I restriction sites) upstream of *kanMX4* (see Figure 1). Primer S2 consists of 40–45 nucleotides of the non-coding strand downstream of the last sense codon followed by 19 nucleotides <u>ATCGATGAATTCGAGCTCG</u>-3′ (*Cla*I, *Eco*RI, *Sac*I restriction sites) downstream of *kanMX4*. The potential problem encountered with duplicated genes can be minimized by selecting target guide sequences 5′ and 3′ of the ORF since, in most cases, only the coding sequences are conserved. Restriction enzyme sites may also be included at the 5′-extremities of the primers used to amplify the marker module. This facilitates subsequent cloning of the PCR product. However, non-homologous DNA sequences at the ends of the transforming DNA with SFH regions may lead to a decrease in integration frequencies. A convenient cloning vector for PCR fragments (pUG7) allowing precise cleavage at the vector-insert boundary has been described (Güldener *et al.*, 1996).

Figure 3. Growth of *kanMX4* transformed *S. cerevisiae* cells and background growth of presumably abortive transformants. The left picture shows a Petri plate (YPD, 2% yeast extract, 1% peptone, 2% glucose, with 200 mg liter⁻¹ geneticin) 2 days after plating 2×10^8 competent FY1679 (diploid S288C derivative) transformed with 2 µg of *kanMX4* SFH-PCR product. Cells of four big colonies (labeled 1–4) and four colonies of intermediate size (labeled A–D) were picked and restreaked on two YPD-geneticin plates. These plates, after 2 days incubation at 30°C, are shown in the middle and right of the figure, respectively. Depending on the host strain, the transformation plates may show a lawn after 1 or 2 days. In that case the whole lawn is replica plated on 200 mg ml⁻¹ geneticin and transformants appearing then are restreaked. Most suppliers of geneticin specify the activity of the batches and the concentration has to be adjusted to 200 mg active geneticin per liter of medium.

Protocol 2. *Saccharomyces cerevisiae* transformation and selection on geneticin

1. Denature carrier DNA (10 µg µl^{-1} shared herring sperm DNA) by boiling for 10 min. Place immediately on ice after boiling.
2. Add 5 µl of carrier DNA to each of the PCR-DNA containing Eppendorf tubes.
3. Add 50 µl lithium acetate (LiAc)-treated competent *S. cerevisiae* cells (Gietz and Woods, 1994) to each tube, vortex briefly and incubate for 20 min at 30°C.
4. Pipette 300 µl polyethylene glycol (PEG)/LiAc solution into each tube, vortex briefly and incubate for 20 min at 30°C.
5. Heat shock the transformation tubes at 42°C for 20 min (the duration must be optimized for every strain). The addition of dimethyl sulfoxide to 10% (v/v) final concentration prior to the heat shock improves the efficiency of transformation up to 10-fold. During the heat shock period, fill one culture tube per transformation with 3 ml of YPD (2% (w/v) yeast extract, 1% (w/v) peptone, 2% (w/v) glucose).
6. Centrifuge the transformation tubes at 5000 rpm for 1 min in a microfuge, remove the PEG/LiAc supernatant and resuspend the cells in 1 ml YPD.
7. Transfer the cell suspension into the prepared YPD culture tubes and incubate cultures for 2–3 h at 30°C with shaking (i.e. for one doubling time).
8. Take 1.5 ml aliquots of each culture and transfer them into Eppendorf tubes, centrifuge at 5000 rpm for 1 min in a microfuge, remove 1.1 ml of the supernatant and resuspend the cells in the remaining volume of supernatant.
9. Plate the suspension on YPD plates containing 200 mg liter^{-1} geneticin (watch for suppliers specification of activity and adjust amount accordingly).
10. Incubate plates at 30°C for 2–3 days. After this period, a few big (3–4 mm in diameter) as well as many small (0.2–2 mm in diameter) colonies are seen on the plates. The small ones are the background of abortive transformants that arise, most probably, from cells that carry non-integrated *kanMX* DNA and produced sufficient aminoglycoside-phosphotransferase to inactivate geneticin during a few rounds of division.
11. Purify transformed cells from the background by streaking out cells from each big colony on YPD-geneticin. Only those clones that can form **colonies from single cells** are putative positive integrants. Note: Do not grow putative transformants as patches on the second selection plate.

used for verification (see Protocol 3 for experimental procedure) is based on a technique for rapid analysis of the mating type of *S. cerevisiae* cells (Huxley *et al.*, 1990). Purification of chromosomal DNA is not necessary. Instead, whole *S. cerevisiae* cells are used. Figure 4A demonstrates the rationale of this method. The criteria for selection of the sequences of the verification primers are summarized in the legend to this figure. An example of the PCR verification of a targeted ORF deletion in diploid *S. cerevisiae* cells is shown in Figure 4B. The data obtained with such analytical PCR experiments are probably more reliable than data obtained by Southern blot analysis because the focus is on the verification of both novel DNA joints rather than on mobility shifts of restriction fragments. This especially holds for verification of replacements of long ORFs. In rare cases, e.g. targeting of members of highly conserved gene families in subtelomeric regions, it may be necessary to verify marker integrations by Southern hybridization to electrophoretically separated chromosomes.

Protocol 3. PCR analysis of geneticin-resistant transformants

1. Transfer single colonies from the restreaked putative transformants into PCR reaction tubes. Do this by picking cell material corresponding to 1 mm in diameter of a single colony with a sterile yellow pipette tip and patch the cells on the wall at the bottom of the tube. Include also one sample with wild-type cells as control.

2. Heat cell pellets in a microwave for 1 min (maximal power). Alternatively, resuspend cells in 50 µl of Zymolyase (20 U ml⁻¹) and incubate for 10 min at room temperature. Collect the cells by centrifugation at 5000 rpm for 1 min in a microfuge, remove the supernatant, and heat the cells for 5 min at 92°C. Note: This method is a little bit more time-consuming but leads to higher amounts of PCR product compared with the "microwave" procedure.

3. Add 25 µl PCR reaction mix [10 mM Tris-HCl (pH 8.3), 50 mM KCl, 2.5 mM $MgCl_2$, 0.2 mM dNTP, 1 µM primers V1, V2, K2 or V3, K3, V4, 1–2 U *Taq* DNA polymerase] to each pellet, vortex, and run PCR. Note: The reaction suspension must be kept on ice until the PCR is started because proteases secreted by *S. cerevisiae* cells might inactivate the DNA polymerase.

4. Run PCR:
   120 s, 94°C (initial denaturation)
   30 s, 94°C (denaturation)
   30 s, 50°C (annealing)   this sequence is repeated 30 times
   90 s, 72°C (elongation)
   Samples are then cooled to 2°C

5. Centrifuge the samples at 10 000 rpm for 1 min in a microfuge and analyze 10 µl of each supernatant by agarose gel electrophoresis.

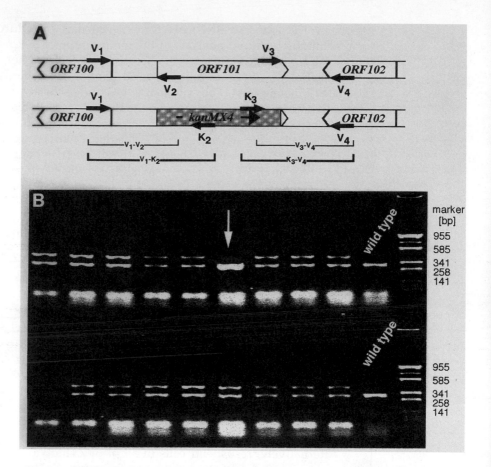

Figure 4. PCR analysis of targeted heterozygous ORF deletions. (A) Genomic context after replacement of one of the two ORF copies by *kanMX4*. Two PCR primers (V1 and V4) are designed to bind 300–400 bp outside of the target locus, two other primers (V2 and V3) to bind within the target locus, and two primers (K2 and K3) to bind within the *kanMX4* module. These primers should be about 20 nucleotides long and the calculated melting temperature should be at least 55°C. In addition, primers V1 to V4 must be checked for strong homology elsewhere in the genome. If necessary, V1 and V4 have to represent sequences further upstream or downstream, respectively. PCR experiments are performed with microwave-treated colonies (see Protocol 3) or isolated genomic DNA from transformants and wild-type cells using combinations of three of the primers (either V1, V2, K2 or K3, V3, V4). Correct integration of the marker DNA is verified by the appearance of two PCR products of predictable length as indicated by the brackets: one product is diagnostic for the wild-type allele (V1–V2 or V3–V4 amplification product) and a second product is diagnostic for the mutated allele (V1–K2 or K3–V4 amplification product). Incorrect transformants will only yield the amplification products of the wild type allele. (B) Agarose-gel electrophoresis of PCR products. In this example, 18 geneticin-resistant transformants and two wild-type colonies were analysed using the primer combination K3, V3, V4. One transformant (marked with the white arrow) carries an incorrectly targeted *kanMX4* marker. The size marker is pUC19 DNA cleaved with *Sau*3A. The concentration of agarose is 1.5%.

D. Gene Targeting Using LFH-PCR Amplified *kanMX4*

Successful targeting of SFH-PCR products depends on perfect homology between the short ends of the transforming DNA and the target locus. Thus, sequence polymorphisms known to exist in different *S. cerevisiae* strains will decrease or prohibit homologous recombination when PCR-made molecules with short target guide sequences are used. Therefore, as an alternative to SFH-PCR, PCR strategies were developed to synthesize selection markers with long flanking homologies (Wach, 1996; Amberg *et al.*, 1995; Maftahi *et al.*, 1996; Mallet and Jacquet, 1996). The long flanking homology technique (LFH-PCR) that we have introduced uses dual-step PCR to produce sufficient quantities of gene targeting cassettes. In the first PCR, two pairs of primers and target cell genomic DNA (or a cosmid clone) are combined in the same reaction tube. One primer pair (L1 and L2) is used to amplify the 5'-region directly upstream of the target gene's second codon (5'-PCR fragment) and the other primer pair (L4 and L3) to amplify the DNA region immediately downstream of its last codon (3'-PCR fragment). The sequences of the 3'-halves (19–22 bases) of the primers L2 and L3 are derived from the *S. cerevisiae* target locus. But, these primers also carry 5'-extensions (20–25 bases) homologous to *kanMX*, thus generating at one end of each fragment short overlapping homologies to this selection marker. In the second PCR with *kanMX* DNA as template, one strand of the 5'-PCR fragment and one strand of the 3'-PCR fragment then serve as long primers to produce the ORF targeting cassette. In this complex reaction, several PCR fragments are produced, i.e. marker with 5'- and 3'-extension (LFH-PCR fragment) but also *kanMX* modules carrying only 5'- or 3'-extensions. The amount of the LFH-PCR fragment can be increased by addition of an excess of the two outer primers (L1 and L4). The yield of the ORF targeting cassette is not as high as with SFH-PCR. However, the longer homology regions lead to improved transformation efficiencies (e.g. in *Saccharomyces* strains with low levels of sequence heterogeneity). A detailed description of the experimental procedure can be found elsewhere (Wach, 1996).

◆◆◆◆◆◆ III. OTHER HETEROLOGOUS MODULES AND THEIR POTENTIAL APPLICATIONS

Figure 5 outlines the structure of members of a series of selection and reporter modules cloned in pFA6a. Reporter modules still carrying the *Pac*I site are also available as clones in pFA6b and pFA6c (altered reading frames in the multiple cloning site; see Figure 1). *Pac*I and *Pme*I are useful 8 bp restriction sites that were used in the cloning of the modules and which are preserved in some (but not all) of the constructs. Not shown in this figure are *LacZ-kanMX* modules (Wach *et al.*, 1994). The diverse potential of these modules for efficient construction of a variety of gene mutations is summarized in Figure 6.

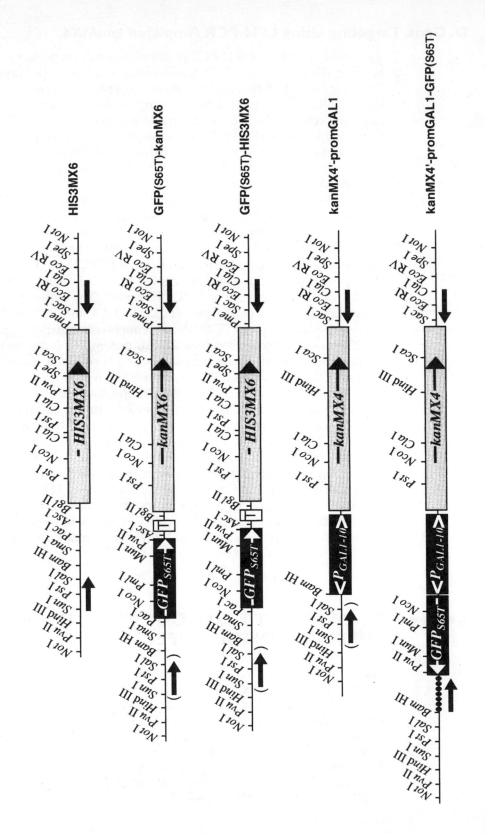

HIS3MX6

GFP(S65T)-kanMX6

GFP(S65T)-HIS3MX6

kanMX4'-promGAL1

kanMX4'-promGAL1-GFP(S65T)

The plasmid pFA6a-*HIS3MX6* (Wach *et al.*, 1997) is similar to pFA6a-*kanMX4* except that the coding sequence of the *kan*[r] gene has been replaced by the coding sequence of the *Schizosaccharomyces pombe HIS3* homolog (Erickson and Hannig, 1995) and that there is a *PmeI* site 3' in this module which is lacking in *kanMX4* but is present in *kanMX6* (Wach, 1996). With less than 5% incorrectly targeted transformants, *HIS3MX* functions almost as reliably as the geneticin resistance marker *kanMX*. We have analyzed the *HIS3* locus in some of the His[+] transformants that are rare false positives by genomic PCR and did not find any evidence for gene conversion of the mutated endogenous *his3* allele by the *S. pombe* homolog. The *HIS3MX6* module allows, for example, constructions of homozygous ORF deletions by two consecutive transformations in both cases using the same primer pair as shown at the bottom of Figure 6. This experimental route is necessary when the ORF deletion in the haploid causes sterility or when the diploid strain is difficult to sporulate in the first place. A certain background of *kanMX4* replacements by *HIS3MX6* cannot be avoided as a result of the second transformation. Therefore, pre-screening of His[+] and geneticin-resistant cells is recommended keeping in mind that the geneticin selection does work on full medium (YPD) but not on high salt-containing minimal medium (Hadfield *et al.*, 1990).

Figure 5. Map of heterologous markers and reporters cloned in pFA6 vectors. For further details of the pFA6 plasmid backbone see Figure 1 and Wach *et al.* (1994). The construction of the heterologous *HIS3MX6* module (*his3*-complementing ORF from *S. pombe* plus promotor and terminator of the *A. gossypii TEF* gene) and the double modules *GFP(S65T)-kanMX6* and *GFP(S65T)-HIS3MX6* are described in Wach *et al.* (1997). The *kanMX4'-promGAL1* module was constructed by integrating the *S. cerevisiae* divergent *GAL1-10* promotor (Johnston and Davis, 1984) as an *EcoRI*(blunt)-*BamHI* fragment into *BglII*(blunt) and *BamHI* cleaved pFA6a-*kanMX4* (see Figure 1). The prime is added to *kanMX* in this and the following module in order to remind investigators about the inversed transcription orientation of *kanMX* with respect to the genomic ORF, the expression of which will be driven by the *GAL1* promotor. The sequence of the Crick strand at the end of the *GAL1* promotor is as follows: 5'-*AGGAGAAAAAAC*CCCGGATCC (original yeast sequence in italics, *BamHI* site underlined). The *kanMX4'-promGAL1-GFP(S65T)* module was constructed in two steps. First, the *GFP(S65T)* ORF was amplified with the following PCR primers: 5'-GGAAGATCTTATA *ATG AGT AAA GGA GAA GAA CTT* (*BglII* site underlined, the first seven codons of *GFP* in italics) and 5'-CGCGGATCC AGC ACC (TGG) AGC ACC ACC *TTT GTA TAG TTC ATC CAT GCC* (*BamHI* site underlined, sequence of hinge region coding for five (optional, six) amino acids written as anti-codons, and the sequence of the non-coding strand for the last seven codons of *GFP* in italics). Second, the *BglII-BamHI* cleaved PCR product was inserted into *BamHI* cleaved *kanMX4'-promGAL1*. Grey arrows show the recommended sites for plasmid-specific sequences in the chimeric PCR primers. The recommended primer binding site downstream of *GAL1* in *kanMX4'-promGAL1* may be moved closer to the *GAL1* sequence in order to shorten the 5' non-transcribed sequence of the targeted ORF. The recommended primer binding site downstream of GFP in *kanMX4'-promGAL1-GFP(S65T)* may include the *BamHI* site, thereby prolonging the hinge region Gly-Gly-Ala-(Pro)-Gly-Ala by glycine and serine.

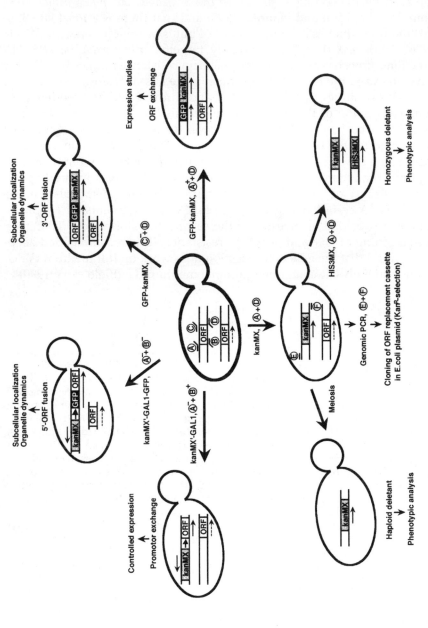

Figure 6. Alternative applications of PCR-based gene targeting. Starting from a diploid cell of a reference strain and four types of gene-specific chimeric primers of 60–65 bases (A to D) several gene alterations can be constructed within 2–3 weeks. The chimeric primers selected for the different pathways in functional analyses of novel genes are schematically shown in the center diploid cell. The black parts represent *S. cerevisiae* sequences and the gray parts module-specific sequences. Depending on the application, primers A or B carry the start codon (designated as A⁺ or A⁻ and B⁺ or B⁻, respectively). Primer C always lacks the stop codon (C⁻). Primer D may or may not contain the stop codon. Regarding the primers, lines drawn parallel to the ORF represent sequences homologous to the target ORF as the 5′-end of the primer, and lines diverging from the ORF represent sequences at the 3′-end of the primer that are homologous to the module to be integrated.

We also cloned the GFP gene from *Aequorea victoria* into pFA-plasmids carrying *HIS3MX6* or *kanMX6* as selection module. The GFP reporter consists of the 0.7 kb GFP(S65T) coding sequence (Heim *et al.*, 1995) lacking the ATG. This sequence was cloned together with the *S. cerevisiae ADH1* terminator (labeled with T in Figure 5) upstream of the selection modules. PCR-synthesized 2.4 kb long GFP-marker modules flanked by short homologies of 40–45 bp were successfully targeted to the carboxy-terminus of several *S. cerevisiae* genes (Wach *et al.*, 1997). The principle of this procedure is outlined in the upper right part of Figure 6. We established that only about 10% of the transformants carried inactivating PCR-introduced mutations in the GFP reporter. With such a small number of clones carrying inactivating GFP mutations, characterization of GFP fusion proteins of unknown localization can be done with confidence by inspection of only a few independent clones. An additional application of these GFP double modules is the replacement of coding sequences by GFP (right part of Figure 6). The principle of this strategy was used, for example, to determine the promotor strength from all genes of *S. cerevisiae* chromosome VIII (M. Johnston and J. H. Hegemann, personal communication).

A targeting module is now available for experiments which aim to place a gene under the control of a regulatable promotor at the genomic level (left part of Figure 6). This *kanMX4'-promGAL1* double module carries at one end sequences homologous to the *S. cerevisiae GAL1-GAL10* locus. However, the yield of correctly targeted modules is still high, despite the presence of *S. cerevisiae GAL1* sequences close to one end of the SFH-PCR product. This is presumably because the SFH is separated by 20 or more base pairs from the *GAL1* promotor sequence (see legend to Figure 5).

For GFP fusions to the amino-terminal end of proteins (upper left part of Figure 6), we have constructed a GFP reporter under control of the *GAL1* promotor. In this module the last codon (Lys) of the GFP ORF is fused to the coding sequence of a hinge region, as indicated by dots in Figure 5. Care should be taken in designing the chimeric PCR primers using this and the *kanMX4'-promGAL1* module because of the inverted orientation of *kanMX* with respect to the targeted genomic ORF. For details see legend to Figure 5.

Plasmid vectors have been described that allow, after cloning of ORFs, the expression of N- or C-terminal GFP-fusion proteins in *S. cerevisiae* (Kahana *et al.*, 1995; Niedenthal *et al.*, 1996). Because of genetic instability of plasmids, targeted integration of the GFP reporter to genomic loci could be advantageous if genetically stable GFP fluorescence is desired, for example, in studying organelle dynamics with the help of fluorescence video microscopy.

◆◆◆◆◆◆ IV. PLASMID REQUESTS

Requests for pFA plasmids. Send plasmid request to Peter Philippsen (FAX: int-41-61-2672118; or E-mail: Sekretariat4@UbaClu.UniBas.ch). The following plasmids (see Figures 1 and 5 for details) are available for scientific research: pFA6a-*kanMX4*, pFA6a-*kanMX6*, pFA6a-*HIS3MX6*, pFA6a-*GFP(S65T)-kanMX6*, pFA6a-*GFP(S65T)-HISMX6* (all *GFP*-containing plasmids are also available in frames b or frames c), pFA6a-*kanMX4'-promGAL1*, and pFA6a-*kanMX4'-promGAL1-GFP(S65T)*. Investigators who plan to use one or more of the plasmids for commercial purposes should please state this fact in their request.

For plasmids containing the *GFP(S65T)* variant, a Howard Huges Medical Institute material transfer agreement has first to be signed. To obtain this document contact Roger Y. Tsien, Howard Huges Medical Institute, Cellular and Molecular Medicine, University of California, San Diego, 9500 Gilman Drive, La Jolla, CA 92093-0647 at FAX: int(1)619-534 5270 and mention that you want to use the pFA plasmids with *GFP(S65T)* registered on A. Wach and P. Philippsen. We may not ship plasmids before you have sent us a copy of the signed Howard Huges Medical Institute material transfer agreement.

Acknowledgements

The experimental work was supported by a grant from the University of Basel to P.P. and by grants from the Swiss Federal Office for Education and Science to A.W. and P.P. (BBW no. 95.0191-12 and 93.0078). K.P. and S.t.H. acknowledge the support of Markus Aebi, ETH Zürich, in whose lab the *kanMX'-promGAL1* was constructed.

References

Amberg, D. C., Botstein, D. and Beasley, E. M. (1995). Precise gene disruption in *Saccharomyces cerevisiae* by double fusion polymerase chain reaction. *Yeast* **11**, 1275–1280.

Baudin, A., Ozier, K. O., Denouel, A., Lacroute, F. and Cullin, C. (1993). A simple and efficient method for direct gene deletion in *Saccharomyces cerevisiae*. *Nucleic Acids Res.* **21**, 3329–3330.

Eberhardt, I. and Hohmann, S. (1995). Strategy for deletion of complete open reading frames in *Saccharomyces cerevisiae*. *Curr. Genet.* **27**, 306–308.

Erickson, F. L. and Hannig, E. M. (1995). Characterization of *Schizosaccharomyces pombe his1* and *his5* cDNA. *Yeast* **11**, 157–167.

Fairhead, C., Llorente, B., Denis, F., Soler, M. and Dujon, B. (1996) New vectors for combinatorial deletions in yeast chromosomes and for gap-repair cloning using "split-marker" recombination. *Yeast* **12**, 1439–1457.

Gietz, R. D. and Woods, R. A. (1994). High efficiency transformation with lithium acetate. *Molecular Genetics of Yeast, A Practical Approach* (J. R. Johnston, ed.), pp. 121–134. IRL Press, Oxford.

Güldener, U., Heck, S., Fielder, T., Beinhauer, J. and Hegemann, J. H. (1996) A new efficient gene disruption cassette for repeated use in budding yeast. *Nucleic Acids Res.* **24**, 2519–2524.

Hadfield, C., Jordan, B. E., Mount, R. C., Pretorius, G. H. and Burak, E. (1990) G418-resistance as a dominant marker and reporter for gene expression in *Saccharomyces cerevisiae*. *Curr. Genet.* **18**, 303–313.

Heim, R., Cubitt, A. B. and Tsien, R. Y. (1995) Improved green fluorescence. *Nature* **373**, 663–664.

Horton, R. M. (1995) PCR-mediated recombination and mutagenesis. SOEing together tailor-made genes. *Mol. Biotechnol.* **3**, 93–99.

Huxley, C., Green, E. D. and Dunham, I. (1990) Rapid assessment of *S. cerevisiae* mating type by PCR. *Trends Genet.* **6**, 236.

Jimenez, A. and Davies, J. (1980) Expression of transposable antibiotic resistance elements in *Saccharomyces*. *Nature* **287**, 869–871.

Johnston, M. and Davis, R. W. (1984) Sequences that regulate the divergent *GAL1-Gal10* promoter in *Saccharomyces cerevisiae*. *Mol. Cell. Biol.* **4**, 1440–1448.

Kahana, J. A., Schnapp, B. J. and Silver, P. A. (1995) Kinetics of spindle pole body separation in budding yeast. *Proc. Natl. Acad. Sci. USA* **92**, 9707–9711.

Laengle-Ronault, F. and Jacobs, E. (1995) A method for performing precise alterations in the yeast genome using a recyclable selectable marker. *Nucleic Acids Res.* **23**, 3079–3081.

Lorenz, M. C., Muir, R. S., Lim, E., McElver, J., Weber, S. C. and Heitman, J. (1995) Gene disruption with PCR products in *Saccharomyces cerevisiae*. *Gene* **158**, 113–117.

Maftahi, M., Gaillardin, C. and Nicaud, J.-M. (1996) Sticky-end polymerase chain reaction method for systematic gene disruption in *Saccharomyces cerevisiae*. *Yeast* **12**, 859–868.

Mallet, L. and Jacquet, M. (1996) Intergenic flip flop, a method for systematic gene disruption and cloning in yeast. *Yeast* **12**, 1351–1357.

Manivasakam, P., Weber, S. C., McElver, J. and Schiestl, R. H. (1995) Microhomology mediated PCR-targeting in *Saccharomyces cerevisiae*. *Nucleic Acids Res.* **23**, 2799–2800.

McElver, J. and Weber, S. (1992) Flag N-terminal epitope overexpression of bacterial alkaline phosphatase and flag C-terminal epitope tagging by PCR one-step targeted integration. *Yeast* **8** (Suppl.), S627.

Niedenthal, R. K., Riles, L., Johnston, M. and Hegemann, J. H. (1996) Green fluorescent protein as a marker for gene expression and subcellular localization in budding yeast. *Yeast* **12**, 773–786.

Oka, A., Sugisaki, H. and Takanami, M. (1981) Nucleotide sequence of the kanamycin resistance transposon Tn903. *J. Mol. Biol.* **147**, 217–226.

Steiner, S. and Philippsen, P. (1994) Sequence and promoter analysis of the highly expressed TEF gene of the filamentous fungus *Ashbya gossypii*. *Mol. Gen. Genet.* **242**, 263–271.

Steiner, S., Wendland, J., Wright, M. C. and Philippsen, P. (1995) Homologous recombination as the main mechanism for DNA integration and cause of rearrangements in the filamentous ascomycete *Ashbya gossypii*. *Genetics* **140**, 973–987.

Wach, A. (1996) PCR-synthesis of marker cassettes with long flanking homology regions for gene disruptions in *S. cerevisiae*. *Yeast* **12**, 259–265.

Wach, A., Brachat, A., Poehlmann, R. and Philippsen, P. (1994) New heterologous modules for classical or PCR-based gene disruptions in *Saccharomyces cerevisiae*. *Yeast* **10**, 1793–1808.

Wach, A., Brachat, A., Alberti-Segui, C., Rebischung, C. and Philippsen, P. (1997) Heterologous *HIS3* marker and GFP reporter modules for PCR-targeting in *Saccharomyces cerevisiae*. *Yeast* **13**, 1065–1075.

6 Studying Essential Genes: Generating and Using Promoter Fusions and Conditional Alleles

Michael J. R. Stark
Department of Biochemistry, University of Dundee, Dundee, UK

◆◆

CONTENTS

List of Abbreviations

Cs⁻	Cold sensitive
5-FOA	5-Fluoroorotic acid
PCR	Polymerase chain reaction
Ts⁻	Temperature sensitive
YCp	Yeast centromeric plasmid
YIp	Yeast integrative plasmid
YFG	"Your favorite gene"

◆◆◆◆◆◆ I. INTRODUCTION

One of the first questions that a yeast researcher will normally ask when faced with analysing a new gene is whether or not it is essential. Chapter 5 describes the use of targeted gene disruption to provide an answer to this question. However, if complete loss of function is lethal to the yeast cell, then other approaches are needed to study the gene's function, since it is difficult to carry out either biochemical or genetic studies with small numbers of dead yeast cells. Following tetrad dissection, when spores deleted for the gene in question are examined microscopically, they may frequently be found to have germinated but to have ceased growth either before dividing or after just a very few rounds of cell division. Although the appearance of such germinated spores can sometimes be informative about the lethal defect, generally it is not. More robust approaches are therefore necessary to learn about the function of essential genes in yeast. This chapter will address two such approaches: the depletion of a gene product by promoter shut-off and the generation of conditional mutant alleles. The latter approach in particular constitutes a powerful way of achieving a functional analysis of an essential yeast gene. In what follows, the essential yeast gene to be studied will be referred to as "*YFG1*" ("your favorite gene") following the nomenclature of Sikorski and Boeke (1991).

A. Fusions to Regulated Promoters

In promoter shut-off experiments, the endogenous gene is replaced by a copy which has been fused to a regulatable promoter. Typically, the galactose-inducible *GAL1–GAL10* promoter (Johnston and Davis, 1984) has been used although, as discussed below, it is not necessarily the best choice. By switching growth conditions to those in which the promoter is repressed (or no longer induced), the effect of depletion of the gene product from the cell can be studied. While this may be helpful in understanding the function of the gene, if the gene product is a particularly stable protein it may persist at levels sufficient to carry out its function for several generations. Thus a promoter shut-off experiment will not necessarily enable the investigator to observe rapid loss of function and the cells could adapt during depletion of the gene product so as to obscure the real effect. Conditional alleles are therefore the preferred way of functional analysis. However, by combining the use of a regulated promoter with a protein-destabilizing genetic element, rapid loss of function may nonetheless be achieved (Park *et al.*, 1992).

B. Conditional Alleles

Conditional alleles typically confer either a temperature-sensitive (Ts⁻) or cold-sensitive (Cs⁻) phenotype such that sufficient gene function remains to support growth at normal temperatures (24–30 °C), but gene function is lost at higher or lower growth temperatures, respectively. Thus mutant cells can be grown under permissive conditions and then shifted to the non-permissive ("restrictive") temperature to examine what happens as the gene function is lost. Conditional mutants can also provide the basis for some of the "smart genetic screens" described in Chapter 16: for example, synthetic lethal, extragenic and multicopy suppressor screens which can identify other genes whose products function in the same process as the protein of interest. This chapter will describe approaches based on the localized mutagenesis method of Muhlrad *et al.* (1992) for selecting conditional alleles starting with virtually any essential yeast gene. Random mutations can be generated using the polymerase chain reaction (PCR) under conditions promoting misincorporation of nucleotides during amplification, followed by generation of a library of mutant strains by *in vivo* gapped plasmid repair and plasmid shuffling. This library can then be screened for clones conferring a conditional growth phenotype. While the emphasis of this chapter is on conditional mutations, the method is of general applicability for the production of any type of mutation given a suitable selectable phenotype (e.g. resistance to an inhibitor, osmosensitivity etc.). Finally, the chapter will consider analysis of conditional mutants once they have been generated.

◆◆◆◆◆◆ II. USING REGULATED PROMOTERS TO STUDY ESSENTIAL GENES

The *GAL1–GAL10* promoter has been used for countless studies of gene function and is available on a wide variety of yeast vectors such as pBM150 (Johnston and Davis, 1984). The promoter is repressed during growth on glucose and induced on galactose, enabling modulation of the expression level by around 1000-fold. To examine what happens when expression of *YFG1* is switched off, the coding sequence of *YFG1* can be fused to the *GAL* promoter and a strain generated in which this is the sole source of *YFG1* function. This could be achieved, for example, by the following steps: (i) production of a diploid strain made heterozygous for a *yfg1Δ* gene knockout by one-step gene disruption (see Chapter 5); (ii) transformation of the diploid with the *GAL–YFG1* fusion on a YIp or YCp vector; and (iii) selection of cells in which the *GAL–YFG1* fusion and *yfg1Δ* gene knockout co-segregate (on galactose-containing medium) following tetrad analysis. The gene fusion can then be switched off by transferring cells to glucose-containing medium and the effects observed or extracts prepared for biochemical studies.

However, the *GAL* promoter is not necessarily ideal for performing this type of experiment. The promoter, when induced, is very strong and will in almost all cases lead to overexpression of the Yfg1p protein, which may itself affect the behaviour of the cells. Overexpression may either be detrimental (possibly lethal) to the cell, or may simply lead to such a high level of protein that it takes too long for the levels to decline after addition of glucose. In some cases, very low levels of expression may suffice to complement the *yfg1Δ* knockout and cells reliant on the *GAL–YFG1* fusion remain capable of normal growth even on glucose medium. For this reason a much weaker, regulated promoter may be more useful and the *MET3* promoter (Mountain *et al.*, 1991, Figure 1A) has recently been used successfully by several groups for modulating the levels of a variety of gene products in yeast cells (Amon *et al.*, 1994; Black *et al.*, 1995; Spang *et al.*, 1995). This promoter is much weaker than *GAL* and yet is tightly repressed by addition of 2 mM methionine to the growth medium; for example, a *MET3–URA3* construct transformed into a *ura3* mutant strain failed to confer 5-FOA sensitivity on methionine-supplemented medium (Mountain *et al.*, 1990). Use of the *MET3* promoter therefore enables more rapid depletion of Yfg1p following promoter shut-off and avoids the possible complications of overexpression when the promoter is on. An

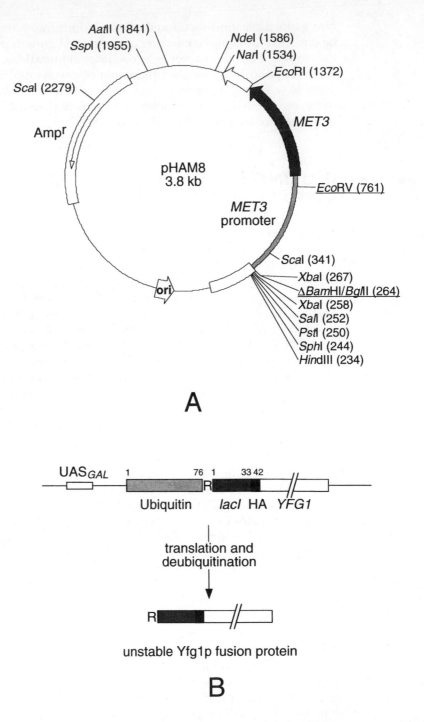

Figure 1. Tools for protein depletion experiments. (A) pHAM8 is a convenient source of the *MET3* promoter developed by Mountain *et al.* (1991). All the *cis*-acting signals required for tightly regulated expression of downstream genes are contained within the 497 bp *BglII-EcoRV* fragment of the *MET3* promoter. (B) The ubiquitin fusion system for regulated expression of constitutively unstable proteins (Park *et al.*, 1992).

example of using the *MET3* promoter to study the function of an essential gene has been reported in detail by Black *et al.* (1995). Again, the basis of the method is to generate a strain in which a *MET3–YFG1* fusion, either integrated or on a YCp vector, is the sole source of *YFG1*. This is achieved using methionine-free medium such that the *MET3–YFG1* fusion is expressed in the desired strain. Cells can then be supplemented with methionine to observe the effect of Yfg1p depletion. Even using the *MET3* promoter, however, it may still take much longer to deplete cells of a stable protein than to inactivate a conditionally mutated gene product. If a suitable antibody is available or *YFG1* has been epitope-tagged, then loss of the gene product following promoter shut-off can be readily followed by Western blot analysis.

A recently developed strategy for making promoter shut-off experiments more useful involves the introduction of a protein destabilizing element into the regulatable *YFG1* construct (Park *et al.*, 1992). Thus, if the *YFG1* product is constitutively unstable, not only will the steady state level be dramatically lowered but it will be rapidly lost from the cell once transcription ceases, thereby avoiding problems associated both with overexpression and slow depletion. To date, this approach has used *GAL* promoter fusions and involves fusing a genetic cassette consisting of the ubiquitin gene, an arginine codon, 31 codons of the *Escherichia coli lacI* gene and an epitope tag to the amino-terminal coding region of *YFG1* (Figure 1B). The chimeric protein is rapidly processed in yeast by a deubiquitinating enzyme to reveal an amino-terminal arginine residue, which destabilizes the remainder of the fusion polypeptide (Bachmair and Varshavsky, 1989). Destabilization is brought about by multiple ubiquitin conjugation onto the *lacI*-encoded region in response to the amino-terminal arginine, followed by ubiquitin-dependent proteolysis. Clearly, for this method to be successful the protein of interest must be capable of tolerating an amino-terminal extension of some 40 residues. Fusion constructs in which deubiquitination reveals a methionine residue (which confers stability) can be used as a control in these experiments. In this way, a half-life as short as 2–3 min can be conferred on the protein of interest (Park *et al.*, 1992). This approach has proved a successful alternative to the generation of conditional alleles in a variety of recent investigations (e.g. Althoefer *et al.*, 1995; Benli *et al.*, 1996; Park *et al.*, 1992).

◆◆◆◆◆◆ III. GENERATING CONDITIONAL ALLELES

A. Conditional Alleles: Ts⁻ *versus* Cs⁻

Will any essential gene necessarily yield Ts⁻ or Cs⁻ alleles? In one study where a yeast strain was mutagenized and Ts⁻ mutations on chromosome I were systematically identified, Harris and Pringle (1991) showed that despite using a range of mutagens with different specificities, many essential genes failed to yield Ts⁻ alleles. In such cases this may be due to the nature of the gene product; for some proteins it may be particularly difficult to generate changes that affect function without causing complete inactivation. However, by focusing on specific essential genes in isolation and using PCR to generate a large, unbiased spectrum of mutations, it is likely that most essential genes will ultimately yield the desired mutant allele. The method described in this chapter has been used successfully in this and many other laboratories to obtain Ts⁻ alleles encoding a range of proteins including enzymes, structural proteins and components of multiprotein complexes (e.g. Connelly and Hieter, 1996; Evans and Stark, 1997; Geissler *et al.*, 1996; Lewis and Pelham, 1996; MacKelvie *et al.*, 1995; Stirling *et al.*, 1994). In this laboratory, we have to date a 100% record of obtaining Ts⁻ alleles in a variety of yeast genes.

Regarding the analysis of conditional alleles, it is probably easier to use Ts⁻ strains than Cs⁻ mutants. Wild-type strains grow rather slowly below 20 °C and it can sometimes be more difficult to obtain a clear distinction between wild-type and a Cs⁻ mutant than with a Ts⁻ strain. Ultimately, it is the difference between the restrictive temperature of the mutant and the maximum (for Ts⁻ alleles) or minimum (for Cs⁻ alleles) growth temperature of the wild-type strain which counts and the greater this difference, the more clear-cut the results will be. Thus most laboratory strains of yeast fail to grow above 38 °C and while 37 °C can be used successfully as the restrictive temperature for a Ts⁻ mutant, a more severe allele that fails to grow above 35 °C might be more useful. Similarly, a Cs⁻ strain that cannot grow below 20–23 °C would be easier to handle than one with a restrictive temperature of 16 °C. Folklore has it that Cs⁻ alleles are harder to find: they may require a more restricted type of alteration to the protein, for example an alteration specifically at the interface of a protein–protein interaction. However, they are potentially just as useful as Ts⁻ mutations and just as different Ts⁻ alleles can be of great use (see below), the ability to obtain Cs⁻ as well as Ts⁻ alleles could be particularly valuable.

B. Generating a Suitable Yeast Strain for Mutant Screening

Figure 2 shows the overall scheme devised by Muhlrad *et al.* (1992) for generating a library of mutant alleles of an essential gene. The starting point for the scheme is the production of a haploid yeast strain in which the gene in question has been deleted from the genome, but which is kept alive by a wild-type copy of the gene on a yeast centromeric plasmid (YCp) carrying the *URA3* gene. Vectors such as YCplac33 (Gietz and Sugino, 1988) and pRS316 (Sikorski and Hieter, 1989) are suitable for constructing the required plasmid carrying *YFG1*. The yeast strain is best generated by tetrad dissection of a diploid strain which has been made heterozygous for a complete *yfg1Δ* knockout by one-step gene disruption (Chapter 5) and then transformed with the *YFG1–URA3* –YCp plasmid, screening for haploid progeny in which the null allele and the plasmid cosegregate. Such progeny should fail to papillate when patched onto 5-FOA medium (Sikorski and Boeke, 1991), indicating that the strain is fully dependent on the plasmid-borne wild-type *YFG1* allele. The marker used to generate the knockout allele is not important, although clearly *URA3* must be reserved to enable 5-FOA counterselection in the later step (Figure 2). If for some reason *URA3* cannot be used, *LYS2* and the Lys2⁺ counterselective agent α-aminoadipic acid could be used instead (Sikorski and Boeke, 1991).

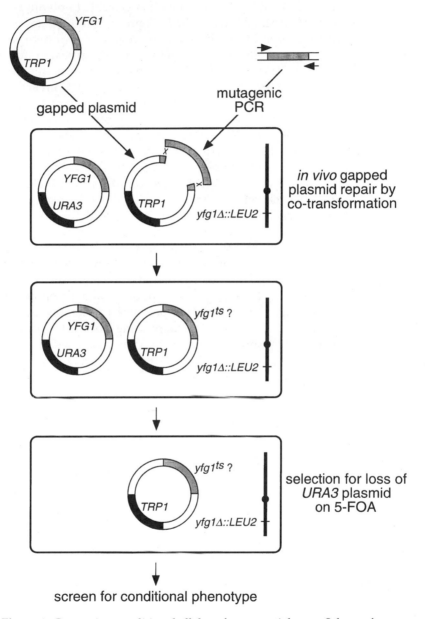

Figure 2. Generating conditional alleles of an essential gene. Scheme for generating conditional alleles of your favourite essential gene (*YFG1*) by mutagenic PCR and gapped plasmid repair *in vivo*. For the purpose of this example, the chromosomal locus has been deleted with *LEU2* and the gapped plasmid carries *TRP1* but other markers may be used as appropriate. The only requirement is that the resident plasmid in the starting yeast strain should carry the *URA3* marker to facilitate selection for its loss in the final step.

C. Primer Design and Plasmid Gapping Strategy

Figure 3 shows the strategy which underlies the *in vivo* gapped plasmid repair step. The region of the gene to be mutated is defined by two sites (A and B in Figure 3) which ideally will enable removal of the bulk of the coding region of the gene. While it is possible to obtain mutations in the regions of overlap between the PCR product and the gapped plasmid, we have never found mutations in these regions in our own work. If nothing is known about which regions of the gene are critical for function, then the larger the section of the coding region excised the better (because this will maximize the region available for mutagenesis). Conversely, if restriction sites A and B lie outside the coding region, then it may be possible to identify mutations that do not change the protein but which affect the promoter activity or mRNA stability, and while these might be of interest, they may not be as useful as mutations that generate an altered protein. Mutagenesis is therefore best limited to the coding region by suitable choice of sites and primers. Restriction sites A and B clearly must not cut the vector backbone or remaining regions of the gene and if no suitable sites are available, they must be introduced (or perhaps a vector site removed) by site-directed mutagenesis, a procedure for which many rapid and easy-to-use kits are now commercially available. Another option would be to replace the coding region with a short oligonucleotide linker containing a unique restriction site when constructing the YCp.

The design of the PCR primers follows standard procedures which aim at optimizing the melting temperature (preferably at least 65°C) and avoiding primer dimers or weak priming from other sites in the template. Several of the commonly used DNA analysis packages contain facilities for doing this, or alternatively a freeware or shareware program such as

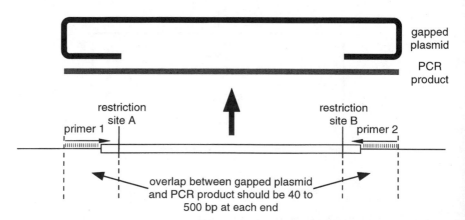

Figure 3. Primer design and plasmid gapping strategy. Primers should be designed that flank the gene to be mutated such that the entire coding region is amplified. For *in vivo* gapped plasmid repair, the regions of overlap between the gapped plasmid and the PCR product should be at least 40 bp at each end and preferably greater. The availablilty of the sites (A, B) used to generate the gapped plasmid will therefore play some role in determining the selection of PCR primers.

Amplify (written by Bill Engels, University of Wisconsin at Madison) can be used. Primers of around 17 bases work well. For the method to work there must clearly be sufficient overlap between the gapped plasmid and PCR fragment for efficient homologous recombination to occur. As with gene knockouts (Chapter 5), this overlap should preferably be at least 40 bp at each end and if one overlap is short (e.g. under 100 bp) it will help if the other is longer. We have not rigorously tested the minimum requirements, but in one case (MacKelvie et al., 1995) the overlaps were 46 and 374 bp while in another instance (Evans and Stark, 1997) they were 148 and 187 bp; both instances yielded large libraries of potentially mutant genes. If the template plasmid and the gapped plasmid are identical, then vector sequences can be used for PCR priming. The largest PCR fragment we have used was 2.5 kb, but there is no reason why longer genes could not be mutated using one set of primers. For very large genes, it may be better to split them into two or more regions if problems are encountered with the PCR step. This might also reduce the chance of obtaining multiple mutations in each conditional allele.

D. Mutagenic PCR

A library of *yfg1* mutant alleles can be rapidly and easily generated by PCR amplification of *YFG1* under conditions that favor misincorporation of dNTPs into the PCR product. This is achieved by using biased dNTP ratios and by including $MnCl_2$ in the amplification reaction (Protocol 1). Although other methods of mutagenesis have been described previously in procedures similar to that presented here (e.g. Sikorski and Boeke, 1991), PCR has the advantage that it is easy, safe and can generate an unbiased range of mutations. Furthermore, mutagenesis is localized to the gene of interest in a way not possible with chemical mutagenesis of the gene present on a plasmid. Finally, in some cases (e.g. where a specific domain is evident in a protein), mutagenesis by PCR can be localized just to the region encoding this domain rather than the whole gene by suitable design of primers. Protocol 1 suggests some typical conditions both for the composition of the PCR reaction and the cycling conditions, but ultimately the best conditions need to be determined empirically. The type of enzyme used for PCR is clearly of importance; because the aim is to generate mutations, a non-proofreading enzyme such as *Taq* polymerase should be used. We have found it best to use a range of perhaps three different $MnCl_2$ concentrations, optimizing the cycling conditions for a good yield of product at each $MnCl_2$ concentration. The dNTP and $MnCl_2$ concentrations can greatly affect the yield of mutants and also the average number of mutational changes per allele. Because it is normally better to work with alleles containing single point alterations, conditions that minimize misincorporation while still yielding a few conditional mutants (e.g. at a level of 1–2 %) are generally preferable. Ultimately, it may be necessary to carry out a few attempts to determine conditions that generate conditional alleles at a reasonable frequency, sequence a few of these to determine the overall level of mutational change and then refine the process as required.

Protocol 1. PCR mutagenesis

1. Set up mutagenic and non-mutagenic (control) PCR reactions in parallel. Standard mutagenic conditions are as follows:
   (i)   ratio of (dTTP + dCTP) : (dATP + dGTP) = 5 : 1 (or can try 1 : 5)
   (ii) include 0.1–0.5 mM MnCl$_2$ in the reaction
   (iii) use double the normal level of *Taq* polymerase in the reaction (i.e. 0.05 U/µl).
2. For each template–primer combination, optimize the following:
   (i) the ratio of MgCl$_2$: MnCl$_2$ in the reaction. This can greatly affect the frequency of mutations obtained and the PCR product yield
   (ii) the PCR cycle conditions for denaturation, annealing and extension.
3. Typical conditions (for a 100 µl reaction) are as follows:

	Stock solution	Final conditions	Volume (µl)
Primer 1	200 ng/µl	4 ng/µl	2
Primer 2	200 ng/µl	4 ng/µl	2
Template[a]	10 ng/µl	0.1–0.2 ng/µl	1–2
Reaction buffer[b]	10×	1×	10
MgCl$_2$[b]	25 mM	2 mM	8
dNTP mix[c]	see above	see above	2
Water	–	–	68–72
MnCl$_2$	10 mM	0.3–0.5 mM	3–5[d]
Taq polymerase	5 U/µl	0.05 U/µl	1

[a] template is a circular plasmid carrying the relevant gene
[b] some PCR buffers already include MgCl$_2$
[c] diluted from 100 mM stocks
• 50× stock "mutagenic" dNTP mix: 25 mM dTTP, dCTP, 5 mM dATP, dGTP
• 50× stock "normal" dNTP mix: 25 mM each dNTP
[d] mutagenic reaction only

4. Mix, heat at 94°C for 5 min and then add the *Taq* polymerase. Mix, add oil (100 µl), pulse spin briefly in a microfuge and then carry out PCR using the preoptimized settings, for example:
   heat to 94°C at ≥ 1°C/s, hold for 45–60 s
   cool at 0.5°/s to 55°C, hold for 2.0–2.5 min
   heat to 72°C at ≥ 1°C/s, hold for 2.5–5.0 min
   repeat for 30 cycles, hold at 72°C for 20 min for the last cycle, cool to 4°C.

E. Gapped Plasmid Repair *in vivo*

The next step is to use the PCR product to generate a library of *YFG1* alleles which can be screened for conditionality. In principle, this could be done by cloning the PCR product into a YCp, introducing the ensuing library of plasmids into the yeast strain generated as above and following the latter part of the procedure shown in Figure 2. However, it is far easier to generate the required plasmids by *in vivo* recombination in yeast and thereby circumvent an extra step using *E. coli*. This is achieved by co-transformation of the PCR fragment and a gapped plasmid whose ends are homologous to the amplified product. When both DNA molecules enter a competent yeast cell, high-efficiency homologous recombination between the PCR fragment and the gapped plasmid leads to reconstruction of a circular plasmid carrying the entire gene. Protocol 2 shows

---

Protocol 2. Gapped-plasmid repair, plasmid shuffling and selection

1. The host strain is a suitable yeast carrying a wild-type copy of the gene in question on a *URA3–CEN* plasmid (YCp) and a gene deletion in the chromosome.
2. Prepare gapped plasmid (a YCp carrying the gene but with a different selectable marker e.g. *TRP1*) by digestion and gel purification using a product such as GeneClean II (Bio101) or Nucleon (Scotlab).
3. Check and PCR product (Protocol 1) for purity by gel electrophoresis after ether extraction of the oil. If a clean band of the expected size is obtained then it can be used following removal of the unused primers, otherwise it should be gel purified as above.
4. Add the following to an aliquot (50 µl) of yeast cells made competent by lithium acetate treatment (Chapter 4):
   10 µl PCR product (roughly 1–5 µg). Use non-mutated product as a control to show that the Ts$^-$/Cs$^-$ clones resulted from the mutagenic PCR
   50–100 ng gapped plasmid
   50 µg sheared single-stranded salmon sperm DNA
5. Mix briefly, then add 300 µl PEG buffer. Mix well. Shake gently at 26°C for 30–45 min and then heat at 42°C for 15 min.
6. Spin 3000 rpm for 2 min and resuspend in 100 µl TE buffer.
7. Plate out on medium selective for the gapped plasmid. Include alongside gapped plasmid alone and a zero DNA sample as controls. Presence of the PCR fragment should stimulate the transformation frequency 5–10 fold.
8. Patch out 500–1000 transformants onto selective plates containing 5-FOA (Sikorski and Boeke, 1991) and then when these have grown, plate out again on 5-FOA medium. Screen these strains on selective plates at 23–26°C and either 35–37°C (to screen for Ts$^-$) or 14–20°C (to screen for Cs$^-$).
9. *Very important:* check any potential mutant alleles by isolating the plasmid, retransformation into the starting strain and using 5-FOA to select for loss of the wild-type allele on the *URA3* YCp:
   • TE buffer: 10 mM Tris-HCl (pH 8.0), 0.1 mM EDTA
   • PEG buffer: 40% polyethylene glycol 4000, 0.1 M lithium acetate in TE buffer (filter sterilized).

---

suitable conditions for achieving this and clearly requires yeast cells that are highly competent for transformation (see Chapter 4). Although Protocol 2 suggests the use of lithium-acetate-treated cells, there is no reason why electroporation could not be used instead. Transformants should be selected at the normal growth temperature just for the gap-repaired plasmid (i.e. include uracil in the plates), thereby allowing for loss of the *URA3–YFG1*–YCp from some fraction of cells in each colony. An important control in this process is to transform the competent cells with the gapped vector alone, which will generate a background level of transformants by recircularization or by gap-repair using the null allele in the genome. Typically, presence of the PCR fragment should stimulate transformation frequency at least 5–10-fold, indicating that efficient gap-repair using the incoming fragment has occurred. Ideally, the number of transformants should be between 500 and 5000 for a reasonable probability of obtaining mutants.

F. Plasmid Shuffling and Selection for Mutant Alleles

By this stage, a library of yeast transformants will have been generated, but because each contains the URA3–YFG1–YCp as well as the gap-repaired plasmid, and because most conditional alleles will be recessive, any effect of a mutant yfg1 will not be evident. To uncover any mutant phenotypes, the URA3–YFG1–YCp must first be lost using 5-FOA counterselection or "plasmid shuffling" (Sikorski and Boeke, 1991), as described in Protocol 2. A minimum of 500 transformants should be patched out onto 5-FOA medium (Sikorski and Boeke, 1991) selecting for the gap-repaired plasmid to enable growth of cells in which the gap-repaired plasmid has become the sole source of YFG1 function in the cell. Because it is likely that a significant fraction of gap-repaired plasmids will encode unconditionally dysfunctional yfg1 alleles, some of the transformants will not generate 5-FOA-resistant progeny at this stage. The temperature chosen for this step is important, because it will define the permissive conditions for any conditional mutations and alleles not supporting growth at this temperature will be lost at this stage. While 26 °C might be typically the best temperature to use, it could be advantageous to try slightly lower temperatures for Ts⁻ screens or slightly higher ones for Cs⁻ screens. A second round of growth on 5-FOA medium ensures complete loss of the URA3–YFG1–YCp.

All 5-FOA resistant colonies should next be replica plated or restreaked onto several plates selecting both for the gap-repaired plasmid and the genomic knockout; one of these is grown under the permissive conditions and the others at a range of potentially restrictive temperatures for both Ts⁻ and Cs⁻ alleles. In this way, colonies that fail to grow specifically on any of the latter plates can be identified as containing candidate conditional yfg1 alleles.

G. Verification of Mutants

For each candidate mutant, the plasmid should be recovered from a small liquid culture grown at the permissive temperature. In this way, following preparation of DNA and recovery of the candidate mutant alleles in E. coli, substantial quantities of DNA can be prepared and the alleles verified. Each recovered plasmid should be re-transformed into the starting strain, shuffling out the URA3–YFG1–YCp on 5-FOA medium and retesting for conditionality as above. Shuffling should be done using at least four or five primary yeast transformants and testing for conditionality carried out using several Ura⁻ derivatives of each, so that a clear and representative result can be obtained. Verification of the plasmids is critical because not every strain that appears Ts⁻ or Cs⁻ initially will consistently continue to do so. Once conditional alleles have been verified in this manner, the site of mutational alteration can be determined by DNA sequencing. Using a suitable series of primers, several alleles can be characterized on a single sequencing gel. Sequencing has two benefits: firstly it confirms that the coding region has been mutated and secondly it demonstrates how many mutations are present. As discussed above, sometimes the

PCR conditions may be sufficiently mutagenic to yield multiple changes and so sequencing one or more alleles at the first opportunity enables the procedure to be repeated using less error-prone conditions for the PCR step. Where multiple missense mutations have been obtained, the mutation responsible for the conditional phenotype may be identified by subcloning (to produce alleles containing single point changes) or by site-directed mutagenesis of a wild-type copy. However, it is not always the case that a single mutation will alone be responsible for the conditional phenotype (e.g. Evans and Stark, 1997) and so repeating the mutagenesis may be preferable as the means of obtaining conditional phenotypes resulting from single point changes.

A final aspect to verification of the mutant alleles concerns their phenotype. The screen described above is of necessity a growth/no growth test performed on plates, but the conditional mutants obtained may show considerable differences in their behavior. In particular, they may differ in how rapidly the mutant protein is inactivated at the restrictive temperature. This may be critical, especially where the possible effects of mutations on the cell division cycle is being examined and where arrest in the first cell cycle following temperature shift is usually needed to obtain clear-cut results. Furthermore, mutations that are conditional on plates may not have such a clear-cut effect in liquid medium, which will be used for much of the future work with them. Thus candidate mutant alleles should also be checked in small-scale liquid culture and those that cease proliferation most rapidly are likely to be of greatest use. In some instances, prior knowledge of the gene function may suggest what sort of phenotype is to be expected following rapid loss of function. In such a case, cells can be examined at the restrictive conditions and alleles identified that seem to confer the expected effect. This could be especially important where a protein has several functions, only one of which is currently of interest to the investigators.

H. Strategies for Integrating Mutant Alleles at their Genomic Locus

The above procedure should reliably yield conditional yeast strains in which the sole source of the mutant protein is encoded by a YCp. While such strains may be very useful, ultimately it is probably best to generate strains in which the mutant allele is integrated, preferably at its normal genomic locus. The copy number of a YCp plasmid can vary somewhat and so the only way of ensuring a stable, single copy of the mutant allele is by integration into the genome. Several possible strategies are available for achieving this, all of which require that the correct integration event be verified by Southern blotting or PCR. It may also be beneficial to verify the integrated mutant allele by PCR amplification of the altered site followed by "cycle sequencing" using one of the commercially available kits, thereby demonstrating presence of the mutation in the genome. Finally, it is worth remembering that a mutation which was Ts⁻ on a YCp plasmid at 1–4 copies per cell may be lethal when integrated in single-copy.

1. Ectopic integration

The mutant allele can be subcloned onto a yeast integrative plasmid (YIp) and integrated at the site of the YIp marker gene. This can be done in a strain heterozygous for a *yfg1* knockout and haploid segregants in which the integrated copy of the gene is the sole source of *YFG1* function identified following tetrad analysis. Alternatively, if a suitable marker is available, the integration could be done in the haploid plasmid-dependent mutant strain obtained following the original mutagenesis. Following transformation, transformants that have lost the plasmid and thus rely on an integrated copy of the *yfg1* conditional allele can be identified following nonselective growth and replica plating. This latter approach avoids possible gene conversion of the integrated allele as the cell contains no wild-type copy of *YFG1* at any stage in the process. Both strategies may result in multiple integrations of the YIp and this should be checked (e.g. by Southern blotting).

2. One-step replacement of a *yfg1Δ* null allele

In this method, a diploid heterozygous for a *yfg1::URA3* null allele is generated and transformed to 5-FOA resistance using a linear fragment of the conditional *yfg1* allele which extends to either side of the region replaced by *URA3*. Transformants should have replaced the knockout with the incoming DNA and, on tetrad analysis, should yield 2:2 segregation of the conditional phenotype. Failure to do so may imply gene conversion of the mutant allele while in the diploid cell, which can occur at a remarkably high frequency (Schneider *et al.*, 1995). Indeed, gene conversion may be the main reason why difficulties may be encountered when trying to integrate conditional alleles in any method where cells contain both mutant and wild-type copies of *YFG1* at some stage. As an alternative, the replacement can be done in a *yfg1::URA3* haploid cell in which the conditional allele is present on a YCp carrying a different (i.e. not *URA3*) marker. This strain can be readily generated by tetrad analysis of a diploid heterozygous for the *yfg1::URA3* null allele and carrying the relevant YCp plasmid because such strains will have an overt conditional phenotype. The strain should papillate colonies on 5-FOA medium in which spontaneous gene conversion of the knockout allele using the YCp plasmid *yfg1* allele has occurred, but transformation with the mutant *yfg1* fragment may increase the efficiency of *yfg1::URA3* replacement. As an alternative, if a *yfg1Δ* using a marker other than *URA3* is already available, co-transformation of a DNA fragment carrying the *yfg1* conditional allele and any other YCp vector with a suitable marker (e.g. at a 100:1 DNA ratio) should enable YCp transformants in which the *yfg1* null allele has been replaced with the mutant copy to be identified by screening for loss of the knockout marker gene (Luke *et al.*, 1996). This variation could be applied to either the haploid or the diploid versions of the procedure.

3. Pop-in, pop out

For this procedure, a YIp plasmid carrying *URA3* and the conditional *yfg1* allele is constructed and integrated at the *YFG1* locus in a haploid strain, resulting in tandem wild-type and mutant copies of *YFG1* separated by the vector sequences. 5-FOA selection can then be used to screen for spontaneous excision of the YIp plasmid by homologous recombination. This should occur in some isolates so as to leave the site of the mutation in the remaining genomic copy (Struhl, 1983). Candidate strains will thus appear as Ts⁻ (or Cs⁻) 5-FOA-resistant derivatives.

4. Integration of a marked *YFG1* fragment

Methods (2) and (3) above result in strains in which the conditional allele is not marked and so it can only be followed in subsequent genetic crosses by its conditional phenotype. In the case of method (1), a marked allele is a consequence of the method itself. Where it is desirable to mark the mutant allele at its genomic locus, this can be achieved by insertion of a suitable marker gene 3' to the gene of interest and transforming a haploid strain with a linear fragment encompassing the whole *yfg1* mutant sequence, the adjacent marker and more distal sequences homologous to the *YFG1* locus. A caveat here is that clearly the site of integration of the marker must not affect the function either of *yfg1* or any downstream gene. Although integration of the marker could occur with a recombination event between the site of the *yfg1* mutation and the marker gene, in practice a sufficiently high proportion of events which incorporate both the mutant *yfg1* and the marker can be found among the transformants. Such marked *yfg1* mutants can readily be converted to unmarked strains by a variant of the pop-in, pop-out procedure (Evans and Stark, 1997).

◆◆◆◆◆◆ IV. ANALYSIS OF CONDITIONAL MUTANTS

Having generated one or more conditional mutants, the way is paved for analysing the effect of the mutations and then using them in further experimental strategies. More often than not, the investigator will have some idea of what effect loss of function may have and will have experimental assays set up to monitor some facet of cell biology in which the *YFG1* product plays a role. Otherwise, examination of the morphology of cells under the microscope following shift to the restrictive temperature is a good starting point. The mutant allele may block cell cycle progression or lead to defective morphological development, in which case it may be useful to examine the effect of the mutation in synchronous cultures. Two methods are commonly used for this: α-factor block-release and centrifugal elutriation. The former has the advantage that it is straightforward and requires no specialist equipment, although the latter method simply selects small cells at the beginning of the cell cycle and does not require perturbation with mating pheromone. Immunofluorescence microscopy using probes for DNA, actin, tubulin and other intracellular components may also be of great use in uncovering the cellular defect under restrictive conditions (Pringle *et al.*, 1991, see Chapter 12). If an antibody is available that recognizes the *YFG1* product (or if the gene has been epitope-tagged) then its abundance or intracellular localization can be examined following temperature shift. Where the polypeptide is known to engage in protein–protein interactions, the effect of the mutation on these can also be tested (e.g. by co-immune precipitation, see Chapter 12, or by building the mutant allele into the two-hybrid assay, see Chapter 15).

It is worth pointing out that the phenotype of a conditional mutation may depend on how far beyond the minimum restrictive temperature the strain is shifted and so it may be important to try more than one set of conditions. Equally, different alleles may have different phenotypes; intragenic complementation between different alleles would suggest that the protein product has different functional domains and that each complementing mutation has affected a different function of the protein. The availability of different conditional alleles is also of great use when they are used for further genetic screens (see Chapter 16). Thus dosage or extragenic suppressors that simply bypass a mutation may nonspecifically suppress any conditional allele (or a deletion) of *YFG1*. In comparison, the ability to identify suppressor mutations of either type that show specificity for one (or a subset) of mutations implies that the suppression is more specific and makes it more likely that the components identified in the screen have a direct involvement with the *YFG1* product. Likewise, allele specificity of synthetic–lethal interactions gives increased confidence that the genetic interaction represents a direct and meaningful biochemical one.

Acknowledgements

Thanks are due to Paul Andrews, David Evans and Doug Stirling for reading the manuscript and for helpful comments.

References

Althoefer, H., Schleiffer, A., Wassmann, K., Nordheim, A. and Ammerer, G. (1995). Mcm1 is required to coordinate G2-specific transcription in *Saccharomyces cerevisiae*. *Mol. Cell. Biol.* **15**, 5917–5928.

Amon, A., Irniger, S. and Nasmyth, K. (1994). Closing the cell cycle circle in yeast: G2 cyclin proteolysis initiated at mitosis persists until the activation of G1 cyclins in the next cycle. *Cell* **77**, 1037–1050.

Bachmair, A. and Varshavsky, A. (1989). The degradation signal in a short-lived protein. *Cell* **56**, 1019–1032.

Benli, M., Doring, F., Robinson, D. G., Yang, X. P. and Gallwitz, D. (1996). Two GTPase isoforms, Ypt31p and Ypt32p, are essential for Golgi function in yeast. *EMBO J.* **15**, 6460–6475.

Black, S., Andrews, P. D., Sneddon, A. A. and Stark, M. J. R. (1995). A regulated *MET3–GLC7* gene fusion provides evidence of a mitotic role for *Saccharomyces cerevisiae* protein phosphatase 1. *Yeast* **11**, 747–759.

Connelly, C. and Hieter, P. (1996). Budding yeast SKP1 encodes an evolutionarily conserved kinetochore protein required for cell cycle progression. *Cell* **86**, 275–285.

Evans, D. R. H. and Stark, M. J. R. (1997). Mutations in the *Saccharomyces cerevisiae* type 2A protein phosphatase catalytic subunit reveal roles in cell wall integrity, actin cytoskeleton organisation and mitosis. *Genetics* **145**, 227–241.

Geissler, S., Pereira, G., Spang, A., Knop, M., Soues, S., Kilmartin, J. and Schiebel, E. (1996). The spindle pole body component Spc98p interacts with the gamma-tubulin-like Tub4p of *Saccharomyces cerevisiae* at the sites of microtubule attachment. *EMBO J.* **15**, 3899–3911.

Gietz, R. D. and Sugino, A. (1988). New yeast–*Escherichia coli* shuttle vectors constructed with *in vitro* mutagenised yeast genes lacking six-base pair restriction sites. *Gene* **74**, 527–534.

Harris, S. D. and Pringle, J. R. (1991). Genetic analysis of *Saccharomyces cerevisiae* chromosome I: on the role of mutagen specificity in delimiting the set of genes identifiable using temperature-sensitive-lethal mutations. *Genetics* **127**, 279–285.

Johnston, M. and Davis, R. W. (1984). Sequences that regulate the divergent *GAL1–GAL10* promoter in *Saccharomyces cerevisiae*. *Mol. Cell. Biol.* **4**, 1440–1448.

Lewis, M. J. and Pelham, H. R. B. (1996). SNARE-mediated retrograde traffic from the golgi complex to the endoplasmic reticulum. *Cell* **85**, 205–215.

Luke, M. M., Dellaseta, F., Di Como, C. J., Sugimoto, H., Kobayashi, R. and Arndt, K. T. (1996). The SAPs, a new family of proteins, associate and function positively with the SIT4 phosphatase. *Mol. Cell. Biol.* **16**, 2744–2755.

MacKelvie, S. H., Andrews, P. D. and Stark, M. J. R. (1995). The *Saccharomyces cerevisiae* gene *SDS22* encodes a potential regulator of the mitotic function of yeast type 1 protein phosphatase. *Mol. Cell. Biol.* **15**, 3777–3785.

Mountain, H. A., Heiber, M., Korch, C. and Byström, A. S. (1990). A *URA3* gene fusion for isolating methionine-specific regulatory genes. *Yeast* **6**, S275.

Mountain, H. A., Byström, A. S., Larsen, J. T. and Korch, C. (1991). Four major transcriptional responses in the methionine/threonine biosynthetic pathway of *Saccharomyces cerevisiae*. *Yeast* **7**, 781–803.

Muhlrad, D., Hunter, R. and Parker, R. (1992). A rapid method for localized mutagenesis of yeast genes. *Yeast* **8**, 79–82.

Park, E. C., Finley, D. and Szostak, J. W. (1992). A strategy for the generation of conditional mutations by protein destabilization. *Proc. Natl. Acad. Sci. USA* **89**, 1249–1252.

Pringle, J. R., Adams, A. E. M., Drubin, D. G. and Haarer, B. K. (1991). Immunofluorescence methods for yeast. *Methods Enzymol.* **194**, 565–602.

Schneider, B. L., Seufert, W., Steiner, B., Yang, Q. H. and Futcher, A. B. (1995). Use of polymerase chain reaction epitope tagging for protein tagging in *Saccharomyces cerevisiae*. *Yeast* **11**, 1265–1274.

Sikorski, R. S. and Hieter, P. (1989). A system of shuttle vectors and yeast host strains designed for efficient manipulation of DNA in *Saccharomyces cerevisiae*. *Genetics* **122**, 19–27.

Sikorski, R. S. and Boeke, J. D. (1991). *In vitro* mutagenesis and plasmid shuffling – from cloned gene to mutant yeast. *Methods Enzymol.* **194**, 302–318.

Spang, A., Courtney, I., Grein, K., Matzner, M. and Schiebel, E. (1995). The Cdc31p-binding protein Kar1p is a component of the half bridge of the yeast spindle pole body. *J. Cell Biol.* **128**, 863–877.

Stirling, D. A., Welch, K. A. and Stark, M. J. R. (1994). Interaction with calmodulin is required for the function of Spc110p, an essential component of the yeast spindle pole body. *EMBO J.* **13**, 4329–4342.

Struhl, K. (1983). The new yeast genetics. *Nature* **305**, 391–397.

7 Ty Mutagenesis

David J. Garfinkel[1], M. Joan Curcio[2] and Victoria Smith[3]
[1]ABL-Basic Research Program, Movable Genetic Elements Section, Gene Regulation and Chromosome Biology Laboratory, Frederick, Maryland, [2]Wadsworth Center for Laboratories and Research, New York State Department of Health, Albany, New York and [3]Department of Molecular Biology, Genentech Inc., South San Francisco, California, USA

◆◆

CONTENTS

List of Abbreviations

AI	Artificial intron
5-FOA	5-Fluoroorotic acid
LTR	Long terminal repeat
PCR	Polymerase chain reaction
pol	Polymerase
SC	Synthetic complete medium (*see Appendix II: Table 2*)
SD	Synthetic dextrose
YAC	Yeast artificial chromosome
YEPD	Yeast extract peptone dextrose (*see Appendix II: Table 1*)

◆◆◆◆◆◆ **I. INTRODUCTION**

The use of transposable elements as insertional mutagens is a powerful tool for genetic analysis. This approach, first developed in bacteria, has also proved to be an effective technique for tagging genes in nematodes, *Drosophila*, mice, and higher plants. Using the retrotransposons Ty1 and Ty2 (see Boeke and Sandmeyer, 1991 and Garfinkel, 1992 for reviews), which are native to *Saccharomyces cerevisiae*, methods for transposon mutagenesis have been developed over the past 10 years (Garfinkel *et al.*, 1988; Rinckel and Garfinkel, 1996). Two critical discoveries have combined to make Ty1 and Ty2 elements useful as insertional mutagens. First, it was demonstrated that retroelement transposition could be stimulated by expressing Ty1 from an efficient promoter (Boeke *et al.*, 1985). Fusing a Ty1 or Ty2 element to the *GAL1* promoter results in galactose-inducible transposition of the element at levels greater than one event per cell. Second, it was demonstrated that Ty1 and Ty2 elements could accommodate selectable markers within their genomes (Boeke *et al.*, 1988; Garfinkel *et al.*, 1988). For example, Ty1 elements carrying *TRP1*, *HIS3*, and the bacterial *Neo* genes have been transcribed from *URA3*-based *GAL1* expression plasmids and shown to transpose into chromosomal targets. Such transposition events were identified as cells that retain the selected Ty marker when the plasmid is lost [plasmid loss is conveniently selected because Ura⁻ cells are resistant to 5-fluoroorotic acid (FOA) (Boeke *et al.*, 1984)]. Earlier versions of the Ty mutagenesis technique have been presented in detail elsewhere (Garfinkel and Strathern, 1991; Strathern *et al.*, 1994; Garfinkel, 1996).

In this chapter, three recent improvements in Ty mutagenesis are presented. First, markers have been developed that only function after transposition; thus cells that have had a transposition event can be directly selected (Curcio and Garfinkel, 1991; Dalgaard et al., 1996). Second, an *I-DmoI* restriction site, which is unique to the yeast genome, has been engineered adjacent to the Ty1 marker gene (Dalgaard et al., 1996). This restriction site can be used in a variety of manipulations that simplify physical mapping and identifying sequences adjacent to the Ty insertion. Third, a technique called genetic footprinting has been developed that allows identification and phenotypic analysis of Ty1-induced mutations within essentially any gene in a population of cells (Smith *et al.*, 1995, 1996).

◆◆◆◆◆◆ II. INSERTIONAL MUTAGENESIS USING Ty1 ELEMENTS MARKED WITH THE RETROTRANSPOSITION INDICATOR GENES *his3-AI* AND *ade2-AI*

A. *his3-AI* and *ade2-AI*

The most recent step in the development of Ty1 as an insertional mutagenesis tool is selectable markers that only function after transposition; thus, cells that have had a transposition event can be directly selected (Protocols 1 and 2). Two retrotransposition indicator genes, *his3-AI* and *ade2-AI*, are currently available (Curcio and Garfinkel, 1991; Dalgaard *et al.*, 1996). The key feature of these alleles is an artificial intron (AI) that has been inserted into the *HIS3* and *ADE2* coding sequence in the wrong orientation for it to be spliced from the *HIS3* or *ADE2* transcript. Cells carrying these alleles are His⁻ or Ade⁻. The *his3-AI* or *ade2-AI* genes have been inserted into a pGTy1 plasmid so that the antisense strand of the marker gene is transcribed as a part of the Ty1 RNA. In this orientation, the AI in *his3-AI* or *ade2-AI* can be spliced from the Ty1 transcript. Reverse transcription and transposition of the spliced transcript results in a Ty1 insertion carrying a functional *HIS3* or *ADE2* allele. With the combined abilities to induce high levels of Ty1 transposition and to select for cells that have had transposition events, marked Ty1 elements can be used efficiently as insertional mutagens and in transposon-tagging protocols.

Protocol 1. Galactose induction of pGTy1 plasmids marked with *his3-AI* or *ade2-AI* in liquid

1. Streak your strain carrying the marked pGTy1 plasmid for single colonies on SC-URA [synthetic complete medium lacking uracil; detailed descriptions of media are presented elsewhere (Sherman, 1991; Appendix II)].
2. Using single colonies as inocula, grow 10 ml overnight cultures in SC-URA liquid with shaking at 30°C. The cultures should grow to saturation at concentrations of about 5×10^7/ml.
3. Centrifuge cells and resuspend at 5×10^6/ml in SC-URA+GAL (synthetic complete medium lacking uracil with galactose as a carbon source).
4. Incubate at 20°C with shaking for 24–48 h.
5. Centrifuge cells and resuspend in sterile water and spread on SC-HIS or SC-ADE plates at 30°C. (In our strains, 1–5% of the cells become His⁺ or Ade⁺. This number is sensitive to the efficiency and timing of galactose induction which can vary from strain to strain.)
6. In some cases, plate the mutagenized cultures under conditions that select for both His⁺ or Ade⁺ and the desired phenotype. In other cases, the first plate selects for His⁺ or Ade⁺ events, and then the resulting colonies are assayed for the desired phenotype.
7. Obtain cells that have lost the pGTy1 vector from before genetic or physical characterization. The *URA3*-based vector is readily lost during growth without selection. Direct selection for cells that have lost the plasmid can be accomplished by growth on medium containing 5-FOA. FOA can be incorporated into either supplemented SD minimal or SC media. We usually use 50 mg of uracil and 850–1000 mg of FOA per liter of media.

Protocol 2. Galactose induction of pGTy1 marked with *his3-AI* or *ade2-AI* on plates

1. Plate your strain carrying the marked pGTy1 plasmid for single-cell colonies on SC-URA+GAL. This can be done either by dilution or streaking for single cells.
2. Incubate at 20°C for 4 days to allow colonies to form.
3. Replica-plate the colonies to SC-HIS or SC-ADE and the condition required for the mutation of interest.
4. Pick papillae that result from cells that satisfy both the transposition and mutant phenotype, insuring that they represent independent events. [In one example, ~0.1% of the His⁺ colonies were also resistant to canavanine. About 73% of these *can1^R* mutants were caused by Ty*his3-AI* or Ty*HIS3* insertions (Rinckel and Garfinkel, 1996)]. Clonally purify the mutants.

1. Advantages of *his3-AI*

A major advantage of the *his3-AI* gene is that almost all His⁺ events result from transpositional integration of one copy of the spliced element. Because splicing occurs with an efficiency of about 15%, this minimizes the possibility of obtaining cells containing two Ty1*HIS3* elements.

2. Problems with *his3-AI*

Because splicing of the Ty1*his3-AI* transcript is inefficient, mutations may be caused by an unspliced Ty1*his3-AI* element. However, because an unspliced element can give rise to His⁺ events (Curcio and Garfinkel, 1991), linkage between a mutant phenotype and His⁺ papillation at 20°C may allow one to isolate a mutant caused by an insertion of Ty1*his3-AI*.

3. Advantages of *ade2-AI*

A major advantage of the *ade2-AI* system over the *his3-AI* system is that because the number of transposition events per induced cell is nine times lower, most (>90%) Ade⁺ prototrophs contain only a single marked Ty1 insertion. Moreover, the efficiency of *ade2-AI* splicing is five times higher than that of *his3-AI*. Both factors significantly reduce the fraction of selected colonies that contain a second marked Ty1 element with an unspliced marker gene. The *ade2-AI-D* indicator gene also contains the site for a rare-cutting endonuclease, *I-DmoI*, that is not present in the normal yeast genome (Dalgaard *et al.*, 1996). Therefore, there is a unique restriction site for *I-DmoI* within the single marked Ty1*ADE2-D* insertion that can be used for physical mapping and cloning. *I-DmoI* digestion of chromosome preparations results in fragmentation of the chromosome that harbors the Ty1*ADE2-D* insertion, which is easily detected by pulse-field gel electrophoresis. *I-DmoI* digestion in combination with partial digestion with *NotI*, *SfiI* or other rare-cutting restriction enzymes can be used for fine-structure mapping. Because the sequence of the entire yeast genome is known, Southern hybridization analyses using the *ADE2* gene as an "end" probe for one of the *I-DmoI*-generated chromosome fragments will determine the exact genomic location of the Ty1*ADE2-D* within the resolution of restriction mapping.

To map the Ty1*ADE2-D* insertion to the nucleotide, the *I-DmoI* site has been used to clone sequences flanking the Ty1*ADE2-D* element without the need for a rescue vector typically used in transposon-tagging protocols. Chromosomal DNA and plasmid pBST KS+D1 (M.J. Curcio, unpublished results), which contains an *I-DmoI* site adjacent to a polylinker, are digested with *I-DmoI* and then ligated together. The ligated chromosomal/vector mixture is then digested with an enzyme that cuts in the polylinker of the ligated pBST KS+D1 and in genomic sequences flanking the Ty1 insertion. Because the *I-DmoI* site is asymmetric, only the chromosomal fragment that contains the 3′ LTR of the Ty1*ADE2-D* element will retain the plasmid origin of replication. The digested chromosome/vector DNA is ligated under conditions that favor circularization. The resulting plasmid should contain the 3' LTR and adjacent genomic DNA, and can be recovered by transformation of *Escherichia coli* for subsequent sequence analysis with a Ty1 LTR primer.

4. Problems with *ade2-AI*

One disadvantage of the pGTy1*ade2-AI-D* system is that a relatively high fraction (5%) of Ade⁺ prototrophs arises by gene conversion of the *ade2-AI-D* gene on the plasmid, or an *ade2* allele in the genome. Gene conversion of the plasmid results in an unstable Ade⁺ phenotype that can be detected on synthetic complete medium (SC) that is limiting for adenine (6 mg liter⁻¹). Because *ade2* strains have a red coloration in the absence of sufficient exogenous adenine, unstable Ade⁺ colonies are pink. Gene conversion of a chromosomal *ade2* allele can be entirely avoided by using a strain with an *ade2* deletion (Aparicio *et al.*, 1991).

B. Insertional Mutagenesis with pGTy Elements Marked with *his3-AI* or *ade2-AI-D*

The simplest approach is to induce transposition of the marked pGTy1 element, select for His⁺ or Ade⁺ cells, and then select or screen for your favorite phenotype (Protocols 1 and 2). Galactose inductions can be performed in liquid or solid media. Note that Ty1 transposition occurs most efficiently at lower temperatures; therefore, we usually induce cultures with galactose at 20°C. Following galactose-induction in liquid synthetic complete medium minus uracil (SC–URA+GAL) for 24–48 h at 20°C, approximately 1–5% of the cells become His⁺ or Ade⁺. Classical genetic crosses are then performed to determine whether the mutation causing the phenotype for which you selected is linked to the *HIS3* or *ADE2*-marked element.

◆◆◆◆◆◆ III. EXAMPLES OF Ty MUTAGENESIS

A. Pheromone Resistance

An extensive analysis of α-factor-resistant mutants has been undertaken using cells mutagenized with Ty1*HIS3*. This selection is advantageous as a test system because at least 12 genes have been previously identified as essential for pheromone response. Seven of these genes have been tagged using Ty1 mutagenesis among less than 20 analyzed (Weinstock *et al.*, 1990; Mastrangelo *et al.*, 1992).

B. Thermoresistance Genes

In a screen for genes essential for growth at high temperature (another screen with several potential targets), three expected target genes have been tagged, but two new genes have also been identified (Kawakami *et al.*, 1992).

C. Uncloneable Loci

Ty mutagenesis has also been used to tag a suppressor gene called *SSM4* that had been known to suppress the thermosensitivity of an *rna14* mutation, but which proved to be uncloneable by plasmid-mediated complementation of the recessive mutant phenotype (Mandart *et al.*, 1994). It has been subsequently learned that plasmid clones containing *SSM4* are toxic in *E. coli*. Therefore, Ty mutagenesis can be an effective technique for tagging genes that cannot be cloned using standard approaches.

◆◆◆◆◆◆ IV. Ty1-MEDIATED FRAGMENTATION OF YEAST ARTIFICIAL CHROMOSOMES

A. Strategy

The pGTy1*ade2-AI-D* mutagenesis system can also be used to isolate strains that sustain a Ty1*ADE2-D* insertion onto a yeast artificial chromosome (YAC) (Dalgaard *et al.*, 1996). The strategy for isolating insertions on a YAC relies on the fact that if a single Ty1*ADE2-D* element transposes onto a YAC carried in a yeast strain, the *ADE2* marker will cosegregate with the *URA3* and *TRP1* YAC markers (Protocols 3 and 4). To isolate strains containing a Ty1*ADE2-D* insertion on a YAC, transposition is induced on galactose-containing medium, and then Ade$^+$ prototrophs are selected. Following a period of nonselective growth, colonies are subject to selection against the *URA3* gene expression by growth on FOA medium. If the Ty1*ADE2-D* element is present on the *URA3*-marked YAC, then the strain will become Ade$^-$ when the YAC is lost and will develop pink coloration on FOA medium.

Protocol 3. Transfer of a YAC to strain JC1474 by *kar1*-mediated chromoduction

1. Make fresh lawns of strains JC1474 and strain AB1380 with the YAC of interest by growing the strains overnight in YEPD and SC-URA-TRP medium at 30°C, respectively.
2. Spread 0.2 ml of each strain onto the same YEPD plate for mating, and grow for approximately 14 h at 30°C.
3. Replica-plate the mating to a SC-URA-TRP+CYH (3 mg liter^{-1} cycloheximide) plate. After 3–4 days, pick cycloheximide-resistant (cyh^R) papillae and clonally purify single colonies on SC-URA-TRP+CYH.
4. Pick independent cyh^R colonies and confirm the transfer of the YAC to strain JC1474 by checking all the relevant phenotypes of the chromoductants, including the nonmating phenotype of the *kar1* recipient strain.

Protocol 4. Isolation and *I-Dmo*I fragmentation of YACs sustaining Ty1*ADE2-D* insertions

1. Transform YAC-containing strains with plasmid pGTy1*ade2AI-D*. Select and maintain the *HIS3*-marked plasmid and *URA3*- and *TRP1*-marked YAC on SC-URA-TRP-HIS medium.
2. Pellet 0.5 ml of a saturated culture grown in SC-URA-TRP-HIS and resuspend the pellet in 5 ml of SC-URA-TRP-HIS+GAL.
3. Grow for 24 h at 20°C with shaking.
4. Spread 400 µl from each culture onto a SC-URA-TRP-ADE plate. About 40–100 colonies will appear after incubation at 30°C for several days. Because most transposition events occur after galactose-induction has ended (Dalgaard *et al.*, 1996), the majority of Ade⁺ colonies will represent independent Ty1*ADE2-D* transposition events.
5. Pick several hundred colonies onto YEPD plates and incubate for 2 days at 30°C. To calculate the number of colonies you will need to analyze, assume that the fraction of Ty1*ADE2-D* insertions on the YAC will be the same as the fraction of the genome that is represented by the YAC. The size of the yeast genome is about 14 Mb. Therefore, a 750 kb YAC represents approximately 5% of the total genome of the strain.
6. Replica-plate to a fresh YEPD plate to go through another round of non-selective growth. The second YEPD plate is designated the master plate.
7. Replica-plate the master YEPD plate to two plates; the first contains SC+FOA+ADE (6 mg liter⁻¹; limiting ADE) and the other contains SC-URA-TRP-ADE. Cell patches that turn pink on FOA-limiting ADE medium with extended incubation at 30°C represent putative Ty1*ADE2-D* insertions onto the YAC. This is because the Ade⁻ (pink) phenotype (loss of a Ty*ADE2-D* insertion) is cosegregating with the Ura⁻ (FOAᴿ) phenotype (loss of the YAC). This can be a subtle phenotype for large stable YACs, so incubations can extend up to 7 days. Cell patches should remain white on the SC-URA-TRP-ADE. Isolates with these characteristics typically contain a Ty1*ADE2-D* insertion on the YAC.
8. Isolates of interest should be retested for cosegregation of the Ura⁺, Trp⁺ and Ade⁺ phenotypes conferred by the tagged YAC, and independent segregation of the His⁺ phenotype conferred by the pGTy1*ade2-AI-D* plasmid. Cells from the master YEPD plate are streaked for single colonies on a fresh YEPD plate. Incubate these plates for 2–3 days at 30°C.
9. Replica-plate the YEPD plate to SC-URA-TRP, SC-ADE, and SC-HIS media. The majority of the colonies should be His⁻, demonstrating independent segregation of the pGTy1*ade2-AI-D* plasmid. Strains containing a Ty1*ADE2-D* insertion on the YAC will contain a small fraction of Ade⁻, Ura⁻, Trp⁻, His⁻ colonies but no Ade⁻, Ura⁺, Trp⁺ colonies. Pick a Ade⁺, Ura⁺, Trp⁺, His⁻ colony from the YEPD plate onto a SC-URA-TRP-ADE plate to maintain the marked YAC. This strain should contain a Ty1*ADE2-D* transposition event on the YAC and lack the pGTy1*ade2-AI-D* plasmid.
10. To map the Ty1*ADE2-D* insertion by fragmenting the YAC at the site of Ty1*ADE2-D* insertion with *I-DmoI*, prepare intact chromosomes of the YAC strains in agarose plugs (Dalgaard *et al.*, 1996), and then incubate the DNA with 10 units of *I-DmoI* in 1 × *I-DmoI* buffer A at 50°C for 3 h [*I-DmoI* can be obtained from Intermountain Scientific, P.O. Box 380, Kaysville, Utah 84037 (tel: 801-547-5047; to order: 800-999-2901; domestic fax: 800-574-7892; international fax: 801-547-5051)]. Separate yeast chromosomes by pulsed-field gel electrophoresis using standard protocols (Dalgaard *et al.*, 1996). Southern blot analysis may be performed using TRP1, URA3, and ADE2 probes to identify the orientation of the Ty1*ADE2-D* insertion, and size of the YAC fragments, because it is possible, albeit unlikely, that a YAC may contain a preexisting *I-DmoI* site.

B. pGTy Plasmids and YACs

Because the *URA3* marker is present on most YACs, a *HIS3*-marked plasmid containing the *GAL1*-promoted Ty1*ade2-AI-D* element has been constructed. Therefore, the strain containing the YAC must contain *his3* and *ade2* mutations in order to maintain the pGTy1*ade2-AI-D* plasmid and select for Ty1*ADE2-D* transposition events. The widely used YAC libraries of Larin *et al.* (1991) and Kusumi *et al.* (1993) are present in strain AB1380, which does not have these markers. However, the YACs can be transferred by chromoduction to a karyogamy mutant (*kar1*) strain JC1474 (*MATa leu2Δ1 trp1Δ63 ura3-52 ade2Δ:hisG his3Δ200 lys2-801 cyh1^R kar1*; ATCC accession no. 200993), which has the appropriate markers (Protocol 3). Alternatively, the new YAC library of Haldi *et al.* (1996) is in a yeast host strain with *ade2-101* and *his3-6* alleles, which can be used directly.

C. *I-DmoI* Mapping

Fragmentation of YACs with *I-DmoI* at the site of Ty1*ADE2-D* insertion is a facile alternative to traditional YAC-fragmentation methods. The *I-DmoI* site is likely to be very rare in mammalian genomes. Fragmentation at a unique *I-DmoI* site generates two YAC fragments, both of which are available for physical analysis. Furthermore, sequences flanking the Ty1*ADE2-D* insertion can be cloned directly (see section II.A.3 above), circumventing the purification of YAC fragments by pulsed-field gel electrophoresis.

The procedure presented in Protocol 4 has been applied to the analysis of two different mouse-derived YACs, which contain inserts of 340 kb or 680 kb (Dalgaard *et al.*, 1996). These analyses show that YACs can be used efficiently as targets for Ty1 transposition. The frequency of Ty1*ADE2-D* elements that have inserted onto the YAC rather than a yeast chromosome was 4.8% and 5.2%, respectively, which is at least as high as that expected if Ty1 target-site selection were random. However, the *I-DmoI*-based mapping of 20 Ty1*ADE2-D* insertions into the 680 kb YAC shows that, like the targeting of Ty1 insertions into yeast chromosomes, Ty1 transposition into YACs occurred at a small number of preferred sites. Quite surprisingly, one of these regions is a highly significant hotspot of 50 kb that corresponds to the transcription unit of the mouse *Steel* locus. It is not understood what constitutes a favorable target for Ty1 integration in mouse DNA, but our knowledge of what controls Ty1 integration into yeast chromosomes suggests that defined chromatin domains, the presence of unique structures within the DNA, or the presence of mouse repetitive elements transcribed by RNA polymerase III (pol III) may influence the integration specificity (Curcio and Morse, 1996; Devine and Boeke, 1996). Certainly, more YACs will need to be analyzed by Ty1-mediated chromosomal fragmentation in order to understand what characteristics of the *Steel* gene render it a favorable target for Ty1 insertion, and whether mammalian transcription units are generally favorable targets.

◆◆◆◆◆◆ V. PROBLEMS WITH Ty MUTAGENESIS

A. Insertions are not Random

One limitation to using Ty1 as an insertional mutagen is that insertions are not random throughout the yeast genome or within YACs containing mammalian DNA (see above). This nonrandomness can be observed both at the level of gene size targets and at the level of site preference within a gene. Ty1 elements appear to have two types of insertion sites in yeast chromosomes: genes transcribed by pol III are preferred targets (Devine and Boeke, 1996), while genes transcribed by pol II are used less efficiently. A variety of studies suggest that pol II-driven genes are used as targets about 10- to 100-fold less efficiently than would be expected if Ty1 insertions are random. Inclusion of mutations in genes that alter target site preferences, such as *RAD6* (Liebman and Newman, 1993), may be of some use. However, *rad6* mutants have pleiotrophic phenotypes that may bias or limit the outcome of a mutant search. In practical terms, this means that one should deal with a large enough pool (>10^7) of Ty1 insertions to insure that insertions into genes that are poor targets are included. Studies using genetic footprinting (see below), however, suggest that Ty1 elements can insert into many pol-II-driven genes (Smith *et al.*, 1995, 1996). It remains possible that some genes will be very poor targets for Ty1 insertions and difficult to mutagenize by this approach.

B. Ty1 Insertions can be Complex

The majority of Ty1 insertions, including those carrying selectable marker genes, are simple insertions of the element into the target sequence accompanied by a 5 bp duplication of the target site. Insertions rarely involve multimeric or rearranged Ty1 elements, or concomitant deletion of target sequences. Multimeric Ty1 insertions have been detected among events that cause the normally silenced *HMLα* locus to be expressed (Weinstock *et al.*, 1990; Mastrangelo *et al.*, 1992). Ty1 monomers at the same positions do not result in this release from the silencing mechanism. Multimeric insertions have also been found as a minor class of Ty insertions that activate the promoterless *his3-Δ1* gene (Weinstock *et al.*, 1990) and insertions that inactivate *CAN1* (Wilke and Liebman, 1989). Such rare multimeric or rearranged Ty1 insertions can present problems for cloning strategies based on isolating sequences adjacent to the marker gene. Similarly, Ty1 insertions associated with a deletion can cause problems in identifying the gene responsible for the mutant phenotype. In a screen for Ty1-induced mutations causing temperature sensitivity, two deletions of the *PET18* locus have been identified that appear to be similar to the spontaneous deletions in this interval caused by recombination between endogenous Ty elements (Kawakami *et al.*, 1992). A smaller deletion associated with a temperature-sensitive phenotype has been obtained in a cluster of endogenous solo-LTR elements. Deletions associated with Ty1 integration have also been reported for the *CYC1* region of chromosome X (Sutton and Liebman, 1992). In these selected cases, Ty1 integration can be accompanied by recombination with adjacent Ty1 elements resulting in deletions. These deletions emphasize the importance of comparing the structure of the mutant and parent strains at the insertion sites.

◆◆◆◆◆◆ VI. GENETIC FOOTPRINTING ANALYSIS: USING Ty1 MUTAGENESIS TO MEASURE PHENOTYPE AND FITNESS

A. Rationale

Genetic footprinting is a genomic strategy for elucidating information regarding the biological role of DNA sequences in *S. cerevisiae* (Smith *et al.*, 1995, 1996). A galactose-inducible Ty1 element is used as a mutagenic tool *in vivo*, to generate a large population of cells harboring insertion mutations at diverse sites. Representative samples of this population are then subjected to growth under different selections. The consequences of Ty1 insertions on cell fitness, for any particular gene, are analyzed using DNA isolated from the selected cells and from the original mutagenized population (the "time zero" population). Ty1 insertions are detected by polymerase chain reaction (PCR) using a labeled primer specific to the gene of interest and an unlabeled primer specific to the Ty1 element. The pattern of amplified products obtained from the selected DNA is then compared with that obtained from time zero DNA; depletion of PCR products following selection implies a growth disadvantage for the cells harboring Ty1 insertions in the gene of interest, and indicates a contribution to fitness by that gene in the selection.

Genetic footprinting analysis provides a means to rapidly investigate gene function. A single Ty1 mutagenesis on a pool of 10^{10}–10^{12} cells provides material for analysis of hundreds to thousands of genes. Selections can also be performed *en masse* on representative mutant cell pools. Other advantages include the analysis of many different mutations for any particular gene, allowing, in many cases, a detailed investigation of functional boundaries. Furthermore, the analysis is quantitative and can reveal even very subtle growth disadvantages that result from Ty1 insertion mutations. Finally, this methodology provides a means to investigate in some detail the Ty1 insertion patterns in particular genes of interest. One disadvantage of this method is that it does not generate an isolated mutant strain of any particular gene for further analysis. However, this does not appear to be a serious limitation, considering the variety of techniques available to introduce mutations into any yeast gene (see Chapters 5, 6 and 16).

B. Induction of Ty1 Mutagenesis

In the 2μ-based plasmid pBTy1 (*URA3*), the Ty1 element is regulated by the *GAL1* promoter and includes a unique 16 bp polylinker sequence (Smith *et al.*, 1995). A sufficiently large population of growing cells (at least 1×10^9) is initially transferred to galactose medium to ensure generation of a diverse mutant cell population. A small amount of sucrose (0.1%) added at the beginning of the mutagenesis assists in maintaining cell growth in the transition from glucose to galactose medium. It does not need to be added in subsequent changes of media. The mutagenesis can be monitored by plating cells on medium containing canavanine to monitor the increase in mutations at *CAN1*. For example, by day 4 of the mutagenesis procedure, plating 10^7 cells should yield 250–350 canavanine-resistant colonies. With the yeast strain used in our study (GRF167, Boeke *et al.*, 1985), we have observed a population doubling time of 4–5 h during Ty1 mutagenesis. Cells are harvested on the fourth day, after which 18–20 population doublings have elapsed. Independently performed mutagenesis yields very similar patterns of Ty1 insertion for all genes analyzed (Smith *et al.*, 1995, 1996).

Analysis of DNA from individual mutant colonies indicated that ≥ 90% of these colonies have six new Ty1 insertions. The number of insertions can be estimated by Southern blot analysis using an oligonucleotide probe complementary to the polylinker sequence in the Ty1 element (each full length Ty1 element will contain two polylinker sequences). It is possible, given this mutation frequency, that a single cell may contain two or more insertions that can result in different phenotypes. This is not a problem, as each PCR analysis is performed on DNA isolated from approximately 7×10^7 independent cells, and up to several hundred (or more) independent cells contribute to the pattern of Ty1 insertions obtained. However, because Ty1 inserts preferentially into noncoding regions, which has far fewer phenotypic consequences, the average number of mutagenic insertions may be more in the order of 0.2 per cell.

Protocol 5. Induction of Ty1 mutagenesis for genetic footprinting analysis

1. Inoculate a 10 ml overnight culture (SC-URA, 2% glucose) with a freshly-streaked colony of a *GAL ura3* yeast strain containing the pBTy1 plasmid.
2. Generate a glucose pre-culture of at least 1×10^9 cells; e.g. inoculate 100 ml SC-URA (2% glucose) with 2 ml of the overnight culture, allow the cells to reach a density of $1–2 \times 10^7$ cells/ml.
3. Collect cells by centrifugation, wash once with sterile water, and transfer to SC-URA medium (≥ 800 ml) containing 2% galactose and 0.1% sucrose, at 25°C.
4. Maintain cell growth for 18–20 population doublings at 25°C (typically 4 days) by transferring cells (≥ 1×10^9) to fresh SC-URA (2% galactose medium). Keep the cell density from exceeding 2×10^7 cells/ml during this time. Cells can be grown in a flask or fermenter. The cell density can reach 4×10^7 cells/ml at completion of the mutagenesis.
5. On completion of the Ty1 mutagenesis, harvest cells by centrifugation. Resuspend several aliquots of cells in 25% glycerol (e.g. 2×10^8 cells per 1 ml tube) and store at –80°C. These cells will be used for selections. DNA is isolated from the remaining cells (Struhl et al., 1979). The cell pellets can be stored for several weeks at –20°C or –80°C prior to isolation of DNA.

C. Selections

At least 4×10^8 mutagenized cells are used for each selection. The frozen cells are thawed on ice and added to the appropriate selection medium (now containing uracil). To assess phenotypes that may result from freezing, one aliquot of cells is grown in rich-glucose medium following mutagenesis without freezing. DNA isolated from these cells is used to confirm phenotypes obtained from cells frozen prior to selection.

The selections themselves can be of any nature, the requirement being that it must be possible to perform them on a scale of at least 4×10^8 cells. Selections that have been performed for genetic footprinting analysis include growth in rich-glucose medium, rich-lactate medium, minimal-glucose medium, growth in medium containing caffeine, medium containing NaCl, growth at high temperature, and the ability to mate with cells of opposite mating type (Smith *et al.*, 1996). The selection should not be so severe as to result in nonspecific cell death. Ideally, footprints should be detected for susceptible mutants, but the bulk of the cell population should grow at a wild-type rate. Pilot-scale analyses can be performed to determine an appropriate severity of selection. Significant nonspecific cell death will be apparent by microscopic examination of the cells, and will also result in more sampling variability in the patterns of Ty1 insertion obtained by PCR analysis. Factors such as cell density (logarithmic growth phase), temperature, and pH should be maintained (for selections where these conditions are not being investigated), to avoid inadvertently applying other selective pressure.

The number of population doublings used for a selection will determine the sensitivity of detection of growth disadvantages. For example, to examine quantitative growth deficits to within a few per cent of the wild-type population growth rate for genes on chromosome V (Smith *et al.*, 1996), mutagenized cells have been subjected to selection in rich-glucose medium for up to 60 population doublings. Typically, however, time-points of 15 to 18 population doublings have been analyzed, and growth deficits as low as 10% (i.e. mutant growing at 90% of the population growth rate) are detectable. DNA isolated from an earlier timepoint (e.g. 10 population doublings) during the selection is useful in confirming quantitative estimates of the growth rate deficit. The sensitivity of detection of growth disadvantages is somewhat lower for selections that do not involve many population doublings (e.g. mating).

D. PCR Analysis and Oligonucleotide Design

The Ty1-specific oligonucleotide (AGAGCTCCCGGGATCCTCTAC-TAAC) includes the 16 bp polylinker sequence that distinguishes this Ty1 element from endogenous Ty1 and solo-δ (LTR) elements. This primer anneals to a sequence located 326 bp into the Ty1 LTR, thus all PCR products corresponding to Ty1 insertions are > 326 bp. The melting temperature is 72°C. This primer is not labeled.

Gene-specific oligonucleotides are typically located 500–900 bp from the initiator ATG, and directed upstream toward the initiator ATG. This permits analysis of several hundred base pairs of coding sequence in a single PCR, while also including regions upstream of the gene. Ty1 insertions that are sufficiently distant from the gene to have no distinguishable effects on fitness generate PCR products that serve as reference points when comparing the Ty1 insertions patterns obtained from different selections. Their presence also confirms that a footprint is the result of genuine depletion of mutant cells from the population under selection, rather than a failed PCR. For small genes (300 bp or less), primers should be located closer to the stop codon (400–500 bp from the initiator ATG) to ensure that peaks corresponding to insertions in the coding sequence will be of a suitable molecular weight for clear resolution on polyacrylamide gel analysis.

Some genes, especially those located in the vicinity of tRNA genes, have very strong upstream site preferences for Ty1 insertions (Devine and Boeke, 1996). This can lead to the generation of artifact peaks in the PCR, possibly due to exhaustion of some of the reagents. Using fewer cycles of PCR (23 instead of 30) can reduce the appearance of artifact peaks. Alternatively, primers specific to these genes can be designed in the opposite orientation: located in the first 100–500 bp of the gene, avoiding the

---

Protocol 6. PCR analysis

PCR reactions are performed in a thermal cycler (e.g. Perkin Elmer 9600). The final reaction volume is 50 μl.

1. Place the 96-well reaction tubes or plate on ice/water. Add 25 pmol of fluorescein-labeled gene-specific oligonucleotide to each well.
2. Prepare a master mix of buffer, nucleotides, Ty1-specific primer, and Taq DNA polymerase. The final concentrations in each well are: 1 × buffer, 250 μM each dNTP, 25 pmol Ty1 primer, 2 U Taq DNA polymerase. Add the appropriate volume of master mix to each well.
3. Add 1 μg of DNA (2 μg of DNA from mated cells) to the top of each well in a 2 or 4 μl droplet. Turn the plate around, if necessary, to find a clean area on each well which has not had contact with other liquid.
4. Centrifuge the plate briefly to mix the DNAs into the PCR reaction. Adding and mixing the DNA in this way for every set of reactions insures reproducibility between different PCRs.
5. Place the plate on the thermal cycler once it has reached a temperature of ≥90°C. Cycle as follows: 93°C, 1 min; then 10 cycles of 92°C for 30 s, 67°C for 45 s, 72°C for 2 min; then 20 cycles of 92°C for 30 s, 62°C for 45 s, 72°C for 2 min.
6. For size fractionation and product detection, remove 5–10 μl of each reaction to a new 96-well plate. Add 4 μl of formamide loading dye (5:1 formamide: 50 mM EDTA, blue dextran) and place the plate at 80°C for 15 minutes.
7. Load 4 μl (the remaining volume) on a 4.75% denaturing polyacrylamide gel, mounted in an Applied Biosystems 373A automated DNA sequencer (or equivalent). Molecular weight size standards (e.g. Applied Biosystems ROX-labeled Genescan-1000 and Genescan-2500) can also be loaded in each well (if labeled with a different fluorophore), or in spare lanes. Electrophorese in 1 × TBE using standard conditions for a sequencing gel. Collect all data in the size range of 300–2000 bp (if using an Applied Biosystems sequencer, use Genescan DNA fragment analysis software).

Ty1-preferred site, and directed downstream toward the stop codon. These "reverse" primers typically generate a more reliable survey of peaks corresponding to Ty1 insertions in the coding sequences of genes with very strong upstream Ty1 site preferences.

Gene-specific oligonucleotides are designed with a melting temperature of 69–74°C as determined by the nearest neighbor method (Breslauer et al., 1986; Rychlik and Rhoads, 1989). Suitable programs for designing these primers are Oligo (National Biosciences), Primer (Whitehead Institute) and primer design software for yeast genome analysis available at the SGD web site (http://www-genome.stanford.edu). The primers are labeled at their 5' end with fluorescein. We have found that fluorescein amidites from a variety of different sources are all effective. The oligonucleotides do not need to be purified or accurately quantitated. For example, we successfully used primers synthesized in 96-well arrays (Lashkari et al., 1995) on a 20 nM scale, without purification or any more quantitation than a general estimate that was applied to all oligonucleotides in the plate. It is important that the fluorescein labeling is reasonably efficient, as the sensitivity of detection is otherwise greatly reduced. When estimating the concentration of gene-specific primers, it is also safer to err on the side of using slightly too much oligonucleotide (25–35 pmol per reaction). This avoids having an excess of the Ty1-specific oligonucleotide in the reaction, a condition that might favor formation of Ty–Ty PCR products from adjacent Ty1 elements.

Single base resolution is not necessary, unless there is interest in mapping insertion sites more precisely. Thus, small gels can be used for fractionation of PCR products. If a suitable fluorescent detector is not available, oligonucleotides can be 5'-labeled with ^{32}P. Ideally, gels should then be imaged using a phosphorimager, to detect and quantitate effectively the large dynamic range of PCR product intensities.

E. Interpretation of data

Footprints are apparent as zones of depletion of PCR products encompassing, typically, the region corresponding to the coding sequence of the gene of interest that is under selection. Growth defects should be inferred from the depletion of several peaks. Caution should be used in giving too much emphasis to individual PCR product peaks: single peaks that do not behave as expected may represent tolerated Ty1 insertions, or PCR artifacts. In addition, the consequences of sampling of the cell population will manifest as variation in the peaks detected between different DNA samples and even in independent PCR analyses of the same DNA samples. This is especially true for many genes that are not favored targets for Ty1 transposition. For these genes, lower-intensity peaks corresponding to insertions in the coding sequence may represent as few as 1–10 cells. Use of a second gene-specific primer, located 100–200 bp from the first primer, is often useful in confirming growth deficits, and allows for a more detailed evaluation of the behavior of individual peaks. If only one primer is used to analyse any particular gene, there is a 2–3% chance that the data generated are unreliable (based on the frequency with which independent primers produced apparently credible but discordant data for genes on chromosome V; Smith *et al.*, 1996).

Gene-specific primers should be analyzed by PCR on DNA isolated from cells that have not been induced for Ty1 transposition, to check for Ty1-independent background. A few background peaks do not usually present a problem for genetic footprinting analysis, but if a large amount of background is apparent, a new independent gene-specific primer should be synthesized. Similarly, synthesis of a new independent primer is generally effective when signal is not detected on PCR analysis of mutagenized and selected DNA samples. However, usable data may not be obtained for a small fraction of genes, even with two or three independent primers (2.6% of genes analyzed on chromosome V). These genes may be very poor targets for Ty1 transposition, or the consequences of Ty1 insertions to cell growth may have been too severe to permit detection of these insertions in the time-zero population. Similarly, footprinting data for genes in which very weak PCR products or only a few peaks of low intensity are detected should be interpreted cautiously, as sampling effects will be more severe.

To obtain quantitative estimates of more subtle growth deficits, PCR data from DNA isolated from one or two or more different timepoints in a selection are compared with the time zero PCR and with each other. Changes in peak area (peak heights also provide a good estimate) can be related to the growth deficit of the mutant strain [% signal remaining = $\frac{1}{2}n(F-1/F)$; where n is the number of population doublings, and F is the fold-increase in generation time (e.g. for a mutant growing at 80% of the population rate, $F=1.25$)]. However, different numbers of products are being amplified in these reactions: for example, if depletion is occurring, a larger number and/or intensity of independent PCR products will be present in the time-zero sample, relative to the selected DNA samples. Reduced competition for PCR reagents can lead to more efficient amplification of the remaining targets in the selected DNA samples. This effect is usually not severe, but more accurate estimates of growth deficits are obtained if the PCR data are normalized to some degree. This can be done by normalizing the signal produced in the different PCRs relative to insertions upstream of the coding sequence, if these insertions do not appear to have consequences on fitness. Generally, a multiplier of 1–2 × (sometimes more) is applied to the time-zero PCR data.

Alternatively, an internal standard can be used to normalize PCR data (Smith *et al.*, 1996). Restriction-enzyme-digested yeast genomic DNA fragments can be cloned into a vector containing the LTR region of the Ty1 element, such that PCR amplification of this library DNA with the Ty1-specific primer and any labeled gene-specific primer will result in a predictable pattern of products. These products are identified by a PCR reaction that includes only the library DNA. The library DNA is added at a fixed concentration to each selected DNA PCR and the time zero PCR, and the library-specific peaks are used to normalize the PCR product peaks from the different reactions relative to each other. This internal standard can assist in obtaining a more accurate estimate of growth rate deficit, although given sampling and other effects, the numbers obtained for growth rate deficits are typically + 5%.

Acknowledgements

Research was sponsored in part by the National Cancer Institute, DHHS, under contract with ABL (D.J.G.), the Wadsworth Center Molecular Core Facility and NIH Grant GM52072 (M.J.C.), and the NIH and Human Frontiers Science Program Organization (V.S.).

References

Aparicio, O. M., Billington, B. L. and Gottschling, D. E. (1991). Modifiers of position effect are shared between telomeres and silent mating-type loci in *S. cerevisiae*. *Cell* **66**, 1279–1287.

Boeke, J. D. and Sandmeyer, S. B. (1991). Yeast transposable elements. In *The Molecular and Cellular Biology of the Yeast* Saccharomyces: *Genome Dynamics, Protein Synthesis, and Energetics* (J. Pringle, E. Jones and J. Broach, eds), pp. 193–261. Cold Spring Harbor Laboratory Press, Cold Spring Harbor, NY.

Boeke, J. D., Lacroute, F. and Fink, G. R. (1984). A positive selection for mutants lacking orotidine 5′-phosphate decarboxylase activity in yeast: 5-fluoroorotic acid resistance. *Mol. Gen. Genet.* **197**, 345–346.

Boeke, J. D., Garfinkel, D. J., Styles, C. A. and Fink, G. R. (1985). Ty elements transpose through an RNA intermediate. *Cell* **40**, 491–500.

Boeke, J. D., Xu, H. and Fink, G. R. (1988). A general method for the chromosomal amplification of genes in yeast. *Science* **239**, 280–282.

Breslauer, K. J., Frank, R., Blocker, H. and Marky, L. A. (1986). Predicting DNA duplex stability from the base sequence. *Proc. Natl Acad. Sci. USA* **83**, 3746–3750.

Curcio, M. J. and Garfinkel, D. J. (1991). Single-step selection for Ty1 element retrotransposition. *Proc. Natl Acad. Sci. USA* **88**, 936–940.

Curcio, M. J. and Morse, R. H. (1996). Tying together integration and chromatin. *Trends Genet.* **12**, 436–438.

Dalgaard, J. Z., Banerjee, M. and Curcio, M. J. (1996). A novel Ty1-mediated fragmentation method for native and artificial yeast chromosomes reveals that the mouse *Steel* gene is a hotspot for Ty1 integration. *Genetics* **143**, 673–683.

Devine, S. and Boeke, J. D. (1996). Integration of the yeast retrotransposon Ty1 is targeted to regions upstream of genes transcribed by RNA polymerase III. *Genes Dev.* **10**, 620–633.

Garfinkel, D.J. (1992). Retroelements in microorganisms. In *The Retroviruses*, Vol. 1. (J. A. Levy, ed.), pp. 107–158. Plenum, New York.

Garfinkel, D. J. (1996). Insertional mutagenesis by Ty elements in *Saccharomyces cerevisiae*. In *Methods in Molecular Biology* (I. Evans, ed.), pp. 227–237. Humana Press, Totowa, NJ.

Garfinkel, D. J. and Strathern, J. N. (1991). Ty mutagenesis in *Saccharomyces cerevisiae*. In *Methods in Enzymology: Guide to Yeast Genetics and Molecular Biology* (C. Guthrie and G. R. Fink, eds), pp. 342–361. Academic Press, New York.

Garfinkel, D.J., Mastrangelo, M. F., Sanders, N. J., Shafer, B. K. and Strathern, J. N. (1988). Transposon tagging using Ty elements in yeast. *Genetics* **120**, 95–108.

Haldi, M. L., Strickland, C., Lim, P., Van Berkel, V., Chen, X.-N., Noya, D., Korenberg, J. R. *et al.* (1996). A comprehensive large-insert yeast artificial chromosome library for physical mapping of the mouse genome. *Mamm. Genome* **7**, 767–769.

Kawakami, K., Shafer, B. K., Garfinkel, D. J., Strathern, J. N. and Nakamura, Y. (1992). Ty element-induced temperature-sensitive mutations of *Saccharomyces cerevisiae*. *Genetics* **131**, 821–832.

Kusumi, K., Smith, J. S., Segre, J. A., Koos, D. S. and Lander, E. S. (1993). Construction of a large-insert yeast artificial chromosome library of the mouse genome. *Mamm. Genome* **4**, 391–392.

Larin, Z., Monaco, A. P. and Lehrach, H. (1991). Yeast artificial chromosome libraries containing large inserts from mouse and human DNA. *Proc. Natl Acad. Sci. USA* **88**, 4123–4127.

Lashkari, D., Hunicke-Smith, S., Norgren, R., Davis, R. W. and Brennan, T. (1995). An automated multiplex oligonucleotide synthesizer: development of high-throughput, low-cost DNA synthesis. *Proc. Natl Acad. Sci. USA* **92**, 7912–7915.

Liebman, S. W. and Newman, G. (1993). A ubiquitin-conjugating enzyme, RAD6, affects the distribution of Ty1 retrotransposon integration positions. *Genetics* **133**, 499–508.

Mandart, E., Dufour, M.-E. and Lacroute, F. (1994). Inactivation of *SSM4*, a new *Saccharomyces cerevisiae* gene, suppresses mRNA instability due to *rna14* mutations. *Mol. Gen. Genet.* **245**, 323–333.

Mastrangelo, M. F., Weinstock, K. G., Shafer, B. K., Hedge, A. M., Garfinkel, D. J. and Strathern, J. N. (1992). Disruption of a silencer domain by a retrotransposon. *Genetics* **131**, 519–529.

Rinckel, L. A. and Garfinkel, D. J. (1996). Influences of histone stoichiometry on the target site preference of retrotransposons Ty1 and Ty2 in *Saccharomyces cerevisiae*. *Genetics* **142**, 761–776.

Rychlik, W. and Rhoads, R. E. (1989). A computer program for choosing optimal oligonucleotides for filter hybridization, sequencing and *in vitro* amplification of DNA. *Nucleic Acids Res.* **17**, 8543–8551.

Sherman, F. (1991). Getting started with yeast. In *Methods in Enzymology: Guide to Yeast Genetics and Molecular Biology* (C. Guthrie and G. R. Fink, eds), pp. 3–20. Academic Press, New York.

Smith, V., Botstein, D. and Brown, P. O. (1995). Genetic footprinting: a genomic strategy for determining a gene's function given its sequence. *Proc. Natl Acad. Sci. USA* **92**, 6479–6483.

Smith, V., Chou, K., Lashkari, D., Botstein, D. and Brown, P. O. (1996). Functional analysis of the genes of yeast chromosome V by genetic footprinting. *Science* **274**, 2069–2074

Strathern, J. N., Mastrangelo, M. Rinckel, L. A. and Garfinkel, D. J. (1994). Ty insertional mutagenesis. In *Molecular Genetics of Yeast: A Practical Approach* (D. Rickwood and B. D. Hames, eds), pp. 111–119. IRL Press, Oxford University, Oxford.

Struhl, K., Stinchcomb, D. T., Scherer, S. and Davis, R. W. (1979). High-frequency transformation of yeast: autonomous replication of hybrid DNA molecules. *Proc. Natl Acad. Sci. USA* **76**, 1035–1039.

Sutton, P. R. and Liebman, S. W. (1992). Rearrangements occurring adjacent to a single Ty1 yeast retrotransposon in the presence and absence of full-length Ty1 transcription. *Genetics* **131**, 833–850.

Weinstock, K. G., Mastrangelo, M. F., Burkett, T. J., Garfinkel, D. J. and Strathern, J. N. (1990). Multimeric arrays of the yeast retrotransposon Ty. *Mol. Cell. Biol.* **10**, 2882–2892.

Wilke, C. M. and Liebman, S. W. (1989). Integration of an abberant retrotransposon in *Saccharomyces cerevisiae*. *Mol. Cell. Biol.* **9**, 4096–4098.

8 Transcript Analysis

Alistair J. P. Brown[1], L. Michael Furness[2] and David Bailey[3]
[1]*Department of Molecular and Cell Biology, Institute of Medical Sciences, University of Aberdeen, Foresterhill, Aberdeen, UK,* [2]*Central Research, Molecular Sciences, Pfizer, Sandwich, Kent, UK* [3]*Incyte Pharmaceuticals, Porter Drive, Palo Alto, California, USA*

◆◆

CONTENTS

Introduction
mRNA abundance measurements
mRNA structure
mRNA stability
Future perspectives

List of Abbreviations

AE	Anchoring enzyme
EST	Expressed sequence tag
GFP	Green fluorescent protein
ORF	Open reading frame
RT-PCR	Reverse transcriptase-polymerase chain reaction
SAGE	Serial analysis of gene expression
YPD	*See Appendix II: Table 1*

◆◆◆◆◆◆ I. INTRODUCTION

mRNA abundance measurements provide a useful tool to study gene regulation. Currently, most groups use such measurements to investigate the expression of a small number of specific genes. However, with the advent of genomic sequencing projects and the completion of the yeast genome sequence, we have entered an era in which genomic transcript analyses are being exploited to generate global pictures of gene regulation. This chapter will guide the reader towards classical methods for the analysis of specific mRNAs, and then describe new technologies that are being developed for gene expression analysis.

While this chapter focuses mainly on mRNA abundance measurements, methods for the analysis of transcript structure and stability are also discussed. By necessity, genome-scale transcript analyses are focusing mainly on the open reading frames (ORFs) themselves, i.e. coding *sequences*. However, the secondary and tertiary *structures* formed by these sequences and their 5′-leader and 3′-trailer regions exert strong influences on gene expression at post-transcriptional levels. Hence these issues are an integral part of "transcript analysis".

◆◆◆◆◆◆ II. mRNA ABUNDANCE MEASUREMENTS

A. A Historical Perspective

In the 1960s, in the absence of methods for direct analysis of mRNAs, cell-free translation systems provided an indirect approach to measure the abundance of a specific mRNA. The amount of a specific protein produced during the cell-free translation of an mRNA population was assumed to reflect the abundance of the cognate mRNA in that population (for standard *in vitro* translation methods, see Sambrook *et al.*, 1989). This was then superseded by dot-blotting, which provided a more direct approach to mRNA quantitation (for standard dot-blotting methods, see Sambrook *et al.*, 1989). This simple approach can provide a useful survey of the expression of specific mRNAs in several RNA populations at the same time. However, dot-blotting suffers several disadvantages. First, it does not give an indication of the integrity of the mRNA being analyzed. Second, it can be difficult to quantitate low-abundance mRNAs accurately because of relatively high backgrounds. These backgrounds are caused by low-level hybridization to other RNAs and to the genomic DNA that frequently contaminates RNA preparations. Therefore, while dot (or slot)-blotting remains a rapid and convenient approach for the analysis of medium- and high-abundance mRNAs, this method has been superseded by Northern blotting, which remains the most widely used method for mRNA quantitation (see section II.B).

A variety of other approaches to mRNA quantitation are available. RNA mapping methods (such as S1 analysis or primer extension, for example) are most frequently exploited for structural analysis of transcripts, but these methods have also been used for mRNA quantitation (section III). RT-PCR provides an alternative approach to mRNA quantitation (Dallman and Porter, 1991). In principle this is a rapid and sensitive method, but it is not used frequently for the analysis of yeast mRNAs, and hence is not discussed further here. Reporter genes are frequently used to study gene regulation as described elsewhere in this volume. This sensitive and accurate approach is now being applied on a genomic scale (section II.C).

New technologies are currently being developed for rapid, genome-scale transcript analysis. Hybridization array technology (section II.E) has been dependent on the development of new methods for the generation of high-density arrays of target sequences and the quantitation of large numbers of individual signals following hybridization of fluorescently labeled probes to these arrays. On the other hand, serial analysis of gene expression (SAGE; section II.D), which has been used to characterize the "yeast transcriptome" of about 15 000 mRNAs, exploits automated DNA sequencing technologies.

B. Northern Analysis

Northern blotting is the most widely used method for mRNA quantitation because it is direct and sensitive, it allows reasonably accurate quantitation, it permits mRNA length estimation, and it tests the integrity of the RNA preparation being analysed. Detailed descriptions of Northern blotting procedures can be found elsewhere (Sambrook *et al.*, 1989). Briefly, RNA samples are subjected to electrophoresis under denaturing conditions, blotted onto a nylon membrane, hybridized with a specific probe, and the signals quantitated in some way. Relative mRNA levels are then calculated once the membrane has been stripped and reprobed for a loading control, most frequently the *ACT1* mRNA (Moore *et al.*, 1991).

Northern blots have been used (a) to test whether a transcript of the appropriate length is expressed from a new ORF, and hence to validate genome sequencing (Fairhead and Dujon, 1994); (b) to study the regulation of specific mRNAs in response to a variety of experimental conditions (environmental changes, cell cycle regulation or phenotypic effects of specific mutations); or (c) to analyse kinetic changes in mRNA levels (for example to measure mRNA half-lives; see section IV). Most frequently, the ultimate purpose of a Northern blot is to compare the level of a particular mRNA in different RNA populations. Several points are worth noting if this is to be achieved accurately (Figure 1).

Northern analysis has now been extended from the analysis of small families of yeast mRNAs to a larger, genomic scale. The first of these studies generated a transcript map of chromosome III (Yoshikawa and Isono, 1990) even before the sequence of the chromosome was published (Oliver *et al.*, 1992). The map was made by constructing an ordered set of chromosome III clones which were used to probe poly(A)-containing mRNA from yeast cells growing on rich medium (YPD). The identification of 155 mRNAs correlated well with the ORFs revealed by the chromosome III sequencing project (Tanaka and Isono, 1993). mRNA levels were categorized qualitatively on the basis of the intensity of autoradiographic signals (Yoshikawa and Isono, 1990). Nevertheless, the data are consistent with the more quantitative results of more recent Northern-based projects that have followed genome sequencing. mRNAs from chromosomes I, VI and XI covering a total of 980 kbp have been analyzed (Barton and Kaback, 1994; Fairhead and Dujon, 1994; Richard *et al.*, 1997; Naitou *et al.*, 1997). 296 of the 475 predicted genes in these regions were shown to be expressed at the mRNA level under the various growth conditions studied.

A related project was undertaken as part of EUROFAN Phase 1 (Oliver, 1996). Rudi Planta (The Netherlands) led a group including four other laboratories (Alistair Brown, UK; Esperanza Cerdan, Spain; Steve Oliver, UK; Johan Thevelein, Belgium) to establish mRNA levels for 1000 ORFs of unknown function (Planta, 1997). The aim was to provide data on the regulation of these ORFs and thereby some clues as to their broad area of function. The primary approach, which started with chromosome XIV ORFs, is outlined in Protocol 1. Multiplexed Northern analyses were used (Figure 2) because neither dot-blotting nor RT-PCR proved sufficiently sensitive or quantitative.

Several important observations have been made by these multiplexed Northern analyses. First, the majority of yeast ORFs are expressed, and hence the yeast genome is densely packed with *expressed* genes (on average, one about every 2.1 kbp; Dujon and Goffeau, 1996). In general, short

---

- Detailed procedures for Northern analysis are published elsewhere (Sambrook *et al.*, 1989).
- The *ACT1* mRNA (YFL039c) provides a useful loading control for medium- to high-abundance mRNAs in many types of experiment, and in particular for the analysis of kinetic changes in mRNA levels over short time courses, or for the comparison of groups of mRNAs under the same growth conditions (Figure 1). However, it should be noted that *ACT1* mRNA levels are not constant under all experimental conditions (Delbruck and Ernst, 1993), and therefore it is important to choose an appropriate loading control for the experimental conditions under analysis.
- The *ORC5* mRNA (YNL261w) is a good control for low-abundance mRNAs. This gene, which encodes an essential component of the origin of DNA replication complex, appears to be expressed constitutively at about 1% of the *ACT1* mRNA.
- In some situations well-defined control mRNAs, which remain at constant levels under the experimental conditions, are not available (e.g. when RNAs from different growth phases are being compared). In these situations, measuring mRNA levels relative to the ribosomal RNAs is often the best approach (Swoboda *et al.*, 1994).
- Significant errors in relative mRNA abundance measurements are introduced if unequal amounts of total RNA samples are analyzed. This is the case even when internal loading controls are used (e.g. the membrane is reprobed for *ACT1*). Therefore, approximately equal amounts of each RNA preparation must be analyzed.
- Simple and convenient hybridization solutions have been developed. Consistent data can be obtained by performing hybridizations on nylon membranes in 7% SDS, 1 mM EDTA, 0.5 M Na_2HPO_4, pH 7.2 at temperatures above 40°C. Membranes can then be washed in standard SSC-based solutions containing 0.1% SDS (Sambrook *et al.*, 1989). This hybridization solution can be stored at room temperature.
- A variety of methods can be used to quantify signals on Northern blots. Autoradiographic images can be misleading: small change in mRNA level (combined with small errors in RNA loading) can yield what appear to be dramatic differences in autoradiographic signals. Hence, although non-radioactive detection methods might be desirable, accurate quantitation is most frequently achieved by direct two-dimensional radioimaging of ^{32}P signals (Swoboda *et al.*, 1994). Phosphorimaging is now the most sensitive method, allowing quantitation of mRNAs at about 1% of the level of the *ACT1* mRNA.

Figure 1. Technical approaches to Northern analysis.

ORFs (of less than 100 codons) do not correspond to functional genes (Fairhead and Dujon, 1994). Intergenic regions are frequently short (about 0.5 kbp; Dujon and Goffeau, 1996). Most yeast genes are expressed at low levels (about 70% are at levels less than about 5% of the *ACT1* mRNA; Planta, 1997), and only a small number of genes (roughly 1%) are expressed at levels significantly higher than *ACT1* (Fairhead and Dujon, 1994; Planta, 1997). About 30% of genes show some glucose-regulation and about 25% vary in response to nitrogen (Planta, 1997). Highly expressed genes are dispersed over the genome, but genes located near telomeres are generally expressed at low levels. With a few notable exceptions such as *GAL1/10/7*, co-regulated genes are dispersed in the genome. Therefore, multiplexed Northern analyses have yielded interesting findings that have implications for our understanding of the global organization, expression and nuclear compartmentation of the yeast genome. However, while this approach is sensitive and accurate, it is labour intensive. Therefore, new technologies have been developed for genome-wide transcript analysis (sections II.D and II.E).

Protocol 1. Approach to genome-scale Northern analysis in EUROFAN Phase 1

1. The yeast strain FY73 (B. Dujon; Institut Pasteur, Paris) is being used in this study (Planta, 1997). FY73 is the haploid *MATα* parent of the diploid strain FY1679 being used for the functional analysis of the yeast genome in EUROFAN Phase 1.

2. Northern blots are prepared using equal loadings of RNA samples isolated during growth under three "steady-state" conditions (growth in rich glucose medium, rich glycerol–ethanol medium and minimal glucose medium), and several "transient" conditions (gluconeogenic to glycolytic growth, carbon starvation, nitrogen starvation, heat-shock and osmotic stress, Figure 1).

3. Probes ranging in length from about 200 to 350 bp are generated by PCR-amplification of each target ORF directly from genomic DNA.

4. The PCR primers are generally of 15 to 16 nucleotides in length, and are designed from the genome sequence at the high stringency setting using OLIGO Primer Analysis Software (Cambio, Cambridge, UK).

5. The PCR products are purified on Wizard™ PCR preps DNA purification columns (Promega, Madison, USA).

6. The PCR products (25 ng) are then radiolabeled to high specific activities (over 5×10^8 dmp µg^{-1} DNA) by random-prime labeling (Feinberg and Vogelstein, 1983), and repurified on Sephadex G50 columns.

7. Northern membranes are hybridized with heat-denatured probes for 15 h at 65°C in 7% SDS, 1 mM EDTA, 0.5 M Na$_2$HPO$_4$, pH 7.2.

8. Membranes are washed for 15–30 min in $0.5 \times$ SSC containing 0.1% SDS: twice at room temperature, and twice at 60–65°C. The signals are then quantified by phosphorimaging.

9. The relative level of each target mRNA is compared under the different growth conditions. Having corrected for variations in the specific activities of the probes, the absolute level of each target mRNA is then compared with that of the *ACT1* mRNA. The relative levels of the *ACT1*, *HSP12*, *RPL25*, *CAR1*, *PCK1* and *ORC5* mRNAs are used as controls for the various growth conditions.

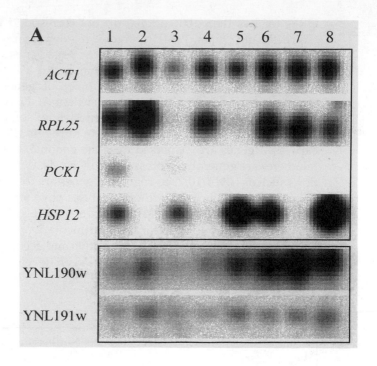

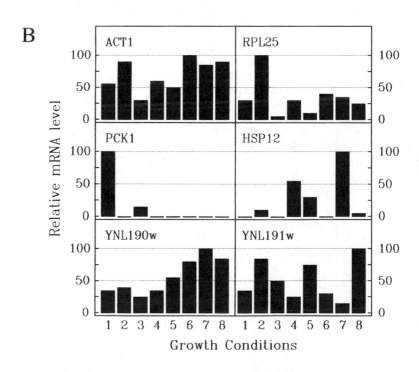

C

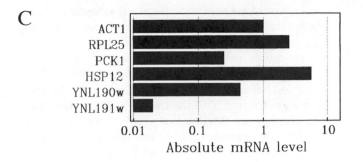

Figure 2. Multiplexed Northern analysis. (A) Approximately equal amounts of RNA prepared under various growth conditions were subjected to Northern analysis and probed with PCR-generated fragments specific for the *ACT1* (YFL039c), *RPL25* (YOL127w), *PCK1* (YKR097w), *HSP12* (YFL014w), YNL19w and YNL191w mRNAs. Low-resolution phosphorimages of the resultant signals are shown. (B) The signals in (A) were quantitated by phosphorimaging. The growth conditions were as follows. Yeast cells were grown exponentially on YPGE (1% yeast extract, 2% bactopeptone, 2% glycerol, 1% ethanol) [1, gluconeogenic growth], then transferred to YPD (1% yeast extract, 2% bactopeptone, 2% glucose) for 60 min [2, glucose induction], and grown to stationary phase in this medium for 24 h [3, carbon starvation]. Cells were grown exponentially in GYNB (2% glucose, 0.65% yeast nitrogen base *with* NH_4^+, without amino acids) [4, amino acid starvation]. Cells were grown exponentially in GYNB (2% glucose, 0.65% yeast nitrogen base *without* NH_4^+, without amino acids) for 2 h [5, nitrogen starvation]. NaCl (1.4 M) was added to the cells growing exponentially on GYNB containing NH_4^+[4], and harvested after 1 h [6, osmotic shock]. Cells were grown exponentially at 25ºC in GYNB (2% glucose, 0.65% yeast nitrogen base with NH_4^+, without amino acids) [7, control], and then subjected to a heat-shock at 36ºC for 30 min [8, heat-shock]. (C) The maximum level of each mRNA was measured relative to the ACT1 mRNA and expressed as the absolute mRNA level.

C. Reporter Genes

Reporter genes, such as those encoding β-galactosidase, luciferase and green fluorescent protein (GFP), provide a rapid and convenient approach towards gene analysis in yeasts. A general discussion of reporters is provided in Chapter 9 by Koloteva *et al.* In this chapter, the discussion is limited to the use of reporter genes for transcript analysis. The principle is simple: the reporter is fused to the gene of interest *in vitro*, and the activity of the reporter protein is used as an indirect measure of the transcriptional activity of the target gene *in vivo*. While this has proven an efficient experimental approach, it should be noted that the degree to which a reporter reflects the *actual* level of the wild-type transcript is limited by several factors.

The nature of the gene fusion is important. For example, the expression of a *lacZ* fusion that includes promoter, 5′-leader, coding or 3′-trailer sequences from the target gene might be subject to multiple levels of regulation at the levels of transcription, translation, mRNA degradation or protein turnover. In this situation, the interpretation of β-galactosidase data would be complex: β-galactosidase activities need not reflect transcript levels accurately because of possible regulatory effects at post-transcriptional levels. In contrast, those fusions that only include 5'-promoter sequences should accurately reflect relative transcription rates under different experimental conditions (Figure 3). This assumes, of course, that all of the sequences responsible for the regulation of the target gene are located 5' to the transcriptional start site. This is not the case for all yeast genes (Mellor *et al.*, 1987; Purvis *et al.*, 1987; Fantino *et al.*, 1992). A further complication is the differential stabilities of wild-type and reporter mRNAs (Mellor *et al.*, 1987; Purvis *et al.*, 1987). Also, the inherent stability of β-galactosidase in yeast limits its usefulness as a reporter of kinetic changes in gene repression. To summarize therefore, several problems preclude the use of reporters for the accurate determination of *absolute* transcript levels for the corresponding wild-type yeast genes. Nevertheless, reporters can provide a useful indication of transcriptional activity (i.e. whether an mRNA is expressed at high or low levels), and can yield accurate information on gene regulation.

Given their ease of use and the accuracy with which they can reveal changes in gene expression, it is not surprising that reporters are being exploited in genome scale analyses of transcriptional regulation. The collaborative approach taken by the Hegemann and Johnston laboratories has been to integrate GFP at chromosomal loci and use green fluorescence as an indicator of promoter activity at these loci (Niedenthal *et al.*, 1997).

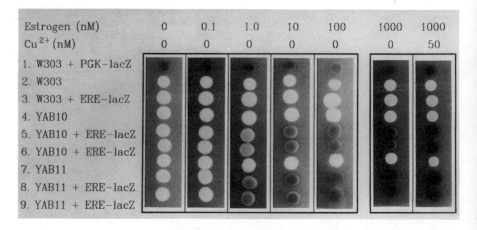

Figure 3. Using *lacZ* to follow transcriptional induction in yeast. A *CUP1–HuER* fusion, in which a human estrogen receptor cDNA was fused to the *CUP1* promoter, was integrated at the *leu2* locus in the yeast strain W303-1B. This generated the strains YAB10 (which carries a single *CUP1–HuER* copy) and YAB11 (which carries two tandem *CUP1–HuER* copies). W303-1B, YAB10 and YAB11 were then transformed with a *lacZ* fusion driven by a promoter carrying an estrogen responsive element (*ERE–lacZ*). *ERE–lacZ* expression is shown to be dependent on both the presence of the human estrogen receptor and estrodiol in the growth medium. Expression levels increase in response to increasing estrogen concentrations and increased levels of the receptor (through increased *CUP1–HuER* copy number, and *CUP1–HuER* induction by copper ions). A constitutively expressed *PGK1–lacZ* fusion is included as a control.

A

- PCR-amplify GFP-*HIS3* cassette using primers with complimentarity to ORF flanking sequences.

- Transform yeast with PCR-amplified GFP-*HIS3* cassette.
- Select for histidine prototrophs.
- Confirm disruption of target ORF using diagnostic PCR.

- Grow GFP-*HIS3* transformants under different conditions.
- Quantify GFP fluorescence by flow cytometry.

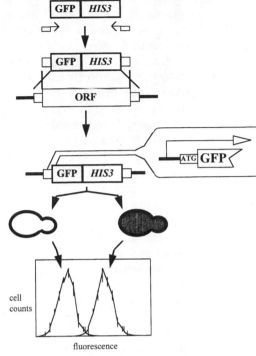

B

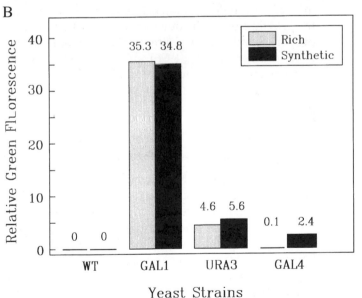

Figure 4. Monitoring promoter activity using GFP fusions. (A) Diagram of the GFP–*HIS3* integration strategy used to generate in-frame GFP fusions at specific ORFs in yeast. Details of this approach are described by Niedenthal *et al.* (1997). (B) Quantitative analysis of fluorescence in wild-type yeast cells and cells containing GFP fusions at the *GAL1*, *GAL4* or *URA3* loci. Fluorescence in two independent transformants was analysed during growth on raffinose/galactose in rich medium or synthetic medium. Adapted from Niedenthal *et al.* (1997) with permission from Professor Hegemann and John Wiley & Sons Ltd.

Briefly, a GFP-*HIS3* cassette is PCR-amplified with a 5′-oligonucleotide primer that includes 45 nucleotides of homology immediately 5′ to the start codon and a 3′-primer with homology immediately 3′ to the stop codon in the target ORF (Figure 4). This PCR product is used to transform a diploid *his3/his3* yeast, and strong-growing histidine prototrophs are selected. Diagnostic PCR is used to confirm the replacement of one of the two alleles of the target ORF with the GFP-*HIS3* cassette. GFP levels are then compared in living cells under different growth conditions by flow cytometry (Niedenthal *et al.*, 1997). GFP fusions have now been generated in a diploid background for all of the 269 genes on chromosome VIII, and their expression is currently being analyzed during glycolytic and gluconeogenic growth in minimal media and in rich medium (R. K. Niedenthal *et al.*, unpublished). The data to date indicate that the promoters for about 1% of the genes show high transcriptional activity under these conditions, but most (60%) are expressed at low levels displaying signals lower than for *URA3*. This is entirely consistent with the genomic Northern data described above (section II.B).

The sensitivity of the GFP reporter allows the relative fluorescence of individual cells to be compared, and this has revealed considerable population heterogeneity in GFP expression levels even when a single copy of the reporter is integrated at a specific locus. Population heterogeneity is masked in other common expression assays which test the population as a whole. The basis for this population heterogeneity is not clear, but it might lie in the physiological heterogeneity and asynchronous nature of the growing cells. The development of GFP mutants with altered spectral properties now allows simultaneous quantitative analysis of different genes (Anderson *et al.*, 1996), an approach that would allow GFP levels to be measured relative to an internal control.

D. cDNA Technologies

1. Conventional cDNA library screening

Early difficulties with RNA handling led to the development of alternative approaches for transcript analysis based on the production of cDNA libraries from target mRNA population. The screening of large, unamplified cDNA libraries permits the accurate quantitation of specific cloned sequences, and consequently the abundance of the corresponding mRNAs. For high quality libraries containing $5–10 \times 10^6$ clones, mRNAs expressed at an abundance of as little as 1 in 10^6 can be quantitated fairly accurately. The abundance of several mRNAs in the same population can be achieved by reprobing the same filters (Schwinn *et al.*, 1995). Although time-consuming, comparative expression studies can be performed in this way.

2. Expressed sequence tags

The first attempts to develop more efficient systems for gene expression analyses in multicellular organisms used high-throughput sequencing methods. In this approach, sequencing runs of about 500 nucleotides on individual cDNA clones are used to assemble databases of ESTs (expressed sequence tags, or partial cDNA sequences). Each tag corresponds to a single transcript, the cumulative data providing an "electronic Northern" (i.e. the relative abundance of an mRNA is based on the proportion of times its sequence arises).

These EST catalogs have proved very useful for transcript analysis in the context of specific genomes (e.g. the human genome; Adams *et al.*, 1991) and are of growing importance in the bacterial and fungal fields for complete genome mapping and sequencing (Fleischmann *et al.*, 1995). However, the sensitivity of the technique when used for transcript analysis depends on the number of individual clones sequenced. Using conventional automated sequencing methods, transcript images of some 5 to 10 000 ESTs can be obtained rapidly from any cDNA library. While sufficient to identify the most abundant mRNAs, this level of sequencing gives poor statistical coverage of minor transcripts. Therefore, modified approaches such as SAGE have been developed to increase the definition of the electronic Northern.

3. Serial analysis of gene expression

Briefly, SAGE involves the sequencing of concatomeric arrays of short cDNA tags each of which defines a specific mRNA within the population (Velculescu *et al.*, 1995). The abundance of a specific mRNA is then defined by the proportion of times its tag is sequenced. In the EST approach, one mRNA hit is obtained per sequencing reaction. SAGE allows more than 10 hits to be generated per reaction, thereby increasing the efficiency with which DNA sequence information can be used to generate a transcript profile (Velculescu *et al.*, 1995; Adams, 1996). The first step is to prepare double-stranded cDNA from the mRNA population of interest, cleave the cDNA with a 4 bp restriction enzyme, and purify the 3'-fragments (Figure 5). Each mRNA molecule is then represented by a specific 3' cDNA fragment whose 5' end is defined by the 4 bp restriction enzyme ("anchoring enzyme"). Around 10 bp of sequence information from this defined position is sufficient to identify the corresponding mRNA (Velculescu *et al.*, 1995). To generate this sequence information, a linker is attached to the 5'-ends of all the 3' cDNA fragments. This linker contains a site for a type IIS restriction enzyme (tagging enzyme) which cleaves a set distance away from this recognition site, leaving the linker attached to a short tag (i.e. the 5'-end of the 3' cDNA fragments). Following sequential ligation, PCR-amplification and cloning, concatomeric arrays of these tags are created for DNA sequencing (Figure 5). The information-rich sequence comprises an array of tags separated by well-defined primers. The proportion of times each sequence tag arises reflects the relative abundance of the corresponding mRNA.

A potential limitation of this technique stems from its reliance on oligo(dT) for priming first strand cDNA synthesis (Figure 5). Poly(A) tails are known to shorten as mRNAs age, and deadenylated but functional mRNAs can accumulate in the yeast cytoplasm. Also, in other systems

- Synthesize ds cDNA with biotinylated oligo(dT) primer

- Cut with anchoring enzyme (AE)
- Bind to streptavidin beads

- Divide in two
- Ligate to linkers 1 and 2

- Cut with tagging enzyme (TE)
- Blunt ends on cDNA tags

- Ligate tags
- PCR-amplify using primers 1 and 2

- Cut off linkers with anchoring enzyme
- Gel-purify dimeric tags
- Ligate into concatomeric arrays
- Clone and sequence

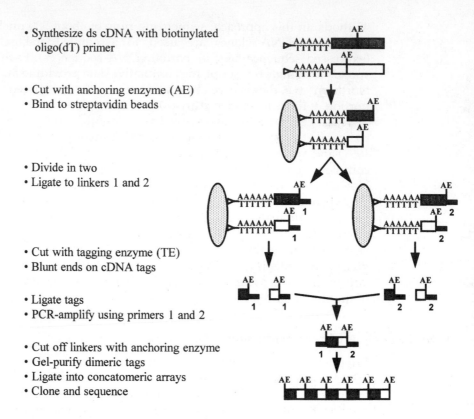

Figure 5. Serial analysis of gene expression (SAGE). The strategy for SAGE is adapted from Velculescu *et al.* (1995). Two double-stranded cDNAs (ds cDNA) are represented by shaded and open rectangles, with their biotinylated ends (▷—) created using a biotin-5'-(T)$_{18}$ primer for first-strand cDNA synthesis. Recognition sites for the anchoring enzyme are represented by AE (a 4 bp site likely to lie in each cDNA sequence), and a streptavidin bead by the large stippled oval. Linkers 1 and 2 (numbered black rectangles) carry sites for the tagging enzyme, a type IIS restriction enzyme which cleaves at a defined distance from the recognition site, thereby releasing a fragment of cDNA (tag) attached to the linker (see text).

some functional mRNAs are not polyadenylated (e.g. histone mRNAs). Deadenylated or non-polyadenylated mRNAs are likely to be under-represented by sequence tags and, as a result, their abundance would be underestimated. This limitation applies to any cDNA-based technique that exploits oligo(dT) primers (see above).

SAGE was first used to analyse a human pancreatic mRNA population (Velculescu *et al.*, 1995). The sequence tags were screened against the data-bases to identify the corresponding mRNAs, but it was not possible to identify all tags because the human genome sequence is incomplete. In contrast, the availability of the complete yeast genome sequence means that, in principle, it is possible to identify the ORFs corresponding to all identified tags. In practice, the analysis of over 60 000 yeast tags has revealed some that do not correspond to annotated ORFs (Velculescu *et al.*, 1997). This is possibly because ORFs shorter than 100 codons, some of which might be functional, have not been annotated in the yeast genome databases. Velculescu and co-workers (1997) have demonstrated that SAGE provides a powerful approach to yeast transcript profiling. Their data indicate that about 76% of yeast genes are expressed during exponential growth in YPD, and that gene expression levels range from about 0.3 to over 200 mRNA molecules per cell. Relating expression levels to genome location, the data reinforce the observation that telomeric regions are underexpressed and that highly expressed genes are distributed across the rest of the yeast genome. Hence their SAGE data are in agreement with those generated by multiplex Northern analyses.

E. Hybridization Array Technology

Despite the valiant efforts of some groups, reporter studies and Northern analyses are generally viewed as too labour intensive for genomic-scale projects. Therefore, in parallel with SAGE, less labour-intensive technologies are being developed for rapid transcript profiling. These technologies involve the immobilization of target sequences in an array on a support. This array is then hybridized to probes derived from the mRNA populations of interest. Hybridization signals are then detected and the resulting expression profiles compared *in silico* (Figure 6). This approach is proving useful in mammalian systems for the identification of sequences that are differentially regulated in response to specific signals (Bernard *et al.*, 1996; Chee *et al.*, 1996; Schena *et al.*, 1996). However, it has the potential to be even more powerful in yeast because the complete genome sequence is known. Non-redundant arrays representing all yeast genes can now be used to quantitate rapidly *all* of the changes in mRNA level that occur in response to a specific stimulus or mutation (DeRisi *et al.*, 1997; Wodicka *et al.*, 1997).

Initially, progress in hybridization array technology was driven by the application of robotics to pick large numbers of individual cDNAs clones into a microtiter plate format (Lennon and Lehrach, 1991). This format permitted the recovery of cDNA clones for gene isolation, functional analysis and systematic sequencing. However, high-density arrays are being developed to increase the efficiency of the approach. In situations where genome sequence information is available, alternative forms of array have been generated using specific PCR products or oligonucleotides.

Different supports have been used for high-density arraying applications, but all such supports must resist deformation to enable efficient image analysis. Although both nylon filters (Takashi *et al.*, 1995; Zhao *et al.*, 1995) and plastic sheets (Matson *et al.*, 1995; Timofeev *et al.*, 1996; Weiler and Hoheisel, 1996) have been used successfully, the best supports have been glass or SiO_2, partly because of their low backgrounds (i.e. their high signal-to-noise ratios). The use of glass matrices for immobilizing cDNAs, PCR products and oligonucleotides has been described in detail elsewhere (Schena *et al.*, 1995; Schena, 1996). This approach appears to be robust and is applicable to a wide range of hybridization targets. Although arraying densities are limited by the spotting robots that

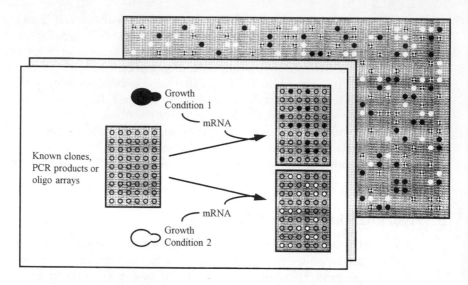

Figure 6. Transcript profiling using hybridization array technology. The principle of hybridization array technology involves the immobilization of DNA (cDNA clones, PCR products or oligonucleotides) in high-density arrays on an appropriate support (nylon, plastic, glass or SiO_2) (see text). These arrays are then hybridized with first-strand cDNA synthesized from the mRNA populations of interest. Often these cDNA probes are labeled with different fluorescent groups, allowing double-probing of arrays. The signals are then detected and the transcript image analysed, thereby revealing well-defined differences in the transcript profiles of the two mRNA populations. This is illustrated in the back panel where black dots represent transcripts present only under growth condition 1, white dots represent transcripts present only under growth condition 2, and stippled dots represent transcripts present under both conditions.

currently have minimum tolerances of 200 µm, new methods based on ink-jet technology have the potential to improve this by an order of magnitude (D. Bancroft, 1997, personal communication).

Higher densities can also be achieved using chip-based matrices in which oligonucleotide targets are synthesized directly on the surface of silicon wafers by photolithography (Jacobs and Fodor, 1994; Chee *et al.*, 1996; McGall *et al.*, 1996). Using this advanced technology, the company Affymetrix (Santa Clara, CA, USA: http://www.affymetrix.com:80/products.html) can construct microarrays of oligonucleotides each of 20–25 nucleotides in length, achieving densities of about 10^6 oligonucleotides per cm^2. To increase the accuracy of the output, multiple oligonucleotides (about 20) are arrayed for each target gene. Even so, all yeast genes can be represented on four chips. The company Synteni (Palo Alto, CA, USA: http://www.synteni.com/backgrnd.htm) is using an alternative approach, arraying PCR products on glass slides at densities of up to 10^5 sequences per 4 cm^2 area. Clearly the availability of such high-density arrays will facilitate the rapid analysis of very large numbers of sequences.

Having generated a "genome chip" the next step is to hybridize the immobilized sequences on the chip with the probe of interest. Optimizing chip-based hybridizations in very small volumes is challenging and depends on many parameters (Southern, 1996), but the key is to maximize both the sensitivity and the fidelity of the system. In the case of transcript profiling, the probe is generally a labeled first-strand cDNA synthesized from the mRNA population of interest. Several labeling methods have been tested by various groups, but fluorescently labelled probes are likely to become the most popular (Schena *et al.*, 1995). One reason for this is that two cDNAs can be labeled with different fluorescent markers to allow a direct comparison between different mRNA populations. Both cDNAs are hybridized simultaneously with the genome chip, and differential gene expression is detected on the basis of the fluorescent signals for each target sequence on the chip resulting from differential amounts of hybridization with each of the fluorescent cDNA probes (Figure 5; Schena *et al.*, 1996). Such differential-labeling strategies provide the opportunity for transcript profiles to be internally controlled using "standard" cDNA hybridization probes, thereby increasing their sensitivity and accuracy. This is important because, as described above, most yeast genes are expressed at very low levels (less than 1 mRNA molecule per cell). The low background signals obtained using glass and SiO_2 supports will further enhance the applicability of this approach for the comprehensive analysis of yeast mRNA populations. Hence, hybrid array technology promises to be a very powerful approach to transcript profiling.

The generation of the experimental readout from matrix hybridizations depends upon the accurate resolution and quantitation of large numbers of signals in two-dimensional microscopic arrays. For example, fluorescent signals have been detected using laser scanning devices that can read high-density arrays at high resolution within minutes. Also, sophisticated software is required to process the complex data sets. Software from Imaging Research (St Catherines, ON, Canada: http://imaging.brocku.ca/default.htm#Imaging), Bio Image (Ann Arbor, MI, USA: http://www.bioimage.com/) and Affymetrix (Santa Clara, CA, USA: http://www.affymetrix.com:80/products.html) have been used by some workers (Schena *et al.*, 1996; Shoemaker *et al.*, 1996).

Hybridization array technologies will provide an enormously powerful tool which can be used by the scientific community in a range of applications including transcript profiling, functional analysis DNA resequencing and the diagnosis of heritable disorders (Chee *et al.*, 1996; Shoemaker *et al.*, 1996). With respect to mRNA analysis in yeast, these methods have facilitated the rapid generation and comparison of complete yeast transcriptomes under different growth conditions and in different mutant backgrounds (DeRisi *et al.*, 1997; Wodicka *et al.*, 1997). The data of DeRisi *et al.* (1997), which support and extend the other technical approaches described above, are available on the Internet (http://cmgm.stanford.edu/pbrown/explore/index.html). However, significant issues may limit the general availability of these methods. They depend upon a set of advanced technologies that are unlikely to be available to most research groups in the near future. Nevertheless, such technologies are being developed rapidly by several companies. Yeast genome chips, and the hardware and software required to read them, will soon become commercially available to the scientific community, but the value of the enormous amounts of data they generate will depend upon intelligent analysis and dissemination to the community as a whole (DeRisi *et al.*, 1997).

◆◆◆◆◆◆ III. mRNA STRUCTURE

A. Transcript Mapping

As described above, genome-scale transcript analyses are currently focusing mainly on ORFs, i.e. coding regions. However, 5'-leader and 3'-trailer sequences exert a strong influence on the translation and stability of an mRNA. Hence, the 5'-transcriptional start sites and 3'-polyadenylation sites of a specific mRNA must be mapped to gain a proper understanding of its structure and regulation. Standard transcript mapping procedures involving S1 or RNase digestion, or primer extension can be found elsewhere (Sambrook *et al.*, 1989). Although relatively few yeast genes contain introns, where necessary these transcript mapping methods can also be used to analyse the intron–exon structure of primary transcripts. Such experiments can be facilitated by increasing the abundance of primary transcripts genetically, using conditional RNA processing mutants (Vijayraghavan *et al.*, 1989).

RNA mapping methods have also been used to measure relative mRNA abundances. Standard transcript mapping procedures involving S1 or RNase digestion, or primer extension are described elsewhere (Sambrook *et al.*, 1989). While these methods are more technically demanding than Northern blotting and hence are not widely used for this purpose, they do allow differential quantitation of closely related mRNAs. This has been particularly useful for the analysis of transcripts with different 5'-ends arising from the same locus (Maicas *et al.*, 1990; Rathgen and Mellor, 1990).

B. mRNA Folding

RNAs fold to form secondary and tertiary structures. Such structures can strongly influence the translation and degradation of an mRNA depending on their location. Although some yeast mRNAs carry stable 5'-secondary structures (e.g. *PMA1*), most yeast 5'-leaders are relatively free of secondary structures because they generally inhibit translation initiation (Cigan and Donahue, 1987). The simplest approach to testing mRNA structure formation is to use well-established computer programs such as FOLD to predict RNA secondary structures and their thermodynamic stability (Zuker and Steigler, 1981; Jaeger *et al.*, 1993). Short-range secondary structure predictions (< 250 bases) are generally thought to be reasonably accurate, but long-range predictions (> 500 bases) must be interpreted with caution. it is desirable to confirm RNA structures experimentally. either enzymatic or chemical probes can be used to dissect such structures *in vitro*, and these methods are reviewed by Jaeger *et al.* (1993).

C. 3'-Poly(A) Tail and 5'-Cap

The 3'-poly(A) tail and 5'-cap play important roles in mRNA translation and degradation, and therefore methods have been developed to probe the polyadenylation and capping status of an mRNA. Polyadenylated mRNA can be fractionated from non-polyadenylated RNA using standard affinity chromatography techniques involving oligo(dT)-cellulose or poly(U)-sepharose (Sambrook *et al.*, 1989). In addition, methods have been developed for the measurement of poly(A) tail length (Muhlrad *et al.*, 1994). Changes in poly(A) tail length cannot be resolved by Northern analysis of the mRNA molecule as a whole, and hence the 3'-end of the mRNA must be analysed in isolation. Briefly, the 3'-end of the target mRNA is cleaved off using RNase H digestion following hybridization with a specific antisense oligonucleotide. The cleavage products are then size-fractionated on polyacrylamide gels, and detected by Northern analysis with a 3'-probe. The capping status of an mRNA is tested by fractionating capped mRNA molecules from uncapped mRNAs by immunoprecipitation with polyclonal antibodies specific for the 5'-cap structure (Muhlrad *et al.*, 1994). The distribution of an mRNA between capped and uncapped RNA populations can then be studied by Northern analysis.

◆◆◆◆◆◆ IV. mRNA STABILITY

The stability of an mRNA fundamentally influences its abundance, and thereby the rate at which its cognate gene is expressed. Constitutive genes expressed at high levels frequently encode stable mRNAs, whereas regulated genes often encode unstable mRNAs. Indeed, an mRNA must have a short half-life if it is to respond rapidly to changes in transcription rate (Ross, 1995). Yeast mRNA half-lives range from about 1 to over 100 min, and the stabilities of some mRNAs are regulated in response to specific environmental stimuli (Mercado *et al.*, 1994; Cereghino and Scheffler, 1996). Yeast is proving to be an attractive and tractable experimental system for the dissection of eukaryotic mRNA degradation pathways because some fundamental aspects of these pathways appear to be conserved from yeast to humans (Ross, 1995; Beelman and Parker, 1995).

Methods for the measurement of yeast mRNA half-lives have been described in detail elsewhere (Parker *et al.*, 1991; Brown, 1994). Four main approaches are available: (a) the analysis of *in vivo* radiolabeling kinetics, or the measurement of mRNA decay kinetics following (b) addition of transcriptional inhibitors, (c) induction of the temperature-sensitive RNA polymerase II mutation, *rpb1-1*, or (d) transcriptional repression using a heterologous promoter such as *GAL1*. The latter three approaches, which are the most popular, depend on accurate and quantitative Northern analysis (section II.B). Methods for the structural analysis of mRNA degradation intermediates are referred to in section III.

◆◆◆◆◆◆ V. FUTURE PERSPECTIVES

The future is exciting with regard to transcript analysis in yeast. Most groups will continue to investigate the structure and regulation of small numbers of mRNAs using standard analytical tools (Northern blotting, RNA mapping and mRNA stability measurements). However, major advances are being made as a direct result of the development of efficient transcript profiling methods for the comparison of global gene expression patterns in yeast mutants or under different growth conditions. SAGE (section II.D) and hybridization array technology (section II.E) are the major contenders. While SAGE is essentially based on a technology that is already widely available (automated DNA sequencing), its use in the analysis of a single "transcriptome" requires considerable effort. In principle, hybridization array technologies can yield equivalent information more efficiently. Therefore, SAGE will probably be superseded by hybridization array technology once the latter has evolved into a rapid, sensitive, reproducible and generally available methodology. With further developments in silicon-chip systems, their evolution building upon experience in the associated microelectronics industry, hybridization array technologies will become the methods of choice for genome-wide expression profiling.

Acknowledgements

A.J.P.B. received funding from the European Commission as part of EUROFAN in Framework IV. We thank Helene Tournu for the data presented in Figure 2, and Ian Purvis (Glaxo Wellcome) for human estrogen receptor and ERE-*lacZ* constructs used to generate the data in Figure 3. We also thank Johannes Hegeman and Mark Johnston for allowing us to present their unpublished data on chromosome VIII GFP fusions.

References

Adams, M. D. (1996). Serial analysis of gene expression – ESTs get smaller. *BioEssays* **18**, 261–262.

Adams, M. D., Kelley, J. M., Gocayne, J. D., Dubnick, M., Polymeropoulos, M. H., Xiao, H., Merril, C. R. *et al.* (1991). Complementary DNA sequencing: expressed sequence tags and human genome project. *Science* **252**, 1651–1656.

Anderson, M. T., Tjioe, I. M., Lorincz, M. C., Parks, D. R., Herzenberg, L. A., Nolan, G. P. and Herzenberg, L. A. (1996). Simultaneous fluorescence-activated cell sorter analysis of two distinct transcriptional elements within a single cell using engineered green fluorescent proteins. *Proc. Natl Acad. Sci. USA* **93**, 8508–8511.

Barton, A. B. and Kaback, D. B. (1994) Molecular cloning of chromosome I DNA from *Saccharomyces cerevisiae*: analysis of the genes in the *FUN38-MAK16-SPO7* region. *J. Bacteriol.* **176**, 1872–1880.

Beelman, C. A. and Parker, R. (1995). Degradation of mRNA in eukaryotes. *Cell* **81**, 179–183.

Bernard, K., Auphan, N., Granjeaud, S., Victorero, G., Schmitt-Verhulst, A-M., Jordan, B. R. and Nguyen, C. (1996). Multiplex messenger assay: simultaneous, quantitative measurement of expression of many genes in the context of T cell activation. *Nucleic Acids Res.* **24**, 1435–1442.

Brown, A. J. P. (1994). Measurement of mRNA stability. In *Molecular Genetics of Yeast: A Practical Approach* (J. R. Johnston, ed.), pp. 147–159. Oxford University Press, Oxford.

Cereghino, G. P. and Scheffler, I. E. (1996). Genetic analysis of glucose regulation in *Saccharomyces cerevisiae* – control of transcription versus messenger RNA turnover. *EMBO J.* **15**, 363–374.

Chee, M., Yang, R., Hubbell, E., Berno, A., Huang, X. C., Stern, D., Winkler, J. *et al.* (1996). Accessing genetic information with high density DNA arrays. *Science* **274**, 610–614.

Cigan, A. M. and Donahue, T. F. (1987). Sequence and structural features associated with translational initiator regions in yeast – a review. *Gene* **59**, 1–18.

Dallman, M. and Porter, A. C. G. (1991). Semi-quantitative PCR for the analysis of gene expression. In *PCR: A Practical Approach* (M. J. McPherson, P. Quirke and G. R. Taylor, eds), pp. 215–224. Oxford University Press, Oxford.

Delbruck, S. and Ernst, J. F. (1993). Morphogenesis-independent regulation of actin transcript levels in the pathogenic yeast *Candida albicans. Mol. Microbiol.* **10**, 859–866.

DeRisi, J. L., Iyer, V. R. and Brown, P. O. (1997) Exploring the metabolic and genetic control of gene expression on a genomic scale. *Science* **278**, 680–686.

Dujon, B. and Goffeau, A. (1996). The yeast genome project: what did we learn? *Trends Genet.* **12**, 263–270.

Fairhead, C. and Dujon, B. (1994). Transcript map of two regions from chromosome XI of *Saccharomyces cerevisiae* for interpretation of systematic sequencing results. *Yeast* **10**, 1403–1413.

Fantino, E., Marguet, D. and Lauquin, G. J.-M. (1992). Downstream activating sequence within the coding region of a yeast gene: specific binding *in vitro* of RAP1 protein. *Mol. Gen. Genet.* **236**, 65–75.

Feinberg, A. P. and Vogelstein, B. (1983). A technique for radiolabelling DNA restriction endonuclease fragments to high specific activity. *Anal. Biochem.* **132**, 6–13.

Fleischmann, R. D., Adams, M. D., White, O., Clayton, R. A., Kirkness, E. F., Kerlavage, A. R., Bult, C. J. *et al.* (1995). Whole-genome random sequencing and assembly of *Haemophilus influenzae* Rd. *Science* **269**, 496–512.

Jacobs, J. W. and Fodor, S. P. A. (1994). Combinatorial chemistry – applications of light-directed chemical synthesis. *Trends Biotechnol.* **12**, 19–26.

Jaeger, J. A., SantaLucia Jr, J. and Tinoco Jr, I. (1993). Determination of RNA structure and thermodynamics. *Annu. Rev. Biochem.* **62**, 255–287.

Lennon, G. G. and Lehrach, H. (1991). Hybridisation analyses of arrayed cDNA libraries. *Trends Genet.* **7**, 314–317.

Maicas, E., Shago, M. and Friesen, J. D. (1990). Translation of the *Saccharomyces cerevisiae TCM1* gene in the absence of a 5′-untranslated leader. *Nucleic Acids Res.* **18**, 5823–5828.

Matson, R. S., Rampal, J., Pentoney, S. L., Anderson, P. D. and Coassin, P. (1995). Biopolymer synthesis on polypropylene supports: oligonucleotide arrays. *Anal. Biochem.* **224**, 110–116.

McGall, G., Labadie, J., Brock, P., Wallraff, G., Nguyen, T. and Hinsberg, W. (1996). Light-directed synthesis of high density oligonucleotide arrays using semiconductor photoresists. *Proc. Natl Acad. Sci. USA* **93**, 13555–13560.

Mellor, J., Dobson, M. J., Kingsman, A. J. and Kingsman, S. M. (1987). A transcriptional activator is located in the coding region of the yeast *PGK* gene. *Nucleic Acids Res.* **15**, 6243–6259.

Mercado, J. J., Smith, R., Sagliocco, F. A., Brown, A. J. P. and Gancedo, J. M. (1994). The levels of yeast gluconeogenic mRNAs respond to several environmental factors. *Eur. J. Biochem.* **224**, 473–481.

Moore, P. A., Sagliocco, F. A., Wood, R. C. M. and Brown, A. J. P. (1991). Yeast glycolytic mRNAs are differentially regulated. *Mol. Cell. Biol.* **11**, 5330–5337.

Muhlrad, D., Decker, C. J. and Parker, R. (1994). Deadenylation of the unstable mRNA encoded by the yeast *MFA2* gene leads to decapping followed by 5′→3′ digestion of the transcript. *Genes Devel.* **8**, 855–866.

Naitou, M., Hagiwara, H., Hanaoka, F., Eki, T. and Murakami, Y. (1997) Expression profiles of transcripts from 126 Open Reading Frames in the entire chromosome VI of *Saccharomyces cerevisiae* by systematic northern analyses. *Yeast* **13**, 1275–1290.

Niedenthal, R. K., Riles, L., Johnston, M. and Hegemann, J. H. (1997). Green fluorescent protein as a marker for gene expression and subcellular localization in budding yeast. *Yeast* **12**, 773–786.

Oliver, S. G. (1996). A network approach to the systematic analysis of yeast gene function. *Trends Genet.* **12**, 241–242.

Oliver, S. G., van der Aart, Q. J. M., Agostoni-Carbone, M. L., Aigle, M., Alberghina, L., Alexandraki, D., Antoine, G. *et al.* (1992). The complete DNA sequence of yeast chromosome III. *Nature* **357**, 38–46.

Parker, R., Herrick, D., Peltz, S. W. and Jacobson, A. (1991). Measurement of mRNA decay rates in *Saccharomyces cerevisiae*. *Methods Enzymol.* **194**, 415–423.

Planta, R. J. (1997). Transcriptional analysis of novel yeast genes. *Yeast* **13**, S15.

Purvis, I. J., Loughlin, L., Bettany, A. J. E. and Brown, A. J. P. (1987). Translation and stability of an *Escherichia coli* β-galactosidase mRNA expressed under the control of pyruvate kinase sequences in *Saccharomyces cerevisiae*. *Nucleic Acids Res.* **15**, 7963–7974.

Rathgen, J. and Mellor, J. (1990). Characterisation of sequences required for RNA initiation from the *PGK* promoter of *Saccharomyces cerevisiae*. *Nucleic Acids Res.* **18**, 3219–3225.

Richard, G.-F., Fairhead, C. and Dujon, B. (1997) Complete transcriptional Map of yeast chromosome XI in different life conditions. *J. Molec. Biol.* **268**, 303–321.

Ross, J. (1995). mRNA stability in mammalian cells. *Microbiol. Rev.* **59**, 423–450.

Sambrook, J., Fritsch, E. F. and Maniatis, T. (1989). *Molecular Cloning: a Laboratory Manual*, 2nd edn. Cold Spring Harbor Laboratory Press, Cold Spring Harbor, New York.

Schena, M. (1996). Genome analysis with gene expression microarrays. *BioEssays* **18**, 427–431.

Schena, M., Shalon, D., Davis, R. W. and Brown, P. O. (1995). Quantitative monitoring of gene expression patterns with a complimentary DNA microarray. *Science* **270**, 467–470.

Schena, M., Shalon, D., Heller, R., Chai, A., Brown, P. O. and Davis, R. W. (1996). Parallel human genome analysis: microarray-based expression monitoring of 1000 genes. *Proc. Natl Acad. Sci. USA* **93**, 10614–10619.

Schwinn, D. A., Johnston, G. I., Page, S. O., Mosley, M. J., Wilson, K. H., Worman, N. P., Campbell, S. *et al.* (1995). Cloning and pharmacological characterisation of human alpha-1 adrenergic receptors. *J. Pharmacol. Exp. Ther.* **272**, 134–142.

Shoemaker, D. D., Lashkari, D. A., Morris, D., Mittmann, M. and Davis, R. (1996). Quantitative phenotypic analysis of yeast deletion mutants using a highly parallel molecular bar-coding strategy. *Nature Genetics* **14**, 450–456.

Southern, E. M. (1996). DNA chips: analysing sequence by hybridisation to oligonucleotides on a large scale. *Trends Genet.* **12**, 110–117.

Swoboda, R. K., Bertram, G., Delbruck, S., Ernst, J. F., Gow, N. A. R., Gooday, G. W. and Brown, A. J. P. (1994). Fluctuations in glycolytic mRNA levels during the yeast-to-hyphal transition in *Candida albicans* reflect underlying changes in growth rather than response to cellular dimorphism. *Mol. Microbiol.* **13**, 663–672.

Takashi, N., Hashida, H., Zhao, N., Misumi, Y. and Sakaki, Y. (1995). High density cDNA filter analysis of the expression profiles of the genes preferentially expressed in human brain. *Gene* **164**, 219–227.

Tanaka, S. and Isonon, K. (1993) Correlation between observed transcripts and sequenced ORFs of chromosome III of *Saccharomyces cerevisiae*. *Nucleic Acids Res.* **21**, 1149–1153.

Timofeev, E. N., Kochetkova, S. V., Mirzabekov, A. D. and Florentiev, V. L. (1996). Regioselective immobilisation of short oligonucleotides to acrylic copolymer gels. *Nucleic Acids Res.* **24**, 3142–3148.

Velculescu, V. E., Zhang, L., Vogelstein, B. and Kinzler, K. W. (1995). Serial analysis of gene expression. *Science* **270**, 484–487.

Velculescu, V. E., Zhang, L., Zhou, W., Vogelstein, J., Basrai, M. A., Bassett, D. E., Hieter, P. *et al.* (1997). Characterisation of the yeast transcriptome. *Cell* **88**, 243–251.

Vijayraghavan, U., Company, M. and Abelson, J. (1989). Isolation and characterization of pre-messenger RNA splicing mutants of *Saccharomyces cerevisiae*. *Genes Devel.* **3**, 1206–1216.

Weiler, J. and Hoheisel, J. D. (1996). Combining the preparation of oligonucleotide arrays and synthesis of high quality primers. *Anal. Biochem.* **243**, 218–227.

Wodicka, L., Dong, H., Mittmann, M., Ho, M.-H. and Lockhart, D. J. (1997) Genome-wide expression monitoring in *Saccharomyces cerevisiae*. *Nature Biotech.* **15**, 1359–1367.

Yoshikawa, A. and Isono, K. (1990). Chromosome III of *Saccharomyces cerevisiae*: an ordered clone bank, a detailed restriction map and analysis of transcripts suggest the presence of 160 genes. *Yeast* **6**, 383–401.

Zhao, N., Hashida, H., Takashi, N., Misumi, Y. and Sakaki, Y. (1995). High density cDNA filter analysis: a novel approach for large-scale, quantitative analysis of gene expression. *Gene* **156**, 207–213.

Zuker, M. and Stiegler, P. (1981). Optimal computer folding of large RNA sequences using thermodynamics and auxillary information. *Nucleic Acids Res.* **9**, 133–148.

9 Reporter Genes and their Use in Studying Yeast Gene Expression

Nadejda Koloteva, John M. X. Hughes and John E. G. McCarthy
Department of Biomolecular Sciences, University of Manchester Institute of Science and Technology (UMIST), Manchester, UK

◆◆

CONTENTS

List of Abbreviations

ARS	Autonomously replicating sequence
CAT	Chloramphenicol acetyltransferase
eIF	Eukaryotic initiation factor
ELISA	Enzyme-linked immunosorbent assay
FACS	Fluorescence-activated cell sorting
GFP	Green fluorescent protein
GPF	*GAL–PGK* fusion
HIV-1	Human immune deficiency virus-1
IRE	Iron responsive element
IRP	Iron regulatory protein
NLS	Nuclear localization sequence
ORF	Open reading frame
RRE	Rev response element
UAS	Upstream activation sequence

◆◆◆◆◆◆ I. GENERAL PRINCIPLES OF THE USE OF REPORTERS AND EXPRESSION VECTORS IN YEAST

Reporter genes have been established as valuable tools for the investigation of individual steps of eukaryotic gene expression. Typically, the sequences to be analyzed for regulatory function are combined with the reading frame of a reporter gene whose product can be readily assayed quantitatively. The effects of the sequence under study on specific steps of the yeast expression pathway are monitored via the reporter gene activity. The reporter gene systems can have a wide range of applications. In terms of monitoring gene function, these include the identification and characterization of sequences that influence transcription, pre-mRNA processing and mRNA translation and decay. Moreover, reporter systems have been used in studies of both *cis*- and *trans*-acting regulatory elements, and are also useful in studies of protein localization.

In this chapter, we describe a range of reporter gene systems and how they can be applied in the study of yeast gene expression. We have focused on the principles involved, giving illustrative examples, rather than attempting to cover this subject area comprehensively. Moreover, we have restricted the coverage of this short article to reporter genes that encode products subject to convenient and quantitative methods of detection. While genes that complement auxotrophic phenotypes can also be regarded as reporters, their primary application is as selection markers and qualitative indicators. It should be noted that "the yeast" referred to in this chapter is *Saccharomyces cerevisiae*, although at least some of the methods and strategies discussed can be used with other yeast species. This latter point will be returned to briefly in section III. In general, we have tried to cite references that provide guides to the use of particular reporter systems and/or direct the reader to further useful publications.

A. Types of Reporter Gene

Both bacterial and eukaryotic reporter genes have been used in yeast (Table 1). Several bacterial reporters, including three genes that encode readily assayable enzymes [β-galactosidase (*lacZ*), chloramphenicol acetyltransferase (*cat*) and β-glucuronidase (*gus*)], have been used in yeast for several years. However, more recent studies have revealed that genes from eukaryotic sources offer several very useful properties. Different types of eukaryotic enzyme reporters are available (Table 1). Firefly luciferase is an attractive example because it can be quantified by means of a simple yet highly sensitive and accurate assay (Tatsumi *et al.*, 1988). Other reporter genes offer the advantage that they encode secretable products, e.g. the fungal glucoamylase gene and the yeast exo-1,3-β-glucanase gene (*EXG1*; Cid *et al.*, 1994). Secretion of the reporter product allows the assay to be performed without disruption of the yeast host cells. Indeed, exo-1,3-β-glucanase has been assayed in living cell populations using flow cytometry (Cid *et al.*, 1994). Of the bacterial and eukaryotic enzyme reporters, only β-galactosidase and β-glucuronidase have so far been shown to tolerate large N-terminal fusions without major losses in activity. Key issues to be considered when choosing a reporter gene for a particular application are the sensitivity of the product assay and the ease and accuracy with which the assay can be performed. This explains why the genes encoding chloramphenicol acetyltransferase (CAT), and especially luciferase, have been favored choices. As will be seen later, both

Table I. A selection of particularly useful reporter genes for yeast

Reporter	Assay/comments	References
Luciferase (*E. coli*, firefly *Photinus pyralis, Renilla reniformis*)	Luminescent reaction, detectable *in vivo* (intact cells) and quantitatively *in vitro* (cell-free extract) in the luminometer	Tatsumi *et al.* (1988), Vietes *et al.* (1994), Srikantha *et al.* (1996)
Chloramphenicol-3-O-acetyltransferase (CAT) (*E. coli*)	Chromatographic (using ^{14}C-chloramphenicol), photometric assay or CAT-ELISA	Mannhaupt *et al.* 1988
β-glucuronidase (GUS) (*E. coli*)	Spectrophotometric, fluorometric and histochemical analysis of the hydrolytic activity	Marathe and McEwen (1995)
Exo-β-glucanases (*S. cerevisiae, C. albicans*)	Hydrolysis of the fluorogenic substrate (MUG, PNPG) allows cell sorting (FACS) on the basis of enzyme activity	Cid *et al.* (1994)
β-galactosidase (*E. coli*)	Detectable via the generation of colored or fluorescent substrates; quantitative enzyme assay using cell extracts	Miller (1972), Rose *et al.* (1981)
Glucoamylase (*Aspergillus awamori*)	Detection of glucoamylase activity in colonies by staining YPD+starch plates with iodine vapors; glucoamylase measured quantitatively in the culture supernatant	Scorpione *et al.* (1993)
Metallothionein (CUP1p) (*S. cerevisiae*)	Semiquantitative assay *in vivo* by monitoring growth in a range of different copper concentrations	Lesser and Guthrie (1993), Stutz and Rosbash (1994)
Green fluorescent protein (GFP) (*Aequorea victoria*) and corresponding mutants:	Detection *in vivo* using fluorescence microscopy and fluorescent ELISA; suitable for FACS; *in vitro* assays also possible	Chalfie *et al.* (1994), Cormack *et al.* (1996), Yang *et al.* (1996)
EGFP (Phe^{64}→Leu Ser^{65}→Thr)	35 × higher intensity	
RSGFP4 (Phe^{64}→Met Ser^{65}→Gly Gln^{69}→Leu)	4–6 × higher intensity	

ELISA, enzyme-linked immunosorbent assay; MUG, 4-methylumbelliferyl-β-D-glucoside; PNPG, *p*-nitrophenyl-β-D-glucopyranoside; FACS, fluorescence-activated cell sorting types of reporter can be measured with great quantitative accuracy and applied in several ways. The selection of one of the other types of reporter listed in Table 1 is most likely to be motivated by particular requirements, for example the need for stable and active fusion constructs or for a secretable product.

Alternatively, *CUP1* is a reporter gene whose product can be measured (semi-) quantitatively *in vivo* on agar plates and which can also be used for direct selection. *CUP1* encodes a yeast metallothionein homolog that is normally non-essential, but it allows cells to grow on otherwise lethal concentrations of copper by chelating the metal in a dosage-dependent manner. The level of *CUP1* gene expression is assessed by monitoring growth in a range of different copper concentrations, varying from 0.1 to 6 mM $CuSO_4$. For example, *CUP1* has been used successfully as a reporter to study the effects of mutations on the efficiency and regulation of mRNA splicing (Lesser and Guthrie, 1993; Stutz and Rosbash, 1994).

Recent years have seen the addition of a new type of reporter gene that encodes a product whose activity is directly detectable in the cell by virtue of its intrinsic fluorescence. The key example is the gene encoding the green fluorescent protein (*GFP*) from the jellyfish *Aequorea victoria*. GFP has become recognized as a versatile reporter for monitoring gene expression and protein localization in a variety of cells and organisms. The GFP protein emits bright green light (λ_{max} = 510 nm with a shoulder at 540 nm) when excited with ultraviolet or blue light (maximally at 395 nm with a minor peak at 470 nm). The fluorophore of GFP is intrinsic to the primary structure of the protein, and its fluorescence is not dependent on additional gene products, substrates or other factors. Moreover, GFP fluorescence can be monitored noninvasively using the techniques of fluorescence microscopy, flow cytometry and fluorescence enzyme-linked immunosorbent assay (ELISA).

In order to improve the potential applications of GFP, a combinatorial mutagenic strategy was targeted at amino acids 64 to 69, the region which spans the chromophore of *A. victoria* GFP, yielding several different mutants with a red-shifted fluorescence excitation spectrum (Yang *et al.*, 1996). One of these mutants, RSGFP4, retains the characteristic green emission spectrum (λ_{max} = 505 nm), but has a single excitation peak (λ_{max} = 490 nm). The fluorescence properties of RSGFP4 are similar to those of another naturally occurring GFP from the sea pansy, *Renilla reniformis* (Wand and Cormier, 1978). Selective excitation of *A. victoria* GFP and RSGFP4 allows for spectral resolution of each fluorescent signal, providing the means to image these signals independently, for example in a mixed population of cells bearing both genes (Yang *et al.*, 1996). Another red-shifted GFP variant, EGFP, encodes a protein that also has a single, red-shifted excitation peak and fluoresces about 35 times more intensively than wild-type GFP (Cormack *et al.*, 1996).

B. Principles of Expression Vector Construction

Expression vectors for use in yeast require a (modular) expression "cassette" containing signals to initiate and terminate transcription, and elements that support stable propagation in *Escherichia coli* and in yeast. Yeast promoters consist of three elements that regulate the efficiency and accuracy of initiation of transcription: upstream activation sequences (UASs), TATA elements and initiator elements of transcription (Struhl, 1989). UASs, which have some functional similarities to mammalian enhancers, determine the activity and regulation of the promoter through specific binding to transcriptional activators (such as the Gal4 and Gcn4 proteins). Regulatable promoters offer obvious advantages in a gene expression system (Romanos *et al.*, 1992). The most powerful tightly regulated promoters of *S. cerevisiae* are those of the galactose-regulated genes *GAL1*, *GAL7* and *GAL10*, all of which are involved in metabolizing galactose. Several other regulatable yeast promoters, which respond to different signals, have been used for protein expression in yeast, e.g. the promoter of the repressible acid phosphatase gene *PHO5*, which responds to depletion of inorganic phosphate, and the *ADH2* promoter, which is induced by a change from glucose to a non-fermentable carbon source in the growth medium. Several promoters have been used in expression vectors for yeast which show no or only limited inducibility. As we shall see, some of these are useful for obtaining a constitutive level of expression, ranging from very weak (e.g. *TRP1*) through a level 100 times higher (e.g. *TEF1*) to the highest (e.g. *PGK1*).

A further property of promoters to be considered is the nature of their transcription initiation sites. Most promoters have multiple start sites which generate heterogeneous 5′ ends in the synthesized mRNA population. This is an undesirable property for studies of posttranscriptional gene expression, which generally require a defined, single 5′ end. This was a major driving force for the development of a type of fusion promoter that combines strong repressibility with high transcription rates and initiation at primarily a single major site (Oliveira *et al.*, 1993a; see also Kingsman *et al.*, 1990).

Yeast transcriptional terminators are incorporated into yeast expression vectors in order to direct precise definition of the mRNA 3′ end. This can be an important factor influencing the attained level of expression of a gene (Zaret and Sherman, 1982). The combination of U/A-rich "positioning" and "efficiency" elements with polyadenylation signals [$C(A)_n$ or $U(A)_n$] ensures the generation of polyadenylated mRNAs of a uniform length (Guo and Sherman, 1996). A further application of termination sequences in expression vectors is to use them to block potential runthrough transcription initiated within vector sequences outside of the 'expression cassette' (Vega Laso *et al.*, 1993).

In addition to the sequences required for reporter gene expression, elements need to be incorporated that allow stable replication of the vector in both yeast and *E. coli*. This means including a bacterial origin of plasmid DNA replication (commonly derived from pBR322) together with a selectable marker for *E. coli* (e.g. the β-lactamase encoding gene *bla*, which confers resistance to ampicillin), and a yeast origin of DNA replication together with a yeast selectable marker (such as *URA3*, *LEU2* or *TRP1*). Yeast DNA replication origins are of two types. One, derived from the yeast 2μ plasmid, allows stable replication at high copy number; stable

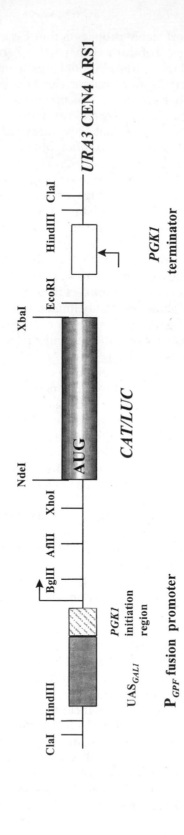

Figure 1. Scheme of the modular expression system vector YCpSUPEX1 (Oliveira *et al.*, 1993a). This scheme illustrates the modular structure of a plasmid based on YCplac33, one of the yeast–*E. coli* shuttle vectors derived from pUC19 (Oliveira *et al.*, 1993a). This version of YCpSUPEX1 contains the P*GPF* fusion promoter, the *PGK1* transcription terminator and either the *cat* or the *LUC* gene as reporter. The restriction sites indicated allow the respective "modules" to be exchanged readily for alternative components. Arrows indicate the sites of transcriptional initiation and termination.

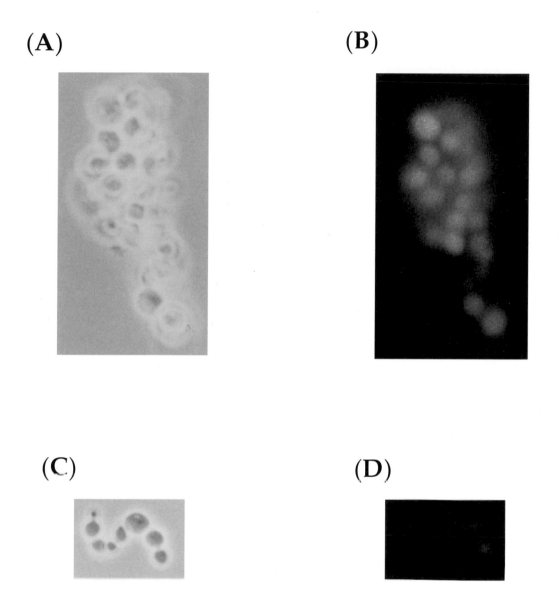

Plate 1. Translational regulation of *GFP* via an RNA-binding protein. Translation of the *GFP* reporter mRNA was regulated via binding of an mRNA binding protein (IRP) to a binding site (IRE) in the 5′UTR (see Figure 3, Chapter 9). Two types of "target" 5′UTR were used: one bearing the wild-type IRE (C and D) and one bearing a deletion mutant derivative of IRE (IREΔC) which shows a greatly reduced binding affinity for IRP (A and B). In all four cases, the IRP gene had been induced in galactose medium for 12 h prior to microscopic examination. The cells were viewed using standard light microscopy (A and C) and fluorescence microscopy (B and D).

plasmid maintenance and copy number control requires *trans*-acting factors, and these are encoded by the endogenous 2μ plasmids in [cir+] strains (nearly all laboratory strains of *S. cerevisiae* are [cir+]). The second type is an origin of chromosomal DNA replication, known as an autonomously replicating sequence (ARS). Plasmids containing an ARS are replicated once per cell cycle but, because of uneven partitioning between mother and daughter on cell division, may accumulate to high copy number under selective conditions. Even partitioning can be conferred on an ARS-containing plasmid by the introduction of a yeast centromeric sequence (CEN), which thus allows stable maintenance of the plasmid at low copy number. It should be emphasized that high copy number often is not a desirable property of a reporter gene vector. ARS–CEN plasmids, therefore, are generally preferred.

An important construction principle for expression plasmids is to include one or more unique restriction sites downstream of the promoter fragment, into which a reporter gene can be introduced. Ideally, sufficient unique restriction sites should be introduced at appropriate positions so that the key components can be assembled or recombined with each other in a "modular" fashion. An illustration of a convenient way to combine these and other important features of reporter gene systems is provided by the YCpSUPEX-series of plasmids (Oliveira *et al.*, 1993a; Figure 1). Many of the examples of reporter gene applications discussed below are based on the use of this vector series.

◆◆◆◆◆◆ II. HOW TO USE REPORTER GENES IN *S. CEREVISIAE*

In the following section we discuss six different categories of reporter gene application, illustrating each strategy with examples using suitably adapted expression constructs.

A. Analysis of Promoter Activity

The most straightforward type of reporter application is that of detecting and characterizing promoter activities. A typical vector designed for this type of work will possess a multiple cloning site upstream of a reporter gene, and no significant promoter activity of its own. Ideally, the vector used should be of the centromeric type, which is maintained in the cell as a single copy, in order to avoid titrating out any transcription factors that influence promoter behavior. Such vectors can be used for detecting promoter activities by inserting fragments of genomic DNA and selecting for expression of the reporter gene activity. Alternatively, the characteristics of an already isolated promoter can be investigated under a range of growth and potential induction/repression conditions. Because the use of this type of vector has been standard practice for several years, we refer the reader to other sources for further details (Mannhaupt *et al.*, 1988; Marathe and McEwen, 1995).

B. Pre-mRNA Processing

The next key step of yeast gene expression that can be monitored with the help of reporter gene systems is nuclear pre-mRNA splicing. By fusing an intron-containing gene of interest to a suitable reporter gene such that expression of the reporter gene is dependent on correct splicing, the requirements for various *cis*- and *trans*-acting factors in splicing may be analysed *in vivo*. Monitoring of the reporter gene activity can reveal the effects not only of individual mutations within introns and in components of the splicing machinery, but also of combinations of mutations, which may provide evidence of interactive suppression. The latter approach is based on the premise that a defect in splicing caused by a mutation within the reporter intron can sometimes be suppressed by a mutation in another gene encoding a product that interacts with the intron. Thus, the effects of the two mutations in combination, to some degree cancel each other out, and the net effect can be measured in terms of expression of the reporter gene. Interactive suppression has been used to demonstrate the requirement for precise intermolecular base-pairing between small nuclear RNAs and conserved sequences at the splice junctions and intron branch point (reviewed by Madhani and Guthrie, 1994) and as the basis for genetic screening for hitherto unknown factors required for splicing (Couto *et al.*, 1987; Newman and Norman, 1991).

The two reporters used in most of these studies are the bacterial *lacZ* and the yeast *HIS4* genes. Neither of these combines the attributes of an ideal yeast reporter, which would allow the quantitative measurement of expression over a wide range *in vivo* and allow direct selection: *lacZ* can be measured quantitatively, but only effectively *in vitro*, and cannot be used for selection, whereas the *HIS4* gene can be selected, but is unsuitable for quantitation. The *CUP1* gene, however, combines selectability with a reasonable degree of quantifiability, and recent reports show that when it is fused to an upstream intron-containing fragment it can be used as a novel reporter system to study the effects of mutations on the efficiency and regulation of splicing (Lesser and Guthrie, 1993; Stutz and Rosbash, 1994).

In one example, a particularly elegant set of reporter constructions were used to demonstrate the function of the Rev protein of human immune deficiency virus-1 (HIV-1) in yeast (Stutz and Rosbash, 1994). Rev protein regulates viral gene expression by enhancing the export of non-spliced or partially spliced viral transcripts to the cytoplasm. This activity requires the interaction of Rev with a highly structured portion of the viral RNA called the Rev response element (RRE). By fusing the *CUP1* coding sequence to an intron-containing fragment upstream and to the RRE downstream, and expressing these fusions in the presence of Rev, a Rev/RRE-dependent pattern of gene expression was reproduced in yeast. Rev activity was monitored for its propensity both to enhance translation of unspliced precursor RNA and to repress translation of spliced RNA by using complementary pairs of reporter gene fusions. The pairs were constructed such that, for one member of each pair, only the spliced RNA contained a translatable, *CUP1* open reading frame (ORF), whereas the reading frame of the unspliced precursor was interrupted by the intron.

Conversely, for the other member of each pair, the unspliced precursor contained a translatable, *CUP1* ORF starting from within the intron, whereas the reading frame of the spliced RNA was interrupted. Thus, the effects of Rev expression could be doubly monitored: both by measuring translation of the unspliced precursor exported to the cytoplasm or, alternatively, of the spliced product. By using this approach, it was shown that Rev protein appears to facilitate the export of unspliced precursor RNA to the cytoplasm by a mechanism that involves interaction with components of the splicing complex.

C. Analysis of Translational Control and mRNA Stability

Posttranscriptional regulation is an important component of eukaryotic gene expression. The translational efficiency of individual cellular mRNAs is controlled primarily by the rate of initiation, which in eukaryotes is thought generally to be mediated via cap-dependent interaction of the 40S ribosomal subunit with the 5' end of each transcript. There is considerable evidence that this initial 5' end-related interaction is followed by migration of the 40S subunit along the 5' untranslated region (5'UTR), during which process the ribosome "scans" the mRNA for the presence of a translational start codon. Several intrinsic structural properties of the 5'UTR can influence this so-called "scanning" process (Figure 2). Thus, restricted length of the 5'UTR, secondary structure, G-rich regions or a poor AUG context can individually, or in combination, reduce the efficiency of translational initiation. Moreover, the presence upstream of the main reading frame of an additional start codon (uAUG) or short reading frame (upstream open reading frame, uORF) can strongly inhibit translation. In order to examine whether a natural 5'UTR contains a potentially modulatory element of the types indicated in Figure 2, or to study the properties of such an element, the whole leader can be inserted upstream

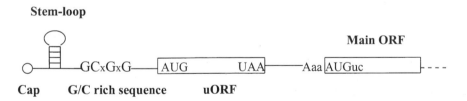

Figure 2. Various structural elements in the 5'UTR of a yeast mRNA can exert posttranscriptional control. The rate of translational initiation at the main reading frame has been shown to be affected by intrinsic structural properties of the 5' untranslated region including its length, secondary structures and/or G/C-rich sequences and the sequence context of the start codon, as well as by upstream AUG codons (uAUGs) or upstream open reading frames (uORFs; see Yoon and Donahue, 1992 for a review). Such elements in the 5'UTR of a yeast mRNA can direct various degrees of translational inhibition. Moreover, their presence can also influence the stability of mRNA (Linz *et al.*, 1997; Oliveira and McCarthy, 1995).

of a reporter gene in a vector of the type shown in Figure 1. Comparison of the expression level of any given 5′UTR with a control leader known not to restrict translation will provide important information about the potential control function of the tested 5′UTR in the cell. It should, however, be stressed that any analysis of this kind must take account of the potential effects of changes in translational efficiency on the stability of the reporter mRNA (Linz *et al.*, 1997). Most importantly, the relationship between translation and mRNA decay varies significantly from reporter to reporter (Linz *et al.*, 1997). For this reason, we recommend paying careful attention to the choice of reporter gene used in such analysis. At the same time, no study of translational control is complete without quantitative assessment of the influence of 5′UTR-directed changes in initiation efficiency upon steady-state mRNA abundance (Oliveira *et al.*, 1993a; Vega Laso *et al.*, 1993; Linz *et al.*, 1997).

The mode of construction of the reporter vector used for this type of work is important. In the YCpSUPEX series of vectors (Oliveira *et al.*, 1993a), a unique *Nde1* restriction site (CAT<u>ATG</u>) functions as the 5′ site for the introduction of the reporter gene. This places no restriction on the sequence downstream of the start codon, i.e. within the coding sequence. Moreover, the only strict requirement for cloning a 5′UTR sequence upstream of this site is the presence of a T at position −1, which is known to be of little significance in terms of start codon selection. Vectors of the YCpSUPEX type have been used in detailed studies of the influence of stem-loop structures at different positions in the 5′UTR on translation (Oliveira *et al.*, 1993a), as well as in other investigations (see below).

Where there is an effect of changes in 5′UTR structure on mRNA stability, the rate of decay can be determined by several methods (Jacobson and Peltz, 1996). One convenient approach involves the use of a mutant strain that has a temperature-sensitive form of RNA polymerase II (*rpbl-1*), allowing transcription to be shut off after a shift to the non-permissive temperature. Alternatively, the use of either the P$_{GAL1}$ or the *GAL-PGK* fusion promoter (P$_{GPF}$, Oliveira *et al.*, 1993a) on the expression plasmid allows transcription of the reporter gene to be shut off specifically by shifting the host from galactose to glucose medium (Jacobson and Peltz, 1996). In either method, Northern blotting can be used to follow the decay of the encoded mRNA subsequent to repression of transcription.

D. mRNA/Protein Interactions

An important principle of specific gene regulation in eukaryotes involves posttranscriptional control mediated by the interaction of an RNA-binding protein with a regulatory sequence placed in the 5′ or 3′UTR. In the best characterized example to date, one of the iron regulatory proteins (IRP1 or IRP2) from higher eukaryotes binds specifically, and with high affinity, to iron responsive elements (IREs) in the target mRNAs (Klausner *et al.*, 1993). More recent work has demonstrated that the IRP/IRE regulatory circuit also works in *S. cerevisiae* (Oliveira *et al.*, 1993b; Koloteva *et al.*, 1997). Insertion of an IRE into the 5′UTR of a reporter gene allows

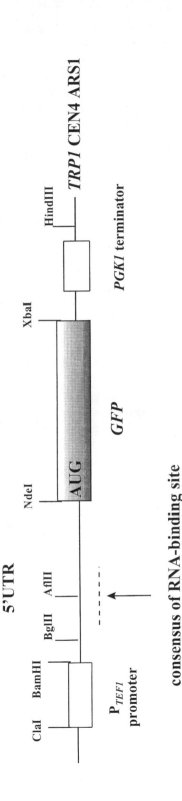

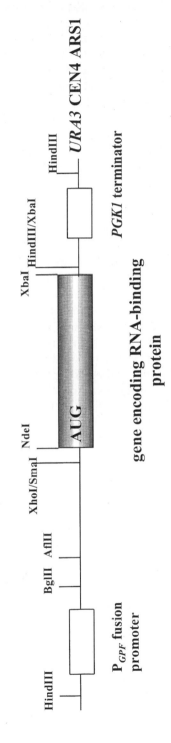

Figure 3. A two-plasmid system for the study of translational regulation via RNA-binding proteins. The vectors are based on the modular expression systems YCpSUPEX (see Figure 1) and pCECY (Vega Laso *et al.*, 1993). In the "target plasmid", transcription is driven by a constitutive promoter (*TEF1*), whereby the 5′UTR of the reporter gene contains the binding site for the RNA-binding protein. The other plasmid directs the inducible expression of the gene encoding the RNA-binding protein. In the case shown, a medium shift from glucose to galactose is used to switch on the hybrid P$_{GPF}$ promoter. The *GFP* gene in the target plasmid provides a convenient means of detecting changes in translation associated with the binding of the encoded protein to the element in its 5′UTR (see Plate 1).

translation to be suppressed by the IRP gene product, which can be inducibly synthesized from a second plasmid maintained in the same cell (Oliveira *et al.*, 1993b; Figure 3). This principle of translational repression in yeast via an RNA-binding protein targeted to the 5'UTR also holds for other RNA-binding proteins, irrespective of whether these proteins normally act as translational repressors (Stripecke *et al.*, 1994). It should also be noted that translational inhibition imposed in this fashion can also affect the stability of the targeted mRNA (Linz *et al.*, 1997). Therefore, as indicated in section II.C, there may need to be corrections made for changes in the abundance of the encoded mRNA.

In the two-plasmid repression system illustrated in Figure 3, the target plasmid carries a reporter gene preceded by a 5' UTR into which the target element can be inserted. Transcription of the reporter gene is driven by a constitutive promoter (P_{TEF1}). The gene encoding a known or putative binding protein specific for the chosen target element is expressed from an inducible promoter (P_{GPF}) in the second plasmid. A particularly attractive version of this type of system makes use of the *GFP* gene (Figure 3 and Plate 1). Because the fluorescent product can be detected in living cell cultures, large numbers of clones can be screened via fluorescence microscopy and/or flow cytometry. Indeed, different levels of translational repression of the *GFP* gene are reflected in distinct fluorescent intensities (Figure 4) and can be used as the basis of separation and/or clone selection using fluorescence-activated cell sorting (FACS).

The two-plasmid system (Figure 3) has at least two potential applications. In the first of these, the system can be used to study the interaction between a known RNA-binding protein and its RNA target in a yeast host (Oliveira *et al.*, 1993b; Koloteva *et al.*, 1997). This kind of study could include analysis of changes in the target element structure and how this influences the binding affinity of the protein, and thus the rate of translational initiation in the test system. The second potential application is to use the system to screen gene banks for genes encoding proteins that specifically bind a given RNA target element. Analogous applications have also been explored within the context of studies of antisense function in yeast (Aygün-Yücel and McCarthy, unpublished data).

E. Subcellular Protein Localization

Several methods are available to determine the localization of proteins or to follow the dynamics of their intracellular transport. These include the formation of fusion proteins with coding sequences of a reporter gene and then monitoring the distribution of these fusions within the cell. One of the most useful reporter genes for use in the monitoring of protein distribution within the cell is that encoding GFP. N- or C-terminal fusions of the *GFP* gene to the gene encoding the protein of interest can be constructed using suitable vectors (Figure 4). The resulting fusion proteins can then be detected in living yeast cells using fluorescence microscopy. For example, we have constructed a *GFP* fusion with the yeast gene encoding the eukaryotic initiation factor 4E (eIF4E) with the intention of determining whether

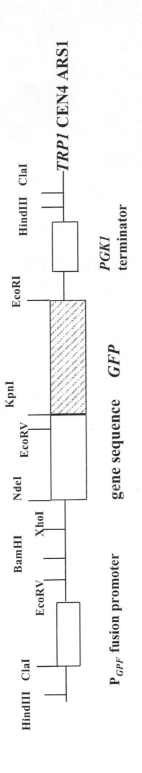

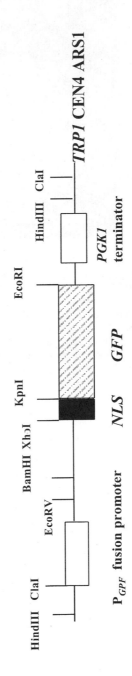

Figure 4. Expression constructs used for studying subcellular protein localization. GFP can be used as a marker for the subcellular distribution of proteins. The schemes shown here illustrate how N-terminal GFP fusions can be constructed using vectors based on the YCpSUPEX series (Figure 1). In general, whole reading frames or segments (domains) of them can be fused in frame to the *GFP* gene. In the example illustrated in Figure 5, the entire yeast *CDC33* reading frame was used. It is advisable to construct a control fusion using a known nuclear localization sequence (NLS), as described in Ptushkina *et al.* (1996). Detection of the expressed fusion proteins is readily achieved via fluorescence microscopy (Figure 5).

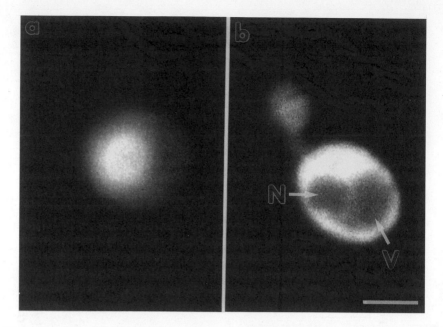

Figure 5. Subcellular distribution of GFP fusion proteins. Two examples of the types of distribution observed with GFP fusion proteins are shown. Fusion of the SV40 T-antigen NLS with *GFP* resulted in expression of a fusion protein that was directed primarily to the nucleus (a). This therefore constitutes a control experiment illustrating the distribution observed with a *GFP* fusion bearing an active NLS. In contrast, a gene fusion constructed between the *CDC33* reading frame encoding yeast eIF4E and *GFP* supported the expression of a fusion protein that localized primarily to the cytoplasm (b). The cell photographed here using fluorescence microscopy was in the process of budding. The location of the nucleus (N) and the vacuole (V), respectively, are indicated on the images.

this protein contains a nuclear localization sequence (NLS; Ptushkina *et al.*, 1996). The behavior of the fusion protein encoded by a construct of this type can be compared with that of a fusion construct in which a synthetic DNA fragment encoding a functional NLS is fused directly to the *GFP* gene (Figure 4). In this particular experiment, there was a clear distinction between the observed distributions of the two types of fusion protein (Figure 5). The primarily nuclear localization of the NLS–GFP fusion is readily discernible. Equally evident, on the other hand, is the inability of the eIF4E protein to direct GFP to the nuclear compartment. Such a result suggests that an NLS sequence is not present in the gene under study. In the case of a positive result, on the other hand, the investigator would have obtained evidence for the existence of an NLS in the tested reading frame.

F. *In Vitro* Synthesized mRNAs as Reporter Templates

An alternative approach to exploring posttranscriptional control is to use *in vitro* synthesized mRNAs as the templates for expression experiments. Their use in cell-free translation extracts is described elsewhere in this volume (see Chapter 19). Here we will focus on the translation of *in vitro* synthesized mRNAs in an almost intact cellular environment. Rather than using expression vectors to generate mRNA within the yeast cell, *in vitro* transcription vectors can be used to synthesize mRNA that can then be introduced into spheroplasts via a form of transient transformation. This approach provides greater control over the sequence and state of modification of the mRNAs that act as templates for translation *in vivo*. Delivery of *in vitro* synthesized transcripts can be achieved either by means of a standard procedure of DNA transformation using polyethylene glycol, or via electroporation (Russell *et al.*, 1991; Everett and Gallie, 1992). The latter procedure allows the amount of mRNA introduced into the spheroplasts to be controlled more precisely, via variables of electroporation such as the applied voltage. Subsequent to introduction of the mRNA which encodes the reporter protein, the reporter activity develops over a period of several hours. Both luciferase and β-glucuronidase can be readily detected in amounts proportional to the quantity of electroporated mRNA for at least 16 h.

A further step in the development of this approach took the form of using different (mutant) cellular backgrounds for the electroporation and expression experiments (Vasilescu *et al.*, 1996). In the example presented here, the ability of the cellular translational machinery to distinguish between 5'-capped and non-capped reporter mRNA was investigated in yeast strains that differed with respect to the form of the cap-binding protein eIF4E that they possessed (Figure 6). This approach therefore allows the experimentalist to examine the influence of selected mutations on the expression of defined mRNAs. The easiest way of generating suitable hosts for such experiments is to use plasmid shuffling in an appropriate disruption mutant in order to substitute mutant genes for the wild-type allele (Vasilescu *et al.*, 1996).

◆◆◆◆◆◆ III. REPORTER ANALYSIS IN OTHER YEAST SPECIES

At least some of the reporters described in this chapter can be exploited in other yeast species. Indeed, there is a general requirement for the development of systems analogous to the ones we have discussed in several organisms. The adaptation of the various types of reporter system would be achieved relatively easily for *Schizosaccharomyces pombe*, in which CAT (Roman *et al.*, 1993), β-galactosidase (Patrikakis *et al.*, 1996), β-glucuronidase (Ach and Gruissem, 1994) and GFP (Atkins and Izant, 1995) have all been shown to function satisfactorily. However, the isolation or construction of a tightly repressible promoter analogous to the *S. cerevisiae* P_{GPF} promoter that would function in this fission yeast must be seen as an urgent priority. Partly driven by increased interest from the pharmaceutical industry, there has recently been considerable activity in developing reporter assays for *Candida* species. This has been complicated by the atypical genetic decoding observed in *Candida* (see Chapter 21). So far, reporter genes that have been used successfully in *Candida* are luciferase

A

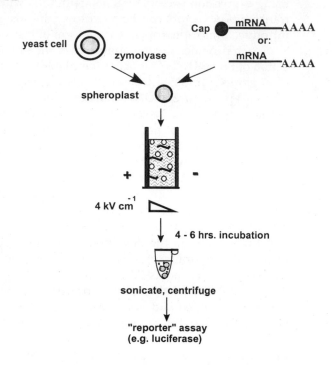

B

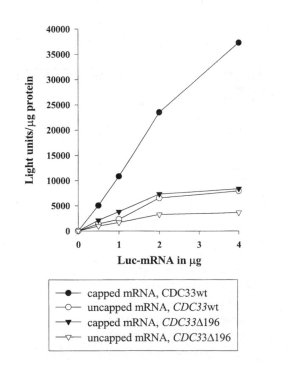

from *Renilla reniformis*, β-galactosidase from *Kluyveromyces lactis* (Leuker *et al.*, 1992; Srikantha *et al.*, 1996), orotidine 5′-monophosphate decarboxylase from *Candida albicans* (Myers *et al.*, 1995) and EGFP. Finally, some progress has also been made with other yeasts, including *Pichia* (Wiemer *et al.*, 1996) and *Hansenula* (Goedecke *et al.*, 1994). The coming years will undoubtedly see rapid further development of a wide spectrum of reporter systems for these and other fungal organisms.

Reporter Genes

References

Ach, R. A. and Gruissem, W. (1994). A small nuclear GTP-binding protein from tomato suppresses a *Schizosaccharomyces pombe* cell-cycle mutant. *Proc. Natl Acad. Sci. USA* **91**, 5863–5867.

Atkins, D. and Izant, J. G. (1995). Expression and analysis of the green fluorescent protein gene in the fission yeast *Schizosaccharomyces pombe*. *Curr. Genet.* **28**, 585–588.

Chalfie, M., Tu, Y., Euskirchen, G., Ward, W. W. and Prasher, D. (1994). Green fluorescent protein as a marker for gene expression. *Science* **263**, 802–805.

Cid, V. J., Alvarez, A. M., Santos, A. I., Nombela, C. and Sanchez, M. (1994). Yeast exo-β-glucanases can be used as efficient and readily detectable reporter genes in *Saccharomyces cerevisiae*. *Yeast* **10**, 747–756.

Cormack, B. P., Valdivia, R. H. and Falkow, S. (1996). FACS-optimized mutants of the green fluorescent protein (GFP). *Gene* **173**, 33–38.

Couto, J. R., Tamm, J., Parker, R. and Guthrie, C. (1987). A *trans*-acting suppressor restores splicing of a yeast intron with a branch point mutation. *Genes Develop.* **1**, 445–455.

Everett, J. G. and Gallie, D. R. (1992). RNA delivery in *Saccharomyces cerevisiae* using electroporation. *Yeast* **8**, 1007–1014.

Goedecke, S., Eckart, M., Janowicz, Z. A. and Hollenberg, C. P. (1994). Identification of sequences responsible for transcriptional regulation of the strongly expressed methanol oxidase-encoding gene in *Hansenula polymorpha*. *Gene* **139**, 35–42.

Guo, Z. and Sherman, F. (1996). 3′-end-forming signals of yeast mRNA. *Trends Biochem. Sci.* **21**, 477–481.

Jacobson, A. and Peltz, S. W. (1996). Interrelationships of the pathways of mRNA decay and translation in eukaryotic cells. *Annu. Rev. Biochem.* **65**, 693–739.

Figure 6. Electroporation of yeast spheroplasts with *in vitro* transcribed mRNA. The presented data illustrate how a mutant phenotype can be studied in transiently transformed yeast cells (spheroplasts). Reporter mRNA (in this case luciferase) can be synthesized *in vitro* and then introduced into yeast spheroplasts by means of electroporation (A). Capping and polyadenylation are optional modifications. The *S. cerevisiae* host chosen here was either a strain containing the wild-type form of the gene (*CDC33*) encoding the cap-binding protein eIF4E, or a mutant derivative with a deletion in *CDC33* that reduces the affinity of the encoded eIF4E for the mRNA cap (*CDC33Δ196*). Luciferase activities were measured in duplicate experiments 5 h after electroporation with different amounts of mRNA (B). The results clearly demonstrate the reduced ability of the translational apparatus in the mutant strain to distinguish between capped and non-capped mRNAs. An analogous strategy could be applied to the analysis of various mutants that are suspected to influence the posttranscriptional control of yeast gene expression.

Kingsman, S. M., Cousens, D., Stanway, C. A., Chambers, A., Wilson, M. and Kingsman, A. J. (1990). High-efficiency yeast expression vectors based on the promoter of the phosphoglycerate kinase gene. *Methods Enzymol.* **185**, 329–341.

Klausner, R. D., Rouault, T. A. and Harford, J. B. (1993). Regulating the fate of mRNA: the control of cellular iron metabolism. *Cell* **72**, 19–28.

Koloteva, N., Müller, P. P. and McCarthy, J. E. G. (1997). The position-dependence of translational regulation via RNA–RNA and RNA–protein interactions in the 5′ untranslated region of eukaryotic mRNA is a function of the thermodynamic competence of 40S ribosomes in translational initiation. *J. Biol. Chem.* **272**, 16531–16539.

Lesser, C. F. and Guthrie, C. (1993). Mutational analysis of pre-mRNA splicing in *Saccharomyces cerevisiae* using a sensitive new reporter gene, *CUP1. Genetics* **135**, 851–863.

Leuker, C. E., Hahn, A. M. and Ernst, J. F. (1992). Beta-galactosidase of *Kluyveromyces lactis* (Lac4p) as reporter of gene expression in *Candida albicans* and *C. tropicalis. Mol. Gen. Genet.* **235**, 235–241.

Linz, B., Koloteva, N., Vasilescu, S. and McCarthy, J. E. G. (1997). Disruption of ribosomal scanning on the 5′ untranslated region, and not restriction of translational initiation *per se*, modulates the stability of non-aberrant mRNAs in the yeast *Saccharomyces cerevisiae. J. Biol. Chem.* **272**, 9131–9140.

Madhani, H. D. and Guthrie, C. (1994). Dynamic RNA–RNA interactions in the spliceosome. *Annu. Rev. Genet.* **28**, 1–26.

Mannhaupt, G., Pilz, U. and Feldmann, H. (1988). A series of shuttle vectors using chloramphenicol acetyltransferase as a reporter enzyme in yeast. *Gene* **67**, 287–294.

Marathe, S. V. and McEwen, J. E. (1995). Vectors with the *gus* reporter gene for identifying and quantitating promoter regions in *Saccharomyces cerevisiae. Gene* **154**, 105–107.

Miller, J. H. (1972). *Experiments in molecular genetics.* Cold Spring Harbor Laboratory, Cold Spring Harbor, New York.

Myers, K. K., Sypherd, P. S. and Fonzi, W. A. (1995). Use of *URA3* as a reporter of gene expression in *C. albicans. Curr. Genet.* **27**, 243–248.

Newman, A. and Norman, C. (1991). Mutations in yeast U5 snRNA alter the specificity of 5′ splice-site cleavage. *Cell* **65**, 115–123.

Oliveira, C. C. and McCarthy, J. E. G. (1995). The relationship between eukaryotic translation and mRNA stability. *J. Biol. Chem.* **270**, 8936–8943.

Oliveira, C. C., van den Heuvel, J. J. and McCarthy, J. E. G. (1993a). Inhibition of translational initiation in *Saccharomyces cerevisiae* by secondary structure: the roles of the stability and position of stem-loops in the mRNA leader. *Mol. Microbiol.* **9**, 521–532.

Oliveira, C. C., Goossen, B., Zanchin, N. I. T., McCarthy, J. E. G., Hentze, M. W. and Stripecke, R. (1993b). Translational repression by the human iron-regulatory factor (IRF) in *Saccharomyces cerevisiae. Nucleic Acids Res.* **21**, 5316–5322.

Patrikakis, M., Izant, J. G. and Atkins, D. (1996). Comparison of three 3′ non-coding regions in *Schizosaccharomyces pombe* expression vectors: efficiencies of transcription termination and mRNA 3′-end formation. *Curr. Genet.* **30**, 151–158.

Ptushkina, M., Vasilescu, S., Fierro-Monti, I, Rohde, M. and McCarthy, J. E. G. (1996). Intracellular targeting and mRNA interactions of the eukaryotic translation initiation factor eIF4E in the yeast *Saccharomyces cerevisiae. Biochem. Biophys. Acta* **1308**, 142–150.

Roman, D. G., Dancis, A., Anderson, G. J. and Klausner, R. D. (1993). The fission yeast ferric reductase gene *frp1+* is required for ferric iron uptake and encodes a protein that is homologous to the gp91-phox subunit of the human NADPH phagocyte oxidoreductase. *Mol. Cell. Biol.* **13**, 4342–4350.

Romanos, M. A., Scorer, C. A. and Clare, J. J. (1992). Foreign gene expression in yeast: a review. *Yeast* **8**, 423–488.

Rose, M., Casadaban, M. J. and Botstein, D. (1981). Yeast genes fused to β-galactosidase in *Escherichia coli* can be expressed normally in yeast. *Proc. Natl Acad. Sci. USA* **78**, 2460–2464.

Russell, P. J., Hambidge, S. J. and Kirkegaard, K. (1991). Direct introduction and transient expression of capped and non-capped RNA in *Saccharomyces cerevisiae*. *Nucleic Acids Res.* **19**, 4949–4953.

Scorpione, R. C., Soares de Camargo, S., Scheinberg, A. C. G. and Astolfi-Filho, S. (1993). A new promoter-probe vector for *Saccharomyces cerevisiae* using fungal glucoamylase cDNA as the reporter gene. *Yeast* **9**, 599–605.

Srikantha, T., Klapach, A., Lorenz, W. W., Tsai, L. K., Laughlin, L. A., Gorman, J. A. and Soll, D. R. (1996). The sea pansy *Renilla reniformis* luciferase serves as a sensitive bioluminescent reporter for differential gene expression in *Candida albicans*. *J. Bacteriol.* **178**, 121–129.

Stripecke, R., Oliveira, C. C., McCarthy, J. E. G. and Hentze, M. W. (1994). Proteins binding to 5′ untranslated region sites: a general mechanism for translational regulation of mRNAs in human and yeast cells. *Mol. Cell. Biol.* **14**, 5898–5909.

Struhl, K. (1989). Molecular mechanisms of transcriptional regulation in yeast. *Annu. Rev. Biochem.* **58**, 1051–1077.

Stutz, F. and Rosbash, M. (1994). A functional interaction between Rev and yeast pre-mRNA is related to splicing complex formation. *EMBO J.* **13**, 4096–4104.

Tatsumi, H., Masuda, T. and Nakano, E. (1988). Synthesis of enzymatically active firefly luciferase in yeast. *Agric. Biol. Chem.* **52**, 1123–1127.

Vasilescu, S., Ptushkina, M., Linz, B., Mueller, P. P. and McCarthy, J. E. G. (1996). Mutants of eukaryotic initiation factor eIF-4E with altered mRNA cap binding specificity reprogram mRNA selection by ribosomes in *Saccharomyces cerevisiae*. *J. Biol. Chem.* **271**, 7030–7037.

Vega Laso, M. R., Zhu, D., Sagliocco, F., Brown, A. J. P., Tuite, M. F. and McCarthy, J. E. G. (1993). Inhibition of translational initiation in the yeast *Saccharomyces cerevisiae* as a function of the stability and position of hairpin structures in the mRNA leader. *J. Biol. Chem.* **268**, 6453–6462.

Vietes, J. M., Navarro-García, F., Pérez-Díaz, R., Pla, J. and Nombela, C. (1994). Expression and *in vivo* determination of firefly luciferase as gene reporter in *Saccharomyces cerevisiae*. *Yeast* **10**, 1321–1327.

Wand, W. W. and Cormier, M. J. (1978). Energy transfer via protein–protein interaction in *Renilla* bioluminescence. *Photobiochem. Photobiol.* **27**, 389–396.

Wiemer, E. A. C., Lueers, G. H., Faber, K. N., Wenzel, T., Veenhuis, M. and Subraman, S. (1996). Isolation and characterization of Pas2p, a peroxisomal membrane protein essential for peroxisome biogenesis in the methylotrophic yeast *Pichia pastoris*. *J. Biol. Chem.* **271**, 18973–18980.

Yang, T. T., Kain, S. R., Kitts, P., Kondepudi, A., Yang, M. M. and Youvan, D. C. (1996). Dual color microscopic imagery of cells expressing the green fluorescent protein and a red-shifted variant. *Gene* **173**, 19–23.

Yoon, H. and Donahue, T. (1992). Control of translation initiation in *Saccharomyces cerevisiae*. *Mol. Microbiol.* **6**, 1413–1419.

Zaret, K. S. and Sherman, F. (1982). DNA sequence required for efficient transcription termination in yeast. *Cell* **28**, 563–573.

10 Transposon Tagging I: A Novel System for Monitoring Protein Production, Function and Localization

Petra Ross-Macdonald[1], Amy Sheehan[1], Carl Friddle[2], G. Shirleen Roeder[1] and Michael Snyder[1]
[1]*Department of Biology, Yale University, New Haven, Connecticut and [2]Department of Genetics, Stanford University, Stanford, California, USA*

◆◆◆

CONTENTS

List of Abbreviations

β-**gal**	β-Galactosidase
FACS	Fluorescence-activated cell sorting
5-FOA	5-Fluoroorotic acid
GFP	Green fluorescent protein
HA	Influenza virus hemagglutinin epitope
HAT tag	HA/transposon tag
mTn	Mini-Tn3 transposon
ORF	Open reading frame
PCR	Polymerase chain reaction
TR	Terminal repeat
UV	Universal vectorette

◆◆◆◆◆◆ I. INTRODUCTION

Yeast molecular geneticists have access to a dazzling range of experimental approaches, both for characterization of individual genes and for surveying the genome. Gene function is usually first investigated by construction of a null allele, typically by creating a complete deletion or an insertion mutation (Hoekstra *et al.*, 1991b; Rothstein, 1991, Baudin *et al.*, 1993; see Chapter 5). Conditional or hypomorphic mutations and separation-of-function alleles may be created by small in-frame insertions, or by substituting bases to alter one or a small number of amino acids (Asubel *et al.*, 1987; Cunningham and Wells, 1989; Wertman *et al.*, 1992). Such alleles can be particularly informative when dissecting active domains within proteins, or when identifying distinct cellular processes in which a gene product participates. Characterization of a protein may be extended by determining its cellular localization, its biochemical activities, and the other proteins with which it interacts. These analyses generally entail generation of specific antibodies, or construction of an epitope-tagged version that is recognized by a commercially available serum (Tyers *et al.*, 1992, Schneider *et al.*, 1995; see Chapter 12). The timing of protein production during the cell cycle or life cycle provides further clues about function and regulation. Production may be assayed using antibodies, or reporters such as β-galactosidase (β-gal; Burns *et al.*, 1994) or green fluorescent protein (GFP; Niedenthal *et al.*, 1996).

Mutant screens of the yeast genome are useful both for investigating the function of known genes and for discovering new pathways or components. In yeast, it is possible to generate large numbers of mutants using either conventional agents (Lawrence, 1991) or insertional mutagens such as Ty-based elements (Garfinkel and Strathern, 1991; Devine and Boeke, 1994) or bacterial transposons (Hoekstra *et al.*, 1991b; Ross-Macdonald *et al*, 1997). Insertion elements have recently come into particular favor (Burns *et al.*, 1994, Smith *et al.*, 1995, Chun and Goebl, 1996; Mosch and Fink, 1997). Like conventional mutagens, they allow the investigator to make many mutations in a particular gene or throughout the genome. However, they have the additional advantage of tagging the mutant locus with an easily detectable DNA element, greatly facilitating identification and analysis of the mutated gene. Use of an insertional mutagen that creates a reporter construct has the additional advantage of allowing identification of genes through expression screens. For example, genes expressed at different times of development or under different growth conditions may be isolated efficiently by this approach (Burns *et al.*, 1994). Reporter constructs can also be used to search for regulators or targets of a protein of interest (Dang *et al.*, 1994, Irniger *et al.*, 1995; see Chapter 11).

Transposon-based systems available in yeast include *in vivo* Ty transposition (Garfinkel and Strathern, 1991; see Chapter 7), *in vitro* mutagenesis using Ty-based transposons (Devine and Boeke, 1994) and "shuttle mutagenesis" in which yeast DNA is mutagenized in *Escherichia coli* using Tn3-based transposons (Seifert *et al.*, 1986). In the latter two methods, mutagenized DNA is returned to yeast by transformation and homologous recombination. While *in vivo* Ty transposition mutagenesis has been extremely valuable for some purposes (Smith *et al.*, 1995), its usefulness may be limited by strong target bias (Ji *et al.*, 1993). The bacterial transposons used in shuttle mutagenesis systems generally show a distribution of insertion sites that is close to random (e.g. Seifert *et al.*, 1986).

Commercial availability of reagents for *in vitro* Ty-based mutagenesis (Perkin Elmer Ltd) should greatly facilitate use of this system. However, at present the transposons available for shuttle mutagenesis are far more powerful and versatile (Ross-Macdonald *et al.*, 1997; see Figure 1). Shuttle mutagenesis has proved extremely useful both for creating many mutations within a single gene (Seifert *et al.*, 1986, Ross-Macdonald *et al.*, 1997) and for generating mutations in many different yeast genes (Burns *et al.*, 1994, Chun and Goebl, 1996; Mosch and Fink, 1997).

◆◆◆◆◆◆ II. NOVEL TRANSPOSONS FOR MUTAGENESIS IN YEAST

We have developed a multipurpose transposon mutagenesis system that allows monitoring of protein production, function and localization in yeast. A simple mutagenesis procedure creates a large number of insertions throughout a gene, and the *lox/cre* recombination system is then used to modify these insertions. In a single experiment (outlined in Figure 2), it is possible to generate null alleles, reporter constructs, and small in-frame insertions that may be screened for additional phenotypes. In addition, the system can be used to place an epitope tag at many sites in a protein, without the labor of creating individual constructs. The tagged protein is a suitable substrate for all immunodetection techniques.

The system uses two basic types of transposon, designated mTn-3×HA/*lacZ* and mTn-3×HA/GFP (HA, influenza virus hemagglutinin epitope) (Figure 1). These transposons are mini-Tn3 derivatives (mTns), created by extensive modification (Ross-Macdonald *et al.*, 1997) of the transposon of Hoekstra *et al.* (1991a). They contain only the 38 bp terminal repeats and the *res* site of Tn3; all enzymatic functions of transposition are provided *in trans* in *E. coli*. Each mTn contains the *tet* gene for selection in *E. coli*, and the yeast auxotrophic marker *URA3*, which may be selected for in *ura3* strains by growth in medium lacking uracil, or selected against using medium containing 5-fluoroorotic acid (5-FOA) (Sikorski and Boeke, 1991). mTn-3×HA/*lacZ* contains a *lacZ* reporter gene, whereas mTn-3×HA/GFP contains coding sequences for a mutant form of GFP (Ile167Thr) that has enhanced fluorescence (Heim *et al.*, 1994). In each mTn an open reading frame (ORF) extends from the end of the transposon into the reporter's coding region. The *lacZ* reporter gene lacks an ATG initiaton codon or promoter sequences, and thus is usually only expressed when fused in-frame to a yeast protein coding region. The GFP reporter has an ATG codon but lacks promoter sequences; presumably it can be expressed either as a protein fusion or when inserted in the appropriate context of gene regulatory sequences. The transposons contain two *lox* sites, one adjacent to each terminal repeat. At one end of each transposon, the *lox* site is adjacent to a coding region for the cleavage site of Factor Xa protease (Ile-Glu-Gly-Arg). At the other end, *lox* is adjacent to a sequence encoding three copies of the hemagglutinin epitope (the 3×HA tag; Tyers *et al.*, 1992).

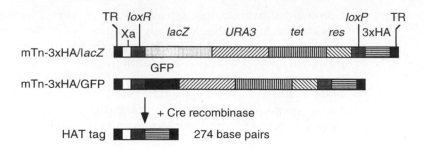

Figure 1. Schematic representation of novel mTns and the derived HAT tag element. mTn-3×HA/*lacZ* contains a truncated *lacZ* gene. mTn-3×HA/GFP contains the coding region for GFP mutant Ile167Thr (Heim *et al.*, 1994). Exposure of these transposons to Cre recombinase catalyzes the formation of a smaller element called the HAT tag, shown below. *loxR*, *loxP*, *lox* sites that serve as targets for Cre recombinase; TR, Tn3 terminal repeats; 3×HA, sequence encoding three tandem copies of the HA epitope; Xa, factor Xa protease cleavage site. Not drawn to scale.

Yeast DNA cloned into an appropriate vector can be mutagenized with a transposon by two bacterial matings, as outlined in Figure 3 and described in section III. Because many cells are matcd, a huge number of insertion alleles can be obtained through a single mutagenesis. The mutagenized DNA is then released from the vector and used to transform yeast, where it replaces the chromosomal copy of the gene by homologous recombination (Figure 2). Transformants are identified by selection for the *URA3* marker on the mTn, and may be screened for phenotypes caused by the mTn insertion. In-frame fusions of yeast genes to the reporter gene can be identified by assays for β-gal (Ross-Macdonald *et al.*, 1995) or GFP production (Niedenthal *et al.*, 1996, Ross-Macdonald *et al.*, 1997).

A yeast strain that carries an in-frame fusion may be used to derive a strain containing a much smaller, in-frame insertion, which we have called a "HAT" tag (HA/transposon tag). This is achieved simply by inducing expression of an exogenous *cre* gene, present on a plasmid under the control of the *GAL* promoter. When exposed to Cre recombinase, excision of all transposon sequences between the two *lox* sites occurs with high efficiency. The remaining mTn sequences result in insertion of 93 amino acids into the encoded protein, including the factor Xa protease cleavage site and the 3×HA tag (Figure 1). The factor Xa site should allow *in vitro* cleavage of the tagged protein with factor Xa protease (this has not yet been attempted). The HAT-tagged protein can be immunodetected using antibodies against the HA epitope. Yeast strains may also be screened for phenotypes caused by the small insertion.

We have tested the new transposon system by mutagenesis of several individual yeast genes (Ross-Macdonald *et al.*, 1997). The HAT tag was successfully used to analyze localization of the Spa2, Arp100 and Sao1 proteins (Ross-Macdonald *et al.*, 1997; M. Snyder, N. Burns and Y. Barral, unpublished data). These proteins localize to sites of polarized cell growth, to the spindle pole body region, and to the nucleus, respectively. The HAT-tagged proteins can also be detected by immunoblot analysis

(M. Snyder, P. Ross-Macdonald, N. Burns and Y. Barral, unpublished data). Mutagenesis of the *BDF1* gene with the mTn that carries a GFP reporter resulted in fusion proteins that localize correctly to the nucleus (Ross-Macdonald *et al.*, 1997). Analysis of several HAT tag insertions in the *SER1* gene indicated that functional domains can be identified by this means, and also demonstrated the ability of small insertions to produce conditional alleles (Ross-Macdonald *et al.*, 1997).

An extant yeast genomic DNA library (Burns *et al.*, 1994) was also mutagenized with each of the novel transposons (section IV). These transposon-insertion libraries can be used as a mutagen that produces tagged genes, or to screen for genes whose expression is differentially regulated by the environment or strain background. These libraries have an advantage over those previously described (Burns *et al.*, 1994, Chun and Goebl, 1996), in that once a strain carrying an in-frame insertion in a gene of interest has been identified, a strain carrying a HAT-tagged version of the protein may be derived. Both the reporter fusion and the tagged protein may be of immediate usefulness in further characterizing the mutagenized gene.

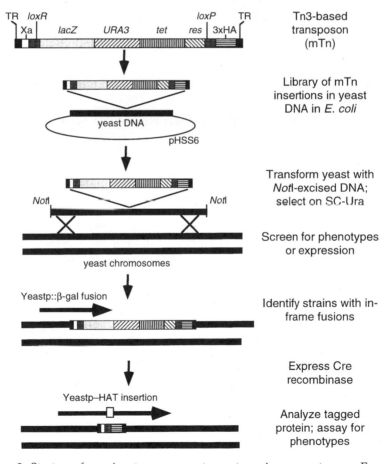

Figure 2. Strategy for using transposons to mutagenize a yeast gene. Example shown uses mTn-3×HA/*lacZ*. Abbreviations and symbols as defined in legend to Figure 1.

◆◆◆◆◆◆ III. MUTAGENIZING A GENE OF INTEREST

A. Generating Transposon Insertions in *E. coli*

An outline of shuttle mutagenesis is shown in Figure 3 and a detailed protocol is presented in Protocol 1. The first step is to clone the gene of interest into a Tn3-free vector, such as the pHSS series described by Hoekstra *et al.* (1991b). This construct is transformed into cells that are producing the Tn3 transposase, TnpA. An F plasmid bearing the desired transposon (Figure 1) is then transferred into these cells by mating. TnpA acts *in trans* to initiate transposition, and a "superplasmid" called a cointegrate is formed between the F plasmid and the pHSS construct. To resolve the cointegrate, a second mating is used to transfer this superplasmid into cells that produce the Tn3 resolvase, TnpR. TnpR acts *in trans* on the *res* sites in the transposon, separating the cointegrate into its two constituents: an F plasmid carrying the transposon, and the pHSS construct, which now also carries the transposon. The mutagenized pHSS construct can be recovered by a standard DNA miniprep procedure and transformed into a standard laboratory host strain. Because in practice each of these steps involves a large population of cells, the final product is a library of pHSS-based plasmids, each carrying a transposon insertion at a different site. It is easy to generate 10^5 independent insertions from a single mutagenesis.

B. Generating Yeast Strains with Reporter Constructs

The mutagenized yeast DNA is excised from the pHSS vector using a rare-cutting restriction enzyme (*Not*I for most vectors). It is then transformed into yeast, selecting for the transposon-borne *URA3* marker. Because the transposon can insert in either orientation or any frame, the first step is to identify strains carrying in-frame fusions in the gene of interest. For transposons containing *lacZ*, this is achieved by a simple color assay for β-gal production. For mTn-3×HA/GFP, other techniques must be used, such as fluorescence-activated cell sorting (FACS; Niedenthal *et al.*, 1996), or direct inspection of cells by fluorescence microscopy (Ross-Macdonald *et al.*, 1997).

C. Generating Epitope-Tagged Strains

Once strains with in-frame fusions to the reporter have been isolated, production of the Cre protein is induced by growth in medium containing galactose. Cre efficiently catalyzes site-specific recombination between the mTn's *lox* sites, excising the body of the transposon. The final chromosomal configuration has a 5 bp duplication generated during transposition in addition to a 274 bp (mTn-3×HA) insertion. This insertion consists of a single *loxR* site and sequences encoding the factor Xa cleavage site, three copies of the HA epitope, and flanking Tn3 terminal repeats (the HAT tag; Figure 1).

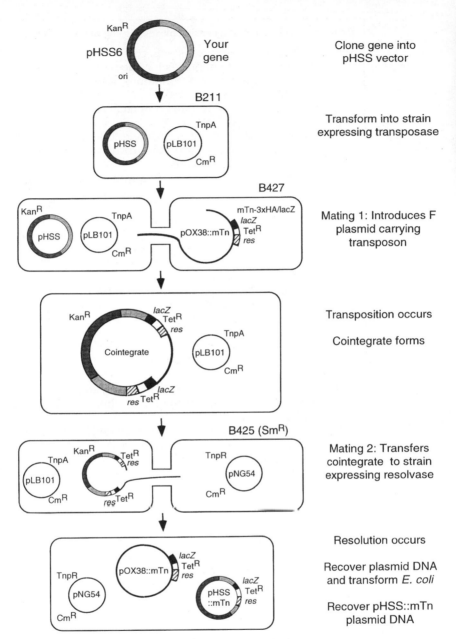

KanR
pHSS6
Your gene
ori

Clone gene into pHSS vector

B211

pHSS
pLB101
TnpA
CmR

Transform into strain expressing transposase

B427

KanR
pHSS
pLB101
TnpA
CmR
pOX38::mTn
mTn-3xHA/lacZ
lacZ
TetR
res

Mating 1: Introduces F plasmid carrying transposon

KanR
lacZ
TetR
res
Cointegrate
pLB101
TnpA
CmR
res TetR
lacZ

Transposition occurs

Cointegrate forms

B425 (SmR)

TnpA
pLB101
CmR
KanR
TetR
res
res TetR
TnpR
pNG54
CmR

Mating 2: Transfers cointegrate to strain expressing resolvase

TnpR
pNG54
CmR
pOX38::mTn
lacZ
TetR
res
pHSS ::mTn
lacZ
TetR
res

Resolution occurs

Recover plasmid DNA and transform E. coli

Recover pHSS::mTn plasmid DNA

Figure 3. Outline of the transposon mutagenesis procedure. A kanamycin-resistant plasmid containing yeast DNA (pHSS) is transformed into an *E. coli* strain containing plasmid pLB101 (B211), which constitutively expresses Tn3 transposase. The transformant is mated to a strain containing an F factor plasmid that carries the mTn (pOX38::mTn) (B427 in this example). pOX38::mTn is transferred into the transformant, where transposition into pHSS leads to formation of a cointegrate containing pHSS, pOX38, and two copies of the mTn (pLB101 is immune to insertion). The cointegrate is transferred by mating into strain B425, which constitutively expresses Tn3 resolvase. Resolution of the cointegrate occurs via the Tn3 *res* site on the mTn. The resulting plasmids are recovered, and pHSS plasmid containing an mTn insertion (pHSS::mTn) is isolated by retransformation into *E. coli*, selecting for kanamycin and tetracycline resistance.

Protocol 1. Shuttle mutagenesis/epitope tagging of a yeast gene

Section I. Transposon mutagenesis in E. coli

1. Clone fragment into pHSS vector (Hoekstra *et al.*, 1991b). Select transformants on LB medium containing Kan (LB+Kan; see Table 1 for information on all antibiotics).
2. Transform this plasmid into competent cells of B211. Select transformants on LB+Kan+Cm.
3. Inoculate a transformant into liquid LB+Kan+Cm. Inoculate strain B426 (or B427 or B428, as preferred) into liquid LB+Tet. Incubate at 37°C overnight with shaking.
4. Subculture each strain at 1:100 dilution into LB. Grow at 37°C for 2–3 h, or until a cell suspension is first visible. Mix 200 µl of each strain in a microfuge tube and incubate at 37°C without agitation for 20 min to 1 h. Plate 100 µl aliquots onto LB+Tet+Kan+Cm (spot 20 µl aliquots of the unmated strains onto this medium as controls) and grow for 1–2 days at 30°C to allow cointegrate formation.
5. Inoculate strain B425 into liquid LB+Cm. Incubate at 37°C overnight with shaking.
6. Elute colonies from cointegrate plates (i.e. put 2 ml of liquid LB onto each plate and scrape cells into a homogeneous suspension). Subculture strain B425 at 1:100 dilution into LB. Dilute an aliquot of the cointegrate eluate into LB to give approximately the same density as the diluted B425 culture. Grow and mate strains as in step 4. Plate 100 µl aliquots onto LB+Tet+Kan+Sm and incubate at 37°C overnight. Perform controls, as in step 4.
7. Elute colonies from the mating. Dilute an aliquot of the mating eluate into LB+Tet+Kan medium to give an almost saturated culture. Grow at 37°C for 1–2 h.
8. Isolate plasmid DNA by a miniprep procedure. Transform about 1/10 of the DNA sample into a standard *recA endA* cloning strain (e.g. DH5). Plate on LB+Tet+Kan.
9. Elute the entire pool of transformants and make a plasmid DNA miniprep as in steps 7 and 8. Make a glycerol frozen stock of the remaining eluate for future use. Plasmid DNA may also be prepared from individual transformants for detailed analysis.
10. Digest the plasmid DNA preparation with *Not*I and transform into yeast, selecting for *URA3*.

Section II. Screening for lacZ *fusions*

1. Patch yeast transformants to SC-Ura (Sherman *et al.*, 1986) (or SC-Ura-Leu if pGAL-*cre* is present in strain).
2. Replica-plate cells to an SC-Ura (or SC-Ura-Leu) plate and onto an SC-Ura plate on which a sterile disc of Whatman 1A filter paper has been placed. Grow overnight at 30°C. Other media or growth conditions can be substituted as desired. For *ade2* strains, test media should contain 80 mg liter^{-1} of adenine.
3. Lift filters from the plates and expose to chloroform vapor in a sealed vessel. We use 15 min exposure; time required may vary between strains.
4. Place filters colony-side up onto X-Gal plates (120 µg/ml 5-bromo-4-chloro-3-indolyl-β-D-galactopyranoside, 0.1 M NaPO₄ (pH 7) and 1 mM MgSO₄ in 1.6% agar) and incubate at 30°C for up to 2 days.
5. Identify transformants carrying productive *lacZ* fusions and recover them from the SC-Ura (or SC-Ura-Leu) plate.

> *Section III. Creating the HAT epitope tag*
> 1. If not already present, transform strains carrying *lacZ* fusions with pGAL-cre, selecting on SC-Ura-Leu.
> 2. Inoculate transformants into 2 ml of SC-Ura-Leu with 2% raffinose as carbon source, and grow to saturation.
> 3. Dilute cultures 1/100 into SC-Leu with 2% galactose as carbon source. As a control, dilute 1/100 into SC-Leu with 2% glucose as carbon source. Grow until saturated.
> 4. To assess the efficiency of excision quantitatively, plate dilutions of each culture onto SC medium. When grown, replica plate to SC-Ura to identify Ura⁻ colonies. Ura⁻ cells derived from the cultures grown with galactose as carbon source should be used.
> 5. Prepare genomic DNA (Philippsen *et al.*, 1991, Burns *et al.*, 1996) and perform PCR (primer sequences in Table 1). Analyze size of products on agarose gel. Purify with Qiaquik (Qiagen) prior to sequencing.

D. Moving Alleles to a Different Strain

If the ability to transfer alleles of interest between strains is required, transposon insertions can be generated and maintained on an autonomously replicating yeast plasmid (G. S. Roeder and J-Y. Leu, unpublished data). A yeast centromere and selectable marker may be cloned into the pHSS construct before the mutagenesis procedure; the plasmid marker is selected for during transformation and all subsequent procedures. The plasmid may be recovered from yeast by standard methods (Rothstein, 1991). An alternative, "plasmid rescue" strategy can be used to recreate a particular HAT-tagged version of a gene in a different strain background. Here, plasmid rescue (section IV.D and Protocol 2, step 8) is used to recover a region containing the transposon insertion, which may be transferred to a new strain background and then reduced by Cre-mediated excision to the HAT-tagged form. Allele rescue by gap repair (Rothstein, 1991) could also be used.

E. Hints and Tips on the Procedure

The only cloning procedure in this protocol occurs at the outset, when the gene of interest must be placed in a vector that lacks Tn3 "immunity" sequences. We have used pHSS6 exclusively and extensively in testing these transposons. While the transposons show little or no site specificity in the cloned gene (Ross-Macdonald *et al.*, 1997), an insertion hotspot is found in the polylinker around the *Eco*RI site in some (but not all) constructs. Therefore, deletion of as much of the polylinker as possible in this cloning step is advised.

Insertions in the polylinker do not represent a drawback for the procedure, because it is unlikely that they will lead to a yeast transformant. However, they do highlight an important point: the final plasmid pool should be large and diverse. In practice, this means ensuring that hundreds (and preferably thousands) of cointegrates are generated in the first mating step, and that this complexity is preserved through the subsequent resolution and transformation steps. Mating reactions may be scaled up to increase numbers, but mating times should be kept brief and constant. The bacterial mating process is sensitive to detergent and agitation.

The final step of shuttle mutagenesis (in which plasmid DNA is prepared from the strain in which resolution has occurred, and then transformed into a standard laboratory host strain from which a further plasmid DNA preparation is made) may seem unnecessary. However, the first preparation also contains the pNG54 plasmid and usually displays a range of products when analyzed by restriction digests with *NotI*, whereas only the expected products are seen after passage through a new host strain. The expected products are a vector band at 2.2 kb and a band whose size corresponds to that of the cloned insert plus that of the transposon. If transposon insertion into the vector has occurred, the converse set of bands may also be seen. To generate a very high quality plasmid preparation without use of phenol, we use the protocol of Morrele (1989). This is a standard alkaline lysis procedure but with 0.5 volumes of 7.5 M ammonium acetate as solution III, and 0.6 volumes of isopropanol to precipitate the DNA.

The epitope-tagging procedures require a *ura3 leu2* yeast host strain capable of inducing the *GAL* promoter. Our strains grow poorly on galactose but give 80–100% excision after a few generations in galactose medium. To save time, the yeast strain is pretransformed with p*GAL-cre* and selection is maintained for the plasmid-borne *LEU2* marker through subsequent steps. Also, selection is maintained for the mTn-borne *URA3* prior to the excision step, as some mutations are deleterious even in the heterozygous state. The frequency of Ura⁻ cells should be 100- to 1000-fold higher in the cultures grown with galactose as the carbon source than in cultures containing glucose.

The location of a HAT tag is determined by size or sequence analysis of the polymerase chain reaction (PCR) product generated from genomic DNA using a primer within the tag and a primer in the mutagenized gene. The primers suggested in Table 1 will generate a product that contains only one copy of the transposon terminal repeat (TR). Owing to self-annealing, PCR products containing both TRs are difficult to generate and impossible to sequence.

The HA triple tag can be detected by the mouse monoclonal antibodies 12CA5 (Boehringer) and 16B12 (101R; BAbCO, Richmond, California). On Western blots of total yeast protein, these antibodies each recognize a single cross-reacting band of about 50 or 125 kD, respectively, and can give a spotty backgound when used for immunofluorescence analysis. Despite this drawback, the 3×HA tag has been used extensively and successfully in yeast. We have found that background immunofluorescence can be eliminated by minimizing the extent of spheroplasting. We also preadsorb primary and secondary antibodies against yeast cells (Burns *et al.*, 1996). A rabbit polyclonal antiserum is also available (101c500; BAbCO) but in our hands this is less reactive. Protocols for yeast immunofluorescence have been presented by Pringle *et al.* (1991; see Chapter 12).

To tag essential genes, a diploid yeast strain must be transformed. After the Cre-mediated excision event, the strain is analyzed by tetrad dissection to determine whether the HAT-tagged gene is functional. It is important to confirm that any phenotype observed is linked to the mutagenized gene and that several excision events give the same phenotype.

Table 1. Reagents required for the mTn-mutagenesis procedure

Strains
- R1123: strain XL1-blue carrying pHSS6 (Hoekstra *et al.*, 1991b)
- B211: strain RDP146 (F⁻ *recA'* (Δ*lac-pro*) *rpsE*; spectinomycin resistant) with plasmid pLB101 (pACYC184 with *tnpA*; active transposase, chloramphenicol resistant) (Hoekstra *et al.*, 1991b)
- B425: strain NG135 (K12 *recA56 gal*-delS165 *strA*; streptomycin resistant) with plasmid pNG54 (pACYC184 with *tnpR*; active resolvase, chloramphenicol resistant) (Ross-Macdonald *et al.*, 1997)
- B227: strain DH5-α carrying p*GAL-cre* (*amp*ᴿ, ori, *CEN*, *LEU2*) (Hoekstra *et al.*, 1991a)
- B426: strain RDP146 with pOX38::mTn-4×HA/*lacZ* (F factor derivative carrying mTn3 derivative; tetracycline resistant) (Ross-Macdonald *et al.*, 1997)
- B427: strain RDP146 with pOX38::mTn-3×HA/*lacZ* (F factor derivative carrying mTn3 derivative; tetracycline resistant) (Ross-Macdonald *et al.*, 1997)
- B428: strain RDP146 with pOX38::mTn-3×HA/GFP (F factor derivative carrying mTn3 derivative; tetracycline resistant) (Ross-Macdonald *et al.*, 1997)

Antibiotics
- Tetracycline HCl, Tet, 3 mg ml⁻¹ stock in 50% ethanol. Use at 3 μg ml⁻¹
- Kanamycin, Kan, 10 mg ml⁻¹ stock in water. Use at 40 μg ml⁻¹
- Chloramphenicol, Cm, 34 mg ml⁻¹ stock in ethanol. Use at 34 μg ml⁻¹
- Streptomycin, Sm, 10 mg ml⁻¹ stock in water. Use at 50 μg ml⁻¹

Oligonucleotide primers
- 5′-3×HA: 5′-CCGTTTACCCATACGATGTTCCTG-3′. Bases 133 to 156 of the HAT tag (sense strand)
- 3′-3×HA: 5′-GAGCGTAATCTGGAACGTCATATGG-3′. Bases 228 to 204 of the HAT tag (antisense strand)

◆◆◆◆◆◆ IV. USING TRANSPOSON-INSERTION LIBRARIES TO MUTAGENIZE OR SCREEN THE YEAST GENOME

A. Available Libraries

Three different transposon-insertion libraries are available from the authors. Each is derived by mutagenesis of a library that contains 18 yeast genome equivalents, composed of 2.5–3.5 kb fragments from a *Sau*3A partial digest of yeast DNA (Burns *et al.*, 1994). This library has been mutagenized with m-Tn3(*LEU2 lacZ*) (Hoekstra *et al.*, 1991b), mTn-3×HA/*lacZ* and mTn-3×HA/GFP (Figure 1; Ross-Macdonald *et al.*, 1997). Each mutagenized library contains 1–2 × 10⁶ insertions. A library made from yeast genomic DNA digested to completion with *Bam*HI and mutagenized with 10⁷ insertions of m-Tn3(*URA3 lac*) (Hoekstra *et al.*, 1991b) has also been described (Chun and Goebl, 1996).

B. Potential Biases in Transposon-Insertion Libraries

We expect that the transposon-insertion procedure will not identify all yeast genes with equal frequency for several reasons. Large yeast genes will be over-represented because they provide larger targets. The ribosomal DNA (rDNA) and 2μ plasmid sequences will also be over-represented in our libraries, whereas in the library of Chun and Goebl (1996), genes residing on large *Bam*HI fragments are likely to be under-represented. Yeast sequences that cannot be cloned in *E. coli* on a vector of moderate copy number (i.e. pHSS6-Sal) will be lacking. Finally, some genes might be under-represented because they are "cold spots" for Tn3 transposition (Wiater and Grindley, 1990).

C. Using Transposon Libraries as Mutagens or Reporters

An outline of the methodology, which is similar for all four libraries, is presented in Figure 2. The mutagenized yeast genomic DNA is released from the plasmid vector by digestion with *Not*I and transformed into the desired yeast host, selecting for the auxotrophic marker carried by the transposon (*URA3* or *LEU2*).

Yeast transformants are screened for a mutant phenotype in the same manner as cells treated with a conventional mutagen. Approximately 30 000 transformants must be screened to ensure 95% coverage of the genome. Transformants may also be screened for reporter activity by β-gal assays (see Protocol 1) or fluorescence analysis (Niedenthal *et al.*, 1996), as appropriate. In this case, approximately 180 000 transformants must be screened for 95% coverage. When a transformant with a phenotype of interest has been identified, genetic analysis is used to determine whether the phenotype segregates with the transposon-borne marker. This is an important consideration, because multiple insertion events or unrelated mutations can occur during transformation. The site of the genomic insertion may then be determined by either plasmid rescue or PCR techniques, as detailed below.

Another potential use for these libraries is screening for downstream targets of proteins of interest such as transcription factors (Dang *et al.*, 1994) or proteases (M. Snyder and H. Friedman, unpublished results). In this approach (described more fully by Burns *et al.*, 1996), the first step is to create a yeast strain with a mutation in the gene of interest, and the wild-type gene on a plasmid with a counter-selectable marker (e.g. *URA3*) so that cells lacking the plasmid (and hence the wild-type gene) can be obtained. The library is transformed into this strain, and transformants are replica-plated onto medium that selects for cells without the plasmid-borne wild-type gene (e.g. 5-FOA), and onto medium selecting for cells that retain the wild-type gene (e.g. SC-Ura). Gene expression in the two cell types may then be analyzed under identical growth conditions, a necessary precaution to avoid the identification of false positives due to environmental effects.

D. Identifying the Insertion Site by Plasmid Rescue

Plasmid rescue using the plasmid YIp5 has previously been described (Burns *et al.*, 1994). We have constructed improved rescue vectors, designated pRSQ2-*LEU2* and pRSQ2-*URA3*. Both are for use with transposons

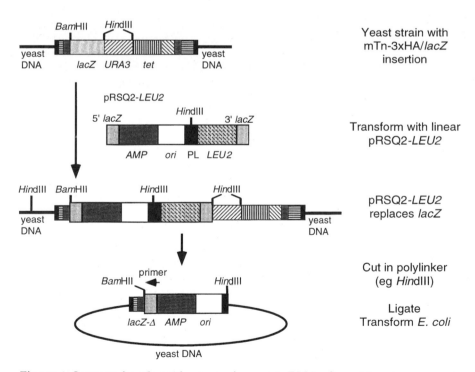

Figure 4. Strategy for plasmid rescue of genomic DNA adjacent to a transposon insertion. In the example, pRSQ2-LEU2 (linearized with *Bam*HI) is used to recover the sequences at the site of mTn-3×HA/*lacZ* insertion. Relevant restriction sites for recovery and plasmid analysis are shown. A primer complementary to *lacZ* sequences is used for sequencing. PL, polylinker region of pRSQ2-*LEU2*.

that contain *lacZ*, and work on the same principle; a detailed protocol is given in Protocol 2. A yeast strain with a transposon insertion (mTn-3×HA/*lacZ* in the example shown in Figure 4) is transformed with linearized pRSQ2 that carries a different auxotrophic marker (pRSQ2-*LEU2* in the example given). Transformation efficiency should be typical for a targeted replacement in your strain background. Transformants are selected for the pRSQ2-borne marker, and the plasmid replaces part of the transposon by homologous recombination between *lacZ* sequences. Yeast DNA is recovered from these transformants by standard means (Philippsen *et al.*, 1991; Burns *et al*; 1996) and cut with one of several possible "recovery" enzymes (usually *Eco*RI or *Hind*III). This digest releases

---

Protocol 2. Plasmid rescue with pRSQ2 vectors

1. Transform the yeast strain carrying the insertion with 1–5 μg of the appropriate pRSQ2 plasmid DNA linearized with *Bam*HI. Select for the pRSQ2 marker.
2. Prepare genomic DNA from transformants (Philippsen *et al.*, 1991; Burns *et al.*, 1996) and cut 5 μg overnight with 5 units of "recovery" enzyme (*Eco*RI, *Hind*III, *Sal*I, *Cla*I, *Xho*I, or *Kpn*I for pRSQ2-*LEU2*; the same enzymes plus *Eco*RV or *Pst*I for pRSQ2-*URA3*) in a total volume of 40 μl.
3. Run 20 μl of the sample on a gel to check digestion. Heat the remainder to 65°C for 25 min to inactivate the restriction enzyme. Add 215 μl of H$_2$O, 25 μl of 10 × ligase buffer and 1 μl of ligase (400 units). To favor intramolecular reactions, the DNA concentration in the ligation should not exceed 10 μg ml^{-1} and can be as low as 2 μg ml^{-1}.
4. After ligation at 16°C for 4–16 h, precipitate DNA by adding 125 μl of 7.5 M NH$_4$Ac and 375 μl of isopropanol, and recover by centrifugation.
5. Wash the DNA pellet with 70% ethanol. Resuspend in 6–20 μl of TE (Asubel *et al.*, 1987). Transform 3 μl into *E. coli*, selecting for resistance to 50 μg ml^{-1} ampicillin.
6. Isolate plasmid from several colonies for each strain. Analyze by double-digestion with *Bam*HI and the 'recovery' enzyme. Desired plasmids display a 2.85 kb band (3.9 kb for *Eco*RI used as recovery enzyme with pRSQ2-*LEU2*) containing vector sequences, plus additional band(s) from genomic DNA. If you get "mystery" plasmids, try a different transformant and/or recovery enzyme.
7. Plasmid DNA preparations may be sequenced by standard techniques using the M13(-40) primer (New England Biolabs). Trust sequence only up to the first site for the recovery enzyme, as chimaerism can occur during circularization.
8. A plasmid that allows the disruption allele to be transferred to other strains can be made by cleaving with one of the following enzymes at step 2: *Avr*II, *Bgl*II, *Bsp*EI, *Eag*I, *Msc*I, *Nae*I, *Nhe*I, *Nru*I, *Pml*I, *Sma*I, *Sna*BI, *Spe*I, *Sph*I or *Xma*I. With these enzymes, a plasmid of >11.7 kb containing sequences both 5′ and 3′ to the transposon insertion is recovered. This plasmid is linearized with the same enzyme to target homologous replacement. The resulting insertion no longer has reporter activity. However, formation of the HAT tag from the derivative of mTn-3×HA/*lacZ* should not be affected.

as a linear segment the bacterial origin of replication, the *amp* gene, the end of the transposon and some adjacent yeast DNA. The fragment is circularized and recovered in *E. coli* as a high copy number plasmid. The plasmid may be sequenced using a primer complementary to the 5' end of the transposon. The plasmid rescue procedure has the advantage of generating an easily sequenced template, and can also be used to move disruption alleles or HAT-tagged alleles into a different strain background (see Protocol 2).

E. Identifying the Insertion Site by Vectorette PCR

The vectorette PCR method was originally described by Riley *et al.* (1990). The procedure is outlined in Figure 5 and Protocol 3. Genomic DNA is prepared by standard methods (Philippsen *et al.*, 1991; Burns *et al.*, 1996) from a strain carrying a transposon, and digested to produce small, blunt-ended fragments. By ligation, an "anchor bubble" is added to each end. A

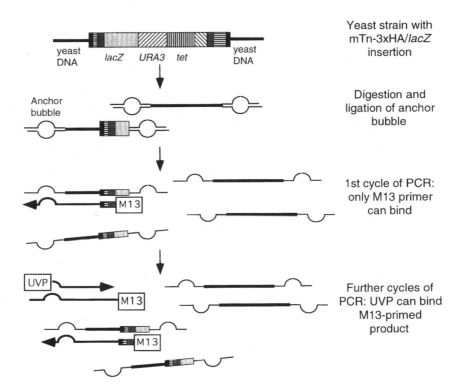

Figure 5. Strategy for vectorette PCR of genomic DNA adjacent to a transposon insertion. Digestion is with any enzyme that cuts frequently and produces blunt ends. The "anchor bubble" is formed by annealing two primers that have a non-homologous central region. A 5' overhang at one end of the anchor bubble ensures that only one copy is added at each end of a restriction fragment, and in a defined orientation. M13, M13(-47) primer, which hybridizes to *lacZ* sequences in the transposon; UVP, universal vectorette primer, identical to a sequence in the non-homologous central region of the anchor bubble.

PCR reaction is performed using a primer complementary to transposon sequences (Protocol 3) and a primer identical (not complementary) to sequences in the bubble (the universal vectorette or UV primer). In the first cycle of PCR, only the transposon primer can bind template and be elongated. In subsequent cycles, the UV primer can bind to these DNA strands. Thus, only a region from the fragment containing the transposon is amplified. The PCR product is then sequenced.

The vectorette PCR method has the advantage of eliminating a cloning step, and has been successful where plasmid rescue has failed (Friddle,

---

Protocol 3. Vectorette PCR

1. Prepare good quality genomic DNA (Philippsen *et al.*, 1991; Burns *et al.*, 1996) from the yeast strain carrying the transposon insertion. Cut 1–3 μg of DNA overnight with 8–10 u of an enzyme that leaves blunt ends, in a total volume of 20 μl. Do not use an enzyme that cuts between the mTn end and the primer binding site (*Alu*I and *Dra*I provide good results).

2. Heat inactivate the enzyme by heating digest to 65°C for 20 min.

3. Add 3 μl of 10× buffer used in restriction digest, 1 μl of annealed anchor bubble (see below), 1 μl of ligase (400 u), 0.5 μl of 5 mM ATP and 24.5 μl of water. Incubate at 16°C for 9–24 h.

4. Set up a 100 μl PCR reaction with the following components: 5 μl of the ligation reaction, 2.5 μl of 20 μM M13(-47) primer (or GFP primer for mTn-3×HA/GFP), 2.5 μl of 20 μM UV primer (see below), 8 μl of 2.5 mM dNTPs, 10 μl of 10× Taq PCR buffer, 71 μl water and 1 μl Taq DNA polymerase (5 u). A "hot start" using Ampliwax (Perkin Elmer) is recommended.

5. Transfer to thermal cycler and denature at 92°C for 2 min. Perform 35 cycles of 20 s at 92°C, 30 s at 67°C and 45–180 s at 72°C (use the longer interval if a product is not apparent), followed by a single cycle of 90 s at 72°C.

6. Run 80 μl of the PCR on a 1–3% SeaKem GTG agarose (FMC) gel. Usually only one band containing 200–400 ng of DNA is seen. Excise all bands individually and recover DNA with Qiaex (Qiagen). Elute DNA with 12 μl of TE (Asubel *et al.*, 1987).

7. Sequence 7 μl of recovered DNA with Sequenase kit (Amersham). Use 200–600 pmol of M13(-47) primer (or GFP primer for mTn-3×HA/GFP), and high specific activity (>1000 Ci/mmol) S^{35} labeled nucleotide (Amersham). Denature sequencing reactions by boiling for 10 min followed by immediate cooling in ice water.
   - Anchor bubble primers: 5′-GAAGGAGAGGACGCTGTCTGTC-GAAGGTAAGGAACGGACGAGAGAAGGGAGAG-3′ and 5′-GACTCTCCCTTCTCGAATCGTAACCGTTCGTACGAGAATCG-CTGTCCTCTCCTTC-3′. To anneal primers to each other, heat an aqueous solution that is 2–4 mM for each primer to 65°C for 5 min, then add MgCl$_2$ to a final concentration of 2 mM and allow to cool slowly to room temperature. Store at –20°C.
   - UV primer: 5′-CGAATCGTAACCGTTCGTACGAGAATCGCT-3′
   - M13(-47) primer: 5′-CGCCAGGGTTTTCCCAGTCACGAC-3′ (bases 177–154 of mTn-3×HA/*lacZ*, antisense)
   - GFP primer: 5′-CATCACCTTCACCCTCTCCACTGAC-3′ (bases 243–219 of mTn-3×HA/GFP, antisense)

---

unpublished observations). It is the only method currently available to determine the insertion site in a yeast genome mutagenized with mTn-3×HA/GFP.

♦♦♦♦♦♦ V. CONCLUSION

Use of transposons allows the rapid construction of a large number of alleles of a gene of interest. Transposon insertion libraries can be used for both mutagenesis and identification of genes regulated by particular growth conditions and strain backgrounds. The new transposons described expand the repertoire of insertions that may be generated to include GFP fusions and epitope tags. We expect that shuttle mutagenesis will continue to be an important tool for characterization of individual genes and their products, and for systematic analysis of the entire yeast genome.

Accession numbers: m-Tn3(*LEU2 lacZ*), U35112; mTn-3×HA/*lacZ*, U54828; mTn-3×HA/GFP, U54830; pRSQ2-*LEU2*, U64693; pRSQ2-*URA3*, U64694; pH556, M84115.

Acknowledgements

We thank Howard Granok and Terry Roemer for comments on this manuscript. This work was supported by N.I.H. grants HD32637 to M.S. and G.S.R., and GM46406 to D. Botstein.

References

Asubel, F. M., Brent, R., Kingston, R. E., Moore, D. D., Seidman, J. G., Smith, J. A. and Struhl K. (1987). *Current Protocols in Molecular Biology*. Wiley Interscience, New York.

Baudin, A., Ozier-Kalogeropoulos, O., Denouel, A., Lacroute, F. and Cullin, C. (1993). A simple and efficient method for direct gene deletion in *Saccharomyces cerevisiae*. *Nucleic Acids Res.* **21**, 3329–3330.

Burns, N., Grimwade, B., Ross-Macdonald, P. B., Choi, E.-Y., Finberg, K., Roeder, G. S. and Snyder, M. (1994). Large-scale characterization of gene expression, protein localization and gene disruption in *Saccharomyces cerevisiae*. *Genes Dev.* **8**, 1087–1105.

Burns, N., Ross-Macdonald, P., Roeder, G. S. and Snyder, M. (1996). Generation, screening and analysis of *lacZ* fusions in yeast. In *Microbial Genome Methods* (K. W. Adolph, ed.). CRC Press, Boca Raton.

Chun, K. T. and Goebl, M. G. (1996). The identification of transposon-tagged mutations in essential genes that affect cell morphology in *S. cerevisiae*. *Genetics* **142**, 39–50.

Cunningham, B. C. and Wells, J. A. (1989). High-resolution epitope mapping of hGH-receptor interactions by alanine-scanning mutagenesis. *Science* **244**, 1081–1085.

Dang, V. D., Valens, M., Bolotin-Fukuhara, M. and Daignan-Fornier (1994). A genetic screen to isolate genes regulated by the yeast CCAAT-box binding protein Hap2p. *Yeast* **10**, 1273–1283.

Devine, S. E. and Boeke, J. D. (1994) Efficient integration of artificial transposons into plasmid targets *in vitro*: a useful tool for DNA mapping, sequencing and genetic analysis. *Nucl. Acids Res.* **22**, 3765–3772.

Garfinkel, D. J. and Strathern, J. N. (1991). Ty mutagenesis in *Saccharomyces cerevisiae. Methods Enzymol.* **194**, 342–361.

Heim, R., Prasher, D. C. and Tsien, R. Y. (1994). Wavelength mutations and post-translational autoxidation of green fluorescent protein. *Proc. Natl Acad. Sci. USA* **91**, 12501–12504.

Hoekstra, M. F., Burbee, D., Singer, J., Mull, E., Chiao, E. and Heffron, F. (1991a). A Tn3 derivative that can be used to make short in-frame insertions within genes. *Proc. Natl Acad. Sci. USA* **88**, 5457–5461.

Hoekstra, M. F., Seifert, H. S., Nickoloff, J. and Heffron, F. (1991b). Shuttle mutagenesis: bacterial transposons for genetic manipulations in yeast. *Methods Enzymol.* **194**, 329–342.

Irniger, S., Piatti, S., Michaelis, C. and Nasmyth, K. (1995). Genes involved in sister chromatid separation are needed for B-type cyclin proteolysis in budding yeast. *Cell* **81**, 269–277.

Ji, H., Moore, D. P., Blomberg, M. A., Braiterman, L. T., Voytas, D. F., Natsoulis, G. and Boeke, J. D. (1993). Hotspots for unselected Ty1 transposition events on yeast chromosome III are near transfer RNA genes and LTR sequences. *Cell* **73**, 1007–1018.

Lawrence, C. W. (1991). Classical mutagenesis techniques. *Methods Enzymol.* **194**, 273–280.

Morrele, G. (1989). A plasmid extraction procedure on a miniprep scale. *BRL Focus* **11**, 7–8.

Mosch, H. U. and Fink, G. R. (1997) Dissection of filamentous growth by transposon mutagenesis in *Saccharomyces cerevisiae. Genetics* **145**, 671–684.

Niedenthal, R. K., Riles, L., Johnston, M. and Hegemann, J. H. (1996). Green fluorescent protein as a marker for gene expression and subcellular localization in budding yeast. *Yeast* **12**, 773–786.

Philippsen, P., Stotz, A. and Scherf, C. (1991). DNA of *Saccharomyces cerevisiae. Methods Enzymol.* **194**, 169–182.

Pringle, J., Adams, A. E. M., Drubin, D. G. and Haarer, B. K. (1991). Immunofluorescence methods for yeast. *Methods Enzymol.* **194**, 565–601.

Riley, J., Butler, R., Ogilvie, D., Finniear, R., Jenner, D., Powell, S., Anand, R., *et al.* (1990). A novel, rapid method for the isolation of terminal sequences from yeast artificial chromosome (YAC) clones. *Nucleic Acids Res.* **18**, 2887–2890.

Ross-Macdonald, P., Burns, N., Malcynski, M., Sheehan, A., Roeder, S. and Snyder, M. (1995). Methods for large-scale analysis of gene expression, protein localization, and disruption phenotypes in *Saccharomyces cerevisiae. Methods Mol. Cell. Biol.* **5**, 298–308.

Ross-Macdonald, P. B., Sheehan, A. S., Roeder, G. S. and Snyder, M. P. (1997). A multipurpose transposon system for analyzing protein production, localization and function in *Saccharomyces cerevisiae. Proc. Natl Acad. Sci. USA.* **94**, 190–195.

Rothstein, R. (1991). Targeting, disruption, replacement, and allele rescue: integrative DNA transformation in yeast. *Methods Enzymol.* **194**, 281–301.

Schneider, B. L., Seufert, W., Steiner, B., Yang, Q. H. and Futcher, A. B. (1995). Use of PCR epitope tagging for protein tagging in *Saccharomyces cerevisiae. Nucleic Acids Res.* **11**, 1265–1274.

Seifert, H. S., Chen, E. Y., So, M. and Heffron, F. (1986). Shuttle mutagenesis: a method of transposon mutagenesis for *Saccharomyces cerevisiae*. *Proc. Natl Acad. Sci. USA* **83**, 735–739.

Sherman, F., Fink, G. R. and Hicks, J. B. (1986) *Methods in Yeast Genetics: A Laboratory Manual*. Cold Spring Harbor Laboratory, Cold Spring Harbor, New York.

Sikorski, R. S. and Boeke, J. D. (1991). *In vitro* mutagenesis and plasmid shuffling: from cloned gene to mutant yeast. *Methods Enzymol*. **194**, 302–329.

Smith, V., Botstein, D. and Brown, P. O. (1995). Genetic footprinting: a genomic strategy for determining a gene's function given its sequence. *Proc. Natl Acad. Sci. USA* **92**, 6479–6483.

Tyers, M., Tokiwa, G., Nash, R. and Futcher, B. (1992). The Cln3-Cdc28 kinase complex of *S. cerevisiae* is regulated by proteolysis and phosphorylation. *EMBO J*. **11**, 1773–1784.

Wertman, K. F., Drubin, D. G. and Botstein, D. (1992). Systematic mutational analysis of the yeast *ACT1* gene. *Genetics* **132**, 337–350.

Wiater, L. A. and Grindley, N. D. F. (1990). Integration host factor increases the transpositional immunity conferred by γδ ends. *J. Bacteriol*. **172**, 4951–4958.

11 Transposon Tagging II: Exploration of Gene Function and Regulatory Networks in Yeast with the Mini-Mu Transposon

Monique Bolotin-Fukuhara

Laboratoire de Génétique Moléculaire, Institut de Génétique et Microbiologie, Université Paris-Sud, Orsay Cedex, France

◆◆

CONTENTS

List of Abbreviations

GFP	Green fluorescent protein
ORF	Open reading frame
RT-PCR	Reverse transcription-polymerase chain reaction
SAGE	Serial analysis of gene expression
SM	Synthetic minimal
2D	Two-dimensional

◆◆◆◆◆◆ I. INTRODUCTION

Deciphering the complete nucleotide sequence of the yeast genome has uncovered more than 6000 open reading frames (ORFs), half of which have as yet no defined biological function. The study of gene function (functional analysis) is the next challenge yeast scientists have to face, going from individual gene analysis to global approaches. As a consequence, new experimental tools must be developed to adapt to the rapid increase in the rate of data accumulation.

Systematic functional analysis of yeast genes requires the combination of various approaches involving the quasi simultaneous study of gene disruption, gene expression and gene product localization. This will require an increase in efficiency while proportionally reducing the amount of experimental work. Following this line of reasoning, it is worth noting that the three classical approaches listed above can all be studied with the help of suitable reporter genes (see Chapters 9 and 10).

If a large-scale analysis is to be undertaken, i.e. many genes having to be studied at the same time, it is not possible to produce and use large numbers of antibodies. Rather, construction of gene fusions is an easy way to monitor both gene expression and gene product localization. Furthermore, if a gene fusion can be constructed in such a way that it interrupts the entire coding region of the target ORF, it may also be a very good tool to study the consequences of a specific gene disruption.

Construction of gene fusions can be done either by *in vitro* ligation or by random insertion using transposon-based systems. Transposon tagging may have advantages over the classical *in vitro* procedure, providing one can make sure that transposition into the target sequences is a random procedure. This strategy is facilitated by the fact that the work is done by a living organism and not by the scientist; this is especially important if experiments have to be scaled up because it reduces considerably the amount of work required. Second, *in vivo* constructions may circumvent several well-known difficulties. Gene fusions are by definition artificial constructions, the stability or expression of which may vary according to the point of insertion of the fusion partner. Unfortunately, we usually have no rationale as to where in the sequence the fusion should be made and how to optimize its expression. Made *in vivo* by transposon tagging, only the correctly expressed fusions will be selected. In addition, the number of insertions one can obtain (see section III), makes it possible to scale up the approach when libraries have to be constructed. Finally, because the reporter gene is inserted within the ORF, thereby interrupting it, the resulting gene fusion can be used to replace the normal ORF and to analyze the disruption phenotype.

Insertional disruption was one of the driving forces behind the establishment of the first yeast transposon tagging system (Daignan-Fornier and Bolotin-Fukuhara, 1988) because many yeast ORFs had yet to be disrupted and analyzed. One may think that this property is not useful any more because a systematic gene disruption study has been undertaken for the whole genome by complete deletion of corresponding ORFs (Chapter 5). Perhaps it is worthwhile to keep in mind, however, the fact that we still need to undertake very detailed studies of yeast ORFs, such as the analysis of proposed functional domains. In such a case, insertional mutagenesis throughout the ORF and the observation of the corresponding phenotypes may be much more important than the complete deletion of the ORF. The necessity to search for interacting genes by selection of suppressors (see Chapter 16) also reinforces the interest of obtaining disruptants by insertion rather than by deletion.

Yeast transposons belonging to the Ty family are not particularly useful for large-scale transposon tagging because: (i) the transposition frequency is not very high, thereby increasing the necessary amount of manipulation; and (ii) insertions are not totally random as was shown by systematic Ty insertions on chromosome III (Ji *et al.*, 1993). These constraints can, however, be easily circumvented because yeast genomic DNA is (or can be) cloned in a variety of ways into many different *Escherichia coli–Saccharomyces cerevisiae* shuttle vectors. This makes possible the direct usage of *E. coli* transposons.

Two types of transposons are currently used for genetic manipulation in *E. coli* and can therefore be adapted to yeast. One, belonging to the Tn family (Tn3/Tn10) has been used to study yeast sequences (Seifert *et al.*, 1986; Huisman *et al.*, 1987; Chapter 10) and was applied to functional analysis by Burns *et al.* (1994). The other is a derivative of the phage Mu called the mini-Mu. Applications of the mini-Mu transposon to yeast gene analysis will be described in this chapter.

◆◆◆◆◆◆ II. THE MINI-Mu TRANSPOSON AND CONSTRUCTION OF DERIVATIVES APPLICABLE TO YEAST

The bacteriophage Mu is a temperate phage of *E. coli* whose properties can be used for genetic manipulation. Insertions into the host DNA take place at non-specific sequences (Mizuuchi and Craigie, 1986) and the choice between the lytic or the lysogenic cycle can be controlled by a temperature-sensitive allele of the *c*-encoded repressor. The two ends of the phage genome are sufficient for transposition and encapsidation of a defective Mu if a wild-type Mu helper, inserted elsewhere in the *E. coli* chromosome, provides the necessary Mu gene products *in trans*.

Casadaban and Chou (1984) have exploited these properties to construct a mini-Mu in which a β-galactosidase fusion protein can be produced if the defective bacteriophage is inserted in-frame within an ORF expressed *in vivo*. The mini-Mu can incorporate relatively large fragments of DNA: up to 39 kb from the *c* extremity (Bukhari and Taylor, 1975). This property, associated with its relatively small size, makes it possible to select insertions into plasmids normally used in gene cloning studies (Castilho *et al.*, 1984).

The use of mini-Mu for insertional mutagenesis, as well as for construction and study of gene fusions, was first developed successfully in *E. coli* (Casadaban and Chou, 1984; Castilho *et al.*, 1984; Groisman *et al.*, 1984; Carlioz and Touati, 1986; Richaud *et al.*, 1987) and related prokaryotic organisms (Dixon *et al.*, 1980; Lee *et al.*, 1980; Csonska *et al.*, 1981; Boulanger *et al.*, 1986). The possibility of applying directly this methodology to yeast was explored by Casadaban and co-workers (discussed in Castilho *et al.*, 1984) but proved unsuccessful.

On the basis of the properties of Mu and the fact that screening for β-galactosidase activity can be performed in yeast (Rose *et al.*, 1981;

Guarente and Ptashne, 1981), a new version of the mini-Mu adapted for functional studies in yeast was developed (Daignan-Fornier and Bolotin-Fukuhara, 1988).

The original mini-Mu construct MudIIPR13 (Figure 1A; described in Richaud *et al.*, 1987) contains the two ends of phage Mu genome that are necessary for transposition and encapsidation: the *cat* gene conferring resistance to chloramphenicol and the '*lac Z* gene (devoid of the first eight codons of the *lacZ* gene (Casadaban *et al.*, 1980). The '*lacZ* gene is placed in-frame with the sequence of one of the Mu ends. Such a construction, if transposed *in-frame* and *in the correct orientation* within any ORF, will create a gene fusion in which the ORF is fused to the reporter gene *lacZ*. This fusion is readily detectable because cells carrying it will be colored blue in the presence of the chromogenic substrate for β-galactosidase (i.e. X-Gal). The point of insertion within the coding sequence can be analyzed by rapid sequencing of the junction using defined oligonucleotide primers.

The next development was to move the system to replicative yeast plasmids. The recovery of these plasmids is easy, as are subsequent manipulations, including DNA sequencing. In addition, overexpression of the gene fusions in yeast multicopy plasmids allows one to monitor the expression of a gene normally expressed too weakly to be easily detectable. Overexpression might, however, create artifacts (such as mis-localization), which prompted us to construct integrative derivatives. Such integrative plasmids are useful in constructing corresponding gene disruptions.

To achieve this dual purpose we inserted into mudIIPR13, in addition to the *LEU2* gene, a 2 μm plasmid origin flanked by two *Sal*I sites to generate the mini-Mu MudIIZZ1 (Daignan-Fornier and Bolotin-Fukuhara, 1988). These sites were later replaced by two *Not*I sites (MudIIZZ4, Dang *et al.*, 1994) (Figure 1B).

These transposition tools can be used to obtain several kinds of information from a single construction: (i) one can obtain a gene fusion with *lacZ* as a reporter, allowing one to measure gene expression and cellular localization of the gene product; (ii) the orientation of the gene fusion confirms the direction of transcription; and (iii) following deletion of the *Not*I cassette (very few *Not*I sites are present in the yeast genome), the plasmid formed is an integrative one and can be used to target the interrupted gene to the chromosome in place of the wild-type sequence.

We have compared the results obtained either by the transposition method described above or by classical yeast molecular biology methods through the detailed study of a 5 kb yeast insert cloned into plasmid pBR322. No discrepancies were found between the two strategies, giving one confidence in the findings obtained (see Daignan-Fornier and Bolotin-Fukuhara, 1989 for details).

The method described is applicable to the analysis of a single plasmid (see section IV) or to a genomic library, i.e. a population of plasmids (see sections VI and VII).

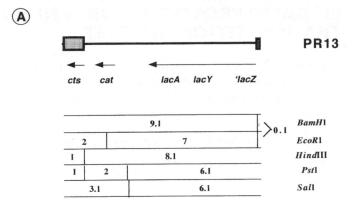

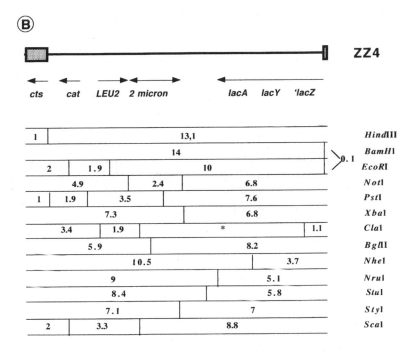

Absent sites: Sal1, Xho1, Sph1, Sma1, Apa1 (and Bgl11 in Δ Not1)

Figure 1. Schematic map of the two mini-Mu transposons MudIIPR13 and MudIIZZ4. (A) The upper part represents the genetic information contained in transposon PR13. *Cts* is the heat-sensitive allele of the c repressor of Mu, *cat* is the gene that encodes chloramphenicol acetyltransferase which confers resistance to chloramphenicol to *E. coli.*, and *lacA*, *lacY* and *lacZ* represent the lactose operon. The direction of transcription is indicated by an arrow. The bottom part is the restriction map of the transposon on the same scale as the genetic map. (B) Corresponding information for the transposon ZZ4. In addition to the genes described in (A), the yeast marker *LEU2* and the 2 μm plasmid origin of replication are indicated. The restriction map is not derived from DNA sequence data and therefore fragment sizes are only approximate. The restriction site *Xba*I is *not* included in the *Not*I fragment. Primers for sequencing are: *lacZ* extremity, 5'CTGTTTCATTTGAAGCGCG3'; *cts* extremity, 5'CCCACCAAATCTAATCCC3'. (*There may be an additional *Cla*I site in this fragment.)

◆◆◆◆◆◆ III. BASIC PROTOCOL FOR MINI-Mu TRANSPOSITION IN YEAST

Protocol 1 describes the different steps one must take to transpose mini-Mu mudIIZZ4 into a yeast plasmid. It is an experimental prototype and can be adapted to other mini-Mu based constructions (see section V).

◆◆◆◆◆◆ IV. APPLICATION OF TRANSPOSON TAGGING TO A SPECIFIC PLASMID

The transposition protocol described in Protocol 1 can be applied to a single plasmid and will produce many independent transductants with up to 10 000 independent transductants being obtained in a single experiment. Below we describe the situations in which it is appropriate to use this method in place of (or in addition to) classical molecular biology protocols.

A typical experiment produces several thousand independent transductants, reflecting several thousand different insertion points. These insertions are essentially random with some hot spots (see Daignan-Fornier and Bolotin-Fukuhara, 1988, 1989 for data and further discussion on this point) and will therefore take place both in the yeast insert and in the vector sequence. This is more than enough to dispose of a large collection of insertional mutagenesis events, each of them occurring on a different plasmid which can therefore be analyzed individually in the *E. coli* host. Alternatively, a more direct approach using plasmid DNA derived from a *pool* of transformed *E. coli* cells takes advantage of selective screening in yeast. The direct approach is recommended because this covers effectively the specific requirements of yeast functional analysis.

A. Selection of Expressed Fusions in Yeast

One can effectively screen for blue colonies after transformation of yeast with a pool of plasmid DNA molecules. This will reveal gene fusions *expressed* in yeast. Because of the large number of *E. coli* transductants obtainable, one can generate a large collection, all differing by the site of

---

Table I. Solutions

- LB liquid medium: 0.5% yeast extract (Difco), 1% Bacto-tryptone (Difco) and 0.5% sodium chloride
- LB + antibiotics: same as above. Add 50 mg liter^{-1} ampicillin (LBA) or 10 mg liter^{-1} of chloramphenicol (LBC) or both (LBAC)
- 1 M MgSO$_4$ stock solution (sterilized by filtration or autoclaving at 110°C, 30 min)
- 1 M CaCl$_2$ stock solution (sterilized as above)
- Chloroform
- Deep-freezing medium for bacteria: same as LB but water is replaced by a mixture of glycerol:water (40:60)

Protocol 1. Basic protocol for mini-Mu transposition

1. Prepare competent *E. coli* cells from a strain that carries a mini-Mu and the Mu helper. Any protocol for preparation of competent cells can be used as long as *cells are grown at 28°C*.
2. Transform the plasmid of interest into the *E. coli* competent cells *taking care not to heat-shock the cells during transformation* because this will induce uncontrolled transposition. The resulting strain is the donor strain.
3. Inoculate an overnight culture of *E. coli* strain M8820 (F-, *araD*139, Δ(*ara-leu*)7697, Δ(*proAB-argF-lacIPOZYA*)XIII, *rpsL*(Sm^R), in 2 ml of LB liquid medium (see Table 1). Grow at 37°C with shaking (around 200 rpm). At the same time, inoculate the donor strain (i.e. the mini-Mu+ plasmid strain constructed in step 2) in 2 ml LBAC liquid medium (see Table 1) and let grow at 28°C with shaking. A temperature below 30°C is necessary to avoid uncontrolled transduction.
4. Next morning, dilute the recipient strain M8820 (1/100) in 10 ml fresh prewarmed LB in a 50–100 ml Erlenmeyer flask. Incubate at 37°C with shaking. Do the same dilution in 2 ml LB for the donor strain but grow at 28–30°C for 2 h with shaking. Use large glass tubes or plastic tubes which are resistant to chloroform.
5. Preheat liquid LB to 60°C and add 4 ml of it to the strain carrying the mini-Mu.
6. Heat shock the culture for 20 min at 42°C. Be careful to use a water bath and conditions in which the temperature will be reached very quickly (e.g. a large tube or small Erlenmeyer flask).
7. At the end of this period add 25 μl of 1 M MgSO$_4$ (sterile) and 10 μl of 1 M of CaCl$_2$ (sterile) to the 10 ml culture of the M8820 strain.
8. Add 100 μl of chloroform to the 6 ml culture of the mini-Mu containing strain and vortex vigorously. Let it rest for 5 min at room temperature and vortex again. Let it sit for 15 min.
9. In 10 ml tubes, add (per tube): 300 μl of phage lysate, *taking care not to bring any chloroform* into the new culture, and 500 μl of treated M8820. Let it sit for 15 min. Plastic tubes that can stand low-speed (4000–6000 rpm) centrifugation are fine. Do not forget the controls: one tube without phage and one tube without M8820 bacteria.
10. Add 3 ml of warmed LB and shake for 2 h at 28°C.
11. Centrifuge 5 min at 4000–6000 rpm to pellet the bacterial cells. Evacuate the supernatant and resuspend the pellet in the residual liquid.
12. Plate the content of each tube on one LBAC plate and let it grow at 28°C overnight.
13. Very tiny colonies should be visible the next morning. An average of 1000–2000 transductants are usually obtained. Because it is possible to make at least 15 plates per experiment, 10 000 to 20 000 transductants should be obtained in a single experiment. It is possible to scale-up this protocol if so required.
14. Let the colonies grow during the day, then harvest the transductants (either one-by-one or as a pool) and store them as follows: scrape the colonies off each plate with a bent glass spreading rod in 600 μl of deep-freezing medium (see Table 1) and transfer the resulting liquid stock to sterile tubes that resist −80°C. At this temperature, colonies will resist practically indefinitely and can be used for successive reinoculation.

insertion of *lacZ*. Advantages of this approach are immediately apparent: (i) the experiment proceeds faster than the one required for individual *in vitro* constructions; (ii) only fusions viable *in vivo* are selected for; (iii) the greater diversity of the population (remember fusions are random along the gene sequence) should make any biological conclusions more sound; and (iv) this method can be used even when useful restriction sites are scarce or when the length of the fragment under study is not optimal.

Plasmids carrying the fusions of interest can be easily rescued from yeast into *E. coli* with current protocols and the inserts immediately localized by restriction mapping (by taking advantage of either the *unique BamHI* or *HindIII* sites of the mini-Mu (see Figure 1) or the one-run sequencing using suitable oligonucleotide primers (see legend of Figure 1 for the sequence of suitable primers).

B. Identification of the Region of Interest

A second useful application of mini-Mu is the delineation of the region able to complement (or suppress) a specific mutation in a large cloned DNA fragment. Considering that the total genome sequence of *S. cerevisiae* is now available, the identification of the location of a yeast insert within a given chromosome is immediate. In most cases, this information is, however, not sufficient to conclude as to which ORF defines the genetic characteristics that were selected. Many libraries contain DNA inserts of 8–12 kb, a size that corresponds usually to several ORFs. In such a case, one has to subclone to identify the ORF responsible for the selected phenotype, an operation that can be time-consuming. In some cases, the absence of suitable restriction sites makes this study all the more difficult. Transposition of a mini-Mu into the plasmid of interest and transformation of the yeast strain with the pool of transductants (one plate of 1000–2000 transductants is usually enough) will rapidly provide yeast transformants that do not present the initial defined characteristics. These transformants most likely carry a non-functional plasmid, probably because of the insertion of the mini-Mu transposon within the part of the sequence required for the genetic function originally being selected. Sequencing the junction will immediately identify the corresponding ORF.

Different types of transposons can be used according to the type of vectors carrying the sequence of interest. Because complementation (or suppression) is obtained in yeast by transforming with genomic libraries built in vectors carrying a selectable marker and a low copy number (i.e. centromeric) or high copy number (2 μm plasmid) origin of replication, the mini-Mu MudIIZZ4 described above is not best suited to this type of analysis. After transposition, it will introduce a second origin of replication. In that case several choices are possible: (i) use MudIIPR13 (which has no yeast replication origin and no yeast selective marker); or (ii) delete the original plasmid for the no longer required yeast markers, or use an integrative derivative of MudIIZZ4 (MudIIZZ5), if the *LEU2* marker is compatible with the genetic markers present in the vector and the yeast strain.

Exploitation of the mini-Mu transposon in yeast has been shown to be efficient, fast and reliable in the two types of genetic screen described above.

◆◆◆◆◆◆ V. HOW TO DEVELOP A NEW MINI-Mu ADAPTED TO SPECIFIC REQUIREMENTS

The transposons described so far were developed in order to answer some of the questions discussed in section II. The parameters chosen (i.e. *lacZ* as a reporter gene, *LEU2* as a selective marker and the 2 μm plasmid replication origin) might not be well adapted to other problems and it may therefore be necessary to modify the present construction. This is quite easy and requires only a plasmid into which one of the already described mini-Mu has been transposed and a RecA-defective *E. coli* strain carrying a Mu helper. Modifications of the mini-Mu are made on this plasmid, taking advantage of all the possible *in vitro* manipulations. A simple and basic protocol to reintroduce a new mini-Mu into the RecA-deficient strain carrying the helper Mu is provided in Protocol 2. It works well provided one pays attention to two possible pitfalls. (i) It is wise to check the phenotype

---

Protocol 2. Construction of new mini-Mu and reintroduction into a Mu-carrying *E. coli* strain

Modifications of the mini-Mu should be carried out on a plasmid into which it has been inserted. If not directly available, the first step is to transpose a strain carrying one mini-Mu into a derivative of pBR322 according to Protocol 1.

1–2. Introduce the plasmid carrying the modified mini-Mu into a RecA *E. coli* strain such as DH1::Mucts (*RecA*1, *gyrA*96, *endA*1, *thr*1-1, *hsdR*17, *supE*44) or JM109:Mucts (same as DH1 plus Δ(*lac-proAB*) *relA[F',traD*36, *proAB,lacI^QZ*ΔM13]). As in Protocol 1, competent cells should be prepared at 28°C and transformation should avoid heat-shock.

3–9. These steps are identical to equivalent steps in Protocol 1 with one important modification: the recipient strain is JM109::Mucts or DH1::Mucts and carries a transposon. Consequently, all steps preceding the heat shock (i.e. step 6) should be carried out at 28°C and not 37°C (both precultures and cultures). The donor strain (carrying the modified mini-Mu) is manipulated as if it were the donor strain of the preceding protocol.

10. Add 3 ml of fresh warmed LB and shake for 2 h at 37°C. This temperature is chosen so as to induce transposition and avoid any lytic cycle.

11. Centrifuge for 5 min at 4000–6000 rpm to pellet the bacterial cells. Evacuate the supernatant and resuspend the pellet in the residual liquid.

12. Plate the tube content on LBC (not LBAC) and incubate at 28°C.

13. At the end of the following day, colonies will appear. Transfer a few hundred clones to LBA and LBAC and compare growth. Some of them will show poor growth on LBAC. Streak them and recheck individual subclones on LBAC and LBC. Keep clones that grow only on LBC (1–2%) and store them at –80°C.

14. These strains can be used as donors *but* the frequency of transposition may vary according to the point of insertion in the *E. coli* chromosome. This frequency should be tested according to Protocol 1 on an already characterized plasmid.

---

of the Rec A-deficient *E. coli* strains and verify the absence of growth on media containing ampicillin (LBA) as well as ampicillin and chloramphenicol (LBAC) before beginning the experiment. These strains do not grow very well and one can easily select secondary mutations. Once verified, the strain should be kept at –80°C as a stock to which one can go back to in case of subsequent problems. (ii) The last step of Protocol 2 describes the reintroduction of the new mini-Mu into the *E. coli* chromosome. After inducing transposition at 37°C, a temperature that does not induce cell lysis, one has to select clones that are chloramphenicol-resistant *and* ampicillin-sensitive. In our hands, these strains appear with a frequency of about 1%. It is important at this stage to keep several of these clones, subclone them and check their transposition efficiency into a well-characterized plasmid according to the procedure described in Protocol 1. This efficiency will depend on the place of insertion of the mini-Mu into the *E. coli* chromosome. We have manipulated the initial mini-Mu several times, including replacing *lacZ* by the gene encoding the green fluorescent protein (*GFP*), changing the selective marker or taking out the replication origin, each time with satisfactory results.

◆◆◆◆◆◆ VI. LIBRARY CONSTRUCTION WITH THE HELP OF A TRANSPOSON

Transposition of an appropriate mini-Mu into a single plasmid can be generalized to a population of plasmids, up to a population large enough to represent the entire yeast genome. The problem is the representativeness of the library and its proper maintenance to avoid any major kind of non-randomness.

The library is constructed in two steps (Figure 2):

(a) *Step 1* is the construction of a genomic library into a suitable vector. A variety of suitable protocols exist with which one can obtain a good representation of the genome. One should, however, avoid using vectors carrying the whole or part of the *lacZ* gene because this might promote homologous recombination with the mini-Mu and lead to rearranged plasmids. Where possible, this library should be constructed directly into the strain that carries the mini-Mu, but this is dependent upon the transformation efficiency of the strain, already lowered by the fact that no heat shock should be performed (as described in Protocol 1, step 2).

(b) *Step 2* involves transposition of the appropriate mini-Mu into the plasmid population. To calculate the size of the library, one should take into account the data obtained in our original study (Daignan-Fornier and Bolotin-Fukuhara, 1988). We have shown that where the vector and the plasmid are about the same size, 20% of the insertions take place into the cloned insert. In addition, only approximately one-sixth of them will be in-frame and in the right orientation with respect to the coding sequence. Finally, despite the fact that the yeast genome is compact (average intergenic regions are rather small), some of the insertions will also take place outside coding sequences. Taking into account all this information, we

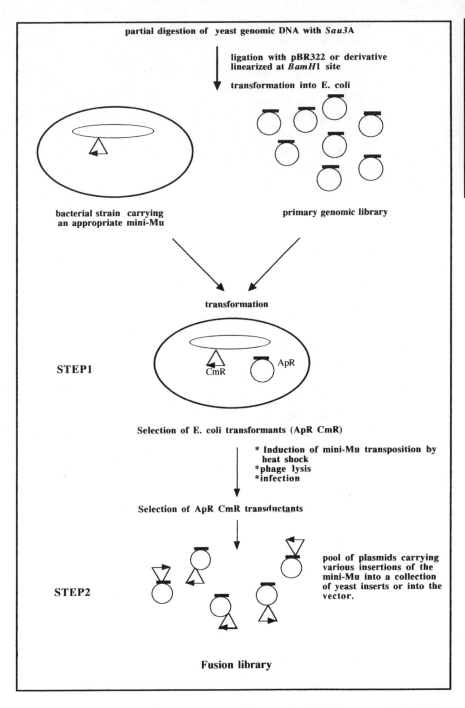

Figure 2. Construction of a transposition library. Small circles are pBR322 plasmids that contain yeast inserts (represented as a thick black line). The mini-Mu is schematized as a triangle, the arrow being the direction of transcription of the *lac* operon.

have estimated that 100 independent transposition events per plasmid should take care of all the known parameters and provide a reasonably representative library. The order of magnitude of standard yeast genomic libraries ranges usually from 8000 to 30 000 independent clones – depending on the average size of the yeast insert – while the transposition libraries should be *two orders of magnitude* larger than these.

A word of caution: libraries can drift to non-randomness very rapidly when propagated through sub-culturing. Consequently, it is recommended to store the original pool of cells in small aliquots at –80°C . In the present case, this is an absolute requirement because the population can evolve at each of the construction steps. To minimize the drift, we recommend one starts with a very high frequency of inserts or alternatively selects only those clones with an insert and separate the genomic library into pools of approximately 1000–2000 clones which could then be individually transposed.

◆◆◆◆◆◆ VII. UTILIZATION OF LIBRARIES

A. Modifications of existing libraries

Construction of libraries is a delicate task which can take time to complete. Transposition as described previously is flexible enough to permit the introduction of a variety of parameters into a basic vector and therefore makes possible and relatively easy the creation of a large variety of libraries depending on one's need. Our original library was constructed in plasmid pBR322. From this initial library it is possible to vary the genetic markers used for selection in yeast, the type of replication origin or the reporter gene because transposition is not the limiting step. Interestingly, with more than half of the insertions being in the plasmid, part of the population will correspond to a classical yeast library with additional markers. One can also introduce a replication origin from another organism, allowing heterologous transformation to take place. The initial library can also be separated into sub-populations such as individual chromosomal libraries. Using this strategy, a chromosomal library dedicated to chromosome III has recently been constructed in order to investigate gene expression of the different ORFs localized on this chromosome (M. Bolotin-Fukuhara, unpublished).

B. Use in Identifying Families of Co-regulated Genes

I. Importance of identifying families of co-regulated genes

The possibility of constructing an exhaustive catalog of yeast gene fusions has interesting applications to the study of global gene expression. Eukaryotic cells adapt to various environmental or genetic conditions by switching on/off a subset of genes usually via transcriptional regulation. This is achieved via the complex interplay of many transcription factors that repress or activate groups of genes. In most cases, however, regulatory factors are initially characterized as regulating one or a few genes with a well-defined biological function. The dependence of other genes upon these same factors is usually examined only if their function is in some way related to that of the known targets. Genes not related – in sequence at least – to the same function will be ignored.

The identification of families of co-regulated genes requires systematic, global and unbiased methods. The identification of regulatory networks will provide a wealth of information: (i) it will allow one to assess global or specific role of a transcription factor; (ii) new ORFs with no biological functions yet defined will be related to precise regulatory networks, allowing us to predict some possible function; and (iii) by identifying known genes as targets, the function of unknown transcription factors can, in turn, be elucidated.

At least 90 different transcription factors have been described in yeast (Svetlov and Cooper, 1995). According to predictive DNA-binding motifs, this number may increase to a few hundred in the total yeast genome.

Several methods – each of them with their own advantages or disadvantages – exist to identify target genes under control of a specific transcriptional regulator and to unravel these regulatory networks. Differential display reverse transcription polymerase chain reaction (dd-RT-PCR) (Bauer *et al.*, 1993) or serial analysis of gene expression (SAGE, Velculescu *et al.*, 1995) are used on mammalian genomes for which no genetic methods are available. However, these methods are not yet really operational for yeast and require considerable expertise for calibration and quantification of the data. Comparison of two-dimensional (2D) gel electrophoresis of yeast proteins in the presence or absence of a given transcription factor might also be used to reveal quantitative differences that will in turn be related to the corresponding genes. The development of protein maps is very promising, but this method may not be readily exploited by the average laboratory. Furthermore, only a portion of the yeast proteins (estimated to be about one-third of the total) are visible by 2D gel analysis of which only a small fraction have been associated to known genes (Boucherie *et al.*, 1995). Computer analysis of the *cis*-acting regulatory sequences upstream of expressed ORFs may be a guide to test *in vivo* regulation as described in Fondrat and Kalogeropoulos (1994). In many cases, these *cis*-acting sequences are not clearly defined and should be considered with caution.

Fusion libraries can be used as an alternative or complementary choice to search for target genes. Because the collection of yeast genes fused to the reporter will carry their own promoter, one can monitor *in vivo* the expression of all the yeast genes in a global way. These libraries are easy to manipulate and every yeast laboratory will be at ease with the basic techniques required to handle them.

It should be emphasized that all methods discussed above do not demonstrate a *direct* action of the factor upon the target genes; they simply try to classify those genes so identified into functional networks.

Method 1

1- Transform a ura3 leu2 yeast strain mutated for the regulatory gene (A) and containing the wild type copy of A on a URA3 centromeric plasmid with the fusion library

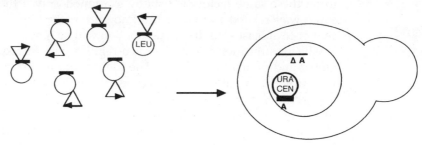

2- Select transformants on SM-URA-LEU and replica-plated on SM-URA-LEU + X-Gal

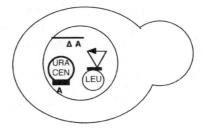

3- Patch blue colonies on SM-URA-LEU and replica-plate on
       SM-LEU   SM-LEU+5FOA

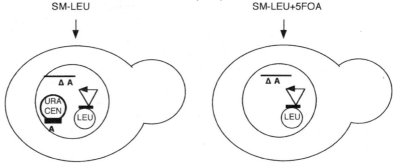

4-Replica-plate each plate (SM-LEU and SM-LEU+5FOA) on SM-LEU+X-Gal

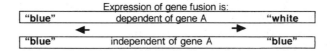

	Expression of gene fusion is:	
"blue"	dependent of gene A	"white
←		→
"blue"	independent of gene A	"blue"

Figure 3. Methodology for selecting regulated gene fusions: search for fusions dependent upon an *activator*. A, the wild-type copy of the activator; ΔA, the deleted copy introduced into the chromosome; URA, the *URA3* gene. The fusion library is schematized as circular plasmids with a triangle. The arrow on the triangle indicates the direction of transcription of the *lac* operon. LEU symbolizes the fact that the mini-Mu carries the *LEU2* gene. After transformation, the plasmids into which the transposon has been integrated confer, to a *leu2* yeast strain, the ability to grow on a minimal medium without leucine. SM (synthetic minimal) medium is described in Dang *et al.* (1994) as well as the concentrations used for X-Gal and 5-fluoroorotic acid (5FOA). For the latter, the appropriate concentration can vary according to strains and vectors and has to be checked before any experiments.

Method 2

1- Transform a *ura3 leu2* yeast strain mutated for the gene encoding the regulator to study (gene A) with the fusion library

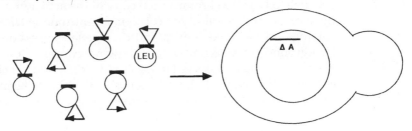

2- Select transformants on SM-LEU and replica-plate on SM-LEU + X-Gal to look for blue colonies

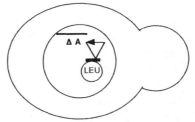

3- Spot blue colonies on SM-LEU medium and cross each colony with (i) a compatible* wild type strain and (ii) the isogenic strain mutated for A. After selection of diploids on appropriate minimal medium, replica-plate on the same medium + X-Gal

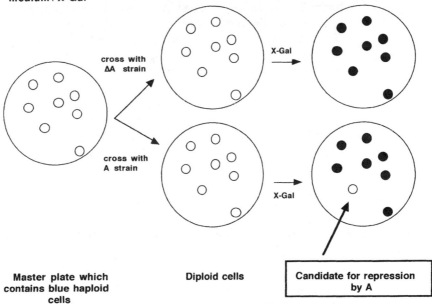

Master plate which contains blue haploid cells

Diploid cells

Candidate for repression by A

Figure 4. Methodology for selecting regulated gene fusions: search for genes under control of a *repressor*. Symbols for the genes used in the screen are the same as in Figure 3. Large circles drawn in part 3 represent Petri dishes while small circles symbolize colonies. These colonies can be either white (hollow circle) or blue (black circle).

2. Methodology to identify co-regulated gene families

Figure 3 illustrates a genetic screen developed to search for genes regulated by a specific activator called A. One transforms a pseudo wild-type cell, deleted for the chromosomal copy of the activator gene, but carrying the corresponding wild-type copy on a centromeric *URA3* plasmid. Blue colonies are selected, which indicate gene expression in a wild-type background. In a second step, following curing of the *URA3* plasmid, expression is again assessed in the cured strain. If the gene fusion is under control of the activator, the color should be white in the presence of X-Gal. This method is well adapted to screen for genes under control of an *activator*, but may not be well adapted to search for targets of a *repressor*. In the latter case, an alternative screen can be used (Figure 4). In this method, a mutant strain is transformed with the library and a collection of blue clones (i.e. expressed in the mutant background) are then crossed to isogenic strains containing the wild-type or mutant allele of the activator. Comparison of the color of these two strains on X-Gal (heterozygous or homozygous for the mutant transcription factor) should reveal clones containing regulated fusions. This second method is also useful when one wants to study two homologous transcription factors (A and B). Instead of crossing the blue colonies with two strains (isogenic A and ΔA strains), diploids from three different crosses should be compared (wild-type AB, AΔB and ΔAB). In this scheme, the distinction between "blue" colonies and "white" colonies is only theoretical. Experimentally, one observes slight changes in coloration that would allow one to screen for negatively controlled gene fusions even in method 1. The reciprocal is also true.

Whatever the method used, the initial screen will provide a set of *candidate* gene fusions for which there is no proof of a real regulatory effect. Screens on plates (or filters) may yield artifactual results depending on the concentration of cells, their growth rate (one has to be especially careful when comparing two strains with very different growth rates) and the uniformity of the medium itself. Further verification is very important and may take different forms. Plasmids carrying the fusions of interest can be extracted from the yeast cells and used to retransform isogenic wild type (A) and deletant (ΔA) strains. Successful candidates may then be monitored by *in vitro* β-galactosidase assay and/or Northern blots and the junction sequenced to identify the regulated gene.

3. An example of a regulatory network: the HAP complex

We have explored this method by searching for genes whose expression is regulated by the HAP transcription complex. The HAP complex is composed of at least four subunits; Hap2p, Hap3p (Hahn and Guarente, 1988) and Hap5p (McNabb *et al.*, 1995) are necessary for DNA binding to the *cis*-acting regulatory sequence, while the Hap4p subunit (Forsburg and Guarente, 1989) is required both for activation and regulation of the complex. The HAP complex is necessary for the expression of the *CYC1* gene on respiratory carbon sources and also acts upon several genes encoding elements of the respiratory chain or of the Krebs cycle (De Winde and

Grivell, 1993, and references therein). As discussed above, no study has been undertaken to see if HAP would regulate other cellular functions.

Because the HAP complex was described as an activator we decided to apply the method described in Figure 3. The initial strain carried a *HAP2* deletion which is sufficient to prevent fixation of the complex on DNA and consequently blocks the regulation. Because such a strain grows only on fermentable media, one can screen regulated gene fusions on galactose (to avoid glucose repression). Using a mini-Mu transposition library and an *in vitro* fusion library, one can obtain many HAP-regulated fusions (see Dang *et al.*, 1994 for details). These experiments led to the following:

(a) About 30 regulated fusions were confirmed, among which only two were obtained more than once. In addition, only one gene so isolated (*CYT1*) was already known to be regulated by HAP, thus confirming the validity of the approach. It appears therefore that the number of genes regulated by the HAP complex is large.

(b) New genes involved in respiratory metabolism were also identified. These included the *SDH3* gene (Daignan-Fornier *et al.*, 1994) which forms part of respiratory complex II and had not previously been identified in yeast. In this specific case, the ability to use the construct for gene disruption facilitated further analysis (see section II). However, the effect of HAP extends beyond the respiratory chain. For example, genes involved in mitochondrial biogenesis such as *RPM2* (the nuclear-encoded subunit of the mitochondrial RNase P) or *PTP1*, involved in mitochondrial transport, were also shown to be regulated by HAP (Dang *et al.*, 1994).

(c) Other cellular functions not directly related to mitochondrial function also appear to be under the control of HAP: for example, gluconeogenic genes, genes involved in hexose transport and phosphorylation, and genes from the glyoxysome and the Krebs cycle were identified (Dang, 1996; Dang *et al.*, unpublished). Unexpectedly, some genes involved in amino acid biosynthesis were also found to be positively regulated by the HAP complex. This is the case for *ASN1*, one of the genes encoding asparagine synthetase (Dang *et al.*, 1996a) and *GDH1*, encoding glutamate dehydrogenase (Dang *et al.*, 1996b). The effect on the latter indicated a cross-regulation between carbon and nitrogen metabolism because *GDH1* is an essential gene for nitrogen metabolism. For both *GDH1* and *ASN1*, regulation is independent of the Hap4p subunit. These data suggest a combinatorial association of the elements of the HAP complex. The DNA-binding part of the complex may associate alternatively with HAP4 or other putative activation components depending on which pathways they are involved in.

This set of results indicates that the HAP complex is involved in controlling a general switch in expression when the cell is transferred from glucose to alternative carbon sources. This effect on energy metabolism is consistent with the observation that some genes from the lipid biosynthetic pathway are negatively regulated by HAP (Dang, 1996; V-D. Dang *et al.*, unpublished observations).

(d) ORFs of unknown function were also uncovered by this screen. The fact that they are part of a large set of genes regulated by the HAP complex allows one to hypothesize a function for them in general energy metabolism.

Similar studies may also be undertaken with other transcription factors and should work *a priori* with any factor provided the gene that encodes it is not essential. In the latter case, conditionally expressed alleles should help to bypass some of the difficulties that can be foreseen.

◆◆◆◆◆◆ VIII. CONCLUSIONS AND PERSPECTIVES

The mini-Mu transposition system described above appears to be a reliable, efficient and rapid way to analyse yeast gene sequences cloned on a plasmid. Because this transposon can be easily manipulated and reintegrated into the *E. coli* chromosome, different types of construction can be achieved by *in vivo* transposition in place of *in vitro* DNA manipulation. The ease of its practical applications is an important advantage in large-scale genomic studies. In particular, the possibility of identifying regulatory networks in yeast by construction of translational fusion libraries is attractive. With this method, a large range of genes controlled by a specific transcription factor can be uncovered and interconnections between networks revealed. Understanding the function and hierarchy of all the transcription factors that are present in this simple eukaryotic cell is going to require significant efforts. Further studies will certainly make use of new sophisticated techniques, probably more exhaustive than the genetic screen presented here. Nevertheless this method of transposon tagging, particularly suited for a global approach and accessible to any laboratory, should be useful for functional analysis of yeast genes.

Acknowledgements

The author would like to thank members of the laboratory who have contributed part of the work presented here. In particular, the contribution of B. Daignan-Fornier was essential to the development of this project, as was the work of V-D. Dang. Support from GREG (65/94) and the EEC (contract BIO2 CT93.00.22 and BIO4 CT 95.00.80) are gratefully acknowledged.

References

Bauer, B., Müller, H., Reich, J., Riedel, H., Ahrenkiel, V., Warthoe, P. and Strauss, M. (1993). Identification of differentially expressed mRNA species by an improved display technique (DDRT-PCR). *Nucleic Acids Res.* **21**, 4272–4280.

Boucherie, H., Dujardin, G., Kermorgant, M., Monribot, C., Slonimski, P. and Perrot, M. (1995). Two-dimensional protein map of *Saccharomyces cerevisiae*: construction of a gene-protein index. *Yeast* **11**, 601–613.

Boulanger, F., Berkaloff, A. and Richaud, F. (1986). Identification of hairy root loci in the T-regions of *Agrobacterium rhizogenes* R1 plasmids. *Plant Molec. Biol.* **6**, 271–279.

Bukhari, A. L. and Taylor, A. L. (1975). Influence of insertions on packaging of host sequences covalently linked to bacteriophage Mu DNA. *Proc. Natl Acad. Sci. USA* **72**, 4399–4403.

Burns, A., Grimwade, B., Ross-MacDonald, P. B., Choi, E-Y., Finberg, K., Roeder, G. S. and Snyder, M. (1994). Large-scale analysis of gene expression, protein localization, and gene disruption in *Saccharomyces cerevisiae*. *Genes Dev.* **8**, 1087–1105.

Carlioz, A. and Touati, D. (1986). Isolation of superoxide dismutase mutants in *E. coli:* is superoxide dismutase necessary for anaerobic life? *EMBO J.* **5**, 623–630.

Casadaban, M. J. and Chou, J. (1984). *In vivo* formation of gene fusions encoding hybrid beta-galactosidase proteins in one step with a transposable Mu-*lac* transducing phage. *Proc. Natl Acad. Sci. USA* **81**, 535–539.

Casadaban, M. J., Chou, J. and Cohen, S. N. (1980). *In vitro* gene fusions that join an enzymatically active beta-galactosidase segment to amino-terminal fragments of exogenous proteins: *E. coli* plasmid vectors for the detection and cloning of translational initiation signals. *J. Bacteriol.* **143**, 971–980.

Castilho, B. A. Olfson, P. and Casadaban, M. J. (1984). Plasmid insertion mutagenesis and *lac* gene fusions with mini-Mu bacteriophage transposons. *J. Bacteriol.* **158**, 488–495.

Csonska, L. N., Howe, M. M., Ingraham, J. L., Pierson III, L. S. and Turnbough, C. L. (1981). Infection in *Salmonella typhimurium* with coliphage MudI (*ApR lac*): construction of *pyr::lac* gene fusions. *J. Bacteriol.* **145**, 299–305.

Daignan-Fornier, B. and Bolotin-Fukuhara, M. (1988). *In vivo* functional characterization of a yeast nucleotide sequence: construction of a mini-Mu derivative adapted to yeast. *Gene* **62**, 45–54.

Daignan-Fornier, B. and Bolotin-Fukuhara, M. (1989). Functional exploration of the yeast (*Saccharomyces cerevisiae*) genome: use of a mini-Mu transposon to analyze randomly cloned sequences. *Yeast* **5**, 259–269.

Daignan-Fornier, B., Valens, M., Lemire, B. and Bolotin-Fukuhara, M. (1994). Structure and regulation of *SDH3*, the yeast gene encoding cytochrome b560 subunit of respiratory complex II. *J. Biol. Chem.* **269**, 15469–15472.

Dang, V-D. (1996) Etude du role physiologique du facteur de transcription HAP2/3/4/5 se fixant au motif CCAAT chez la levure *Saccharomyces cerevisiae*. Thèse de l'Université Paris-Sud.

Dang, V-D., Valens, M., Bolotin-Fukuhara, M. and Daignan-Fornier, B. (1994). A genetic screen to isolate genes regulated by the yeast CCAAT-box binding protein Hap2p. *Yeast* **10**, 1273–1283.

Dang, V-D., Valens, M., Bolotin-Fukuhara, M. and Daignan-Fornier, B. (1996a). Cloning of the *ASN1* and *ASN2* genes encoding asparagine synthetases in *Saccharomyces cerevisiae:* differential regulation by the CCAAT-box binding factor. *Mol. Microbiol.* **22**, 681–692.

Dang, V-D., Bohn, C., Bolotin-Fukuhara, M. and Daignan-Fornier, B. (1996b). The CCAAT-box binding factor stimulates ammonium assimilation in *Saccharomyces cerevisiae*, defining a new cross-pathway regulation between nitrogen and carbon metabolism. *J. Bacteriol.* **178**, 1842–1849.

De Winde, J. H. and Grivell, L. A. (1993). Global regulation of mitochondrial biogenesis in *Saccharomyces cerevisiae*. *Prog. Nucleic Acid Res. Mol. Biol.* **46**, 51–91.

Dixon, R., Eady, R. R., Espin, G., Hill, S., Iaccarino, M., Kahn, D. and Merrick, M. (1980). Analysis of regulation of *Klebsiella pneumoniae* nitrogen fixation (*nif*) gene cluster with gene fusions. *Nature* **280**, 128–132.

Fondrat, C. and Kalogeropoulos, A. (1994). Approaching the function of new genes by detection of their potential upstream activation sequences in *Saccharomyces cerevisiae*: application to chromosome III. *Curr. Genet.* **25**, 396–406.

Forsburg, S. and Guarente, L. (1989). Identification and characterization of HAP4: a third component of the CCAAT-bound HAP2/HAP3 heteromer. *Genes Dev.* **3**, 1166–1178.

Groisman, E. A., Castilho, B. A. and Casadaban, M. J. (1984). *In vivo* DNA cloning and adjacent gene fusing with the mini-Mu *lac* bacteriophage containing a plasmid replicon. *Proc. Natl Acad. Sci. USA* **81**, 1480–1483.

Guarente, L. and Ptashne, M. (1981). Fusion of the *E. coli lacZ* to the cytochrome c gene of *Saccharomyces cerevisiae. Proc. Natl Acad. Sci. USA* **78**, 2199–2203.

Hahn, S. and Guarente, L. (1988). Yeast HAP2 and HAP3: transcriptional activators in a heteromeric complex. *Science* **240**, 317–321.

Huisman, O., Raymond, W., Froehlich, K. U., Errada, P., Kleckner, N., Botstein, D. and Hoyt, A. (1987). A *Tn10-lacZ-kanR-URA3* gene fusion transposon for insertion mutagenesis and fusion analysis of yeast and bacterial genes. *Genetics* **116**, 163–168.

Ji, H., Moore, D. P., Blomberg, M. A., Braiterman, L. T., Voytas, D. F., Natsoulis, G. and Boeke, J. D. (1993). Hotspots for unselected Ty1 transposition events on yeast chromosome III are near tRNA genes and LTR sequences. *Cell* **73**, 1007–1018.

Lee, J. H., Heffernan, L. and Wilcox, G. (1980). Isolation of *ara-lac* gene fusions in *Salmonella typhimurium* LT2 by using transducing bacteriophage MudI (Ap^R *lac*). *J. Bacteriol.* **143**, 1325–1331.

McNabb, D. S., Xing, Y. and Guarente, L. (1995). Cloning of yeast HAP5: a novel subunit of heteromeric complex required for CCAAT binding. *Genes Dev.* **9**, 47–58.

Mizuuchi, K. and Craigie, R. (1986). Mechanism of bacteriophage Mu transposition. *Ann. Rev. Genet.* **20**, 385–429.

Richaud, C., Higgins, W., Mengin-Lecreux, D. and Stragier, P. (1987). Molecular cloning, characterization and chromosomal localization of *dapF*, the *E. coli* gene for diaminopimelate epimerase. *J. Bacteriol.* **169**, 1454–1459.

Rose, M., Casadaban, M. J. and Botstein, D. (1981). Yeast genes fused to beta-galactosidase in *E. coli* can be expressed normally in yeast. *Proc. Natl Acad. Sci. USA* **78**, 2460–2464.

Seifert, H. S., Chen, E. Y., So, M. and Heffron, F. (1986). Shuttle mutagenesis: a method of transposon mutagenesis for *Saccharomyces cerevisiae. Proc. Natl Acad. Sci USA* **83**, 735–739.

Svetlov, W. V. and Cooper, T. (1995). Compilation and characteristics of dedicated transcription factors in *Saccharomyces cerevisiae. Yeast* **11**, 1439–1484.

Velculescu, V. E., Zhang, L., Vogelstein, B. and Kintzler, K. W. (1995). Serial analysis of gene expression. *Science* **270**, 484–487.

12 Immunological Approaches to the Study of Protein Localization in Yeast

Iain M. Hagan and Colin J. Stirling
School of Biological Sciences, University of Manchester, Manchester, UK

◆◆

CONTENTS

Introduction
Antibodies and antigens
Immunodetection techniques
Immunoprecipitation
Indirect immunofluorescence

List of Abbreviations

GFP	Green fluorescent protein
IMAC	Immobilized metal affinity chromatography
ORF	Open reading frame
SPB	Spindle pole body

◆◆◆◆◆◆ I. INTRODUCTION

The organization of the eukaryotic cell is immensely complex with numerous compartments and microenvironments comprising distinct subsets of proteins involved in specialized functions. In the yeast *Saccharomyces cerevisiae*, several membrane-bound organelles have been characterized (Zinser and Daum, 1995) but within each still further layers of structural complexity can exist. For example, some proteins are sequestered to topologically distinct regions of the cell such as the bud neck, apical tip, or nucleolus (Debeus *et al.*, 1994; Chant *et al.*, 1995; TerBush and Novick, 1995), whilst others function in association with multimeric protein complexes (such as the vacuolar proton ATPase; Graham *et al.*, 1995) or supramolecular structures such as the mitotic spindle (Roof *et al.*, 1992). Thus, in order to investigate fully the role of an individual protein it is necessary both to identify its subcellular localization and to identify any biochemical interactions with other factors that may be intrinsic to its function. Antibodies raised against a specific protein can provide immensely powerful research tools for such studies. Our aim in this chapter is to introduce some of the theory behind the most commonly used immunological techniques with specific reference to their applications in yeast gene analysis. Whilst some methods will be provided, our principal aim is to encourage the reader to adapt and modify published methods. The review will not cover the more specialized techniques involved in immunoelectronmicroscopy for which the interested reader should refer to recent reviews (Dubochet, 1995; Binder *et al.*, 1996; McDonald *et al.*, 1996).

◆◆◆◆◆◆ II. ANTIBODIES AND ANTIGENS

A. Sources of Antibodies

The key to success in any immunological approach is the availability of suitable reagent antibodies. For those interested in well-characterized proteins they may be fortunate enough to find a commercial source of a suitable antibody (e.g. antibodies against vacuolar CPY, or mitochondrial porin are available from Molecular Probes, Eugene, Oregon). However, most commercially available reagents are directed against proteins from other species, yet even these can prove useful in yeast when the target protein is highly conserved, e.g. anti-β-tubulin (Chemicon, Temecula, CA) or MAb414 (which cross-reacts with two yeast nuclear pore proteins; BAbCO, Richmond, CA). In other cases, the generosity of colleagues will often provide a characterized antiserum or hybridoma cell line.

B. Raising Reagent Antibodies: Source of Antigen

For the practicalities of raising antibodies and the theoretical background to the immune response the reader is referred to the excellent volume by Harlow and Lane (1988) for details. However, an overview of some of the most important points is relevant here with particular reference to raising antibodies against yeast proteins. The source of antigen is the first and most vital consideration. Using a protein purified from yeast cells might offer the best prospects of raising antibodies that recognize the protein's native conformation. However, this approach can be labor intensive and in many cases is neither practicable nor necessary. Using yeast-derived material also raises the possibility that even low levels of immunogenic contaminant proteins might result in unacceptably high levels of unwanted reactivity in the final antiserum, particularly if the contaminant contains an immunodominant epitope. It is also important to consider the nature of any covalent modifications of the target protein; for example, immunization with a glycoprotein would be likely to raise antibodies specific to sugar moieties which would react with many other glycoproteins in the cell. Such an antiserum would be of little value in the localization of an individual protein. However, it should be noted that antibodies specific for certain covalent modifications (e.g. sugar linkages or phosphotyrosine) have proven invaluable in a variety of functional studies (for examples see Baker *et al.*, 1988; Dean and Pelham, 1990; Vandre *et al.*, 1984).

The immunogen need not necessarily be a single antigen. For example, by raising a panel of monoclonal antibodies against an enriched spindle pole body fraction, Rout and Kilmartin (1990) identified four protein components of the yeast spindle. Similarly, Strambio-de-Castilla *et al.* (1995) raised an antibody against a nuclear envelope fraction that was found to be specific for a component of the nuclear pore complex.

Recombinant DNA technology offers a very practical and efficient route to the purification of antigens for immunization purposes. It is particularly relevant for the yeast geneticist where a gene or open reading frame (ORF) has been identified, but where little may be known about the biochemical properties of its protein product. The construction of gene fusions designed for overexpression of fusion proteins in *Escherichia coli* has been widely used in recent years with various strategies coming into, and out of, fashion (Lavallie and McCoy, 1995). The use of a short oligohistidine tag is currently very much in vogue, with fusion proteins expressed to very high levels (using the T7 promoter), then purified by immobilized metal affinity chromatography (IMAC; see Studier *et al.*, 1990; Crowe *et al.*, 1994). The level of overexpression obtained often results in the bulk of such a fusion protein being found in inclusion bodies that require highly chaotropic reagents to achieve solubilization (e.g. 8 M urea or 6 M guanidine hydrochloride). However, because chromatography steps can be performed under these conditions, the stringency of the metal affinity purification is improved. Even under these conditions we find that IMAC-purified material still contains low levels of contaminating *E. coli* proteins, but these have not proven problematic when raising antibodies against yeast proteins in either sheep or rabbits. If necessary, such contaminants can be eliminated by further purification of the fusion protein by preparative SDS-PAGE. In some cases the overexpression of a fusion protein may be toxic in *E. coli* and it may be necessary to use an expression system that is tightly regulated to prevent constitutive expression (for a review see Studier *et al.*, 1990), but even then the toxicity associated with induction may result in relatively poor yields. It is particularly worth noting that the overexpression of integral membrane proteins in *E. coli* is notoriously problematic and so it is advisable, where possible, to express a fusion carrying only a soluble domain from the membrane protein of interest (Wright *et al.*, 1988), or to raise antibodies against synthetic peptides (Stirling *et al.*, 1992).

C. Synthetic Peptides as Antigens

Synthetic peptides offer two distinct advantages as antigens: they are simple to produce and the antibodies they elicit will be directed against a

Immunological Approaches

203

defined portion of the target protein. This latter feature can be useful for structural/functional analysis, such as the determination of membrane protein topology (Serrano *et al.*, 1993; Kuroda *et al.*, 1996), but also has great potential when studying specific sequence motifs or protein families. For example, Gould *et al.* (1990) raised antibodies against the peroxisomal targeting motif, Ser-Lys-Leu, which in turn cross-react with several peroxisomal proteins. Conversely, when studying a highly conserved protein family it may be possible to identify specific peptide sequences that are unique to one family member which might then direct a monospecific antibody response; for example, Santa Cruz Biotechnology (Santa Cruz, CA) offer several products that discriminate between individual Hsp70s.

Oligopeptides of 9–15 residues can be extremely immunogenic when conjugated in large molar excess to a carrier protein such as keyhole limpet haemocyanin (see Stirling *et al.*, 1992). However, it is not uncommon to raise antibodies that recognize free peptide but which do not react with the intact protein from which the peptide sequence is derived; for example, of five anti-peptide antibodies raised against sequences from Sec61p only one recognizes the full length protein in immunoblots (C. J. Stirling and R. Schekman, unpublished data). This presumably reflects the topological constraints imposed upon the peptide within the intact protein. There are no hard and fast rules governing the choice of a peptide sequence as a potential immunogen but a few simple guidelines have served us well in recent studies. Firstly, the longer the peptide the better. Secondly, peptides corresponding to either the N- or C-terminus of the mature protein will often cross-react with the intact molecule, presumably because the peptide structure more closely mimics that of the free terminal. Thirdly, where antibodies are required to interact with the fully folded protein then one should choose a hydrophilic sequence that would be predicted to be on the surface of a globular protein. Finally, it would seem prudent to search the yeast protein database to ensure that your chosen peptide sequence is unique, but note that this is no guarantee against the presence of structurally related epitopes on other proteins. We strongly recommend that synthetic peptides are purified by HPLC to eliminate potentially immunodominant contaminants. The technicalities surrounding the use of synthetic peptides as antigens is covered in detail by Harlow and Lane (1988), but it may be of interest to note that some companies now offer a combined peptide synthesis and antibody production service at very reasonable cost (e.g. Research Genetics Inc., Huntsville, Alabama).

D. Choice of host system

I. Monoclonal antibodies

Monoclonal antibodies can be raised in either rats or mice and offer several distinct advantages, the most obvious being that they are frequently very highly specific (usually recognizing a single epitope on the target protein), and are available in essentially limitless supply from cultured hybridoma cell lines. Rats have the advantage that the initial pre-bleeds are larger and can be used as polyclonals in their own right, even if the fusion fails – on the down side they are larger and more expensive to maintain. However, raising a panel of monoclonal antibodies is both time-consuming and expensive and should not be undertaken without expert advice. Moreover, it must be stressed that the key to success in raising monoclonals is the existence of a practical, high-throughput screen with which to identify clones of interest rapidly. All too often this approach runs into problems when the intricacies of the screen have not been carefully considered in advance.

2. Polyclonal antibodies

Raising a polyclonal antiserum is technically simpler and, for this reason alone, is often the preferred route. There are a range of alternative hosts available for antibody production (see Harlow and Lane, 1988). Where the antigen is limiting, or where secondary antibody specificity is a consideration, then rat, mouse, or guinea pig are very acceptable. However, rabbits are a cheap and reliable system offering yields of serum that are more commensurate with longer-term studies. Polyclonal antisera can also be made in sheep, where the yield of antiserum exceeds even the most demanding of applications. Whatever the host, it is imperative to screen a test serum (e.g. by western blotting or immunofluorescence) prior to immunization in order to exclude any animals with high background reactivity to yeast proteins. In our experience some 20–50% of both sheep and rabbits must be excluded at this stage. Ideally, at least two animals should be immunized to mitigate against unpredictable variations in the quality and titer of the humoral response. In general the immunization regimes using Freund's complete/incomplete adjuvant yield satisfactory responses following two to four boosts after the initial immunization, but recent reports suggest that alternative adjuvants such as Hunter's Titermax (Sigma-Aldrich, cat. no. H4397) can induce a strong IgG response following a single immunization (Bennett *et al.*, 1992).

E. Antibody Specificity

Most polyclonal antisera will recognize a spectrum of yeast proteins in addition to the specific antigen of interest. The simplest way to examine this spectrum is by western blotting (Protocol 1). By comparing the patterns of staining observed with immune versus pre-immune sera, one can identify the band(s) that have arisen during the immunization regime. However, not all such bands necessarily correspond to the desired antigen. Whilst some may be due to cross-reactivity as a consequence of shared epitopes, others may be a result of antibodies specific to contaminants in the original immunogen, or may even have arisen through the accidental exposure of the animal to unrelated environmental immunogens. One must therefore exercise caution when using pre-immune serum as a control for antigen specificity. In order to identify unequivocally the native form(s) of an antigen it may be useful to compare an extract from wild-type cells with one from cells overexpressing the protein of interest and also, where possible, from null mutant cells. A second source of background that is often overlooked is the secondary serum used to decorate the primary antibody. The problem varies from batch to batch but stems from the fact that all animals are under constant attack from fungal infection and thus produce a huge range of anti-fungal protein antibodies. Even the sera that are marketed as having been affinity-purified on pure immunoglobulin matrices do not necessarily avoid this problem as the pure immunoglobulins, by definition, offer all of the epitopes imaginable (see below). The problem can be circumvented by preincubating secondary antibody with yeast protein extract for 30 min prior to use.

Protocol I. Quantitative western blotting

1. Load whole-cell extracts corresponding to 0.25–1 OD_{600} equivalent of cells. After SDS-PAGE, remove gel and lay on top of two suitably sized pieces of Whatman 3 mm paper (pre-wetted in transfer buffer). Then lay a sheet of prewetted nitrocellulose membrane (Schleicher & Schuell 0.45 mm; BA85, ref. no. 401196, or Amersham HyBond C) over the gel, being careful not to introduce any air bubbles. Now layer three sheets of prewetted Whatman paper over the nitrocellulose membrane. Using a clean pipette (or similar), roll gently over the assembled sandwich to remove air bubbles. Carefully close sandwich and note which side is which. Quickly transfer sandwich into a suitable blotting apparatus and transfer according to manufacturer's instructions (e.g. for the Hoeffer TE22 either 1 h at 300 mA with vigorous stirring and water-cooling or transfer overnight at 100 mA, again with stirring/cooling).

2. After transfer, carefully remove the nitrocellulose membrane and place protein-side up in a clean staining tray. Rinse off excess polyacrylamide with water. Cover with diluted Ponceau S stain (Sigma-Aldrich, cat. no. P7767) and gently agitate for 5–10 min. Pour off stain (which may be reused several times) and destain membrane with H_2O until banding pattern is visible.

3. Examine staining pattern and note any evidence of uneven transfer (usually caused by trapped air bubbles). Cut-off top left-hand corner of blot (to define orientation). ^{14}C-labeled molecular weight standards markers are preferred, but if using unlabeled molecular weight standards then mark their positions on the blot in pencil.

4. Incubate the membrane in blocking solution (20 min to overnight) containing 1% (w/v) fat-free dried milk reconstituted in TBS-NP. Change the blocking solution at least once. Blocking the solution should remove all traces of the Ponceau S stain from the membrane.

5. Incubate with the primary antibody diluted in 1% (w/v) milk/TBS-NP with gentle agitation. Optimum antibody dilution and incubation time should be determined empirically. A 1 h incubation is usually sufficient while longer incubation times (e.g. overnight) may increase the signal but may result in higher backgrounds.

6. Rinse the blot twice in 1% (w/v) milk/TBS-NP, then wash three times (5 min) with agitation. Finally rinse the blot once more.

7. Incubate with 5 µCi ^{35}S-labeled donkey anti-rabbit IgG (Amersham; SJ434) diluted in 10 ml 1% milk/TBS-NP. This diluted antibody solution can usually be reused at least 10 times without significant loss of signal if stored at 4°C with sodium azide (0.02% w/v).

8. Wash filter as in step 7, including two 5 min washes in TBS-NP.

9. Expose the filter to a storage-phosphor/"phosphorimaging" detection system (usually 2 h to overnight), or to conventional pre-flashed X-ray film overnight (or as appropriate).

10. In order to draw valid comparisons between samples, the amount of protein loaded must be carefully controlled. In addition, when analysing variations in any given antigen it is highly desirable to measure levels of some internal control (e.g. α-tubulin). Particular care must also be taken when comparing quantitative data between experiments, especially when probed with different batches of secondary antibody.

Solutions

Quantities shown are per liter of solution

Transfer buffer: Tris-base 2.4 g, glycine 11.25 g, methanol 200 ml, H_2O 800 ml
10×TBS: NaCl 80 g, KCl 2 g, Tris-base 24.2 g. Add 800 ml H_2O, then pH to 7.6, make up to 1000 ml
TBS-NP: 10×TBS 100 ml, H_2O 899 ml, NP40 1 ml. Stir thoroughly to get NP40 into solution
1% milk TBS-NP: As TBS-NP, but add 10 g non-fat dried milk (e.g. Marvel dried skimmed milk "over 99% fat free"). Stir thoroughly before use.

F. Affinity Purification

When an antiserum is to be used for immunolocalization studies, it is essential that it is specific for the antigen(s) of interest with no extraneous reactivities. In most cases this demands affinity purification where the antiserum is incubated with immobilized antigen, then the specifically bound immunoglobulins eluted and concentrated (see Harlow and Lane, 1988, for specific methodologies). The purity of the immobilized antigen is vital to the success of this approach. However, in cases where epitopes are shared then even affinity-purified antibodies will not eliminate cross-reactivity. If such cross-reactivity appears fortuitous (i.e. there is no close homolog of the antigen of interest encoded in the genome), then it would be probable that only a subset of antigen-specific polyclonal antibodies would be responsible for the cross-reactivity. Under these circumstances it might prove fruitful to construct fusion proteins comprising only portions of the protein of interest. Affinity purification using these truncated fusion proteins might then yield a monospecific antibody population. Alternatively, such fusions could be used for immunization purposes to avoid an immunodominant epitope that may be responsible for the cross-reactivity. When studying non-essential proteins, the preadsorption of primary antibody against null mutant cells can also greatly improve signal specificity (Roberts *et al.*, 1991).

G. Epitope Tags

Another approach is to construct a gene fusion that incorporates an amino acid sequence containing an epitope against which there is a well-characterized monoclonal antibody such as HA (Roof *et al.*, 1992) or cMyc (Craven *et al.*, 1996). Not only does this avoid the need to raise a specific antibody, but the usefulness of the relevant monoclonals is already proven. The major disadvantage of using any fusion protein for localization studies is the possibility that the chimera may not report faithfully the distribution of the wild-type protein (Vallen *et al.*, 1992; Biggins and Rose, 1994; Spang *et al.*, 1995). When studying an essential gene, one can at least test for the ability of the fusion to complement a null mutation, thus proving that at least a proportion of fusion protein must be correctly localized. Despite these caveats Snyder and co-workers have elegantly exploited this approach both in the analysis of specific proteins (Page *et al.*, 1994) and as a first step in the functional analysis of large numbers of uncharacterized genes (Ross-Macdonald *et al.*, 1995; see Chapter 10).

◆◆◆◆◆◆ III. IMMUNODETECTION TECHNIQUES

Western blotting of whole-cell extracts is a simple and reliable method for studying the steady-state levels of a particular protein and can be used to monitor changes in protein levels under a variety of conditions (e.g. throughout the cell cycle, during sporulation, under stress, or changes in growth conditions). Whole-cell extracts intended for analysis by SDS-PAGE are best prepared under denaturing conditions (e.g. directly in an SDS-based sample buffer) in order to minimize *in vitro* proteolysis. This routinely involves denaturation by heating samples to 95°C, but one should note that many integral membrane proteins of yeast, such as Hmg1p or Sec61p, will aggregate under these conditions preventing them from even entering a polyacrylamide gel (Stirling *et al.*, 1992). Given that some 10% of yeast genes are predicted to encode membrane proteins (see the *yeast protein database* at http://quest7.proteome.com), then their exclusion from gel analysis represents a major limitation. In our experience many such proteins remain soluble if samples are heated to only 37–50°C for 10 min. There may be other reasons why some proteins might not be detected in gel analysis but which might be observed by *in situ* analysis of fixed cells: for example, extreme sensitivity to proteolysis, heterogeneous gel mobility, or antigen preservation. Such considerations emphasize that the antigen(s) detected in one technique need not necessarily reflect the full spectrum of those observed by another method.

The problems associated with *in vitro* proteolysis during yeast cell breakage under non-denaturing conditions can be minimized by using protease-deficient strains and/or a combination of protease inhibitors (for a review see Jones, 1991). A protease inhibitor cocktail tailored for use in yeast cell extracts is now available from Sigma-Aldrich (St Louis, Missouri; cat. no. P8215).

When coupled to cellular fractionation techniques, western blotting can provide a powerful means with which to define biochemically the localization of a protein within the cell. The principles and practices of cellular fractionation in *S. cerevisiae* have been comprehensively addressed in a recent review by Zinser and Daum (1995). In most cases it is impossible to purify any specific organelle to homogeneity, rather one uses the principle of co-localization with well-characterized organelle markers (Zinser and Daum, 1995) to determine the most probable localization of a novel protein. Solubilization studies using chaotropic salts or detergents may then further refine the localization by distinguishing soluble proteins from either peripheral or integral membrane proteins (Stirling *et al.*, 1992). In this respect it is worth noting that whilst carbonate extraction is often considered to be evidence that a protein is peripherally membrane-associated, there are examples where such treatments may also solubilize integral membrane proteins (Esnault *et al.*, 1994).

The demands for precision in assaying co-localization of two proteins requires accurate quantitation. Quantitative western blotting requires saturating antibody concentrations (determined empirically), but is highly dependent upon the detection method used. It is possible to digitize and "quantify" the signal from any detection method but one must be cautious when correlating this to the actual levels of antigen. This is particularly the case when using enzyme-coupled detection methods where the kinetics of the reaction(s) must be carefully controlled. Some reactions are actually product-inhibited and so are unsuited to quantitation. In our experience the most reliable, generally applicable method for quantitative western blotting is to use a radiolabeled secondary antibody in combination with a storage-phosphor detection system (e.g. the Fujix BAS2000). A range of ^{35}S- or ^{125}I-labeled secondary antibodies are available from a variety of sources (e.g. ^{35}S-labeled donkey anti-rabbit IgG from Amersham International (Slough, England).

◆◆◆◆◆◆ IV. IMMUNOPRECIPITATION

Many proteins function as components of larger oligomeric or multimeric protein complexes. Localizing a protein to such a complex, and the identification of its interacting partner(s), may prove invaluable in determining the biochemical role played not only by any given protein subunit, but also that of the complex itself. The biochemical co-fractionation of a novel protein with a known complex again provides evidence for an interaction (Graham *et al.*, 1995). Moreover, the specificity of antibodies can provide for the very simple purification of a protein from a cell extract and the co-immunoprecipitation of associated proteins can provide compelling evidence for the existence of a complex. There are many examples of co-immunoprecipitation studies identifying complexes involved in a range of processes including transcription (Eisenmann *et al.*, 1992; Swanson and Winston, 1992), mRNA processing (Brown *et al.*, 1996), translation (Deshmukh *et al.*, 1993), protein targeting (Hann *et al.*, 1992), secretory vesicle traffic (Sweet and Pelham, 1993; TerBush and Novick, 1995), and in the spindle pole (Geissler *et al.*, 1996) amongst others. Immunoprecipitation from a radiolabeled cell extract is ideal where potential partner proteins are not known (Deshaies *et al.*, 1991), but where antibodies are available against candidate partners then immunoprecipitates from unlabeled extracts can be analyzed by Western blotting for the presence of co-precipitated antigens (Hann *et al.*, 1992; TerBush and Novick, 1995). No single cell lysis buffer is likely to preserve all protein–protein interactions and so some informed manipulation of conditions may be required (see the cited literature for examples).

The key to any successful immunoprecipitation revolves around the immunochemical qualities of the antibodies. Because antibody binding occurs in free solution, then antibody *affinity* for antigen is important as it determines the proportion of bound antigen at equilibrium in a concentration-dependent manner. Equilibrium is rapidly reached and so incubations with high affinity antibodies can be very short (e.g. 1 h) thus greatly improving specificity. Longer incubations with even low-affinity polyclonal antibodies can give rise to high *avidity* immune complexes in which antigen can be efficiently recovered (although non-specific precipitation can increase). In contrast, a low-affinity monoclonal IgG can form only bivalent interactions (unless the epitope is repeated within the antigen), and so may be of little value in immunoprecipitation. The very high local concentration of antigen immobilized on western blots can promote low-affinity interactions perhaps explaining why some antibodies perform well in blots but not in immunoprecipitations.

Immune complexes are generally harvested using protein A immobilized on Sepharose beads or using formalin-fixed *Staphylococcus aureus* cells. Protein A has a high affinity for rabbit IgG but a lower affinity for sheep IgG for which protein G is recommended (Harlow and Lane, 1988). However, in our experience, immunoprecipitation of antigens using sheep polyclonals is very efficient even with protein A Sepharose, presumably because of the multivalent immune complexes having sufficient avidity for protein A. In addition, certain IgG classes from both rat and mouse have only a low affinity for protein A. Consequently, some monoclonals require decoration with a secondary antibody in order to achieve efficient recovery.

Clearly, one must guard against spurious cross-reactivity when seeking to identify proteins that co-precipitate with your antigen. Where "co-precipitated" bands are also detected in immunoprecipitations from denatured cell extracts then they can be considered to be due to cross-reactivity rather than a protein complex. Alternatively, performing immunoprecipitation after some conventional biochemical fractionation step (e.g. sucrose gradient fractionation) may help to resolve an authentic complex from major cross-reacting antigens. Where possible, null mutant extracts lacking the antigen of interest provide an excellent control. Such extracts can also be engineered for essential proteins using shut-off alleles (e.g. *GAL1*) or perhaps other conditional mutants (see Chapter 6).

Some antibodies might not recognize the native antigen but may be successful in immunoprecipitation of SDS-denatured antigen for example. These can be invaluable in analysing protein biogenesis and modifications and can be used to identify protein complexes by co-immunoprecipitation after chemical crosslinking of whole-cell extracts (Deshaies *et al.*, 1991).

Antibodies that react with native proteins have many other uses. For example, Milne *et al.* (1996) used antibodies in gel mobility supershift experiments to demonstrate direct roles for the Ku80 and Hdf1 proteins in DNA binding. In addition, when available in sufficient quantities such antibodies can be used for large-scale immunoaffinity purification (Brown *et al.*, 1994). Conversely, the use of an immunodepleted cell extract may help to determine the biochemical role played by a specific protein (or complex) in reconstitution studies (Milne *et al.*, 1996; Lewis *et al.*, 1996; Cockell *et al.*, 1995).

◆◆◆◆◆◆ V. INDIRECT IMMUNOFLUORESCENCE

Immunofluorescence offers the immense advantage of localizing native cellular antigens *in situ*. This not only permits localization of proteins to specific compartments but can also reveal further levels of complexity in the organization of the cell which might not be evident upon cell lysis such as the apical tip (TerBush and Novick, 1995) or bud neck (Chant *et al.*, 1995). Immunofluorescence studies have also been instrumental in revealing the temporal redistribution of some proteins in phase with the cell cycle (Nasmyth *et al.*, 1990; Hennessy *et al.*, 1990; Chuang and Schekman, 1996). Comparing specific localization patterns in a range of mutant backgrounds can also shed light on protein function and can serve to elucidate intricate regulatory networks (Hagan and Hyams, 1988, Hagan and Yanagida, 1997). Finally, recent advances in video-enhanced microscopy, coupled with sophisticated image analysis software, have also increased the resolution and sensitivity of immunofluorescence analysis, especially in the imaging of complex three-dimensional structures and in the co-localization of different antigens (Lange *et al.*, 1995).

One important aspect of immunological techniques is that they permit the localization of the native protein and are therefore not subject to the obvious caveats that apply to the localization of fusion proteins, although in some cases this advantage is diminished by the localization of an epitope-tagged version of the protein. However, in contrast to the localization of green fluorescent protein (GFP)-fusion proteins in living cells (see Chapter 9), all immunological methods require that cells are first fixed in order to freeze the internal structures in their native state. The importance of the fixation step cannot be overstressed. It is easy to forget the heated debates that filled the literature until fairly recently arguing for or against the existence of particular structures depending upon the fixation conditions used (for a fascinating discussion see Hopwood, 1985). Whilst many of these debates have been laid to rest by the advent of cryo-electron microscopic techniques (Dubochet, 1995; McDonald *et al.*, 1996) those embarking on protein localization in yeast by immunofluorescence would do well to heed their warnings.

Methods for immunofluorescence in budding yeast have been reviewed by Pringle *et al.* (1991) and this remains an excellent source of information for expert and novice alike. We will therefore concentrate on aspects of fixation and highlight instances where variations in fixation conditions have produced dramatic results. Given the very widespread acceptance of a standard fixation procedure (3.7% formaldehyde) for *S. cerevisiae*, we feel it likely that the more adventurous readers would be rewarded by experimenting with fixation conditions, even with established antisera.

A. Fixation

The fixation procedure must be rapid, in order to maintain spatial order, but it is equally important that epitopes on target proteins are preserved and remain accessible to antibodies. Essentially two methods can be used: one is to employ cross-linking agents, such as aldehydes, to crosslink cellular components covalently; the other is rapid dehydration using cold solvents to induce the *in situ* precipitation of proteins. Numerous variations on these two approaches appear in the published literature, but the latter has been used relatively rarely for *S. cerevisiae*. It is important to stress that the precise fixation conditions will vary for different antigen/antibody combinations and that the optimal conditions must be sought empirically.

1. Aldehyde fixation

Aldehyde fixation is by far the most commonly used method for immuno-fluorescence analysis in yeast. Most protocols are based on the use of either 3.7–5% formaldehyde (Pringle *et al.*, 1991) or a combination of 3% paraformaldehyde (a polymer of formaldehyde that forms formaldehyde on heating) and 1% glutaraldehyde (Kilmartin and Adams, 1984). In either case the fixative(s) can either be added directly to the culture medium, or to cells harvested by centrifugation or filtration (Kilmartin and Adams, 1984). The addition of aldehyde directly to the culture has the advantage that cells are fixed prior to harvesting; this may be critical when examining antigens whose distribution might be altered during the harvesting process, perhaps by the forces involved during centrifugation or in response to nutrient limitation (Pringle *et al.*, 1991; Roberts *et al.*, 1991).

2. Formaldehyde fixation

The use of formaldehyde alone has proven extremely successful in immunofluorescence studies of a wide range of proteins. Formaldehyde has the distinct advantage that it rapidly penetrates cells and thus the fixation process is also rapid. Despite this, most fixation protocols recommend long incubation periods ranging from 30 min to 12–15 h. In studies on vacuolar morphology, long fixation times are positively beneficial (Roberts *et al.*, 1991). However, the exact opposite may also be true. For example, Rout and Kilmartin (1990) initially reported immunofluorescence localization of Spc110p to the spindle pole body (SPB) in spheroplasts fixed with methanol and acetone. Meanwhile, an independent study on the localization of the same protein in formaldehyde-fixed cells revealed diffuse nuclear staining, leading Mirzayan *et al.* (1992) to conclude that Spc110p was a component of the nuclear matrix. In subsequent studies, Kilmartin *et al.* (1993) were able to show that a monoclonal antibody specific for Spc110p gave a punctate staining pattern typical of SPBs in cells that were subjected to a brief formaldehyde fixation protocol (2–5 min), whilst longer treatments (30 min) produced a weak and diffuse staining pattern similar to that reported by Mirzayan *et al.* (1992). The SPB is a densely packed protein complex wherein protein–protein crosslinking efficiency might be expected to be high. It has therefore been suggested that the sensitivity of SPB antigens to aldehyde fixation may be due to the formation of extensive crosslinks which render the organelle impervious to antibody molecules (Kilmartin *et al.*, 1993; Rout and Kilmartin, 1990). Similar observations have been reported for nuclear pore complex proteins (Wente *et al.*, 1992).

Whilst shorter aldehyde fixation times may be successful with some antigens, other antigens have been revealed by treatment of formaldehyde-fixed cells with either cold methanol (–20°C, 6 min) and acetone (–20°C, 30 s; Drubin *et al.*, 1988), or SDS (1–5% w/v SDS for 0.5–5 min; Roberts *et al.*, 1991). The additional methanol/acetone fixation may generally enhance antibody permeability, fix additional antigens, and may also expose previously hidden epitopes through denaturation (see below). This additional step is now in widespread use but its efficacy is not reported in most cases. Treatment of fixed cells with SDS may also serve to improve antibody permeability and denature some antigens, but has the added advantage that cells become more adherent to polylysine-treated microscope slides, thus reducing cell loss during extensive washing procedures (Roberts *et al.* 1991).

3. Dual aldehyde fixation

Whilst formaldehyde is a powerful fixative it must be recognized that its uses are limited: for example, it yields insufficient preservation of fine structure for electron microscopy. In contrast, glutaraldehyde provides excellent preservation of ultrastructure, but the levels required to ensure rapid fixation can result in an unacceptable loss of antibody reactivity. This problem may be overcome by the use of dual fixatives where formaldehyde treatment is quickly followed by the addition of low levels of glutaraldehyde (Kilmartin and Adams, 1984). The theory here is that formaldehyde rapidly penetrates the cell, fixing most cellular structures, but then the glutaraldehyde penetrates more slowly forming additional crosslinks, thus fixing further proteins, and preserving greater structural detail.

A considerable advance in the application of dual aldehyde fixation in yeast was made when it was discovered that treatment of cells with an osmotic stabilizer immediately before fixation resulted in a marked improvement in the preservation of the Golgi apparatus (Preuss *et al.*, 1992). The inclusion of sorbitol during formaldehyde fixation is now also widespread but once again the effect of such treatment is often not discussed. The rather intangible nature of some fixation effects is perhaps best highlighted by the observations that whilst 4% formaldehyde on its own is incapable of routinely preserving cytoplasmic microtubules in fission yeast, *either* dual fixation with 0.2% glutaraldehyde *or* the inclusion of 1.2 M sorbitol (added as prewarmed 2.4 M stock in medium 5 min before fixation) results in the complete preservation of the interphase cytoskeleton (D. R. Drummond, A. J. Bridge and I. M. Hagan, unpublished data).

In general, glutaraldehyde tends to be more detrimental to antigen preservation than is formaldehyde alone. For example, staining of fission yeast cells with the monoclonal antibodies N350 anti-actin (Amersham), MPM1 and MPM2 (Vandre *et al.*, 1984) is abolished if glutaraldehyde is included in the fixation. Therefore one must carefully titrate the glutaraldehyde fixation conditions for each new antigen/antibody combination in order to optimize both structural preservation and antigenicity. Glutaraldehyde can result in high levels of background fluorescence which must be quenched with $NaBH_4$ (Kilmartin and Adams, 1984). Such aldehyde-induced fluorescence is less problematic when glutaraldehyde is used at concentrations of 0.2% and below (I. M. Hagan, unpublished observations).

4. Storage and preparation of aldehydes

Formaldehyde is a monoaldehyde that can react with free amino groups to generate intermediates capable of crosslinking proteins *via* the formation of methylene bridges (Sharma and Sharma, 1965). Moreover, formaldehyde can polymerize to form long chains that may provide for the formation of longer-range crosslinks between proteins. Several mechanisms may contribute to the role of glutaraldehyde as a protein fixative, but they undoubtedly reflect the ability of this bifunctional dialdehyde to form a large range of complex polymers (Hopwood, 1985). The extent of polymerization increases with storage, consequently ultrapure fresh glutaraldehyde is routinely used for electron microscopy because it minimizes artifactual fixation (i.e. crosslinkers are short). Despite this we have generally found the opposite for immunofluorescence using dual aldehyde fixation and get the best results with less pure solutions that have been stored at 4°C for anything up to 3 years. Obviously there is a point after which the degree of polymerization in the glutaraldehyde stock solution may reduce the cross-linking capabilities *in vivo* so a gradual reduction in fixation capacity of old solutions should be guarded against.

> **Protocol 2.** Preparation of fresh formaldehyde
>
> Dissolve powdered paraformaldehyde (Sigma-Aldrich) in PEM buffer (100 mM Pipes, 1 mM EGTA, 1 mM MgSO$_4$, pH 6.9) to a final concentration of 30% (w/v). Place the suspension in a water bath at 65°C for 5 min and then add sodium hydroxide to dissolve the solid (*mixing* in a fume hood; adding approximately 120 µl of 5 N NaOH for every 10 ml of fix). Add alkali slowly until just before all of the formaldehyde has gone into solution at which point the solution is faintly gray and a small pellet is formed upon centrifugation. In this state the formaldehyde will begin to polymerize rapidly upon dilution and excess unbuffered NaOH is avoided.

Whilst complex crosslinking polymers of glutaraldehyde may be beneficial (certainly not detrimental), the opposite would seem likely for formaldehyde. In this case the formation of long polymers might be expected to reduce the rate at which it penetrates cells, thus diminishing the extent of fixation towards the center of the cell. For this reason relatively fresh stocks of 37% (w/w) are preferred; indeed, some workers routinely prepare fresh formaldehyde from paraformaldehyde stocks for immediate use (see Protocol 2). In contrast some researchers routinely use formalin (a solution of formaldehyde stabilized with methanol), in which formaldehyde polymerization can occur. It is not clear whether or not such polymers might be positively advantageous when formaldehyde is used as the sole fixative. We would recommend that all variations on aldehyde fixation be considered.

5. Solvent fixation

This technique is extremely simple and is widely used in several systems, yet it is used relatively rarely in *S. cerevisiae*. Examples of its success in yeast include localization of components of the SPB (see above; Rout and Kilmartin, 1990) and localization of peroxisomal thiolase (Erdmann and Kunau, 1994). One obvious disadvantage of this method is the artifactual shrinking of the fixed cell, but this is compensated by the fact that solvent fixation tends not to affect antigenicity as severely as does aldehyde fixation. Moreover, because the effect of solvent fixation is to precipitate proteins, then the permeability of the fixed cell to antibody is increased such that antibody dilutions can be reduced to as little as one-tenth of those required for combined formaldehyde and glutaraldehyde fixation. Similarly antibody incubation times can be reduced from overnight to 30 min. Good control antibodies for solvent fixation are the Amersham anti-actin antibody (cat. no. N350) and TAT1 anti-α tubulin antibody (Woods *et al.*, 1989).

Fixation of yeast spheroplasts by incubation at room temperature in methanol (5 min) followed by acetone (5 min) is sufficient to preserve peroxisomal structures (Erdmann and Kunau, 1994). Lower temperatures have been used to fix components of the SPB: methanol for 5 min at either

0 or –20°C, followed by 30 s in acetone at room temperature (Rout and Kilmartin, 1990). Lower fixation temperatures may be beneficial because they would be more likely to freeze internal structures rapidly thus preventing their breakdown before solvent fixation. Indeed, fixation in methanol at –80°C gives much better preservation of the *S. pombe* mitotic spindle than is seen at –20°C (I. M. Hagan, unpublished observations). Even greater preservation of ultrastructure can be achieved by plunging very small samples into extremely cold liquids with low heat capacity (such as liquid helium and liquid propane), but this requires specialized expertise and equipment and is not recommended for routine fluorescence microscopy, rather for electron microscopy where precise preservation is a necessity (Tanaka and Kanbe, 1986; Kanbe *et al.*, 1989; Ding *et al.*, 1993; Walther *et al.*, 1992; Baba *et al.*, 1994). However, the evidence from several recent comparative studies in other systems suggests that very low temperature cryofixation can provide superior fixation and immunocytochemical properties when compared with aldehyde fixed samples (Baskin *et al.*, 1996; Hoying *et al.*, 1995). It would therefore seem probable that this approach will continue to gain favor as the technology becomes more accessible.

6. Fixation artifacts

The effects of fixation and permeabilization conditions on the artifactual localization of soluble proteins have been examined extensively and systematically by Melan and Sluder (1992). This particular study highlights the problem that artifactual staining can appear as attractive as any authentic pattern. One can guard against this to some extent by comparing the staining patterns observed in cells fixed with aldehyde versus solvent-fixed cells. Even if similar distributions are obtained there may be qualitative or quantitative differences that favor the use of one approach over another. For example, when fission yeast cells are fixed by formaldehyde and glutaraldehyde, around four to six cytoplasmic tubulin filaments are seen and there is no great variation associated with cell cycle progression. In contrast, when solvent fixation is used this number increases to around eight to 10 and the post-mitotic cells are virtually full of microtubules (Hagan and Hyams, 1988).

The use of GFP to study localization is subject to the same caveats associated with any fusion protein approach (see above). Nonetheless this technique offers a powerful complement to immunolocalization approaches. Not only can one examine the localization of a GFP fusion protein in living cells, and compare this with the immunolocalization pattern for the native protein, but one can also directly compare GFP staining in fixed *versus* unfixed cells to identify any potential changes induced during fixation. The relatively long time taken for GFP to cyclize does, however, limit its uses in the analysis of proteins with short (functional) half-lives. One final word of caution regarding GFP: living cells respond rapidly to environmental stresses and therefore a controlled environment is required for microscopic examination, otherwise cells should be fixed!

Finally, variations in staining pattern can pose important questions relating to reproducibility. For example, staining variation from cell to cell may reflect genuine differences in protein localization or may merely reflect uneven antibody accessibility caused by limited cell wall digestion in some cells. Antibody permeability can be controlled by staining a sample of fixed cells with an unrelated antibody (e.g. against α-tubulin, DNA, etc.). When examining staining patterns of a particular antigen in a variety of mutant strain backgrounds it becomes important to ensure that fixation and antibody staining is carefully controlled between samples. One simple way to achieve this is to process a mixed culture that contains a morphologically distinct strain alongside the strain of interest, e.g. *cdc4* or *cdc34* mutants that accumulate multiple buds. If the signal is uniformly strong in all of the multiply budded cells but weak in all of the cells with normal morphology, one can be fairly confident that the signal is genuinely reduced in that particular genetic background (Yaffe *et al.*, 1996).

B. Post-Fixation: Stuck on Slides?

A final consideration that is often overlooked is the mechanism by which the cells are processed following fixation. There are two possibilities as to how to process fixed samples: either the fixed cells are immobilized within polylysine-coated PTFE wells on microscope slides or the cells are processed in microfuge tubes. Most of the work in budding yeast uses the former whilst researchers studying fission yeast use the latter. The differences may be historical, but the two approaches offer their own specific advantages. For example, if a large number of different yeast strains are to be screened with a readily available monoclonal antibody it is clearly advantageous to have multiple specimens stuck to slides which can be dipped into staining baths. Moreover, if a brief fixation regime is required that might leave the cells fragile, then securing them at an early stage may be beneficial.

In contrast, fixed cells can be incubated with antibodies in microfuge tubes and washed following brief centrifugation. This has the very particular advantage that stained cells can be stored for long periods (up to 4 years in the dark at 4°C) and re-examined at leisure. This is more difficult with slide-mounted specimens where the background fluorescence in the mounting medium accumulates and the medium can dry out. Being able to remount cells is also advantageous when trying to attain an appropriate cell density for efficient examination; it is all too easy to lose cells from the surface of slides during multiple wash steps. Finally, when using an upright microscope to examine specimens mounted on slides, the light must pass through the coverslip and mounting medium and back again which can generate high background. This is markedly reduced if stained cells are mounted on the underside of coverslips. Because cells are mounted only at the final step, then the risk of accumulating debris on the surface of the coverslip is minimized.

Acknowledgements

I.M.H. is supported by the Cancer Research Campaign. C.J.S. is a Lister Institute Jenner Research Fellow.

References

Baba, M., Takeshige, K., Baba, N. and Ohsumi, Y. (1994). Ultrastructural analysis of the autophagic process in yeast: detection of autophagosomes and their characterization. *J. Cell Biol.* **124**, 903–913.

Baker, D., Hicke, L., Rexach, M., Schleyer, M. and Schekman, R. (1988). Reconstitution of *SEC* gene product-dependent intercompartmental protein transport. *Cell* **54**, 335–344.

Baskin, T. I., Miller, D. D., Vos, J. W., Wilson, J. E. and Hepler, P. K. (1996). Cryofixing single cells and multicellular specimens enhances structure and immunocytochemistry for light microscopy. *J. Microsc.* **182**, 149–161.

Bennett, B., Check, I. J., Olsen, M. R. and Hunter, R. L. (1992). A comparison of commercially available adjuvants for use in research. *J. Immunol. Methods* **153**, 31–40.

Biggins, S. and Rose, M. D. (1994). Direct interaction between yeast spindle pole body components: Kar1p is required for Cdc31p localization to the spindle pole body. *J. Cell Biol.* **125**, 843–852.

Binder, M., Hartig, A. and Sata, T. (1996). Immunogold labeling of yeast cells: an efficient tool for the study of protein targeting and morphological alterations due to overexpression and inactivation of genes. *Histochem. Cell Biol.* **106**, 115–130.

Brown, C. E., Tarun, S. Z., Jr, Boeck, R. and Sachs, A. B. (1996). *PAN3* encodes a subunit of the Pab1p-dependent poly(A) nuclease in *Saccharomyces cerevisiae*. *Mol. Cell. Biol.* **16**, 5744–5753.

Brown, J. D., Hann, B. C., Medzihradszky, K. F., Niwa, M., Burlingame, A. L. and Walter, P. (1994). Subunits of the *Saccharomyces cerevisiae* signal recognition particle required for its functional expression. *EMBO J.* **13**, 4390–4400.

Chant, J., Mischke, M., Mitchell, E., Herskowitz, I. and Pringle, J. R. (1995). Role of Bud3p in producing the axial budding pattern of yeast. *J. Cell Biol.* **129**, 767–778.

Chuang, J. S. and Schekman, R. W. (1996). Differential trafficking and timed localization of two chitin synthase proteins, Chs2p and Chs3p. *J. Cell Biol.* **135**, 597–610.

Cockell, M., Palladino, F., Laroche, T., Kyrion, G., Liu, C., Lustig, A. J. and Gasser, S. M. (1995). The carboxy termini of Sir4 and Rap1 affect Sir3 localization: evidence for a multicomponent complex required for yeast telomeric silencing. *J. Cell Biol.* **129**, 909–924.

Craven, R. A., Egerton, M. and Stirling, C. J. (1996). A novel Hsp70 of the yeast ER lumen is required for the efficient translocation of a number of protein precursors. *EMBO J.* **15**, 2640–2650.

Crowe, J., Dobeli, H., Gentz, R., Hochuli, E., Stuber, D. and Henco, K. (1994). 6×His-Ni-NTA chromatography as a superior technique in recombinant protein expression/purification. *Methods Mol. Biol.* **31**, 371–387.

Dean, N. and Pelham, H. R. B. (1990). Recycling of proteins from the Golgi compartment to the ER in yeast. *J. Cell Biol.* **111**, 369–377.

Debeus, E., Brockenbrough, J. S., Hong, B. and Aris, J. P. (1994). Yeast *NOP2* encodes an essential nucleolar protein with homology to a human proliferation marker. *J. Cell Biol.* **127**, 1799–1813.

Deshaies, R. J., Sanders, S. L., Feldheim, D. A. and Schekman, R. (1991). Assembly of yeast Sec proteins involved in translocation into the endoplasmic reticulum into a membrane-bound multisubunit complex. *Nature* **349**, 806–808.

Deshmukh, M., Tsay, Y. F., Paulovich, A. G. and Woolford, J. L. (1993). Yeast ribosomal-protein l1 is required for the stability of newly synthesized 5S ribosomal-RNA and the assembly of 60S ribosomal- subunits. *Mol. Cell. Biol.* **13**, 2835–2845.

Ding, R., McDonald, K. L. and McIntosh, J. R. (1993). 3-dimensional reconstruction and analysis of mitotic spindles from the yeast, *Schizosaccharomyces pombe*. *J. Cell Biol.* **120**, 141–151.

Drubin, D. G., Miller, K. G. and Botstein, D. (1988). Yeast actin-binding proteins: evidence for a role in morphogenesis. *J. Cell Biol.* **107**, 2551–2561.

Dubochet, J. (1995). High-pressure freezing for cryoelectron microscopy. *Trends Cell Biol.* **5**, 366–368.

Eisenmann, D. M., Arndt, K. M., Ricupero, S. L., Rooney, J. W. and Winston, F. (1992). Spt3 interacts with TFIID to allow normal transcription in *Saccharomyces cerevisiae*. *Genes Develop.* **6**, 1319–1331.

Erdmann, R. and Kunau, W. H. (1994). Purification and immunolocalization of the peroxisomal 3-oxoacyl-CoA thiolase from *Saccharomyces cerevisiae*. *Yeast* **10**, 1173–1182.

Esnault, Y., Feldheim, D., Blondel, M. O., Schekman, R. and Kepes, F. (1994). Sss1 encodes a stabilizing component of the Sec61 subcomplex of the yeast protein translocation apparatus. *J. Biol. Chem.* **269**, 27478–27485.

Geissler, S., Pereira, G., Spang, A., Knop, M., Soues, S., Kilmartin, J. and Schiebel, E. (1996). The spindle pole body component Spc98p interacts with the gamma-tubulin-like Tub4p of *Saccharomyces cerevisiae* at the sites of microtubule attachment. *EMBO J.* **15**, 3899–3911.

Gould, S. J., Krisans, S., Keller, G. A. and Subramani, S. (1990). Antibodies directed against the peroxisomal targeting signal of firefly luciferase recognize multiple mammalian peroxisomal proteins. *J. Cell Biol.* **110**, 27–34.

Graham, L. A., Hill, K. J. and Stevens, T. H. (1995). *VMA8* encodes a 32-kDa V1 subunit of the *Saccharomyces cerevisiae* vacuolar H(+)-ATPase required for function and assembly of the enzyme complex. *J. Biol. Chem.* **270**, 15037–15044.

Hagan, I. M. and Hyams, J. S. (1988). The use of cell-division cycle mutants to investigate the control of microtubule distribution in the fission yeast *Schizosaccharomyces pombe*. *J. Cell Sci.* **89**, 343–357.

Hagan, I. M. and Yanagida, M. (1997). Evidence for cell-cycle specific, spindle pole-body mediated, nuclear positioning in the fission yeast *Schizosaccharomyces pombe*. *J. Cell. Science* **110**, 1851–1866.

Hann, B. C., Stirling, C. J. and Walter, P. (1992). *SEC65* gene product is a subunit of the yeast signal recognition particle required for its integrity. *Nature* **356**, 532–533.

Harlow, E. and Lane, D. (1988). *Antibodies: A Laboratory Manual.* Cold Spring Harbor Laboratories, New York.

Hennessy, K. M., Clark, C. D. and Botstein, D. (1990). Subcellular-localization of yeast Cdc46 varies with the cell-cycle. *Genes Develop.* **4**, 2252–2263.

Hopwood, D. (1985). Cell and tissue fixation, 1972–1982. *Histochem. J.* **17**, 389–442.

Hoying, J. B., Chen, S. C. and Williams, S. K. (1995). Interaction of colloidal gold-labelled glucosylated albumin with endothelial cell monolayers: comparison between cryofixation and glutaraldehyde fixation. *Microsc. Res. Tech.* **30**, 252–257.

Jones, E. W. (1991). Tackling the protease problem in *Saccharomyces cerevisiae*. *Methods Enzymol.* **194**, 428–453.

Kanbe, T., Kobayashi, I. and Tanaka, K. (1989). Dynamics of cytoplasmic organelles in the cell cycle of the fission yeast *Schizosaccharomyces pombe*: three-dimensional reconstruction from serial sections. *J. Cell Sci.* **94**, 647–656.

Kilmartin, J. V. and Adams, A. E. (1984). Structural rearrangements of tubulin and actin during the cell cycle of the yeast *Saccharomyces*. *J. Cell Biol.* **98**, 922–933.

Kilmartin, J. V., Dyos, S. L., Kershaw, D. and Finch, J. T. (1993). A spacer protein in the *Saccharomyces cerevisiae* spindle pole body whose transcript is cell cycle-regulated. *J. Cell Biol.* **123**, 1175–1184.

Kuroda, R., Kinoshita, J., Honsho, M., Mitoma, J. and Ito, A. (1996). *In situ* topology of cytochrome b5 in the endoplasmic reticulum membrane. *J. Biochem. (Tokyo)* **120**, 828–833.

Lange, B. M. H., Sherwin, T., Hagan, I. M. and Gull, K. (1995). The basics of immunofluorescence video-microscopy for mammalian and microbial systems. *Trends Cell Biol.* **5**, 328–332.

Lavallie, E. R. and McCoy, J. M. (1995). Gene fusion expression systems in *Escherichia coli*. *Curr. Opin. Biotechnol.* **6**, 501–506.

Lewis, J. D., Gorlich, D. and Mattaj, I. W. (1996). A yeast cap-binding protein complex (yCBC) acts at an early step in pre-mRNA splicing. *Nucleic. Acids Res.* **24**, 3332–3336.

McDonald, K., Otoole, E. T., Mastronarde, D. N., Winey, M. and McIntosh, J. R. (1996). Mapping the 3-dimensional organization of microtubules in mitotic spindles of yeast. *Trends Cell Biol.* **6**, 235–239.

Melan, M. A. and Sluder, G. (1992). Redistribution and differential extraction of soluble proteins in permeabilized cultured cells. Implications for immuno-fluorescence microscopy. *J. Cell Sci.* **101**, 731–743.

Milne, G. T., Jin, S., Shannon, K. B. and Weaver, D. T. (1996). Mutations in two Ku homologs define a DNA end-joining repair pathway in *Saccharomyces cerevisiae*. *Mol. Cell. Biol.* **16**, 4189–4198.

Mirzayan, C., Copeland, C. S. and Snyder, M. (1992). The *NUF1* gene encodes an essential coiled-coil related protein that is a potential component of the yeast nucleoskeleton. *J. Cell Biol.* **116**, 1319–1332.

Nasmyth, K., Adolf, G., Lydall, D. and Seddon, A. (1990). The identification of a second cell cycle control on the HO promoter in yeast: cell cycle regulation of SW15 nuclear entry. *Cell* **62**, 631–647.

Page, B. D., Satterwhite, L. L., Rose, M. D. and Snyder, M. (1994). Localization of the Kar3 kinesin heavy chain-related protein requires the Cik1 interacting protein. *J. Cell Biol.* **124**, 507–519.

Preuss, D., Mulholland, J., Franzusoff, A., Segev, N. and Botstein, D. (1992). Characterization of the *Saccharomyces* Golgi-complex through the cell-cycle by immunoelectron microscopy. *Mol. Biol. Cell* **3**, 789–803.

Pringle, J. R., Adams, A. E. M., Drubin, D. G. and Haarer, B. K. (1991). Immunofluorescence methods in yeast. *Methods Enzymol.* **194**, 565–602.

Roberts, C. J., Raymond, C. K., Yamashiro, C. T. and Stevens, T. H. (1991). Methods for studying the yeast vacuole. *Methods Enzymol.* **194**, 644–661.

Roof, D. M., Meluh, P. B., and Rose, M. D. (1992). Kinesin-related proteins required for assembly of the mitotic spindle. *J. Cell Biol.* **118**, 95–108.

Ross-Macdonald, P., Burns, N., Malczynski, M., Sheehan, A., Roeder, S. and Snyder, M. (1995). Methods for large-scale analysis of gene-expression, protein localization, and disruption phenotypes in *Saccharomyces cerevisiae*. *Methods Mol. Cell. Biol.* **5**, 298–308.

Rout, M. P. and Kilmartin, J. V. (1990). Components of the yeast spindle and spindle pole body. *J. Cell Biol.* **111**, 1913–1927.

Serrano, R., Monk, B. C., Villalba, J. M., Montesinos, C. and Weiler, E. W. (1993). Epitope mapping and accessibility of immunodominant regions of yeast plasma membrane H(+)-ATPase. *Eur. J. Biochem.* **212**, 737–744.

Sharma, A. K. and Sharma, A. S. (1965). *Chromosome Techniques: Theory and Practice*. Butterworths, London.

Spang, A., Courtney, I., Grein, K., Matzner, M. and Schiebel, E. (1995). The Cdc31p-binding protein Kar1p is a component of the half bridge of the yeast spindle pole body. *J. Cell Biol.* **128**, 863–877.

Stirling, C. J., Rothblatt, J., Hosobuchi, M., Deshaies, R. and Schekman, R. (1992). Protein translocation mutants defective in the insertion of integral membrane proteins into the endoplasmic reticulum. *Mol. Biol. Cell* **3**, 129–142.

Strambio-de-Castillia, C., Blobel, G. and Rout, M. P. (1995). Isolation and characterization of nuclear envelopes from the yeast *Saccharomyces*. *J. Cell Biol.* **131**, 19–31.

Studier, F. W., Rosenberg, A. H., Dunn, J. J. and Dubendorff, J. W. (1990). Use of T7 RNA polymerase to direct expression of cloned genes. *Methods Enzymol.* **185**, 60–89.

Swanson, M. S. and Winston, F. (1992). Spt4, spt5 and spt6 interactions – effects on transcription and viability in *Saccharomyces cerevisiae*. *Genetics* **132**, 325–336.

Sweet, D. J. and Pelham, H. R. B. (1993). The *TIP1* gene of *Saccharomyces cerevisiae* encodes an 80 kDa cytoplasmic protein that interacts with the cytoplasmic domain of Sec20p. *EMBO J.* **12**, 2831–2840.

Tanaka, K. and Kanbe, T. (1986). Mitosis in the fission yeast *Schizosaccharomyces pombe* as revealed by freeze-substitution electron microscopy. *J. Cell Sci.* **80**, 253–268.

TerBush, D. R. and Novick, P. (1995). Sec6, Sec8, and Sec15 are components of a multisubunit complex which localizes to small bud tips in *Saccharomyces cerevisiae*. *J. Cell Biol.* **130**, 299–312.

Vallen, E. A., Scherson, T. Y., Roberts, T., van Zee, K. and Rose, M. D. (1992). Asymmetric mitotic segregation of the yeast spindle pole body. *Cell* **69**, 505–515.

Vandre, D. D., Davis, F. M., Rao, P. N. and Borisy, G. G. (1984). Phosphoproteins are components of mitotic microtubule organizing centers. *Proc. Natl Acad. Sci. USA* **81**, 4439–4443.

Walther, P., Chen, Y., Pech, L. L. and Pawley, J. B. (1992). High-resolution scanning electron microscopy of frozen-hydrated cells. *J. Microsc.* **168**, 169–180.

Wente, S. R., Rout, M. P. and Blobel, G. (1992). A new family of yeast nuclear pore complex proteins. *J. Cell Biol.* **119**, 705–723.

Wilkinson, B. M., Critchley, A. J. and Stirling, C. J. (1996). Determination of the transmembrane topology of yeast Sec61p, an essential component of the endoplasmic reticulum translocation complex. *J. Biol. Chem.* **271**, 25590–25597.

Woods, A., Sherwin, T., Sasse, R., MacRae, T. H., Baines, A. J. and Gull, K. (1989). Definition of individual components within the cytoskeleton of *Trypanosoma brucei* by a library of monoclonal antibodies. *J. Cell Sci.* **93**, 491–500.

Wright, R., Basson, M., D'Ari, L. and Rine, J. (1988). Increased amounts of HMG-CoA reductase induce "karmellae": a proliferation of stacked membrane pairs surrounding the yeast nucleus. *J. Cell Biol.* **107**, 101–114.

Yaffe, M. P., Harata, D., Verde, F., Eddison, M., Toda, T. and Nurse, P. (1996). Microtubules mediate mitochondrial distribution in fission yeast. *Proc. Natl Acad. Sci. USA* **93**, 11664–11668.

Zinser, E. and Daum, G. (1995). Isolation and biochemical characterization of organelles from the yeast, *Saccharomyces cerevisiae*. *Yeast* **11**, 493–536.

13 Posttranslational Modifications of Secretory Proteins

F. M. Klis, A. F. J. Ram, R. C. Montijn, J. C. Kapteyn, L. H. P. Caro, J. H. Vossen, M. A. A. Van Berkel, S. S. C. Brekelmans and H. Van den Ende

Fungal Cell Wall Group, Institute for Molecular Cell Biology, University of Amsterdam, BioCentrum Amsterdam, The Netherlands

◆◆◆

CONTENTS

List of Abbreviations

BCA	Bicinchoninic acid
BSA	Bovine serum albumin
ConA	Concanavalin A
CPY	Carboxypeptidase Y
CRD	Cross-reacting determinant
CWP	Cell wall protein
ECL	Enhanced chemiluminescence
Endo-H	Endoglycosidase-H
ER	Endoplasmic reticulum
GPI	Glycosylphosphatidylinositol
HA	Hemagglutinin
HF	Hydrofluoric acid
MES	2-(N-Morpholino)ethanesulfonic acid
OD	Optical density
OT-ase	Oligosaccharyl transferase
PBS	Phosphate buffered saline
PI-PLC	Phosphatidylinositol-specific phosopholipase C
PMSF	Phenylmethylsulfonyl fluoride
RT	Room temperature
YPD	Yeast peptone dextrose (*see Appendix II: Table I*)

METHODS IN MICROBIOLOGY, VOLUME 26
ISBN 0–12–521526–6

◆◆◆◆◆◆ I. CLASSIFICATION OF SECRETORY PROTEINS

Proteins that traverse the secretory pathway and reach the plasma membrane may be targeted to various locations (Table 1). One final destination is the plasma membrane itself. The plasma membrane contains not only many transmembrane proteins such as Fks1p/Cwh53p, a component of β1,3-glucan synthase, and Pma1p, an abundant H⁺-ATPase (Ram *et al.*, 1994; Douglas *et al.*, 1994; Dexaerde *et al.*, 1996), but also a special group of proteins, including the extensively studied Gas1p, which are C-terminally linked to the outer leaflet of the plasma membrane through a so-called glycosylphosphatidylinositol (GPI) anchor (section IV). Between the plasma membrane and the cell wall, periplasmic proteins such as invertase and acid phosphatase may accumulate. These are large soluble proteins that have difficulty in passing through the cell wall owing to their size. The cell wall itself contains some proteins such as the endoglucanase Bgl2p that can be extracted with hot detergents, but most, including α-agglutinin, are covalently linked to β-glucan through a GPI anchor-derived structure (Lu *et al.*, 1995; Kapteyn *et al.*, 1996; see also section V). These proteins can be solubilized enzymatically by using cell wall degrading enzymes (section V). Finally, a limited number of proteins such as chitinase and several heat shock proteins are secreted into the medium (Orlean *et al.*, 1991; see also section VI). Secretory proteins are modified in

Table I. Location and glycosylation of representative cell surface and extracellular proteins

Location	Protein	M_r (kDa)	Modifications
Plasma membrane	Gas1p	115	O-chains Short N-chains GPI anchor
Periplasmic region	Invertase (Suc2p)	110–150	O-chains Variable number of long N-chains
Cell wall			
SDS-extractable	Bgl2p	33	Single, short N-chain
SDS-resistant	α-Agglutinin (Agα1p)	350	O-chains Long N-chains GPI-anchor-derived structure β-glucosylated
Culture medium	Chitinase (Cts1p)	150	Heavily O-glycosylated

various ways during their passage through the secretory pathway. In this chapter, techniques will be discussed to identify and characterize the posttranslational modifications of secretory proteins. It is important to realize that these techniques, although primarily developed for *Saccharomyces cerevisiae*, are in many cases also valid for other Ascomycetes, including the filamentous fungi belonging to this taxonomic group.

◆◆◆◆◆◆ II. GLYCOSYLATION OF SECRETORY PROTEINS

Posttranslational modifications of secretory proteins comprise both proteolytic steps and glycosylation reactions. For example, a subset of secretory proteins such as pro-alpha-factor, the major exoglucanase (Exg1p) and HSP150 are proteolytically processed in a late Golgi compartment by the endoprotease Kex2p, which cleaves carboxyterminally to a basic dipeptide such as lysine-arginine. In this chapter, however, we will focus on the addition of carbohydrates to the protein backbone. Most secretory proteins carry *N*-chains and/or *O*-chains, whereas a subset of secretory proteins also receive a GPI anchor. In all cases, glycosylation of secretory proteins begins in the endoplasmic reticulum (ER) (Herscovics and Orlean, 1993; Lehle and Tanner, 1995).

N-linked glycosylation is an essential process in eukaryotic cells and begins with the stepwise assembly of a dolichol-linked core oligosaccharide ($Glc_3Man_9GlcNAc_2$-PP-Dol) at the membrane of the ER (Herscovics and Orlean, 1993; Lehle and Tanner, 1995). The oligosaccharide moiety is then transferred *en bloc* by the oligosaccharyl transferase complex to asparagine residues in the consensus sequence Asn-Xxx-Ser/Thr. This is immediately followed by a series of trimming reactions resulting in a $Man_8GlcNAc_2$ side-chain, which is characteristic of the ER. In the Golgi this structure is extended by mannose residues resulting in a mature core of about 9–13 mannose residues ($Man_{9-13}GlcNAc_2$). This core is often, but not always, extended by a long backbone of $\alpha1,6$-linked mannose residues carrying multiple short side chains of $\alpha1,2$-linked mannose residues terminating in an $\alpha1,3$-linked mannose residue (Herscovics and Orlean, 1993; Lehle and Tanner, 1995). This outer chain may contain more than 50 mannose residues. In addition, both small and large *N*-chains may be substituted with mannosyl phosphate groups.

In the case of *O*-glycosylation, the first mannose residue is attached to selected serine and/or threonine residues in the ER, and extension of the *O*-linked chains up to a final length of five mannose residues takes place in the Golgi (Herscovics and Orlean, 1993; Lehle and Tanner, 1995). Despite extensive research, no clear rules have yet emerged as to how the cell selects specific serine and threonine residues for *O*-glycosylation (Lehle and Tanner, 1995).

A selected group of secretory proteins also receive a GPI anchor in the ER. Proteins carrying a GPI anchor addition signal at their C-terminus (Figure 1) exchange this signal for a preassembled GPI anchor, which in yeast contains four mannose residues (Sipos *et al.*, 1995). In the Golgi another mannose residue is attached (Figure 2). Some of the GPI anchored proteins, such as Gas1p, Yap3p and Kre1p, become attached to the outer leaflet of the plasma membrane. In contrast, GPI-anchored precursors of cell wall proteins are, in a yet unknown way, released from the plasma membrane, presumably by cleavage within their GPI anchor. Subsequently, they are believed to become linked to a β1,6-/β1,3-glucan heteropolymer through a remnant of their GPI anchor (Kapteyn *et al.*, 1996). The resulting structure represents the major building block of the yeast cell wall (Figure 3). Evidence for a similar structure has also been found in other Ascomycetous fungi (S. Brul, personal communication).

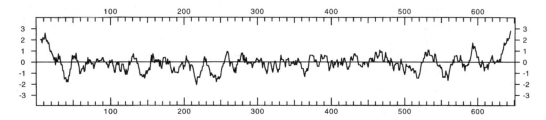

Figure 1. Hydropathy plot of α-agglutinin. α-Agglutinin is a well-characterized GPI-anchored protein found in the cell walls of MATα cells. Note the hydrophobic regions at both ends of the protein. The mature protein possesses 29% serine and threonine residues.

PROTEIN - C=O
  \
   NH-CH$_2$-CH$_2$-PO$_4$⁻
       \
        6
        3 Manα1-2Manα1-6Manα1-4GlcNα1-6Inositol-PO4-LIPID
        /
  Manα1-2/3Manα1

Figure 2. Structure of a GPI anchor in yeast. In the endoplasmic reticulum a transamidase exchanges the C-terminal hydrophobic sequence of the protein for a preassembled GPI anchor. The C-terminal amino acid of the mature protein is coupled to ethanolamine, which in turn is connected to an oligosaccharide followed by an inositol phospholipid. In yeast the conserved core oligosaccharide in the preassembled anchor (6Manα1-2Manα1-6Manα1-4GlcNα1-) is extended with a fourth mannose residue. In the Golgi, a fifth mannose residue is added at either the 2- or 3-position of the fourth mannose residue.

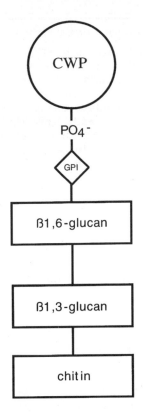

Figure 3. Proposed structure of the major building block of the yeast cell wall. The β1,6-glucan presumably consists of about 140 residues and the β1,3-glucan of about 1500 residues. CWP stands for cell wall protein, whereas the vertical lines that connect the separate components represent covalent linkages. We imagine that the building blocks are kept together by numerous hydrogen bridges between parallel-running β1,3-glucan chains originating from separate building blocks.

◆◆◆◆◆◆ III. DETECTION OF *N*-AND *O*-GLYCOSYLATION IN SECRETORY PROTEINS

A. General

A sensitive way to detect the presence of either *O*- or *N*-chains without discriminating between the two forms is based on the recognition of α-linked mannose residues by the lectin Concanavalin A (ConA). Protocol 1 describes a procedure to detect glycosylated proteins making use of peroxidase-conjugated ConA in combination with chemiluminescence (ECL). Another generally useful tool comes from the temperature-sensitive *sec* mutants. These are secretory mutants, which at the restrictive temperature become blocked at a specific step in the secretory pathway.

> **Protocol 1.** Detection of mannoproteins by peroxidase-conjugated ConA
>
> 1. Separate proteins by SDS-PAGE and transfer them onto a PVDF-filter.
> 2. Block the filter with 3% (w/v) BSA in phosphate buffered saline (PBS) (BSA/PBS) at room temperature (RT) for at least 1 h.
> 3. Rinse the filter twice with PBS for 5 min.
> 4. Add peroxidase-conjugated ConA (Sigma) to BSA/PBS containing 2.5 mM CaCl$_2$ and 2.5 mM MgCl$_2$, to a final concentration of 1 μg/ml.
> 5. Incubate the filter in this solution at RT for 1 h.
> 6. Rinse the filter twice in PBS for 5 min and once for 10 min.
> 7. Develop the filter using ECL-detection reagents (Amersham).

They allow accumulation of the ER form (*sec18*), the Golgi form (*sec7* and *sec14*), and the secreted form (*sec1*, *sec4* and *sec6*) of secretory proteins (Lu *et al.*, 1994).

B. Selected Secretory Proteins

To determine whether a specific secretory protein is *N*- or *O*-glycosylated or both, an epitope-tagged version of the protein (e.g. HA- or myc-tagged), or protein-specific antibodies need to be generated. Alternatively, one can make use of null mutants and strains expressing multiple copies of the gene to identify a selected protein.

1. N-chains

The presence of *N*-chains in secretory proteins can be easily established using endoglycosidase H (Endo-H) (Orlean *et al.*, 1991; Kapteyn *et al.*, 1996; Protocol 2). Treatment of isolated, SDS-boiled proteins with Endo-H results in cleavage of the *N*-chains after the first *N*-acetylglucosamine residue thereby removing almost the entire *N*-chain. The corresponding loss in molecular mass can be detected as increased electrophoretic mobility. Alternatively, tunicamycin can be used, because this drug inhibits the addition of *N*-chains to the protein (Orlean *et al.*, 1991). The presence of outer chains can be detected by studying the protein of interest in an *mnn9* or an *och1* background (Ballou, 1990; Nakayama *et al.*, 1992) and comparing it with its wild-type form. These mutants are defective in the addition of an outer chain to the inner core. As the outer chains can vary largely in size even for a given protein, the loss of outer chains results not only in increased electrophoretic mobility, but also in more discrete bands and, as a consequence, in greater sensitivity in detection levels. Different steps in the construction of the outer chain can be monitored using antibodies directed against α-1,6-linked, α-1,2-linked and α-1,3-linked mannose residues (Ballou, 1991).

Protocol 2. Endo-H digestion of yeast mannoproteins

1. Mix 40 μl protein solution, 10 μl 2% SDS, and 1.5 μl β-mercaptoethanol.
2. Denature the proteins by heating at 100°C for 5 min.
3. Add 150 μl of 50 mM NaAc, pH 5.5, containing 2 mM EDTA and 1 mM PMSF.
4. Add 1 μl pepstatin (0.7 mg ml^{-1} in methanol) and 1 μl leupeptin (0.7 mg ml^{-1} in water).
5. Add 10 μl recombinant Endo-H solution (Boehringer) and incubate overnight at 37°C.
6. Precipitate proteins by adding acetone to a final volume of 2 ml. Keep at −20°C for 2 h.
7. Dry the precipitated proteins and take up in sample buffer.

As mentioned earlier, the precursor form of *N*-chains is a lipid-linked oligosaccharide that is assembled step by step. Intermediate forms of the oligosaccharide moiety can be identified by HPLC allowing the detection of genes involved in particular assembly steps (Burda *et al.*, 1996).

2. O-chains

Secretory proteins often contain serine- and threonine-rich regions, which can be *O*-mannosylated extensively. *O*-chains can therefore contribute considerably to the molecular mass of secretory proteins. For example, chitinase is a heavily *O*-mannosylated protein with a molecular mass of about 150 kDa, whereas the unglycosylated protein has a mass of 60 kDa (Orlean *et al.*, 1991). Because Kre2p/Mnt1p adds the third mannose to the growing *O*-chains (Hill *et al.*, 1992; Haüsler *et al.*, 1992), glycosylation in a *kre2/mnt1* background results in shorter *O*-chains and increased electrophoretic mobility of the secretory protein. Kre2p/Mnt1p is not only involved in *O*-glycosylation, but also in *N*-glycosylation (Hill *et al.*, 1992). When a particular protein is both *N*- and *O*-glycosylated, removal of the *N*-chains by endo-H becomes essential for detection of *O*-chains (section III.B.1). As an additional advantage, resolution in gel electrophoresis will be considerably improved. Alternatively, one may consider analysis in a *pmt1* or *pmt2* background (Lussier *et al.*, 1995; Strahlbolsinger *et al.*, 1993). Both Pmt1p and Pmt2p are involved in the addition of the first mannose residue to serine and threonine residues. Although other protein mannosyl transferases can partially take over their function, *O*-glycosylation is sufficiently disturbed in a deletion strain to distinguish between wild-type and mutant *O*-glycosylation. However, Pmt1p and Pmt2p do not seem to be involved in the *O*-glycosylation of all proteins. Recently, it has been shown that the *O*-glycosylation of Gas1p is only affected in a *pmt4* deletion mutant, whereas the *O*-glycosylation of chitinase was primarily affected in a *pmt1* or *pmt2* background (Gentzsch and Tanner, 1996).

229

◆◆◆◆◆◆ IV. DETECTION OF GPI-ANCHORED PROTEINS

GPI-anchored proteins show a characteristic hydropathy plot characterized by a hydrophobic region at both ends of the molecule (Figure 1). The N-terminal hydrophobic region represents the signal peptide needed to enter the ER where it is removed by a signal peptidase. The carboxy terminal end (the GPI anchor addition signal) is replaced in the ER with a GPI anchor (Figure 2) and this step is carried by a putative transamidase complex (Udenfriend and Kodukula, 1995). GPI-anchored proteins in yeast often contain serine- and threonine-rich regions (Caro et al., 1997). Although there does not exist an absolute consensus sequence for the addition signal of GPI anchors, such a signal is nevertheless easy to recognize and the C-terminal amino acid of the mature protein (the ω amino acid) can be predicted with reasonable accuracy (Udenfriend and Kodukula, 1995). The GPI anchor addition signal is composed of a C-terminal sequence of hydrophobic amino acids preceded by a shorter, more polar region separating it from the ω amino acid. The ω amino acid may vary, but only a limited number of amino acids are used. In yeast asparagine, glycine, serine, and threonine are the preferred amino acids in this position (Caro et al., 1997; Fankhauser et al., 1993). The ω+2 amino acid also is restricted to small amino acids. In silico analysis has identified about 60 potential GPI proteins in the yeast genome (Caro et al., 1997).

To confirm the presence of a GPI anchor, several techniques are available. As a preliminary indication, the observation can be used that GPI-anchored proteins are membrane-bound and resistant to extraction with high salt, urea, and sodium carbonate (Wojciechowicz et al., 1993; Conzelmann et al., 1990). In addition, when GPI-anchored proteins are fractionated using Triton X-114 as a detergent, they are found in the detergent fraction before treatment with PI-PLC, and in the aqueous fraction after this treatment. Cleavage of the GPI anchor by PI-PLC also results in the formation of 2,3-phosphodiester-linked inositol, which is recognized by commercially available, so-called anti-CRD (cross-reacting determinant) antibodies (Oxford GlycoSystems). Radioactive labeling of the GPI anchor with inositol and palmitate is also frequently used (Wojciechowicz et al., 1993; Conzelmann et al., 1990). Labeling the GPI anchor with ethanolamine is less suitable because of relatively low incorporation levels (M. Van Berkel, personal communication; Müller et al., 1996).

Finally, cleavage of both phosphodiester bridges in protein-bound GPI anchors (Figure 2) using hydrofluoric acid (HF) allows the isolation and analysis of their glycan cores (Sipos et al., 1995; Schneider and Ferguson, 1995).

◆◆◆◆◆◆ V. IDENTIFICATION OF β-GLUCOSYLATED CELL WALL PROTEINS

As mentioned above, the yeast genome contains about 60 GPI proteins, some of which are destined for the plasma membrane, whereas the others are directed to the cell wall. Interestingly, known GPI-bound proteins in the plasma membrane contain a dibasic motif just preceding the predicted ω amino acid, which is absent from known GPI proteins in the cell wall (Vossen et al., 1997). Using this and other criteria, it is possible to identify about 20 GPI proteins that are potential plasma membrane proteins, whereas the remaining 40 GPI proteins are presumably incorporated in the cell wall (Caro et al., 1997).

A. Release of Cell Wall Proteins

SDS-resistant cell wall proteins (CWPs) are believed to be covalently linked through a remnant of the GPI anchor to a β1,6/β1,3-glucan complex (Figure 3). Because a phosphodiester bridge is present, this implies

---

Protocol 3. Release of cell wall proteins by laminarinase

1. Harvest about 100 optical density $(OD)_{530}$ units of cells from an early-logarithmic-phase culture.
2. Wash the cells three times with ice-cold buffer: 10 mM Tris-HCl, pH 7.8, 1 mM PMSF.
3. Suspend the cells in 1 ml wash buffer in a capped Pyrex glass tube.
4. Add glass beads (diameter = 0.5 mm) to just below the surface of the suspension.
5. Shake the suspension in the cold room using a Griffin shaker at half maximal speed for 30 min. Check breakage of the cells microscopically.
6. Collect the cell lysate. Wash the glass beads with cold 1 M NaCl and collect the washings until they are clear.
7. Spin the walls down at 1500g for 8 min and wash them 2× with ice-cold 1 M NaCl and 3× with 1 mM PMSF.
8. Add 5 µl extraction buffer/mg (wet weight) of cell walls. This buffer consists of 50 mM Tris-HCl, pH 7.8, 0.1 M EDTA, 2% SDS, to which 2.8 µl mercaptoethanol/ml buffer (40 mM) are added just before use.
9. Heat the cell wall suspension at 100°C for 5 min.
10. Cool down to RT and wash the cell walls 5× with water and 1× with 0.1 M NaAc, pH 5.5, 1 mM PMSF.
11. Resuspend the cell walls in the wash buffer: 2 µl mg^{-1} (wet weight) of walls.
12. Add 0.5 µl of a stock solution of Mollusk laminarinase (Sigma) per mg of walls and incubate at 37°C for 2 h. The stock solution of laminarinase contains 500 mU/ml.
13. Add fresh Mollusk laminarinase: 1 µl of stock solution per mg of walls and incubate again.
14. Add an equal volume of 2× SDS-PAGE sample buffer, and heat at 100°C for 5 min.
15. Spin down at 10 000g for 5 min and analyze the supernatant by SDS-PAGE.

---

that CWPs can be released by using aqueous HF, which specifically cleaves phosphodiester bonds. HF-released CWPs often appear as relatively discrete bands after SDS-PAGE (Kapteyn et al., 1995). This approach is generally useful for isolating CWPs from fungi and also allows determination of their N-terminal sequences and cloning of the corresponding genes (E. Schoffelmeer, personal communication). Figure 3 also explains why (partial) digestion with an endo-β1,6-glucanase or endo-β1,3-glucanase can release CWPs. We have successfully used *Trichoderma harzianum* β1,6-glucanase (Delacruz et al., 1995), a recombinant β1,3-glucanase, and Mollusk laminarinase or *Trichoderma* laminarinase for this purpose (Kapteyn et al., 1996; Van Der Vaart et al., 1995; see also Protocol 3). One should realize that laminarinase preparations contain not only β1,3-glucanase but also β1,6-glucanase activities. We normally terminate β-glucanase digestions before the digestion is complete, because this allows the detection of incompletely digested β-glucan chains that are still linked to the protein. To maximize the yield of liberated CWPs, the cell walls should be extracted with hot SDS after incubation (Protocol 3).

B. Raising Antibodies Against β-Glucans

Figure 3 predicts that CWPs released by β-glucanases may still carry part of the original β-glucan. This can be confirmed by using antisera specifically directed against either β1,3-glucan or β1,6-glucan. As carbohydrates *per se* are not very antigenic, they first have to be coupled to a carrier protein. We have successfully used the following general procedure. First, laminarin (β1,3-glucan) and pustulan (β1,6-glucan) were partially hydrolyzed (Montijn et al., 1994). Oligosaccharides in the range of 5–15 sugar residues long were isolated by gel filtration and, subsequently, covalently conjugated to the lysine residues of bovine serum albumin (BSA) by reductive amination (Roy et al., 1984). Keyhole limpet hemocyanin is probably a useful alternative for BSA. The specificities of the antisera raised in this way can be determined by testing them on CWPs in combination with competition experiments using periodate-treated laminarin or pustulan (Montijn et al., 1994). The antisera can be further characterized by periodate oxidation of the β-glucan carrying proteins used in analysis. Sugars carrying at least two neighboring hydroxyl groups such as found in β1,6-glucan are sensitive to periodate, whereas β1,3-glucan, which lacks sugars with two neighboring hydroxyl groups, is resistant to periodate. The specificity of the antisera can be further improved by affinity purification against β1,3-glucan or β1,6-glucan oligosaccharides immobilized on an epoxy-sepharose column (Kapteyn et al., 1996).

C. Detection of cell wall proteins

Separation of liberated CWPs by SDS-PAGE reveals that they generally have large (apparent) molecular masses of up to 500 kDa and higher depending on the enzyme used, requiring 3–15% gradient gels. Proteins

> **Protocol 4.** Protein assay of isolated cell walls
>
> 1. Resuspend 30 mg wet weight of isolated walls in 0.1 ml 1 N NaOH.
> 2. Boil the suspension for 10 min.
> 3. Neutralize the suspension by adding 0.1 ml 1 N HCl.
> 4. Spin the cell walls down at 10 000g for 5 min.
> 5. Take a 50 µl sample from the supernatant.
> 6. Add 1 ml of BCA protein reagent (Pierce).
> 7. Incubate the mixture at 37°C for 30 min.
> 8. Measure the OD at 562 nm, and estimate the protein content using a calibration curve generated using BSA in 0.5 M NaCl.

can be detected using either silver staining or western blotting. Silver staining is preceded by periodate treatment to increase sensitivity (De Nobel *et al.*, 1989). Western blots can be developed using peroxidase-conjugated ConA (Protocol 1) or antipeptide antibodies. In the latter case, it is essential to pretreat blotted proteins with periodate (50 mM periodate/0.1 M sodium acetate at pH 4.5 for 30 min) to increase the accessibility of the protein to the antibodies (J. C. Kapteyn, personal communication).

Depending on the procedure used to liberate CWPs, one may expect either relatively discrete bands or extended smears (Kapteyn *et al.*, 1996). In general, using HF results in discrete bands with the lowest molecular mass, but the yield may be incomplete. β1,6-Glucanase and laminarinase will also release relatively discrete bands with a slightly higher molecular mass, but will give considerably higher yields. The use of pure β1,3-glucanase results in large smears with extremely high apparent molecular masses.

To assay the protein content of isolated cell walls, the following procedure is used (J. H. Kapteyn and S. S. C. Brekelmans, personal communication; adapted from Smith *et al.*, 1985). Isolated walls are first boiled in 1 N NaOH for 10 min. The suspension is neutralized with 1 N HCl, centrifuged, and a sample from the supernatant is assayed for protein content (Protocol 4).

◆◆◆◆◆◆ VI. SECRETORY PROTEINS FOUND IN THE CULTURE MEDIUM

S. cerevisiae secretes a limited number of proteins into the culture medium and only in low amounts (Bussey *et al.*, 1983). Analysis of medium proteins therefore often requires prior concentration. In general, two methods can be recommended. The method of Ozols (1990) uses deoxycholate in combination with trichloroacetic acid to precipitate medium proteins. For smaller volumes the method by Wessel and Flügge (1984), which uses methanol and chloroform to precipitate proteins and which is faster and more efficient, is recommended. As described in the next section, chitinase can be specifically concentrated by adsorbing it to chitin (Abeijon *et al.*, 1993).

When, as a result of drug treatment or mutant background, the incorporation of CWPs into the wall is affected, non-glucosylated or partially β-glucosylated CWPs are found in the culture medium, where they can be detected as additional protein bands using SDS-PAGE. Alternatively, one can quantify the amount of a specific CWP, such as Cwp1p, in the culture medium using antisera. We have successfully raised a polyclonal antiserum against Cwp1p using a protein-coupled peptide derived from Cwp1p with the sequence: DDGKLKFDDDKYAV (J. H. Vossen *et al.*, personal communication).

◆◆◆◆◆◆ VII. GENERAL DEFECTS IN *N*- AND *O*-GLYCOSYLATION

Defects in *N*-glycosylation may result from defects in genes involved in either the assembly of the dolichol-linked oligosaccharide precursor of *N*-chains (*ALG* genes), the transfer of this chain in the ER to selected asparagine residues [oligosaccharyl transferase (OT-ase) genes], or further processing of this chain in both ER and Golgi (Lehle and Tanner, 1995). Other genes may result indirectly in defective *N*-glycosylation. Interestingly, defective *N*-glycosylation is often correlated with increased sensitivity to aminoglycosides such as hygromycin, gentamycin and kanamycin (Dean, 1995) and to increased resistance to vanadate (Ballou *et al.*, 1991). Such genes can therefore be easily detected. A possible defect in *N*-glycosylation can be confirmed by determining the electrophoretic mobility of invertase using native electrophoresis in combination with enzymatic activity staining *in situ* (Ballou, 1990; and modified by Abeijon *et al.*, 1993), or by western blotting using invertase-specific antibodies. When the addition of *N*-chains to the growing peptide chains in the ER is inefficient, carboxypeptidase Y (CPY) can be studied as a marker protein. This vacuolar protein normally has four *N*-chains, which are not outer-glycosylated. Decreased efficiency in the transfer of *N*-chains leads to the appearance of underglycosylated forms of CPY and a characteristic pattern of up to five discrete bands running closely to each other (te Heesen *et al.*, 1992). For a more detailed discussion of the properties of individual mannan-defective (*mnn*) mutants and their detection one is referred to Ballou (1990).

Interestingly, defects in *O*-glycosylation often result in (partial) resistance to K1 killer toxin, which is indicative of decreased levels of β1,6-glucan in the cell wall (Brown *et al.*, 1994). A more direct approach to detect defects in *O*-glycosylation is based on the analysis of the extracellular protein chitinase, which is heavily *O*-glycosylated and has no *N*-chains (Orlean *et al.*, 1991). Chitinase can be easily concentrated and purified from the culture medium by adsorption to chitin (Abeijon *et al.*,

1993). The fully glycosylated form has a molecular mass of 150 kDa whereas the nonglycosylated form has a mass of 60 kDa (Orlean *et al.*, 1991). In addition, the variation in the lengths of the *O*-chains can be determined. Under mild alkaline conditions, the *O*-chains are released from the protein backbone (Ballou, 1990). They can subsequently be separated by HPLC (Montijn *et al.*, 1994), or by thin-layer chromatography after prior labeling of the cells with [2-^{3}H]-mannose (Abeijon *et al.*, 1993).

N-chains and, more recently, also *O*-chains are known to contain mannosyl phosphate groups. As these groups are responsible for binding the cationic dye Alcian blue, this compound can be used to screen for the loss of mannosyl phosphate groups at the cell surface (Ballou, 1990).

◆◆◆◆◆◆ VIII. GENERAL SCREENS FOR THE IDENTIFICATION OF CELL-WALL-RELATED GENES

The cell wall is a supramolecular structure, and its correct construction depends on a controlled supply of various components in space and time during the cell cycle. It is therefore not surprising that a large group of genes is (in)directly involved in cell wall construction. To detect cell-wall related genes, several general screens have been developed. The most extensively and successfully used screen is based on the disturbing action of Calcofluor White on cell wall construction (Ram *et al.*, 1994). Mutations in cell wall-related genes generally result in hypersensitivity to Calcofluor White (Protocol 5). Other useful general screens are based on the observation that mutations in cell-wall-related genes often result in hypersensitivity to SDS (Kanikennulat *et al.*, 1995) and in increased sensitivity or resistance to cell wall degrading enzymes (Ram *et al.*, 1994). More

---

Protocol 5. Calcofluor White spot assay: Identification of cell-wall-related genes

1. Filter-sterilize a 1% (w/v) Calcofluor White stock solution and store at room temperature.
2. Cool YPD medium down to 70°C, add Calcofluor to a final concentration of 50 μg/ml, and pour the plates. When minimal medium is used, buffer the medium with 1% (w/v) MES and bring the pH to 6.0 with NaOH. Note that the sensitivity to Calcofluor may vary depending on the genetic background.
3. Make a tenfold dilution series of a cell suspension containing 10 OD$_{600}$ units per ml.
4. Spot 3 μl of each dilution in a row on a YPD plate containing 50 μg ml^{-1} Calcofluor: undiluted + 10×, 100×, 1000× and 10 000× diluted.
5. Score growth after 2 days at 30°C. Wild-type cells (S288C) grow well on all spots.

---

specialized tests relating to *N*- and *O*-glycosylation of CWPs and to incorporation of proteins into the cell wall have already been discussed above. Additional screens for further characterization of cell-wall-related genes are discussed in Ram *et al*. (1994).

Acknowledgements

We thank Dr Frans Hochstenbach for critically reading the manuscript. This work was financially supported by the Dutch Ministry of Economic Affairs, GLAXO-Wellcome, Unilever, and by the EC program EUROFAN.

References

Abeijon, C., Yanagisawa, K., Mandon, E. C., Hausler, A., Moremen, K., Hirschberg, C. B. and Robbins, P. W. (1993). Guanosine diphosphatase is required for protein and sphingolipid glycosylation in the Golgi lumen of *Saccharomyces cerevisiae*. *J. Cell Biol*. **122**, 307–323.

Ballou, C. E. (1990). Isolation, characterization, and properties of *Saccharomyces cerevisiae mnn* mutants with non-conditional protein glycosylation defects. *Methods Enzymol*. **185**, 440–470.

Ballou, L., Hitzeman, R. A., Lewis, M. S. and Ballou, C. E. (1991). Vanadate-resistant yeast mutants are defective in protein glycosylation. *Proc. Natl Acad. Sci. USA* **88**, 3209–3212.

Brown, J. L., Roemer, T., Lussier, M., Sdicu, A-M. and Bussey, H. (1994). In *Molecular Genetics of Yeast. A Practical Approach: The K1 Killer Toxin: Molecular and Genetic Applications to Secretion and Cell Surface Assembly* (J. R. Johnston, ed.), pp. 217–232. Oxford University Press, Oxford.

Burda, P., te Heesen, S., Brachat, A., Wach, A., Düsterhoeft, A. and Aebi H. (1996). Stepwise assembly of the lipid-linked oligosaccharide in the endoplasmic reticulum of *Saccharomyces cerevisiae*: identification of the *ALG9* gene encoding a putative mannosyl transferase. *Proc. Natl Acad. Sci. USA* **93**, 7160–7165.

Bussey, H., Steinmetz, O. and Saville, D. (1983). Protein secretion in yeast: two chromosomal mutants that oversecrete killer toxin in *Saccharomyces cerevisiae*. *Curr. Genet*. **7**, 449–456.

Caro, L. H. P., Tettelin, H., Vossen, J. H., Ram, A. F. J., Van Den Ende, H. and Klis, F. M. (1997). *In silicio* identification of glycosylphosphatidylinositol-anchored plasma membrane and cell wall proteins of *Saccharomyces cerevisiae*. *Yeast* **13**, 1477–1489.

Conzelmann, A., Fankhauser, C. and Desponds, C. (1990). Myoinositol gets incorporated into numerous membrane glycoproteins of *Saccharomyces cerevisiae*; incorporation is dependent on phosphomannomutase (SEC53). *EMBO J*. **9**, 653–661.

Dean, N. (1995). Yeast glycosylation mutants are sensitive to aminoglycosides. *Proc. Natl Acad. Sci. USA* **92**, 1287–1291.

Delacruz, J., Pintortoro, J. A., Benitez, T. and Llobell, A. (1995). Purification and characterization of an endo-beta-1,6-glucanase from *Trichoderma harzianum* that is related to its mycoparasitism. *J. Bacteriol*. **177**, 1864–1871.

De Nobel, J. G., Dijkers, C., Hooijberg, E. and Klis, F. M. (1989). Increased cell wall porosity in *Saccharomyces cerevisiae* after treatment with dithiothreitol or EDTA. *J. Gen. Microbiol*. **135**, 2077–2084.

Dexaerde, A. D., Supply, P. and Goffeau, A. (1996). Review: Subcellular traffic of the plasma membrane H+-ATPase in *Saccharomyces cerevisiae*. *Yeast* **12**, 907–916.

Douglas, C. M., Foor, F., Marrinan, J. A., Morin, N., Nielsen, J. B., Dahl, A. M., Mazur, P. *et al.* (1994). The *Saccharomyces cerevisiae FKS1* (*ETG1*) gene encodes an integral membrane protein which is a subunit of 1,3-beta-D-glucan synthase. *Proc. Natl Acad. Sci. USA* **91**, 12907–12911.

Fankhauser, C., Homans, S. W., Thomasoates, J. E., McConville, M. J., Desponds, C., Conzelmann, A. and Ferguson, M. A. J. (1993). Structures of glycosyl-phosphatidylinositol membrane anchors from *Saccharomyces cerevisiae*. *J. Biol. Chem.* **268**, 26365–26374.

Gentzsch, M. and Tanner, W. (1996). The *PMT1* family: protein O-glycosylation in *Saccharomyces cerevisiae* is vital. *EMBO J.* **15**, 5752–5759.

Haüsler, A., Ballou, L., Ballou, C. E. and Robbins, P. W. (1992). Yeast glycoprotein biosynthesis – *MNT1* encodes an alpha-1,2-mannosyltransferase involved in O-glycosylation. *Proc. Natl Acad. Sci. USA* **89**, 6846–6850.

Herscovics, A. and Orlean, P (1993). Glycoprotein biosynthesis in yeast. *FASEB J.* **7**, 540–550.

Hill, K., Boone, C., Goebl, M., Puccia, R., Sdicu, A. M. and Bussey, H. (1992). Yeast *KRE2* defines a new gene family encoding probable secretory proteins, and is required for the correct N-glycosylation of proteins. *Genetics* **130**, 273–283.

Kanikennulat, C., Montalvo, E. and Neff, N. (1995). Sodium orthovanadate-resistant mutants of *Saccharomyces cerevisiae* show defects in Golgi-mediated protein glycosylation, sporulation and detergent resistance. *Genetics* **140**, 933–943.

Kapteyn, J. C., Montijn, R. C., Dijkgraaf, G. J. P., Van Den Ende, H. and Klis, F. M. (1995). Covalent association of beta-1,3-glucan with beta-1,6-glucosylated mannoproteins in cell walls of *Candida albicans*. *J. Bacteriol.* **177**, 3788–3792.

Kapteyn, J. C., Montijn, R. C., Vink, E., Delacruz, J., Llobell, A., Douwes, J. E., *et al.* (1996). Retention of *Saccharomyces cerevisiae* cell wall proteins through a phosphodiester-linked beta-1,3-/beta-1,6-glucan heteropolymer. *Glycobiol.* **6**, 337–345.

Lehle, L. and Tanner, W. (1995). In *Glycoproteins* (J. Montreuil, J. F. G. Vliegenthart and H. Schachter, eds), pp. 475–509. Elsevier, Amsterdam.

Lu, C. F., Kurjan, J. and Lipke, P. N. (1994). A pathway for cell wall anchorage of *Saccharomyces cerevisiae* alpha-agglutinin. *Mol. Cell. Biol.* **14**, 4825–4833.

Lu, C. F., Montijn, R. C., Brown, J. L., Klis, F., Kurjan, J., Bussey, H. and Lipke, P. N. (1995). Glycosyl phosphatidylinositol-dependent cross-linking of alpha-agglutinin and beta 1,6-glucan in the *Saccharomyces cerevisiae* cell wall. *J. Cell Biol.* **128**, 333–340.

Lussier, M., Gentzsch, M., Sdicu, A. M., Bussey, H. and Tanner, W. (1995). Protein O-glycosylation in yeast – the PMT2 gene specifies a second protein O-manno-syltransferase that functions in addition to the PMT1-encoded activity. *J. Biol. Chem.* **270**, 2770–2775.

Montijn, R. C., Van Rinsum, J., Van Schagen, F. A. and Klis, F. M. (1994). Glucomannoproteins in the cell wall of *Saccharomyces cerevisiae* contain a novel type of carbohydrate side chain. *J. Biol. Chem.* **269**, 19338–19342.

Müller, G., Gross, E., Wied, S., Bandlow, W. (1996). Glucose-induced sequential processing of a glycosylphosphatidylinositol-anchored ectoprotein in *Saccharomyces cerevisiae*. *Mol. Cell. Biol.* **16**, 442–456.

Nakayama, K., Nagasu, T., Shimma, Y., Kuromitsu, J. and Jigami, Y. (1992). *OCH1* encodes a novel membrane bound mannosyltransferase – outer chain elongation of asparagine-linked oligosaccharides. *EMBO J.* **11**, 2511–2519.

Orlean, P., Kuranda, M. J. and Albright, C. F. (1991). Analysis of glycoproteins from *Saccharomyces cerevisiae*. *Methods Enzymol.* **194**, 682–697.

Ozols, J. (1990). In *Methods in Enzymology: Amino Acid Analysis* (M. P. Deutscher, ed.), vol. 182, pp. 587-601. Academic Press, London.

Ram, A. F. J., Wolters, A., Ten Hoopen, R. and Klis, F. M. (1994). A new approach for isolating cell wall mutants in *Saccharomyces cerevisiae* by screening for hypersensitivity to Calcofluor White. *Yeast* **10**, 1019–1030.

Roy, R., Katzenellenbogen, E. and Jennings, H. J. (1984). Improved procedures for the conjugation of oligosaccharides to protein by reductive amination. *Can. J. Biochem. Cell Biol.* **62**, 270–275.

Schneider, P. and Ferguson, M. A. J. (1995). Microscale analysis of glycosyl-phosphatidylinositol structures. *Methods Enzymol.* **250**, 614–630.

Sipos, G., Puoti, A. and Conzelmann, A. (1995). Biosynthesis of the side chain of yeast glycosylphosphatidylinositol anchors is operated by novel mannosyl-transferases located in the endoplasmic reticulum and the Golgi apparatus. *J. Biol. Chem.* **270**, 19709–19715.

Smith, P. K., Krohn, R. I., Hermanson, G. T., Mallia, A. K., Gartner, F. H., Provenzano, M. D., Fujimoto, E. K. *et al.* (1985). Measurement of protein using bicinchoninic acid. *Anal. Biochem.* **150**, 76–85.

Strahlbolsinger, S., Immervoll, T., Deutzmann, R. and Tanner, W. (1993). *PMT1*, the gene for a key enzyme of protein *O*-glycosylation in *Saccharomyces cerevisiae*. *Proc. Natl Acad. Sci. USA* **90**, 8164–8168.

te Heesen, S., Janetzky, B., Lehle, L. and Aebi, M. (1992). The yeast *WBP1* is essential for oligosaccharyl transferase activity *in vivo* and *in vitro*. *EMBO J.* **11**, 2071–2075.

Udenfriend, S. and Kodukula, K. (1995). How glycosyl-phosphatidylinositol-anchored membrane proteins are made. *Annu. Rev. Biochem.* **64**, 563–591.

Van Der Vaart, J. M., Caro, L. H., Chapman, J. W., Klis, F. M. and Verrips, T. (1995). Identification of three mannoproteins in the cell wall of *Saccharomyces cerevisiae*. *J. Bacteriol.* **177**, 3104–3110.

Vossen, J. H., Müller, W. H., Lipke, P. N. and Klis, F. M. (1997). Restrictive glyco-sylphosphatidylinositol anchor synthesis in *cwh6/gpi3* yeast cells causes aberrant biogenesis of cell wall proteins. *J. Bacteriol.* **179**, 2202–2209.

Wessel, D. and Flügge, U. I. (1984). A method for the quantitative recovery of protein in dilute solution in the presence of detergents and lipids. *Anal. Biochem.* **138**, 141–143.

Wojciechowicz, D., Lu, C. F., Kurjan, J. and Lipke, P. N. (1993). Cell surface anchorage and ligand-binding domains of the *Saccharomyces cerevisiae* cell adhesion protein alpha-agglutinin, a member of the immunoglobulin superfamily. *Mol. Cell. Biol.* **13**, 2554–2563.

14 Studying Posttranslational Modifications in Yeast: Intracellular Proteins

Michael J. R. Stark
Department of Biochemistry, University of Dundee, Dundee, UK

◆◆◆

CONTENTS

List of Abbreviations

CIAP	Calf intestinal alkaline phosphatase
ES MS	Electrospray mass spectrometry
GST	Glutathione-S-transferase
HAT	Histone acetyl transferase
MAP	Mitogen-activated protein
MALDI MS	Matrix-assisted laser desorption/ionization mass spectrometry
MS	Mass spectrometry
ORF	Open reading frame
PFTase	Protein farnesyl transferase
PGGTase-I	Protein geranylgeranyl transferase I
PP2A	Protein phosphatase 2A
PVDF	Polyvinylidene difluoride
Ub	Ubiquitin
YPD	Yeast Protein Database

◆◆◆◆◆◆ I. INTRODUCTION

For many intracellular proteins, translation on the ribosome is not the end of the story and further posttranslational modifications may occur. Such modifications may be important for the localization or activity of the

Table I. Sequence motifs indicative of potential protein modification

Modification	Motif(s)[a]	Location[b]
Myristoylation	G{EDKRHPYFV}XX(STAGCN){P}	N
Geranylgeranylation	C{DENQ}(LIVM)(LIF),CC, CCXX, or CXC	C
Farnesylation	C{DENQ}(LIVM)[CSMQA]	C
Palmitoylation	C	–
Carboxymethylation	prenyl-C	C
Mak3p N-acetylation	MLRF	N
Phosphorylation	S, T, Y	–

[a] X, any residue; ( ), degenerate residues; { }, excluded residue(s)
[b] N, extreme amino terminus; C, extreme carboxy-terminus; –, anywhere

protein, and protein modification may also form part of a regulatory mechanism controlling the function of the particular polypeptide. For many yeast investigators, the first hint of protein modification may be the detection of multiple forms of a protein by western blot analysis, the presence of a particular sequence motif in the protein of interest or unexpected behavior of the protein during cell fractionation. Table 1 shows several such sequence motifs that may indicate the potential for particular types of protein modification. Phosphorylation is probably the most frequently considered example of posttranslational modification, and the presence of a slower-migrating form of a protein is usually taken to indicate that a protein is most likely to be phosphorylated. However, phosphorylation is by no means the only posttranslational modification to which yeast proteins may be subjected, and well over 100 different modified amino acid residues may be found in specific proteins (Krishna and Wold, 1997). This chapter will attempt to survey the most commonly occurring types of modification that may occur, and point the reader towards techniques for identifying them and examining their function. In addition, it will summarize the application of mass spectrometry to the study of protein modification because this has recently become one of the most powerful methods available for identifying posttranslational modifications.

◆◆◆◆◆◆ II. PHOSPHORYLATION

A. Determining that a Protein is Phosphorylated

When one or more slower-migrating forms of a protein are observed on a western blot, the most likely scenario is that these represent differentially phosphorylated forms. The simplest way to demonstrate this is to show that the extra bands can be converted back to the fastest-migrating form following phosphatase treatment. Typically, the yeast protein sample can be incubated with a non-specific phosphatase such as calf intestinal alkaline phosphatase (CIAP), because the majority of phosphate groups can be removed in this way. There are many examples of the application of this type of approach to yeast proteins (e.g. Donaldson and Kilmartin, 1996; Kuo *et al.*, 1997). However, phosphorylation frequently may not change the apparent mobility of a protein on SDS-PAGE analysis. Therefore, a complementary approach involves *in vivo* labeling of the protein with [^{32}P]orthophosphate followed by affinity isolation and autoradiography, for example by immune precipitation using a specific antibody or after epitope-tagging the protein of interest. Given the universality of phosphorylation as a reversible regulatory mechanism, the phosphorylation state of a protein will frequently vary depending on the growth conditions or phase of the cell division cycle. It may therefore be necessary to bear this in mind when looking at a specific protein of interest and phosphorylation, detectable either by labeling or by mobility shift, may only be evident under particular conditions.

Where phosphorylation can be followed as a mobility shift by western blot analysis, cell extracts can be simply prepared in boiling SDS-PAGE sample buffer (e.g. Stirling and Stark, 1996). However, when extracts are prepared under non-denaturing conditions for phosphorylation studies, it is necessary to include protein kinase and protein phosphatase inhibitors to preserve the phosphorylation state of a protein until it has been isolated (Hardie *et al.*, 1993). Typically, sodium vanadate (100 μM) is used to inhibit tyrosine-specific and dual-specificity phosphatases, while sodium fluoride (50 mM) plus sodium pyrophosphate (5 mM) inhibit most other phosphatases. EDTA (5 mM) inhibits protein kinases (which require Mg^{2+}-ATP) and also inhibits protein phosphatase 2B, which is Ca^{2+}-dependent. Microcystin-LR also inhibits most phosphatases of the PPP family (such as PP1 and PP2A).

B. What Type of Phosphorylation is Present?

Although phosphorylation on serine, threonine and tyrosine residues is by far the most common situation, histidine, lysine, arginine and aspartate are all possible phosphoaccepting residues. Phosphate linked to these latter four residues is characterized by its acid lability (e.g. to 0.1 M HCl at 60°C for 10 min), while β-elimination (e.g. using 3 M NaOH at 120°C for 3 h) removes phosphate from serine, threonine and arginine but not from tyrosine and the other residues (Krishna and Wold, 1997). The investigation of the Sln1p-Ypd1p-Ssk1p phosphorylation system which regulates the Hog1p mitogen-activated protein (MAP) kinase pathway in yeast provides a good example of how yeast proteins phosphorylated on histidine and aspartate can be studied effectively (Posas *et al.*, 1996).

To demonstrate directly the nature of the phosphorylated residue(s) in a ^{32}P-labeled polypeptide, phosphoamino acid analysis can be carried out in the case of the acid-stable phosphoserine, phosphothreonine and phosphotyrosine modifications (van der Geer *et al.*, 1993). This procedure involves acid hydrolysis of the phosphoprotein followed by two-dimensional electrophoresis to resolve the three phosphoamino acids. For proteins containing phosphoserine or phosphothreonine, the catalytic subunit of protein phosphatase 2A (PP2A) is a useful tool for demonstrating phosphorylation. PP2A is a broad-specificity protein phosphatase and has the advantage that it is exquisitely sensitive to the inhibitors okadaic acid and microcystin-LR, either of which can be added to control incubations to demonstrate that the conversion of slower-migrating to faster-migrating forms of a protein (or a ^{32}P-labeled species to an unlabeled form) is really due to dephosphorylation. Note that PP2A can also dephosphorylate phosphohistidine. For phosphotyrosine-containing proteins, commercially available phosphotyrosine-specific protein phosphatases such as PTP1B can be used in the same way, and in fact the combination of both types of protein phosphatase was used in the original demonstration that MAP kinases require both threonine and tyrosine phosphorylation to be active (Anderson et al., 1990). Finally, anti-phosphotyrosine antibodies such as 4G10 can be used both to confirm the presence of this modification in a protein and as a tool to monitor the phosphorylation state under different conditions. Such an approach is particularly useful for examining MAP kinase activation in yeast as in other systems (e.g. Wurgler-Murphy et al., 1997; Zarzov et al., 1996).

C. Identifying Site(s) of Phosphorylation

Basic methods for determining the site(s) of phosphorylation in proteins have been previously described in detail and so will only be summarized briefly here (see, for example, Hardie et al., 1993; Krishna and Wold, 1997; Morrice and Carrey, 1997). Firstly, the phosphoprotein must generally be ^{32}P-labeled, either by isolation following in vivo labeling or after an in vitro phosphorylation reaction. In the former case, this might typically involve use of an affinity-tagged protein or a larger-scale immune precipitation. ^{32}P-labeled peptides are then isolated by HPLC following chemical or enzymatic cleavage of the protein. Each labeled polypeptide can be sequenced both under normal conditions (to demonstrate where it originates in the protein of interest) and using modified conditions (using protein immobilized on PVDF membrane and reagents that permit effective extraction of the highly charged phosphoamino acid derivative). In the latter case, the particular Edman cycle(s) in which ^{32}P is released allow the phosphoamino acid(s) to be placed within the sequence. For unlabeled proteins, mass spectrometry can be used to identify those peptides in a mixture derived from the protein of interest, which lose 80 mass units when dephosphorylated with a protein phosphatase (see above). However, because many peptides will contain multiple phosphorylatable residues, this approach may not definitively identify the phosphorylation site(s). Unlabeled phosphoserine and phosphothreonine can also be identified during peptide sequencing following conversion by β-elimination to derivatives yielding distinctive products during Edman degradation (Krishna and Wold, 1997). Where a candidate phosphorylation site has been identified, mutation to a non-phosphorylatable residue (e.g. serine or threonine to alanine, or tyrosine to phenylalanine) can be used to confirm that the mutant protein can no longer be phosphorylated.

D. Protein Phosphorylation and Protein Function

When phosphorylation sites have been identified in a protein, identifying the protein kinase(s) responsible and determining whether the phosphorylation state at any site is important for function are clearly key questions. Although the sites preferred for many mammalian protein kinases are known, in yeast relatively few of these have been rigorously established; Table 2 lists several cases where consensus sequences have been demonstrated. Otherwise, where there are good candidate yeast homologs of mammalian protein kinases, it is at least possible that they will show similar specificity and this may give clues as to which protein kinase may modify the protein of interest. Where both kinase and substrate can be obtained in pure form (e.g. by expression of recombinant proteins), then experiments can be performed *in vitro* to show that the kinase can modify the candidate substrate at a particular site (Errede *et al.*, 1993). However, to show that this occurs *in vivo* requires null mutations in the relevant kinase gene (or conditional mutations if it is an essential protein kinase), which also begs the question of whether the phosphorylation is direct or via activation of yet another protein kinase. Thus establishing the precise nature of a phosphorylation event may not be at all straightforward, and a combination of both *in vivo* and *in vitro* approaches are likely to be required. Protein phosphatases also have a role to play in controlling the phosphorylation state of proteins and with the increasing number of mutants now available (Stark, 1996), it is feasible to test the effect of inactivating a specific yeast protein phosphatase on the phosphorylation state of a given protein of interest. However, this approach clearly is also open to the caveat of direct versus indirect effects.

Where the protein of interest has a readily assayable function *in vitro*, the role of phosphorylation can be tested by comparing the activity of phosphorylated and non-phosphorylated forms, using protein kinases and protein phosphatases as appropriate to change the phosphorylation state of the isolated protein. This can be coupled with the use of non-phosphorylatable mutant proteins and mutants in which the phosphorylatable residue has been changed to an aspartate (mimicking the negative charge on the phosphate group), although this latter approach may not always work. The mutants have the advantage that they can be

Table 2. Candidate recognition sites for yeast protein kinases

Kinase	Phosphorylation site[a]
Cdc28p (p34^{cdc2} homolog)	(ST)PX(KR)
cAMP-dependent protein kinase	RRX(ST)ϕ
Casein kinase II (Cka1p, Cka2p)	SXX(DE) [(DE)(DE)(DE)]
Pkc1p (protein kinase C)	[RXX] (ST)X(KR)
Snf1p	ϕXRXXSXXXϕ

[a]X, any residue; ( ), degenerate residues; ϕ, hydrophobic; [ ] optional residues. The phosphorylated residue is underlined

tested for function *in vivo* as well as *in vitro*, assisting in the demonstration of the significance of the phosphorylation event. In either case, the situation is made more complex where multiple phosphorylations, possibly by more than one protein kinase, are involved.

◆◆◆◆◆◆ III. LIPID MODIFICATIONS

A. Myristoylation

N-myristoylation of proteins in yeast is carried out by the *NMT1* gene product (myristoyl CoA:protein N-myristoyl transferase), which is an essential gene (Duronio *et al.*, 1989). This modification occurs exclusively on N-terminal glycine residues and the substrate specificity of the enzyme has been studied in detail, leading to a generalized consensus sequence (Table 1) such that the possibility of N-myristoylation can be predicted knowing the amino acid sequence encoded by an open reading frame (ORF) (Towler *et al.*, 1988). The Yeast Protein Database (YPD; Payne and Garrels, 1997) currently lists eight known N-myristoylated yeast proteins, including the PPZ protein phosphatases (Clotet *et al.*, 1996), the B subunit of yeast calcineurin (Cyert and Thorner, 1992) and several GTP-binding proteins including Gpa1p (Song *et al.*, 1996).

Proof of myristoylation requires labeling of yeast cells with [^{3}H]myristate followed by recovery of the protein in question so that incorporation of the label can be demonstrated, and several groups have described a procedure for achieving this (Cyert and Thorner, 1992; Song *et al.*, 1996; Zhu *et al.*, 1995). Cells are first treated with cerulenin for 15 min to block endogenous fatty acid synthesis, then labeled with [^{3}H]myristate for up to 1.5 h. The cells can then be broken by lysis with glass beads and a protein extract prepared. Because N-myristoylated proteins can be membrane-associated, it is important to ensure that proteins are completely solubilized at this stage. Recovery of the protein of interest requires that an antibody be available or that the protein has been epitope-tagged, so that an immune precipitation can be performed. The [^{3}H]-labeled protein can be examined by fluorography following SDS-PAGE gel electrophoresis of the affinity-isolated material. A western blot should be carried out in parallel on part of the sample to ensure that the labeled band co-migrates with the protein in question. It is also possible to release the fatty acid from the protein by acid hydrolysis, and to verify that the fatty acid is myristate using reverse-phase HPLC (Simon and Aderem, 1992). This can be important because myristate may be converted to palmitate before incorporation into the protein *in vivo*. Furthermore, whereas palmitate can be released from proteins by acid hydrolysis or hydroxylamine, N-myristate is not susceptible to release using the latter reagent (Towler *et al.*, 1988). An alternative to using antibodies is to fuse the amino-terminal region of the protein of interest to glutathione-S-transferase (GST) so that the potentially labeled fusion protein can be examined following affinity isolation on glutathione agarose. A $(His)_6$ tag might also be useful in the case of a very insoluble protein, because using this approach, the affinity isolation step could be performed on fully denatured material.

Because the N-terminal glycine is essential for N-myristoylation to take place, mutagenesis (typically to alanine) provides an opportunity both to confirm that a mutant protein is no longer labeled and to test the functional significance of the modification *in vivo*. While N-myristoylation is essential for function in some cases (Song *et al.*, 1996), in others it is apparently not (Zhu *et al.*, 1995).

B. Palmitoylation

The best-known examples of palmitoylated proteins are the small GTPase proteins of the *ras* superfamily. Many of these proteins are palmitoylated on a cysteine residue at the C-terminus, to which the fatty acid is joined by a thioether linkage. However, other categories of protein are also palmitoylated, for example the synaptobrevin homolog Snc1p is palmitoylated on a cysteine residue adjacent to its transmembrane domain (Couve *et al.*, 1995). As with N-myristoylation, demonstration of palmitoylation requires *in vivo* labeling, but in this case using [³H]palmitic acid (Bhattacharya *et al.*, 1995; Couve *et al.*, 1995). Cells are labeled for 2–4 h and then the protein recovered by an affinity method such as immune precipitation. The recovered protein can then be separated by SDS-PAGE, the labeled protein visualized by fluorography and a western blot run in parallel to ensure co-migration of the target protein. Since palmitoylated proteins behave essentially as integral membrane proteins even when they do not also have a transmembrane domain (Towler *et al.*, 1988), care is needed to ensure proper solubilization of the modified protein, for example by boiling in SDS buffer and then diluting the protein extract before immune precipitation (Couve *et al.*, 1995). Incorporated label should be releasable by hydroxylamine treatment or alkali, and by fixing the gel in either methanol alone or methanol plus 0.2 M NaOH it should be possible to see loss of the palmitate label (Molenaar *et al.*, 1988). While it may be possible to detect some proteins in this way, it is likely that in many cases it will be necessary to overproduce the protein (e.g. using the *GAL1* promoter) in order to obtain sufficient incorporation of the label (Molenaar *et al.*, 1988). As with myristoylation, it should also be possible to make specific mutations which can be used both to show that the mutant protein is no longer modified and to test the functional requirement for palmitoylation *in vivo*. Thus *SNC1* encodes a single cysteine which, when mutated to serine, blocks labeling by [³H]palmitic acid and leads to reduced protein stability (Couve *et al.*, 1995). Also, loss of palmitoylation in the *RAS2* C318S mutant affects some Ras2p functions and blocks plasma membrane targeting of the protein (Bhattacharya *et al.*, 1995).

IV. PRENYLATION

Prenylation is the attachment of a lipophilic isoprenoid group to a cysteine residue in a protein via a thioether linkage (Zhang and Casey, 1996). The isoprenoid modification is either farnesyl (C_{15}) or geranylgeranyl (C_{20}) and is added to the C-terminus of the protein. As in other eukaryotes, prenylation of yeast proteins is frequently signaled by the presence of a "CaaX box" at the extreme C-terminus (Table 1), where C is the modified cysteine, the "a" residues are aliphatic and X is one of either C, S, M, Q, A (farnesylation) or L, I, F (geranylgeranylation), as shown by Caplin *et al.* (1994). However, modification at related C-terminal sequences such as -CC, -CXC or -CCXX is possible (Schafer and Rine, 1992); examples of this in yeast are Ypt1p, Sec4p and the yeast casein kinase I homologs Yck1p and Yck2p (Jiang *et al.*, 1993; Vancura *et al.*, 1994). After modification, the three C-terminal residues are processed proteolytically and the now C-terminal prenylcysteine is carboxymethylated (see below). The -CC and -CXC motifs are not processed after prenylation and, in the former case, the C-terminal prenylcysteine is not carboxymethylated (Zhang and Casey, 1996). The YPD currently lists 12 known or predicted geranylger-anylated proteins and 24 farnesylated proteins in *Saccharomyces cerevisiae*; the former include small GTPase proteins of the rho and rab/ypt families and casein kinase I, while the latter include ras superfamily members, the γ subunit of a heterotrimeric G-protein (Ste18p) and the mating pheromone **a**-factor. Three enzymes catalyse these modifications (Jiang *et al.*, 1993; Omer and Gibbs, 1994; Schafer and Rine, 1992): protein farnesyl transferase (PFTase, consisting of Ram2p and Ram1p), protein geranyl-geranyl transferase I (PGGTase-I, consisting of Ram2p and Cdc43p) and PGGTase-II (Bet2p and Bet4p/Mad2p). PFTase and PGGTase-I modify CaaX-box proteins while PGGTase-II is specific for the -CC and -CXC motifs.

While the sequence motifs that signal protein prenylation are therefore quite predictable, not all such proteins are prenylated *in vivo* (Schafer and Rine, 1992). In *S. cerevisiae*, *in vivo* labeling of prenyl groups has proved ineffective because the cells are unable to take up [³H]mevalonate (Mitchell and Deschenes, 1995). This has led to the development of an alternative procedure which involves expression of the protein of interest in yeast as a [³⁵S]-labeled GST fusion, followed by complete digestion with proteases and analysis of the products by reversed-phase HPLC (Mitchell and Deschenes, 1995). Alternatively, *in vitro* assays can be used to demon-strate prenylation of a protein of interest (Caplan *et al.*, 1992; Caplin *et al.*, 1994; Jiang *et al.*, 1993). In these cases, the candidate protein is expressed in *Escherichia coli* (leading to an unmodified product) and, if necessary, purified or affinity isolated. Crude *S. cerevisiae* protein extracts can be used as a source of the transferase enzymes (Caplan *et al.*, 1992; Goodman *et al.*, 1990), or alternatively these can be generated by bacterial co-expression of the relevant polypeptides (Caplin *et al.*, 1994). In addition to the substrate and the source of enzyme, the *in vitro* assay requires either [³H]geranylgeranyl pyrophosphate or [³H]farnesyl pyrophosphate as the isoprenoid donor. Modification can be measured either by incorporation of label into acid-precipitable material or by running out the reaction products on an SDS gel. If required, the labeled prenyl group can be released from the protein by methyl iodide cleavage and then its identity checked by reverse-phase HPLC (Newman *et al.*, 1992).

◆◆◆◆◆◆ V. UBIQUITINATION

Modification of many proteins by ubiquitination targets them for degradation by the proteasome and thus many short-lived proteins in yeast are modified in this way (Hochstrasser, 1995; Jentsch, 1992). Ubiquitin (Ub) is a highly-conserved 76-residue polypeptide which can be joined to the ε-amino group of certain lysine residues in proteins via an isopeptide bond in a reaction involving three classes of enzyme termed E1 (Ub-activating enzyme), E2 (Ub-conjugating enzyme) and E3 (Ub ligase). Ubiquitin itself can be ubiquitinated on lysines 48 and 63, allowing multiubiquitin chains to be assembled and it is these that are most effective at directing proteasomal degradation. While some proteins may be constitutively ubiquitinated, in other instances the modification is regulated, for example to target proteins for degradation at certain stages of the cell division cycle.

Several approaches are available for demonstrating the ubiquitination of yeast proteins. In the simplest of these, the protein of interest can be immune precipitated and the immune complexes examined by Western blotting with commercial anti-Ub antibodies. The Ub-conjugated material will appear as multiple, slower-migrating species recognized by both the anti-Ub and the anti-protein antibodies (Yoon and Carbon, 1995). Alternatively, copper-inducible Ub and myc-epitope-tagged Ub constructs can be used (Ellison and Hochstrasser, 1991). In this instance, the protein of interest is immune precipitated from cells overexpressing either of these forms of Ub and visualized by western blotting with antibodies that recognize the protein itself; the Ub-conjugated forms migrate more slowly when the Ub contains the myc-tag and therefore can be identified as such. The myc-tag in this construct also apparently inhibits the subsequent proteolysis of the conjugated protein (Ellison and Hochstrasser, 1991). Given that the Ub-conjugated protein is unstable, radiolabeling is frequently needed to increase the sensitivity of detecting the Ub-conjugated forms in immune precipitates.

Another engineered form of Ub has been described recently that may be of particular use in examining protein ubiquitination in yeast (Willems *et al.*, 1996). In this case, Ub has been modified with both the myc epitope and $(His)_6$ tags for ease of detection and/or isolation of the Ub-conjugated protein. Two mutations within the Ub sequence itself also enhance the usefulness of this construct. Firstly, the K48R mutation blocks the formation of polyubiquitin chains, leading to reduced loss of the conjugated target protein through proteolysis by the proteasome. Secondly, the G76A mutation inhibits the removal of Ub by deubiquitinating activities present in yeast cells, rendering the modification more stable. Thus proteins modified with this Ub derivative are more readily detectable. Again, expression of the tagged and mutated Ub is copper-inducible via the *CUP1* promoter.

Finally, it is possible to study Ub conjugation using *in vitro* assays that rely either on crude yeast extracts or on at least partially purified components (Deshaies *et al.*, 1995; Kornitzer *et al.*, 1994; Zachariae and Nasmyth, 1996). The substrate can be affinity-purified protein of interest or a GST fusion protein, and either the Ub or the substrate can be radiolabeled to aid detection of the Ub-conjugated forms. Such assays are probably more useful for examining the requirements of the conjugation reaction (e.g. the need for substrate phosphorylation or the effect of mutations in known E2- or E3-encoding genes on the activity) rather than for demonstrating simply whether or not the protein of interest is modified by ubiquitination. Where the site(s) of Ub addition can be identified, mutation of the key lysine residues can be used to assess the effects of blocking ubiquitination and hence stabilizing the protein.

◆◆◆◆◆◆ VI. OTHER MODIFICATIONS

A. Acetylation

The majority of yeast proteins are modified by N-acetylation, while some proteins are acetylated elsewhere in the polypeptide chain. An example of the second group are the histones, which in yeast just as in other eukaryotic cells are acetylated on the ε-amino group of certain lysine residues. At least three activities in yeast are responsible for N-acetylation: Nat1p and Ard1p form one of these (Park and Szostak, 1992), while Nat2p is an N-terminal methionine-specific acetyltransferase (Kulkarni and Sherman, 1994) and Mak3p specifically acetylates proteins that begin with the sequence met-leu-arg-phe- (Tercero *et al.*, 1993). Protein N-acetylation usually becomes evident when attempts to determine the amino-terminal sequence of the protein fail to yield data. Because N-acetyl groups in proteins are small and cannot be specifically labeled *in vivo*, the best way to demonstrate this modification involves isolation of the amino-terminal peptide after chemical or enzymatic fragmentation of the protein, followed by mass spectrometry.

Histone acetyl transferases (HATs) fall into two groups: HAT A enzymes are localized in the nucleus and they acetylate core histones in nucleosomes, while HAT B enzymes are cytoplasmic and acetylate free histones, a step that might be involved in chromatin assembly during DNA replication (Brownell and Allis, 1996). In yeast, Gcn5p has been identified as a HAT A enzyme (Kuo *et al.*, 1996) while Hat1p and Hat2p are responsible for the major HAT B function (Parthun *et al.*, 1996). Both activities can be monitored by *in vitro* assays, in which transfer of label from [³H]acetyl coenzyme A to a suitable histone substrate catalysed by crude extracts or purified enzyme is followed. In principle, any yeast protein that was thought to be acetylated could be investigated in this way by looking for *in vitro* transfer of [³H]acetyl groups from [³H]acetyl coenzyme A to a recombinant, affinity-isolated protein of interest. If labeling was found, then just as with the yeast histones H3 and H4 (Parthun *et al.*, 1996), the site(s) of modification could be determined by purifying a labeled peptide and subjecting it to Edman degradation, monitoring the cycle in which the radioactivity was released as well as the amino acid sequence generated.

B. Carboxymethylation

Almost all proteins that are subjected to prenylation are subsequently modified by carboxymethylation of the C-terminal prenylcysteine (Schafer and Rine, 1992). In *S. cerevisiae* this is carried out by Ste14p. To demonstrate carboxymethylation, cells can be labeled with S-adenosyl-[methyl-³H]methionine and the protein of interest recovered by immune precipitation (Deschenes *et al.*, 1989; Hrycyna *et al.*, 1991). Overproduction of the protein may be needed to give a sufficiently strong signal, and double-labeling with [³⁵S]methionine can be used to facilitate identification of the labeled protein (Hrycyna *et al.*, 1991). Carboxymethylation results in the generation of an alkali-labile methyl group. After gel electrophoresis, gel slice vapor phase assays are carried out to demonstrate release of [³H]methanol from the protein of interest. Because *STE14* is a non-essential gene (Sapperstein *et al.*, 1994), a *ste14* knockout strain could be useful as a control. The use of a bacterially expressed Ste14p fusion protein for *in vitro* studies of protein carboxymethylation in yeast has also been described recently (Hrycyna *et al.*, 1995).

It is likely that some yeast proteins are carboxymethylated on residues other than prenylcysteine. Such carboxymethylation events are unlikely to be Ste14p-dependent. For example, mammalian PP2A catalytic subunit is carboxymethylated on the C-terminal leucine residue (Favre *et al.*, 1994). Conservation of the C-terminal residue in all PP2A catalytic subunits, including the yeast PP2As, makes it quite likely that it is also modified in this way. In principal, carboxymethylation of this or any other yeast protein could be studied following S-adenosyl-[methyl-³H]methionine labeling exactly as discussed above for Ste14p-dependent events.

C. Disulfide bonds

Any protein that possesses multiple cysteine residues can potentially contain disulfide bonds between pairs of cysteine residues and it may therefore be important to demonstrate both the existence of such bonds and the location in the sequence of the participating residues. Several methods have been described for achieving the former aim using relatively large amounts of purified protein, but in principle these can be adapted to much smaller amounts of protein either by using antibodies to detect the protein or by using radiolabeled protein isolated using an affinity method such as His$_6$-tagging and Ni^{2+}-chelate chromatography. Clearly, it is critical that the protein does not become reduced during isolation. To determine the presence of disulfide bonds, the protein sample is first fully denatured and then free thiol groups modified using iodoacetamide, which does not change the net charge of the protein. In a second step, the protein is then progressively reduced either using a range of increasing DTT concentrations or increasing reaction times, and then newly revealed thiol groups are alkylated with iodoacetate, adding one negative charge per thiol group. The modified samples can be separated by gel electrophoresis. When disulfide bonds are present, their progressive reduction results in a series of bands of different mobility in the samples. The total number of different species visible less 1 corresponds to the number of disulfide bonds (Hollecker, 1997).

To locate the disulfide bonds within the protein sequence, the protein must be fragmented using proteases, separated into peptide fragments by HPLC and then cystine-containing peptides identified. This can often be done on the complete digest by comparing the HPLC traces of reduced and non-reduced samples, but this might require the analysis of individually separated peptides. Cystine-containing peptides thus identified can be subjected to peptide sequencing to locate them within the protein sequence (Gray, 1997). Alternatively, mass spectrometry of reduced and non-reduced samples should reveal the required information: peptides containing an internal –S–S– bond will decrease in mass by 2 Da, while peptides linked by such a bond will generate subfragments equal to the original mass plus 2 Da. Lipari and Herscovics (1996) describe a recent example of the application of these techniques to a yeast protein, in this case a sample of recombinant α1,2-mannosidase.

◆◆◆◆◆◆ VII. MASS SPECTROMETRY AND THE STUDY OF POSTTRANSLATIONAL MODIFICATIONS

Mass spectrometry (MS) is increasingly becoming a powerful tool for the analysis of protein sequence and posttranslational modification. A recent article provides a good summary of the range of techniques currently available for protein sequence and protein modification analysis, including basic details of the methods and approaches (Jensen *et al.*, 1997). In the case of *S. cerevisiae*, the fact that the sequence of the entire complement of gene products can be predicted following completion of the genome sequence makes this a particularly useful technique. Two MS methods are particularly relevant for protein modification analysis: electrospray MS (ES MS) and matrix-assisted laser desorption/ionization MS (MALDI MS), which is usually combined with a time-of-flight analyser. The latter method is not really quantitative, but has the advantage that it requires less rigorous sample preparation and can use smaller amounts of material. These MS methods permit mass determination of polypeptides in the 500–150 000 Da range with greater than 0.1% accuracy.

For studying protein modifications by MS, several possibilities are available, some of which have already been mentioned above. Thus mass determination of a purified protein can be used to check whether the amino-terminal methionine has been removed and to test whether extra mass corresponding to known posttranslational modifications is present. However, peptide mapping by MS may be more useful, in which mixtures of peptide fragments generated by enzymatic or chemical cleavage are analysed. This allows the detection of modified peptides that do not correspond in mass to those predicted from the protein sequence, assisting both in the identification of the specific modification and also in the localization of the modification within the protein sequence. Finally, comparison of mass spectra of the peptide mixture before and after treatment to remove the particular modification can permit rapid identification of the modified peptide within the sequence, as discussed in section II.C in the case of protein phosphorylation. ES MS can also be coupled to HPLC separation in a technique that monitors loss of the phosphate group as a phosphopeptide elutes from the HPLC column and enters the mass spectrometer (Jensen *et al.*, 1997). Finally, MS is useful when modified peptides can be isolated and analysed. While MS is still a "high-tech" procedure, generating results that often require considerable experience to interpret them, it is clearly going to become a powerful tool for studying posttranslational modification of proteins from yeast or any other organism.

Acknowledgements

Thanks are due to Nick Morrice, Grahame Hardie and Carol MacKintosh for helpful comments.

References

Anderson, N. G., Maller, J. L., Tonks, N. K. and Sturgill, T. W. (1990). Requirement for integration of signals from two distinct phosphorylation pathways for activation of MAP kinase. *Nature* **343**, 651–653.

Bhattacharya, S., Chen, L., Broach, J. R. and Powers, S. (1995). Ras membrane targeting is essential for glucose signaling but not for viability in yeast. *Proc. Natl Acad. Sci. USA* **92**, 2984–2988.

Brownell, J. E. and Allis, C. D. (1996). Special HATs for special occasions: linking histone acetylation to chromatin assembly and gene activation. *Curr. Opin. Genet. Devel.* **6**, 176–184.

Caplan, A. J., Tsai, J., Casey, P. J. and Douglas, M. G. (1992). Farnesylation of Ydj1p is required for function at elevated growth temperatures in *Saccharomyces cerevisiae. J. Biol. Chem.* **267**, 18890–18895.

Caplin, B. E., Hettich, L. A. and Marshall, M. S. (1994). Substrate characterization of the *Saccharomyces cerevisiae* protein farnesyltransferase and type-I protein geranylgeranyltransferase. *Biochim. Biophys. Acta* **1205**, 39–48.

Clotet, J., Posas, F., de Nadal, E. and Arino, J. (1996) The NH_2-terminal extension of protein phosphatase PPZ1 has an essential functional role. *J. Biol. Chem.* **271**, 26349–26355.

Couve, A., Protopopov, V. and Gerst, J. E. (1995). Yeast synaptobrevin homologs are modified posttranslationally by the addition of palmitate. *Proc. Natl Acad. Sci. USA* **92**, 5987–5991.

Cyert, M. S. and Thorner, J. (1992). Regulatory subunit (*CNB1* gene product) of yeast Ca^{2+}/calmodulin-dependent phosphoprotein phosphatases is required for adaptation to pheromone. *Mol. Cell. Biol.* **12**, 3460–3469.

Deschenes, R. J., Stimmel, J. B., Clarke, S., Stock, J. and Broach, J. R. (1989). RAS2 protein of *Saccharomyces cerevisiae* is methyl-esterified at its carboxyl terminus. *J. Biol. Chem.* **264**, 11865–11873.

Deshaies, R. J., Chau, V. and Kirschner, M. (1995). Ubiquitination of the G1 cyclin Cln2p by a Cdc34p-dependent pathway. *EMBO J.* **14**, 303–312.

Donaldson, A. D. and Kilmartin, J. V. (1996). Spc42p: a phosphorylated component of the *S. cerevisiae* spindle pole body (SPB) with an essential function during SPB duplication. *J. Cell. Biol.* **132**, 887–901.

Duronio, R. J., Towler, D. A., Heuckeroth, R. O. and Gordon, J. I. (1989). Disruption of the yeast N-myristoyl transferase gene causes recessive lethality. *Science* **243**, 796–800.

Ellison, M. J. and Hochstrasser, M. (1991). Epitope-tagged ubiquitin. A new probe for analyzing ubiquitin function. *J. Biol. Chem.* **266**, 21150–21157.

Errede, B., Gartner, A., Zhou, Z. Q., Nasmyth, K. and Ammerer, G. (1993). MAP kinase-related FUS3 from *S. cerevisiae* is activated by STE7 *in vitro*. *Nature* **362**, 261–264.

Favre, B., Zolnierowicz, S., Turowski, P. and Hemmings, B. A. (1994). The catalytic subunit of protein phosphatase 2A is carboxyl-methylated *in vivo*. *J. Biol. Chem.* **269**, 16311–16317.

Goodman, L. E., Judd, S. R., Farnsworth, C. C., Powers, S., Gelb, M. H., Glomset, J. A. and Tamanoi, F. (1990). Mutants of *Saccharomyces cerevisiae* defective in the farnesylation of Ras proteins. *Proc. Natl Acad. Sci. USA* **87**, 9665–9669.

Gray, W. R. (1997). Disulfide bonds between cysteine residues. In *Protein Structure: A Practical Approach* (T. E. Creighton, ed.), pp. 165–186. Oxford University Press, Oxford.

Hardie, D. G., Campbell, D. G., Caudwell, F. B. and Haystead, T. A. J. (1993). Analysis of sites phosphorylated *in vivo* and *in vitro*. In *Protein Phosphorylation: A Practical Approach* (D. G. Hardie, ed.), pp. 61–85. Oxford University Press, Oxford.

Hochstrasser, M. (1995). Ubiquitin, proteasomes, and the regulation of intracellular protein degradation. *Curr. Opin. Cell Biol.* **7**, 215–223.

Hollecker, M. (1997). Counting integral numbers of residues by chemical modification. In *Protein Structure: A Practical Approach* (T. E. Creighton, ed.), pp. 151–164. Oxford University Press, Oxford.

Hrycyna, C. A., Sapperstein, S. K., Clarke, S. and Michaelis, S. (1991). The *Saccharomyces cerevisiae* STE14 gene encodes a methyltransferase that mediates C-terminal methylation of **a**-factor and RAS proteins. *EMBO J.* **10**, 1699–1709.

Hrycyna, C. A., Wait, S. J., Backlund, P. S. and Michaelis, S. (1995). Yeast STE14 methyltransferase, expressed as TrpE-STE14 fusion protein in *Escherichia coli*, for *in vitro* carboxymethylation of prenylated polypeptides. *Methods Enzymol.* **250**, 251–266.

Jensen, N. O., Shevchenko, A. and Mann, M. (1997). Protein analysis by mass spectrometry. In *Protein Structure: A Practical Approach* (T. E. Creighton, ed.), pp. 29–57. Oxford University Press, Oxford.

Jentsch, S. (1992). The ubiquitin-conjugation system. *Ann. Rev. Genet.* **26**, 179–207.

Jiang, Y., Rossi, G. and Ferro-Novick, S. (1993). Bet2p and Mad2p are components of a prenyltransferase that adds geranylgeranyl onto Ypt1p and Sec4p. *Nature* **366**, 84–86.

Kornitzer, D., Raboy, B., Kulka, R. G. and Fink, G. R. (1994). Regulated degradation of the transcription factor Gcn4. *EMBO J.* **13**, 6021–6030.

Krishna, R. G. and Wold, F. (1997). Identification of common post-translational modifications. In *Protein Structure: A Practical Approach* (T. E. Creighton, ed.), pp. 91–116. Oxford University Press, Oxford.

Kulkarni, M. S. and Sherman, F. (1994). *NAT2*, an essential gene encoding methionine N-alpha-acetyltransferase in the yeast *Saccharomyces cerevisiae. J. Biol. Chem.* **269**, 13141–13147.

Kuo, M.-H., Brownell, J. E., Sobel, R. E., Ranalli, T. A., Cook, R. G., Edmondson, D. G., Roth, S. Y. *et al.* (1996). Transcription-linked acetylation by Gcn5p of histones H3 and H4 at specific lysines. *Nature* **383**, 269–272.

Kuo, M.-H., Nadeau, E. T. and Greyhack, E. J. (1997). Multiple phosphorylated forms of the *Saccharomyces cerevisiae* Mcm1 protein include an isoform induced in response to high salt concentrations. *Mol. Cell. Biol.* **17**, 1819–1832.

Lipari, F. and Herscovics, A. (1996). Role of the cysteine residues in the α1,2-mannosidase involved in N-glycan biosynthesis in *Saccharomyces cerevisiae*. The conserved Cys340 and Cys385 residues form an essential disulfide bond. *J. Biol. Chem.* **271**, 27615–27622.

Mitchell, D. A. and Deschenes, R. J. (1995). Characterization of protein prenylation in *Saccharomyces cerevisiae. Methods Enzymol.* **250**, 68–78.

Molenaar, C. M., Prange, R. and Gallwitz, D. (1988). A carboxyl-terminal cysteine residue is required for palmitic acid binding and biological activity of the ras-related yeast YPT1 protein. *EMBO J.* **7**, 971–976.

Morrice, N. A. and Carrey, E. A. (1997). Peptide mapping. In *Protein Structure: A Practical Approach* (T. E. Creighton, ed.), pp. 29–57. Oxford University Press, Oxford.

Newman, C. M., Giannakouros, T., Hancock, J. F., Fawell, E. H., Armstrong, J. and Magee, A. I. (1992). Post-translational processing of *Schizosaccharomyces pombe* YPT proteins. *J. Biol. Chem.* **267**, 11329–11336.

Omer, C. A. and Gibbs, J. B. (1994). Protein prenylation in eukaryotic microorganisms: genetics, biology and biochemistry. *Mol. Microbiol.* **11**, 219–225.

Park, E. C. and Szostak, J. W. (1992). ARD1 and NAT1 proteins form a complex that has N-terminal acetyltransferase activity. *EMBO J.* **11**, 2087–2093.

Parthun, M. R., Widom, J. and Gottschling, D. E. (1996). The major cytoplasmic histone acetyltransferase in yeast: links to chromatin replication and histone metabolism. *Cell* **87**, 85–94.

Payne, W. E. and Garrels, J. I. (1997). Yeast Protein Database (YPD): A database for the complete proteome of *Saccharomyces cerevisiae. Nucleic Acids Res.* **25**, 57–62.

Posas, F., Wurgler-Murphy, S. M., Maeda, T., Witten, E. A., Thai, T. C. and Saito, H. (1996). Yeast HOG1 MAP kinase cascade is regulated by a multistep phosphorelay mechanism in the SLN1-YPD1-SSK1 "two-component" osmosensor. *Cell* **86**, 865–875.

Sapperstein, S., Berkower, C. and Michaelis, S. (1994). Nucleotide sequence of the yeast *STE14* gene, which encodes farnesylcysteine carboxyl methyltransferase, and demonstration of its essential role in **a**-factor export. *Mol. Cell. Biol.* **14**, 1438–1449.

Schafer, W. R. and Rine, J. (1992). Protein prenylation. Genes, enzymes, targets, and functions. *Ann. Rev. Genet.* **26**, 209–237.

Simon, S. M. and Aderem, A. (1992). Myristoylation of proteins in the yeast secretory pathway. *J. Biol. Chem.* **267**, 3922–3931.

Song, J., Hirschman, J., Gunn, K. and Dohlman, H. G. (1996). Regulation of membrane and subunit interactions by N-myristoylation of a G protein alpha subunit in yeast. *J. Biol. Chem.* **271**, 20273–20283.

Stark, M. J. R. (1996). Yeast protein serine/threonine phosphatases: multiple roles and diverse regulation. *Yeast* **12**, 1647–1675.

Stirling, D. A. and Stark, M. J. R. (1996). The phosphorylation state of the 110 kDa component of the yeast spindle pole body shows cell cycle dependent regulation. *Biochem. Biophys. Res. Commun.* **222**, 236–242.

Tercero, J. C., Dinman, J. D. and Wickner, R. B. (1993). Yeast MAK3 N-acetyltransferase recognizes the N-terminal 4 amino acids of the major coat protein (gag) of the L-A double-stranded RNA virus. *J. Bacteriol.* **175**, 3192–3194.

Towler, D. A., Gordon, J. I., Adams, S. P. and Glaser, L. (1988). The biology and enzymology of eukaryotic protein acylation. *Annu. Rev. Biochem.* **57**, 69–99.

van der Geer, P., Luo, K., Sefton, B. M. and Hunter, A. (1993). Phosphopeptide mapping and phosphoamino acid analysis on cellulose thin-layer plates. In *Protein Phosphorylation: A Practical Approach* (D. G. Hardie, ed.) pp. 31–59. Oxford University Press, Oxford.

Vancura, A., Sessler, A., Leichus, B. and Kuret, J. (1994). A prenylation motif is required for plasma membrane localization and biochemical function of casein kinase I in budding yeast. *J. Biol. Chem.* **269**, 19271–19278.

Willems, A. R., Lanker, S., Patton, E. E., Craig, K. L., Nason, T. F., Mathias, N., Kobayashi, R. *et al.* (1996). Cdc53 targets phosphorylated G1 cyclins for degradation by the ubiquitin proteolytic pathway. *Cell* **86**, 453–463.

Wurgler-Murphy, S. M., Maeda, T., Witten, E. A. and Saito, H. (1997). Regulation of the *Saccharomyces cerevisiae HOG1* mitogen-activated protein kinase by the *PTP2* and *PTP3* protein tyrosine phosphatases. *Mol. Cell. Biol.* **17**, 1289–1297.

Yoon, H. J. and Carbon, J. (1995). Genetic and biochemical interactions between an essential kinetochore protein, Cbf2p/Ndc10p, and the *CDC34* ubiquitin-conjugating enzyme. *Mol. Cell. Biol.* **15**, 4835–4842.

Zachariae, W. and Nasmyth, K. (1996). TPR proteins required for anaphase progression mediate ubiquitination of mitotic B-type cyclins in yeast. *Mol. Biol. Cell* **7**, 791–801.

Zarzov, P., Mazzoni, C. and Mann, C. (1996). The SLT2 (MPK1) MAP kinase is activated during periods of polarized cell growth in yeast. *EMBO J.* **15**, 83–91.

Zhang, F. L. and Casey, P. J. (1996). Protein prenylation: molecular mechanisms and functional consequences. *Annu. Rev. Biochem.* **65**, 241–269.

Zhu, D., Cardenas, M. E. and Heitman, J. (1995). Myristoylation of calcineurin B is not required for function or interaction with immunophilin-immunosuppressant complexes in the yeast *Saccharomyces cerevisiae. J. Biol. Chem.* **270**, 24831–24838.

15 Two-Hybrid Analysis of Protein–Protein Interactions in Yeast

John Rosamond
School of Biological Sciences, University of Manchester, Manchester, UK

◆◆

CONTENTS

List of Abbreviations

HA	Hemagglutinin
3-AT	3-amino-1,2,4-triazole
UAS	Upstream activating sequence

◆◆◆◆◆◆ I. INTRODUCTION

Protein–protein interactions underpin a wide range of complex cellular processes and in many cases make a critical contribution to the regulation of those processes. Consequently, a key question in the study of the cellular function of any novel protein is to identify other proteins with which it interacts. There are many physical, biochemical and immunological methods available to identify and quantify such interactions (Phizicky and Fields, 1995). These have recently been complemented by genetic-based methods of which one, the yeast two-hybrid system (Chien *et al.*, 1991), was developed as a sensitive genetic tool to identify protein–protein interactions *in vivo*. In this chapter, I will describe the background to the two-hybrid system, the properties of the basic elements used to define interactions using the two-hybrid system and briefly review some of the potential future developments of this technology.

METHODS IN MICROBIOLOGY, VOLUME 26
ISBN 0–12–521526–6

Initiation of transcription in yeast, as in most eukaryotes, requires the concerted action of several molecular processes. All yeast genes are preceded by a loosely conserved TATA-box that influences both the start site for transcription and the intrinsic basal level of gene expression. In addition to the TATA-box, other *cis*-acting elements include upstream activating sequences (UAS), which act as binding sites for specific transcription factors. Interaction of a transcription factor with its cognate UAS acts to enhance expression of the adjacent structural gene. The two-hybrid system takes advantage of the specificity of this interaction between a transcription factor and its UAS as well as relying on other properties of the transcription factor itself.

Eukaryotic transcription factors are modular proteins containing discrete, highly defined functional domains within the protein (Keegan *et al.*, 1986; Hope and Struhl, 1986; Ma and Ptashne, 1987). Experiments with the *GAL4*-encoded transcription factor Gal4p established at least two, separable domains: one domain (the DNA-binding domain) was required to bind the protein to the DNA at the specific UAS (UAS_G) upstream of the genes regulated by *GAL4*; the second domain (the activation domain) was required to activate transcription, probably by direct interaction between that domain of Gal4p and the transcriptional apparatus itself. Both domains of Gal4p are required for effective transcriptional regulation so that a protein containing only the DNA-binding domain or the activation domain will not activate transcription. However, several experiments have demonstrated that the two domains need not be localized within the same protein molecule providing they could be brought into an appropriate configuration at the UAS. Using a Gal4p derivative that contained the DNA-binding domain and a region that binds Gal80p (but lacked the activation domain), Ma and Ptashne (1987) showed that this protein was unable to activate the transcription of a reporter gene containing an upstream UAS_G. However, when the protein was co-expressed in yeast with a Gal80p protein fused to the Gal4p activation domain, interaction between the Gal80p activation-domain fusion and the Gal4p DNA-binding protein at UAS_G resulted in transcription of the adjacent reporter gene. In analogous experiments, Fields and Song (1989) showed that transcriptional activation of a reporter gene in yeast could be used specifically to monitor the association between two proteins if one was expressed as a fusion with a DNA-binding domain and the other as a fusion with an activation domain. Using Snf1p fused to the Gal4p DNA-binding domain and Snf4p fused to the Gal4p activation domain, they showed that the known interaction between Snf1p and Snf4p could effectively reconstitute functional Gal4p transcription factor activity and direct the expression of a *lacZ* reporter with an upstream UAS_G. These observations served as the basis for the two-hybrid system to detect protein–protein interactions *in vivo*. As shown in Figure 1, the assay of a simple reporter gene can be used as an output to measure the interaction between two known proteins fused independently to DNA-binding and activation domains (Iwabuchi *et al.*,

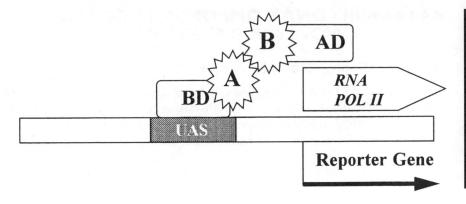

Figure 1. Schematic representation of the two-hybrid system in which protein A, fused to the DNA-binding domain (BD), interacts with protein B fused to the transcriptional activator domain (AD), to reconstitute active transcription factor activity from the UAS adjacent to a reporter gene.

1993), or to screen for interactions between a known protein in one fusion and a genomic or cDNA library as the other fusion.

All current versions of the two-hybrid system that have been developed in several laboratories are based on these experimental findings. Although the several versions differ in their specific components, all comprise the same three basic constituents:

- yeast vectors for the expression of DNA-binding domain fusions, usually referred to as "bait" fusions;
- vectors for the expression of cDNA or genomic DNA fused to a transcription activator domain, usually referred to as "prey" fusions; and
- yeast strains carrying reporter gene(s) with appropriate UAS sites for the DNA-binding domain.

The basic protocol to test for interactions between two proteins can be summarized as follows:

(a) constructing a plasmid encoding a protein fusion between a DNA-binding domain and a protein under test;
(b) constructing a plasmid encoding protein fusions between an activation domain and either the second protein under test or a gene library;
(c) introducing both plasmids into a yeast cell carrying an appropriately regulated reporter gene; and
(d) assaying the transformed cells for reporter gene activity.

The following sections describe the properties of several specific components for each of these steps and illustrate the relative merits and demerits of the various alternatives.

◆◆◆◆◆◆ III. DNA COMPONENTS OF THE TWO-HYBRID SYSTEM

A. DNA-Binding Domain Vectors

All DNA-binding domain vectors for the two-hybrid system are based on either Gal4p or LexAp. A list of potential vectors is given in Table 1 with their relevant properties, selectable markers, utilizable restriction sites and the reading frame from the DNA-binding domain through the cloning sites.

The Gal4p DNA-binding domain is the N-terminal region of the protein comprising residues 1–147. This portion of the Gal4p protein is sufficient to ensure both the efficient localization of fusion proteins to the yeast cell nucleus and to maintain high-affinity binding to the well-characterized binding site for Gal4p. In the plasmids pAS2 and pGBT9, the DNA-binding-domain fusion is constitutively expressed from the yeast *ADH1* promoter at a high level (pAS2) or at a relatively low level from a truncated version of the promoter (pGBT9). This is an important consideration in cases where the fusion protein is toxic to the cell, in which case pGBT9 can be used more effectively. However, this has to be balanced against the observation that some interactions will only be detected with the higher levels of expression obtained with pAS2. As with many aspects of the two-hybrid system, there is no immediately obvious way to predict from the outset which vector will be better for any particular fusion. In addition to the Gal4p DNA-binding domain, pAS2 also contains an HA-epitope tag that enables the expression of fusion proteins to be monitored by Western blotting using commercially available antibodies, and the dominant *CYH2* gene that confers sensitivity to cycloheximide. This can be used after screening to select for loss of the "bait" plasmid to test for direct transcriptional activation by the activation-domain plasmid alone. Although these two genetic determinants serve a useful purpose, they do introduce some unwanted complications. For example, the HA-epitope tag can act to increase non-specific interactions with some activation-domain fusions, so that many DNA-binding domain fusions derived from pAS2 have an associated low level of endogenous reporter gene activity. In addition, the *CYH2* gene can be toxic when overexpressed. When mating is used to produce co-transformants (see below), using pAS2 reduces the efficiency of diploid formation by 5–10 fold relative to vectors lacking *CYH2*.

A further potential difficulty in using vectors based on *GAL4* can occur when using DNA fragments from an organism that can complement *gal4* and *gal80* mutations in the host yeast strain, which will result in a significant number of false positives. In these cases, vectors such as pEG202 (Gyuris *et al.*, 1993; Zervos *et al.*, 1993) provide an alternative in the form of the *LexA* coding sequence. *LexA* does not have a nuclear localization signal, but is expressed at high level from the yeast *ADH1* promoter and so enters the nucleus and can occupy LexAp binding sites (LexA$_{op}$) upstream of reporter genes. There appear to be no native yeast proteins that can interact with LexA$_{op}$ so that background levels of reporter gene activation are minimal.

Table I. Properties of DNA-binding domain vectors

Plasmid	DNA-binding domain	Markers	Multiple cloning sites	Reference	Genbank accession
pAS2	GAL4 (1–147)	amp^R TRP1 CYH2 HA-tag	NdeI SfiI NcoI SmaI BamHI SalI	Harper et al. (1993)	U30497
pGBT9	GAL4 (1–147)	amp^R TRP1	EcoRI SmaI BamHI SalI PstI	Bartel et al. (1993)	U07646
pEG202	LexA (1–202)	amp^R HIS3	EcoRI BamHI SalI NcoI NotI XhoI	Gyuris et al. (1993)	N/A

Reading frames through the multiple cloning sites are as follows:

pAS2: Gal4 (1–147)   CAT ATG GCC ATG GAG GCC CCG GGG ATC CGT CGA C

pGBT9: Gal4 (1–147)   GAA TTC CCG GGG ATC CGT CGA CCT GC¹ G

pEG202: LexA (1–202) GAA TTC CCG GGG ATC CGT CGA TGG CCG CCG CTC GAG

N/A = not available.

B. Activation-Domain Fusion Vectors

Several different vector systems have been developed to express fusion proteins from genomic and cDNAs with a transcriptional activation domain at the amino terminus of the fusion protein (Table 2). Because activation domains are largely interchangeable, various proteins have been used, although most vectors are based on the strong Gal4p activator (residues 768–881), the equally strong activator from the herpes simplex virus VP16 protein or the weak bacterial activator B42. In most cases, expression of the activation-domain fusion is constitutive from the *ADH1* promoter (high level in pACT2, low level in pGAD424). In pJG4-5, B42 expression is galactose-inducible via the *GAL1* promoter; this vector is designed specifically for use as a partner with the *LexA* vector pEG202 (Table 1).

C. Activation-Domain Libraries

As with any genetic screen, the overall success is critically dependent on the quality and complexity of the gene library being screened. Additionally, it goes without saying that the easiest way to get a library for two-hybrid screening is to pick up the telephone and ask a colleague if they have what you need! A large number of libraries have been constructed and used to screen for interaction; from simple unicellular organisms such as yeast to complex multicellular eukaryotes such as plants and humans. Many of these libraries are now commercially available (e.g. from Clontech Inc.). In some cases, these libraries have been made by partial digestion of genomic DNA with a restriction enzyme such as *Sau*3A and ligated into a compatible restriction site (e.g. *Bam*HI or *Bgl*II) in one of the activation-domain vectors. Unlike conventional gene libraries, this is an ineffective way to produce fusion libraries; for an enzyme such as *Sau*3A, whose recognition site occurs on average every 256 bp, an in-frame fusion will occur about every 1.5 kb throughout the genome.

Several strategies have been used to improve the quality of a library by increasing the frequency of in-frame fusions. One route is to engineer the activation-domain vector to produce derivatives in which the frame of the unique cloning site is altered by one or two nucleotides with respect to the activation-domain coding sequence. This, in effect, requires the construction and screening of three separate libraries with a concomitant increase in the number of transformants to be analysed, but significantly increases the probability of productive interactions. An extension of this approach has been to combine it with the use of several different enzymes to generate the genomic fragments (James *et al.*, 1996). Genomic DNA for library construction was isolated from a yeast strain deleted for the *GAL4* and *HIS3* genes, then digested partially with one of five enzymes (*Aci*I, *Hin*PI, *Mae*II, *Msp*I and *Taq*I). Each of these enzymes produces fragments whose termini are compatible with *Cla*I, which enabled their cloning into derivatives of pGAD424 containing the *Cla*I site in a different reading frame with respect to the *GAL4* coding sequence. This increased the frequency of in-frame fusions to about one every 97 bp.

Table 2. Activation-domain vectors

Plasmid	Activation domain	Markers	Multiple cloning sites	Reference	Genbank accession
pACT2	Gal4 (768–881)	amp^R *LEU2*	*NdeI SfiI NcoI SmaI BamHI EcoRI XhoI*	Li *et al.* (1994)	U29899
pGAD424	Gal4 (768–881)	amp^R *LEU2*	*EcoRI SmaI BamHI SalI PstI*	Bartel *et al.* (1993)	U07647
pJG4-5	B42	amp^R *TRP1*	*EcoRI XhoI*	Gyuris *et al.* (1993)	N/A

Reading frames through the multiple cloning sites are as follows:
pACT2: Gal4 (768–881)-HA tag CAT ATG GCC ATG GAG GCC CCG GGG ATC CGA ATT CGA AGC TCG AGA GAT CT
pGAD424: Gal4 (768–881) GAA TTC CCG GGG ATC CGT CGA CCT GCA G
pJG4-5: B42-HA tag GAA TTC
N/A, not available.

An alternative and complementary approach makes use of a technique that has been widely used to generate DNA fragments for library construction as part of large-scale sequencing projects. Rather than cutting DNA with restriction enzymes, high-molecular-weight DNA was initially fragmented by hydrodynamic shearing or by ultrasonication to produce molecules with a defined, limited size range, the precise value of which can be controlled by varying the shearing parameters (Povinelli and Gibbs, 1993). The termini of these fragments can be repaired by a combination of Mung bean exonuclease, T4 DNA polymerase and the Klenow fragment of DNA polymerase to generate blunt-ended molecules that can either be cloned directly into the activation-domain vector, or converted to cohesive termini with short oligonucleotides before cloning. This latter option provides yet a further refinement that increases library quality by reducing the frequency of non-recombinant molecules. By partially filling-in the cohesive ends of the vector with Klenow polymerase (e.g. by converting the *Bam*HI cohesive sequence GATC to GA by treating with Klenow polymerase, dATP and dGTP) and by using oligonucleotides at the ends of the genomic fragments that have an unpaired complementary sequence (in this case TC), recombinant formation can be enhanced while simple vector religation is virtually abolished. This type of approach has been used to construct libraries within the EUROFAN project (see Chapter 1), the systematic project for the functional analysis of the yeast genome, and for the EC-funded TAPIR programme (Two-hybrid analyses of proteins involved in RNA metabolism) to provide an exhaustive screen of functional interactions of proteins involved in RNA processing in yeast, where the libraries are estimated to contain an in-frame fusion equivalent to every 24 bp through the yeast genome (Fromont-Racine *et al.*, 1997).

◆◆◆◆◆◆ IV. INTERACTION SCREENING

A. Yeast Strains

Several yeast strains have been developed to meet the specific requirements of hosts for two-hybrid screening and the properties of some of these are given in Table 3. In general, all strains contain:

- mutations in the chromosomal *GAL4* and *GAL80* genes;
- auxotrophic markers for transformation;
- reporter genes downstream of appropriate transcription factor binding sequences.

In many of the strains used originally for two-hybrid screening, such as GGY1::171 (Gill and Ptashne, 1987; Table 3), interaction between two proteins was reported solely by transcription of the *GAL1-lacZ* reporter, which required one to assay for β-galactosidase activity in transformed

Table 3. Yeast strains used for two-hybrid screening

Strain	Genotype	Reporter	Transformation markers	Reference
GGY1::171	MATa leu2-3,112 his3-200 met tyr1 ura3-52 ade2 gal4Δ gal80Δ URA3::GAL1-lacZ	lacZ	his3 leu2	Gill and Ptashne (1987)
Y187	MATα ura3-52 his3-200 ade2-101 trp1-901 leu2-3,112 met gal4Δ gal80Δ URA3::GAL1$_{UAS}$–GAL1$_{TATA}$–lacZ	lacZ	trp1 leu2	Harper et al. (1993)
HF7c	MATa ura3-52 his3-200 ade2-101 lys2-801 trp1-901 leu2-3,112 gal4-542 gal80-538 LYS2::GAL1$_{UAS}$–GAL1$_{TATA}$–HIS3 URA3::GAL4$_{17mer}$–CYC1$_{TATA}$–lacZ	HIS3 lacZ	trp1 leu2	Feilotter et al. (1994)
EGY48	MATα ura3 his3 trp1 LexA$_{op(x6)}$–LEU2	LEU2	his3 trp1 ura3	Estojak et al. (1995)

cells (see below). Although this remains an essential component of most screens, it was unnecessarily tedious and wasteful for high-complexity library screens. This has been improved by incorporating a second reporter gene such that interactions *in vivo* convert an auxotrophic strain to prototrophy. The markers for this are typically *GAL1–HIS3* or *LexA–LEU2*, and these greatly ease the complexity of any screen because potential interacting clones can be identified first by selection on medium lacking histidine or leucine. Although the use of auxotrophic and *lacZ* reporters decreases the incidence of false positives, the expression of *GAL1–HIS3* can be leaky so that some of the strains can in fact be phenotypically prototrophic for histidine. This can be reversed by including the anti-metabolite 3-amino-1,2,4-triazole (3-AT; Yocum *et al.*, 1984) in the medium although this in turn has several disadvantages, one of these being that the optimum 3-AT concentration will vary for each individual DNA-binding domain construct. Alternatively, the strains HF7c and CG-1945 (Table 3) can be used without 3-AT, because these contain a *GAL4*-dependent *HIS3* reporter with a tightly regulated promoter such that these strains are phenotypically auxotrophic for histidine (Feilotter *et al.*, 1994).

Auxotrophic selection for *HIS3* or *LEU2* provides a useful first-step screen for interactions, but the *lacZ* reporter provides an essential role in quantifying the interaction. *LacZ* expression is routinely measured by assaying β-galactosidase activity for which there are several published protocols (Transy and Legrain, 1995). Usually this is done by assaying hydrolysis of the chromogenic substrate X-gal by replica-plating His+ (or Leu+) colonies either directly onto medium containing X-gal or onto filters before lysis, and assay *in situ*. While the X-gal medium provides a simple, convenient method that is amenable to large-scale screens, it is less sensitive than the filter-lift method and can give more problems with background color development. The time taken for color development is also considerably more variable and generally longer (up to 6 days) on medium. However, neither of these methods provides a direct quantitative assay of β-galactosidase activity. This can best be done by liquid assay of whole-cell extracts using *o*-nitrophenyl-β-D-galactopyranoside, chlorophenol red-β-D-galactopyranoside or chemiluminescent substrates (Campbell *et al.*, 1995).

B. Screening Methods

The two-hybrid screen requires co-expression of two fusion proteins within a single yeast cell (Figure 2), and there are three ways to engineer that:

- sequential transformation of the DNA-binding- and activation-domain constructs;
- co-transformation of both constructs;
- transformation of independent strains by the two constructs, then mating the transformants to form a diploid in which the interaction is assayed.

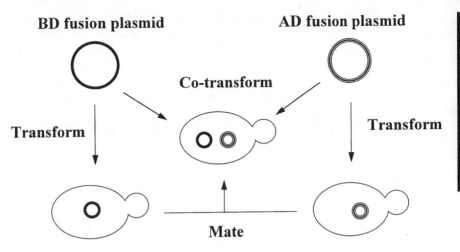

BD fusion plasmid

AD fusion plasmid

Co-transform

Transform

Transform

Mate

Figure 2. Summary of methods to obtain co-expression of fusion proteins in a single yeast cell.

Co-transformation (introducing both DNA-binding-domain and activation-domain fusions in a single transformation) and sequential transformation (transforming with the DNA-binding-domain construct then subsequently transforming with the activation-domain fusion) using published transformation protocols (Gietz *et al.*, 1992; see Chapter 4) are both satisfactory for most applications to study direct interactions involving the domains of two known proteins, where both plasmid constructs contained specific DNA sequences. Both methods can be, and have been, used successfully to screen libraries for novel interacting partners. However, both are inherently inefficient in generating the large numbers of co-transformants needed exhaustively to screen complex libraries, for which yeast mating provides an excellent solution (Finley and Brent, 1994, 1995). In this method, the DNA-binding-domain fusion and the activation domain fusion are transformed initially into different strains, one *MAT*a and the other *MAT*α. These cells can be mated and the interaction assayed in the resulting diploid cell, an effective means of producing co-transformants that has been used to examine specific interactions between sites of proteins, to screen libraries and to define complete sets of interactions amongst the proteome of a small genome (Bartel *et al.*, 1996).

◆◆◆◆◆◆ V. FUTURE DEVELOPMENTS

The impact of the two-hybrid method for defining and characterizing protein–protein interactions has been remarkable and although it represents only one of the many available methods for studying such interactions, its application is enhanced considerably by the explosion in genome sequence information. Despite this though, the method in its current form remains confined to identifying those interactions that can occur between fusion proteins within the context of the yeast nucleus. Not surprisingly then, analogous systems to analyse interactions in other cell types are already being developed, including one for mammalian cells (Luo *et al.*, 1997). Although this is conceptually similar to the yeast system, requiring the co-expression of fusion proteins to regenerate nuclear transcription factor activity to direct the expression of a reporter gene, it offers a useful starting point to develop an homologous system for the analysis of human gene products that are appropriately posttranslationally modified. In addition, by using reporter systems that rely on other physical properties resulting from protein–protein interaction, rather than transcription factor activity, it will be possible to screen for direct interactions in cell compartments other than the nucleus.

The two-hybrid method in its original format is constrained in detecting only direct binary interactions. This is a major limitation and a significant advance has been the development of analogous systems for the detection of ternary interactions (so-called three-hybrid methods). These allow the detection of interactions in which the DNA-binding-domain fusion protein does not directly contact the activation-domain fusion protein without an intermediate effector protein, or in which a protein will only bind to a complex site formed as a result of the interaction between two other proteins (Zhang and Lautar, 1996). The potential of this development is enormous, because the effector molecule that docks the DNA-binding- and activation-domain fusions together at the UAS does not itself need to be a protein but can instead be a small synthetic molecule or even a molecule of RNA (SenGupta *et al.*, 1996; Licitra and Liu, 1996). The recent use of a three-hybrid assay to screen for interactions in mammalian cells can only enhance the range of applications for this technology (Liberies *et al.*, 1997).

Finally, we are currently witnessing extraordinary advances in genome research, with complete microbial sequences already in the public databases and the prospect of the genome of at least one multicellular organism being completed imminently. When combined with chip-based technologies for gene identification and analysis (see Chapter 8), the prospect of using two-hybrid technology as a tool to define the complete interaction set for the proteome of an organism becomes an intriguing possibility. It is probably significant that this approach has already been applied successfully to a small bacteriophage genome (Bartel *et al.*, 1996) and is currently being used to define complete interaction sets for complex cellular processes such as pre-mRNA splicing in yeast. These analyses serve not only to confirm the interactions defined previously by other technologies, but also to open up entirely new and unexpected interactions that serve to highlight novel overlaps between discrete cellular processes.

References

Bartel, P., Chien, C-T., Sternglanz, R. and Fields, S. (1993). Using the two-hybrid system to detect protein–protein interactions. In *Cellular Interactions in Development: A Practical Approach* (D. A. Hartley, ed.), pp. 153–179. Oxford University Press, Oxford.

Bartel, P., Roecklein, J. A., SenGupta, D. and Fields, S. (1996). A protein linkage map of *Escherichia coli* bacteriophage T7. *Nature Genet.* **12**, 72–77.

Campbell, K. S., Buder, A. and Deuschle, U. (1995). Interactions between the amino-terminal domain of p56^{lck} and cytoplasmic domains of CD4 and CD8-alpha in yeast. *Eur. J. Immunol.* **25**, 2408–2412.

Chien, C. T., Bartel, P. L., Sternglanz, R. and Fields, S. (1991). The two-hybrid system: a method to identify and clone genes for proteins that interact with a protein of interest. *Proc. Natl Acad. Sci. USA* **88**, 9578–9582.

Estojak, J., Brent, R. and Golemis, E. (1995). Correlation of two-hybrid affinity data with *in vitro* measurements. *Mol. Cell. Biol.* **15**, 5820–5829.

Feilotter, E., Hannon, G. J., Ruddell, C. J. and Beach, D. (1994). Construction of an improved host strain for two hybrid screening. *Nucleic Acids Res.* **22**, 1502–1503.

Fields, S. and Song, O-K. (1989). A novel genetic system to detect protein–protein interactions. *Nature* **340**, 245–246.

Finley, R. L. and Brent, R. (1994). Interaction mating reveals binary and ternary connections between *Drosophila* cell cycle regulators. *Proc. Natl Acad. Sci. USA* **91**, 12980–12984.

Finley, R. L. and Brent, R. (1995). Interaction trap cloning with yeast. In *DNA Cloning II, Expression Systems: A Practical Approach* (B. D. Hames and D. M. Glover, eds), pp. 169–203. Oxford University Press, Oxford.

Fromont-Racine, M., Rain, J. C. and Legrain, P. (1997). Toward a functional analysis of the yeast genome through exhaustive two-hybrid screens. *Nature Genet.* **16**, 277–282.

Gietz, D., St. Jean, A., Woods, R. A. and Schiestl, R. H. (1992). Improved method for high efficiency transformation of intact yeast cells. *Nucleic Acids Res.* **20**, 1425.

Gill, G. and Ptashne, M. (1987). Mutants of GAL4 protein altered in an activation function. *Cell* **51**, 121–127.

Gyuris, J., Golemis, E., Chertkov, H. and Brent, R. (1993). Cdi1, a human G1 and S phase protein phosphatase that associates with Cdk2. *Cell* **75**, 791–803.

Harper, J. W., Adami, G. R., Wei, N., Keyomarsi, K. and Elledge, S. (1993). The p21 Cdk-interacting protein Cip1 is a potent inhibitor of G1 cyclin-dependent kinases. *Cell* **75**, 805–816.

Hope, I. A. and Struhl, K. (1986). Functional dissection of a eukaryotic transcriptional activator protein, GCN4 of yeast. *Cell* **46**, 885–894.

Iwabuchi, K., Li, B., Bartel, P. and Fields, S. (1993) Use of the two-hybrid system to identify the domain of p53 involved in dimerisation. *Oncogene* **8**, 1693–1696.

James, P., Halladay, J. and Craig, E. A. (1996). Genomic libraries and a host strain designed for highly efficient two-hybrid selection in yeast. *Genetics* **144**, 1425–1436.

Keegan, L., Gill, G. and Ptashne, M. (1986) Separation of DNA-binding from the transcription-activating function of a eukaryotic regulatory protein. *Science* **231**, 699–704.

Li, L., Elledge, S., Peterson, C. A., Bales, E. S. and Legerski, R. J. (1994) Specific association between the human DNA repair proteins XPA and ERCC1. *Proc. Natl Acad. Sci. USA* **91**, 5012–5016.

Liberies, S. D., Diver, S. T., Austin, D. J. and Schreiber, S. L. (1997). Inducible gene expression and protein translocation using non-toxic ligands identified by a mammalian three-hybrid screen. *Proc. Natl Acad. Sci. USA* **94**, 7825–7830.

Licitra, E. J. and Liu, J. O. (1996). A three-hybrid system for detecting small ligand-protein receptor interactions. *Proc. Natl Acad. Sci. USA* **23**, 12817–12821.

Luo, Y., Batalao, A., Zhou, H. and Zhu, L. (1997). Mammalian two-hybrid system: a complementary approach to the yeast two-hybrid system. *BioTechniques* **22**, 350–352.

Ma, J. and Ptashne, M. (1987). A new class of yeast transcriptional activators. *Cell* **51**, 113–119.

Phizicky, E. M. and Fields, S. (1995). Protein–protein interactions: methods for detection and analysis. *Microbiol. Rev.* **59**, 94–123.

Povinelli, C. M. and Gibbs, R. A. (1993). Large-scale sequencing library production: an adaptor-based strategy. *Anal. Biochem.* **210**, 16–26.

SenGupta, D. J., Zhang, B., Kraemer, B., Pochart, P. Fields, S. and Wickens, M. (1996). A three-hybrid system to detect RNA–protein interactions *in vivo*. *Proc. Natl Acad. Sci. USA* **93**, 8496–8501.

Transy, C. and Legrain, P. (1995) The two-hybrid: an *in vivo* protein–protein interaction assay. *Mol. Biol. Rep.* **21**, 119–127.

Yocum, R. R., Hanley, S., West, R. and Ptashne, M. (1984). Use of *lacZ* fusions to delimit regulatory elements of the inducible divergent *GAL1–GAL10* promoter in *Saccharomyces cerevisiae*. *Mol. Cell. Biol.* **4**, 1985–1998.

Zervos, A. S., Gyuris, J. and Brent, R. (1993) Mxi1, a protein that specifically interacts with Max to bind Myc–Max recognition sites. *Cell* **72**, 223–232.

Zhang, J. and Lautar, E. (1996). A yeast three-hybrid method to clone ternary protein complex components. *Anal. Biochem.* **242**, 68–72.

16 "Smart" Genetic Screens

Jesús de la Cruz, Marie-Claire Daugeron and Patrick Linder
Département de Biochimie Médicale, Centre Médicale Universitaire, Genève, Switzerland

◆◆

CONTENTS

List of Abbreviations

5-FOA	5 Fluoro-orotic acid
cs	Cold-sensitive
EF	Elongation factor
eIF	Eukaryotic initiation factor
MAP	Mitogen-activated protein
ORF	Open reading frame
PCR	Polymerase chain reaction
TIF	Translation initiation factor
ts	Temperature-sensitive

◆◆◆◆◆◆ I. INTRODUCTION

The yeast *Saccharomyces cerevisiae* is perhaps the best-studied eukaryotic organism and it serves as a model system for studying a variety of basic processes in the eukaryotic world. It has a relative short generation time, and it grows on different defined media. Moreover, yeast biochemistry and cell biology are well established. Because *S. cerevisiae* can exist as any of three stable cell types, two haploid (MAT**a** and MATα) and one diploid (MAT**a**/MATα), it is an ideal organism on which to perform classical genetics. Its genome is relatively small (approximately 14 000 kb) and only a few genes carry introns. An additional advantage is the high efficiency of transformation, which facilitates the application of a variety of molecular genetic techniques. Moreover, the high rate of homologous recombination allows the efficient replacement of genomic sequences with *in vitro* modified genes. The completion of the genome sequence will facilitate considerably the molecular analysis of known and new genes. Nevertheless, the functions of most of its 6000 or so genes, as well as the interaction of their products, remain to be elucidated.

In classical genetics a gene is defined by a mutation. Such a mutation can reveal a phenotype that gives hints to the function of the wild-type gene product. In order to study a cellular process in detail, it is important to identify its components, define their function, identify all relevant interactions and determine the factors that can regulate it. There exist various approaches to do so. One possibility is to search for a "needle in a haystack", i.e. to screen for a particular phenotype in an unselected population. Also, it is possible to screen a collection of existing conditional-lethal mutants (i.e. temperature-sensitive, ts) for a particular defect. In addition to these general approaches, a more direct way is to start with a given mutation in a gene and, by performing genetic screens, find all possible interacting gene partners. Yeast offers the possibility to search for various changes in phenotype, including growth/non-growth, colony morphology or pigment formation. Compared with generalized screening procedures, where a large population of cells or colonies are individually checked for a phenotype, it is also possible to apply a specific selection for a certain phenotypic trait. In this way, it is possible to identify a rare genetic event in a large population.

A key element in establishing a genetic screen is the starting strain and the formulation of questions that should be answered by potential candidates. For example, if the starting mutant shows a high reversion rate, it will be difficult to screen or select for suppressors because the high numbers of revertants will obscure their presence. Also, searching for genes or mutations that suppress or enhance an existing mutant phenotype may result in different outcomes if the starting mutation is a missense or a nonsense mutation, or a null-allele obtained by the complete replacement of the gene. Moreover, the isolated suppressors may be of different types if the starting mutant produces a highly unstable or an inactive protein.

The application of screens allows, in general, the assignment of genetic interactions among two or more gene products. This may reflect a direct physical interaction of the two gene products or more often a functional interaction without actual physical contact. In this latter case, the gene products may be isofunctional, they may act at different stages in the same metabolic pathway or cellular processes, or they may be components of parallel or cross-talking pathways. Owing to these uncertainties, it is important to use other tools to understand the role of a gene isolated by one of the screens described below. In the first instance, a mutational analysis of the gene isolated is required to detect all possible phenotypes and compare them with those of the starting mutation(s). Finally, it is necessary to use other, complementary, tools based on biochemical or cell biological methods to confirm and extend the primary genetic observations.

In this chapter, we will describe the genetic screens most commonly used for *S. cerevisiae*. Owing to space limitations, it will not be possible to refer to all possible variations of the screens described. Several excellent reviews, as well as descriptions of particular screens and their molecular explanations, have been published (Doye and Hurt, 1995; Guarente, 1993; Huffaker *et al.*, 1987; Rose, 1995). Moreover, for basic procedures, the reader is referred to various published protocols (Ausubel *et al.*, 1994; Guthrie and Fink, 1991; Kaiser *et al.*, 1994).

◆◆◆◆◆◆ II. PLASMID SEGREGATION IN GENETIC SCREENS AND SELECTIONS

Classical genetics uses mutants, revertants or suppressors to study gene function and to identify functionally related genes. The development of molecular genetic methods, which use recombinant DNA transformed into yeast, enabled the elaboration of a variety of different screens. In most of them, a copy of a gene of interest (mutant or wild-type) is cloned on a plasmid, transformed into yeast and then the segregation of the plasmid is monitored. These types of screens are based on the fact that plasmids in yeast are not stably maintained but segregate spontaneously at a frequency of $1–5 \times 10^{-2}$ per generation. Plasmid segregation can be followed either by replica-plating, by selecting for cells that have lost the plasmid (counterselection) or by screening for a colony producing a pigment, induced by the presence or absence of a plasmid-encoded marker. If conditions are found under which plasmid maintenance is essential for survival, no plasmid segregation should be observed. Below, we discuss the genetic markers most commonly used to follow plasmid segregation in a yeast population.

A. Counterselection Using the *URA3* or *LYS2* Genes

Cells harboring a wild-type *URA3* gene metabolize 5-fluoro-orotic acid (5-FOA) into 5-fluoro-uracil, which is toxic to cells (Boeke *et al.*, 1984). Thus, if a culture of yeast cells harboring a *URA3* plasmid is plated on medium containing 5-FOA, only cells that have segregated out the plasmid will be able to form colonies. It is important to realize that this selection does not induce plasmid segregation by itself, but rather selects for plasmid-free cells. For this reason it is important to grow under non-selective conditions before the application of counterselection. Assuming that an essential gene is present on a *URA3* plasmid (either single or multicopy type) and its corresponding chromosomal copy has been inactivated by a mutation, an otherwise wild-type cell cannot lose the plasmid and therefore should not be able to grow on 5-FOA. However, upon suppression of the genomic mutation, segregation again becomes possible and cells will grow on 5-FOA-containing medium. These characteristic features have been widely used in the so-called "plasmid shuffling" technique (Boeke *et al.*, 1987).

As in the case with the *URA3* gene, it is possible to counterselect a *LYS2* gene in the presence of α-aminoadipate (Chattoo *et al.*, 1979). However, the open reading frame (ORF) of the *LYS2* gene is very large compared with the *URA3* gene and thus contains many of the restriction sites present in the polylinkers of most cloning vectors. Moreover, many laboratory strains are prototrophic for lysine and would therefore need to be rendered *lys2* before employment of the *LYS2* plasmid system. Therefore, its application in the plasmid shuffling technique has not been widely used.

Counterselection can also be applied using the *can1*/*CAN1* alleles in the presence of canavanine and the *cyh2*/*CYH2* alleles in the presence of cycloheximide (Mann *et al.*, 1987; Harper *et al.*, 1993). The arginine analog canavanine is toxic to a cell and mutants in the arginine permease-encoding gene, *CAN1*, are resistant to it. Also, cells that are wild-type for the *CYH2* locus, which codes for the ribosomal L29 protein, cannot grow in the presence of cycloheximide, but *cyh2* mutants can.

B. Screening Plasmid Segregation by Colony Color

The products of the *ADE2* and *ADE1* genes act consecutively in the purine biosynthesis pathway. Mutations in these genes lead to an accumulation of a red pigment, derived from the substrate of the Ade2p activity (Fisher, 1969). The red coloration of *ade2* colonies, which can be enhanced by incubating the plates for a few days at 4°C, allows one to follow easily the segregation of plasmids carrying the *ADE2* gene in an *ade2* background. A cell that no longer contains the *ADE2* plasmid will give an entirely red colony, whereas a colony with a mixture of plasmid-containing and plasmid-free cells will have white and red sectors. Cells that cannot lose the plasmid will appear entirely white. A variation of this method is to use a plasmid carrying an ochre suppressor tRNA encoded by *SUP11* or *SUP4*, instead of the *ADE2* gene, in an *ade2-101*^{ochre} background (Moreau *et al.*, 1996; Riles and Olson, 1988).

A major drawback of the accumulation of the red pigment in *ade2* strains is that it leads to reduced growth rate. Thus, in sectoring assays the red sectors can be very small or even undetectable. Moreover, a white colony color may also be the result of mutations affecting respiration, because non-respiring cells do not accumulate the red pigment. An improvement of the white/red sectoring method comes from the combination of the *ade2* and *ade3* alleles (Koshland *et al.*, 1985). The product of the *ADE3* gene acts before that of the *ADE2* gene in the purine biosynthesis pathway and therefore the double mutant, *ade3 ade2*, does not accumulate the red pigment unless the *ADE3* gene is provided on a plasmid. Upon loss of the *ADE3* plasmid, white sectors can be observed. Because the white cells grow better than the red ones, white sectors are easily visible.

C. The *MET15* Gene Allows Counterselection and Color Screening

Recently, the construction of a plasmid carrying the *MET15* gene involved in methionine biosynthesis has been described (Cost and Boeke, 1996).

The *MET15* marker combines the advantages of the two previously described techniques. Mutants in the *MET15* gene become darkly pigmented in the presence of Pb^{2+} ions in the medium and it is possible to counterselect the *MET15* gene in the presence of methylmercury (Singh and Sherman, 1974) .

◆◆◆◆◆◆ III. GENETIC SCREENS TO IDENTIFY NEW GENES AND THEIR FUNCTION

The genetic approach to gene function relies on decreasing or increasing the activity of a given gene product by mutation, depletion or over-expression. Because homologous recombination is very efficient in yeast and gene replacements are easy to perform, additional information can be obtained by creating null alleles. However, in some instances gene disruption is lethal whereas in other cases it does not reveal any phenotype under standard laboratory conditions. In both cases, it is necessary to apply additional methods to reveal the function of the gene; this is done, in the case of an essential gene, by creating loss-of-function conditional alleles and, in the case of a non-essential one, by searching for dominant negative mutations leading to a specific phenotype. Moreover, in order to define all the components of a particular process, it is necessary to characterize additional genes and their products. The classical genetic approach to this is to isolate secondary mutations or genes that affect positively or negatively the phenotype of the starting mutation in the cellular process under study.

In this section, we describe the most commonly used genetic screens to analyse gene function and interaction (Table 1). Some screens start with a particular mutation conferring a phenotype, whereas others start with a wild-type gene.

A. Complementation

Complementation of a mutation represents probably the most straight-forward screen. To be able to complement a mutant allele by its wild-type counterpart, the mutation needs to be recessive. The recessive or dominant character of a mutation is most easily assessed by crossing a mutant with a wild-type strain. If the heterozygous diploid behaves as a wild-type, the mutation is recessive, whereas the mutation is dominant if the diploid strain manifests the same phenotype as the parental mutant strain. To isolate the wild-type allele of a recessive mutation, the mutant strain is transformed with a genomic library and colonies are screened or selected for reappearance of the wild-type phenotype. Plasmid segregation and back-transformation are needed to confirm the plasmid-linked complementation. To prove that the isolated gene represents the wild-type version of the target gene rather than a suppressor, it is cloned on an

Table I. Overview of yeast genetic screens. The main applications and procedures used are described in the text

Approach	Manipulation	Screen for	Expected result	Section
Complementation	Transformation	Growth	Wild-type gene	III.A
Extragenic suppressors	Mutagenesis	Growth	Mutation in related genes	III.B
Multicopy suppressors	Transformation	Growth	Wild-type gene with related function	III.C
Multicopy suppressees	Mutagenesis	Non-growth/non-segregation	Mutation in related genes	III.D
High-dosage phenotype	Transformation	Non-growth	Wild type gene, required in stoichiometric amounts	III.E
High-dosage dominant-negative mutations	Mutagenesis/transformation	Non-growth	Protein and domain function	III.E
Synthetic enhancement	Mutagenesis	Non-growth/non-segregation	Mutation in related genes	III.F
Non-allelic non-complementation	Mutagenesis	Non-growth of diploid cells	Mutation in other genes (heteromeric complex)	III.G
Conditional lethal strain collections		Specific phenotype	Mutant (specific cellular process)	III.H
EUROFAN	Disruption	Phenotype	Phenotypic analysis	III.I
lacZ insertion and fusion libraries	Transformation	β-galactosidase activity	Expression analysis	III.J
Heterologous genes	Transformation/mutagenesis	Growth/non-growth	Heterologous gene/mutation in yeast genes	III.K

integrative plasmid and inserted together with the plasmid marker into the genome of a wild-type strain. After verification of correct integration by polymerase chain reaction (PCR) or Southern analysis, this marked strain is then crossed to the original mutant strain. In the case of allelic complementation, a 2:2 segregation of the wild-type (marked) and the mutated allele will be observed. In the case of non-allelic complementation, which represents a form of suppression, a random segregation of the markers is expected. In general, it is useful to use libraries constructed on low-copy-number plasmids to limit the isolation of non-allelic complementing genes.

B. Extragenic Suppressors

The most classical approach to analyse genetic interactions and to identify new genes whose products are involved in the same cellular process as the starting mutated gene is the isolation of extragenic suppressor mutations. In general, a mutant having a clear and pronounced phenotype (i.e. ts) is used to isolate spontaneous or mutagen-induced mutations suppressing the original mutant phenotype (Figure 1). Often, the strain and the nature of the starting mutation govern the type of suppressors obtained. Dominant or recessive suppressors may be obtained with haploid strains, whereas only dominant ones are obtained with

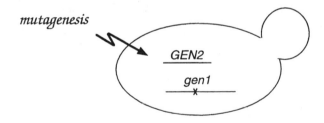

No growth at non-permissive conditions

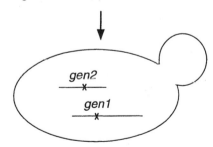

Growth at non-permissive conditions

Figure 1. Isolation of extragenic suppressors. A strain carrying a mutation in the gene of interest (*gen1*) is mutagenized and screened for additional mutations (*gen2*) that allow growth under specific non-permissive conditions.

diploid strains. In general, recessive suppressors are considered to be loss-of-function mutations, whereas dominant suppressors represent gain-of-function mutations. If the starting mutation is a nonsense mutation, the suppressor can be of the informational type, such as a tRNA. Therefore, it is generally advisable to use missense mutations in such screens.

After the isolation of a suppressor mutation, it has to be determined if it is recessive or dominant and if it is encoded by a single nuclear locus different from the original mutant one. To do so, the suppressor strain is crossed with the parental mutant strain, the phenotype of the diploid strain assessed, and the 2:2 segregation of the suppressor phenotype analysed upon tetrad analysis. To determine whether the suppressor is of intragenic or extragenic type, crosses with a wild-type strain are performed. In the case of a revertant or an intragenic suppressor only wild-type spores will be obtained upon tetrad analysis. In the case of an extragenic and unlinked suppressor, the original mutant and the suppressor phenotype will not co-segregate. Spores carrying only the suppressor mutation will also show whether the suppressor mutation by itself has a phenotype, which may be exploited in a follow-up analysis such as cloning of its wild-type counterpart by complementation. Moreover, if a mutagen has been used, it is likely that additional mutations to the one conferring the suppressor phenotype are present in the genome. Multiple back-crosses with the parental strain are therefore required to exclude secondary effects by unrelated mutations.

Analysis of the suppressor mutation in combination with other mutations may reveal the nature of the suppression. For this, the suppressor mutation is combined with other mutations in the gene of interest including its null allele, with mutations in other genes of the same pathway, and with other completely unrelated mutations. Gene- and allele-specificity are often interpreted as a physical interaction of the two gene products, viewed as compensatory conformational changes by the suppressor mutation. Allele-specific but gene-unspecific suppressors are likely to be of the informational type. Gene-specific and allele-unspecific suppressors as well as some gene- and allele-unspecific suppressors can be explained by compensatory mutation in other genes of the same cellular process, such as bypass or epistatic suppressors that make downstream steps of a process independent or less dependent on the upstream steps.

An example of the extragenic suppressor approach is the dissection of the translation initiation machinery using the *GCN4* gene, which encodes a transcription factor that is regulated on the translational level (Hinnebusch and Liebman, 1991). Reduced *GCN4* expression affects many amino acid biosynthetic genes and hence it can be followed by hypersensitivity to 3-amino-triazole, a competitive inhibitor of the *HIS3* gene product. This hypersensitivity was used to isolate suppressor mutations compensating for reduced *GCN4* activity. The wild-type genes of these extragenic suppressors code for translation initiation factors (Greenberg *et al.*, 1986; Hinnebusch and Fink, 1983; Vazquez de Aldana and Hinnebusch, 1994).

C. Multicopy Suppressors (Gene Dosage-Dependent Suppressors)

Increasing the concentration of a gene product in a cell can suppress mutant phenotype(s) owing to mutation in another gene. If the mutation is a point mutation, it is possible to obtain suppression by compensating partially for its loss of function. If the mutation is a null allele, the suppressor is most likely to create a bypass. The overexpression can most easily be achieved by using multicopy plasmids based on the 2 μm plasmid (Figure 2). Alternatively, cDNAs may be cloned behind a strong (inducible) promoter to obtain increased expression. The mRNAs used in the construction of the cDNA library needs to be those expressed in the growth conditions most relevant to the type of gene to be isolated. Thus, when rarely expressed or uncommonly regulated genes are searched, an appropriate library using genomic fragments fused to a strong promoter can be used instead of the cDNA library (Ramer *et al.*, 1992). Also, libraries under the control of a strong inducible promoter circumvent the limitation of some tightly regulated genes that cannot be highly expressed even in high copy number. In both cases, the multicopy or dosage-dependent approach has the clear advantage that the isolation of the suppressor coincides with the cloning of the suppressor gene. Moreover, the possibility of obtaining sequence information of the entire clone by comparing short flanking sequences with the known yeast genome sequence has increased

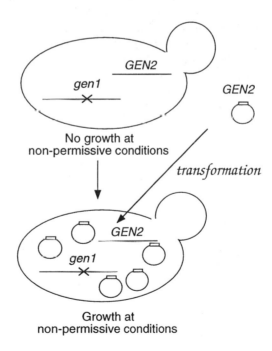

Figure 2. Isolation of multicopy suppressors. A strain carrying a mutation in the gene of interest (*gen1*) is transformed with a multicopy library and transformants are screened for growth under non-permissive conditions, allowing the isolation of *GEN2*.

the popularity and use of the multicopy suppressor approach. In any case, co-segregation of the library plasmid and the suppressor phenotype needs to be confirmed. Also, the plasmid bearing the multicopy suppressor should be retransformed back to the original mutant to demonstrate a direct link between the suppressor effect and the gene present on the plasmid.

One successful application of the multicopy suppressor approach has been the study of translation initiation (Coppolecchia *et al.*, 1993). A strain, harboring as the only eIF4A-encoding genes the ts *tif1-1* allele on an *LEU2* plasmid, was transformed with a *URA3* multicopy genomic library and transformants were selected for growth at 37°C. The candidates were then analysed for extragenic plasmid-bearing suppressors, genomic suppressors, bypass suppressors and plasmids harboring a wild-type *TIF1* gene, by segregation of the plasmids under non-selective (full medium) and non-permissive conditions (37°C). *STM1*, one of the extragenic plasmid-bearing suppressors isolated, showed weak homology to mammalian eIF4B. In a subsequent analysis, it was found that deletion of the *STM1* gene (also called *TIF3*; Altmann *et al.*, 1993) was synthetically lethal (see below) with mutations in some translation initiation factor (*TIF*) genes, including *TIF1*. In addition, polysome analysis and *in vitro* translation experiments confirmed the genetic results, suggesting a role of the *STM1* gene product in translation initiation (Altmann *et al.*, 1993; Coppolecchia *et al.*, 1993).

A second example of this approach was the screen for multicopy suppressors of a *ste20-1* mutant, which resulted in isolation of the *STE5* gene (Leberer *et al.*, 1993). Ste20p is a protein kinase involved in the signal transduction of the mating pheromone response. Stimulation of Ste20p activity leads to the transcriptional induction of several genes whose products are required to mediate various aspects of mating, including the *FUS1* gene involved in the fusion of two cells of opposite mating type. The authors used the histidine auxotrophy of a *his3 ste20-1* strain containing a *FUS1–HIS3* fusion to isolate multicopy suppressors of the *ste20-1* mutation by transforming the mutant strain with a multicopy library. Because the *FUS1–HIS3* fusion is not expressed in the absence of a functional Ste20p, the cells could grow in the absence of histidine only if suppression of the *ste20-1* defect or if secondary mutations, changing the transcriptional control of the *FUS1* gene, occurred. The wild-type *STE20* gene was discarded by restriction analysis. The mating efficiency of the candidates was analysed to eliminate *HIS3* clones amongst the remaining positive candidates, because mating is completely abolished in the *ste20-1* mutant. The *STE5* gene was identified as one of the final positive candidates. Consistent with the partial suppression of the mating deficiency, the restored sensitivity to mating factor confirmed the involvement of *STE5* in the mating factor signaling pathway. The suppression of *ste20-1* by *STE5* was dependent on Ste20-1p, because no suppression could be observed in a *ste20-Δ2* strain, in which the kinase domain of Ste20p had been deleted. Thus, overexpression of *STE5* did not lead to a bypass of the Ste20p requirement. Independent results obtained by the two-hybrid system suggest that Ste5p interacts with other members of the mitogen-activated protein (MAP)-kinase cascade and it functions as a scaffold protein for these proteins in the pheromone signaling pathway (Printen and Sprague, 1994). Thus, suppression of *ste20-1* by the *STE5* gene can be explained by the re-establishment of a functional complex through the elevated Ste5p levels. This example demonstrates how the requirement for a conditional mutation for a suppressor screen can be replaced by the creation of a conditional non-growth phenotype such as, in this case, by using the *FUS1-HIS3* fusion.

Multicopy suppression can also lead to the isolation of bypass suppressors, especially when a null allele is used as the starting mutation. For example, strains lacking the *PGI1* gene, encoding phosphoglucose isomerase, cannot grow on glucose as a sole carbon source, but overexpression of the *GDH2* gene (encoding NAD-dependent glutamate dehydrogenase) allows these mutant cells to grow again on glucose-containing medium. In this case, it has been reported that the growth defect of the *PGI1*-deletion mutants on glucose is due to a depletion of the $NADP^+$ pool. Overexpresssion of the NAD-dependent glutamate dehydrogenase activity leads, by a completely different metabolic pathway, to regeneration of the intracellular levels of $NADP^+$ (Boles *et al.*, 1993).

Multicopy suppression may also lead to the isolation of redundant genes, as in the case of ribosomal protein genes. Many ribosomal proteins of *S. cerevisiae* are encoded by duplicated genes (see Appendix I) whose products do not contribute equally to growth. Strains lacking both copies of a particular ribosomal protein gene are generally inviable, while strains lacking one of the two copies grow either as the wild-type or exhibit a slow growth phenotype. In a slow growing, the introduction of multiple copies of the redundant gene fully restores wild-type growth (Rotenberg *et al.*, 1988).

Finally, it is also possible to reverse the multicopy suppressor phenotype, by introducing a second gene on a multicopy plasmid into the cell (Bi and Pringle, 1996).

Occasionally, increased expression of the mutated gene can also result in suppression of the mutant phenotype. It is therefore important to analyse the effect of overexpression of the mutant gene before using multicopy suppressor analysis. If overexpression of the mutated gene completely or even partially suppresses the mutant phenotype, then undesired transcriptional changes, genomic rearrangements or, if the mutant gene is present on a plasmid, increased plasmid copy numbers are likely to be found. Depending on the type of mutation, it is also feasible to isolate tRNA genes that, when present on multiple copies, can suppress (at least partially) the mutant phenotype (Hohmann *et al.*, 1994). It has also been observed that gene dosage alteration of the elongation factor EF-1α, even if present on a centromeric plasmid, can cause misreading of nonsense mutations (Song *et al.*, 1989).

Although the multicopy suppressor approach is easy to perform and leads to the concomitant cloning of the suppressor gene, in some cases the suppression does not reveal a direct interaction, but rather may reflect two different independent functions that indirectly influence each other. For example, the genes *ZDS1* and *ZDS2* have been isolated as multicopy suppressors in many screens involving mutations in a variety of different cellular processes including replication (*cdc28-1N*; Yu *et al.*, 1996), morphogenesis (*cdc24*; Bi and Pringle, 1996), mRNA capping (*ceg1*; Schwer *et al.*, 1998), cell cycle (*cdc20*; Clark and Burke, cited in Yu *et al.*, 1996), casein kinase II function (*cka2*; McCann and Glover, cited in Yu *et al.*, 1996), translation (*tif1-1*; Schwer *et al.*, 1998) and probably others. However, the genes isolated by these "zillion different screens" (*ZDS*) have not yet been attributed a precise biochemical function. This example confirms the necessity for a follow-up analysis of the isolated suppressors.

D. Multicopy Suppressees (Gene Dosage-Dependent Mutations)

In this section we describe the isolation of mutations in genomic loci that are at least partially suppressed by the presence of multiple copies of a gene of interest (Figure 3). The analysis of the mutated locus can give clues to the function of the overexpressed gene. This type of screen is conceptually based on the multicopy suppressor approach; however, there is one fundamental difference between the two approaches. In the multicopy suppressor method, a mutation in the gene of interest is available and the screen consists of searching for another gene whose overexpression suppresses the mutant phenotype. In the multicopy suppressee

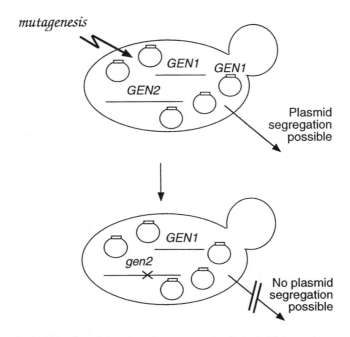

Figure 3. Isolation of multicopy suppressees. A strain with a multicopy plasmid bearing *GEN1* is mutagenized and screened for mutations in other genes (*gen2*) strictly dependent on *GEN1* overexpression.

method, the gene that is overexpressed is the one available and the aim of the screen is to find extragenic mutations, so-called *multicopy suppressee* mutations, that require this high expression for viability or growth. The genetic strategy of the multicopy suppressee method can be very useful for isolating functionally related gene products or interacting partners when other methods (e.g. multicopy suppression) cannot readily be applied because of the lack of a starting mutant phenotype. Two different approaches have been described that differ in their use of either a strong inducible promoter or a high-copy-number plasmid.

The *RAS1* and *RAS2* genes are functionally related and their products are GTPases involved in the cAMP signaling pathway in *S. cerevisiae*. In order to isolate temperature-sensitive mutants in the *RAS2* gene, as well as mutants dependent on high levels of Ras2p, Mitsuzawa *et al.* (1989) placed *RAS2* under the control of an inducible *GAL1* promoter on a plasmid. A *Δras1 RAS2* strain, harboring the *GAL1–RAS2* plasmid, was mutagenized and candidates were isolated as colony formers at 37°C on galactose plates but which failed to grow on glucose plates. By this method, mutations in two genes were identified. One mutated gene was the genomic *RAS2*, whereas the second one corresponded to an unlinked locus, the *CYR1* gene. The *CYR1* gene product is an adenylate cyclase, which is required for the synthesis of cAMP and acts downstream to Ras1p and Ras2p in the signaling pathway.

In a similar approach, but using a multicopy plasmid, Bender and Pringle (1991) isolated mutants that required the presence of multiple copies of the *MSB1* gene for survival. The non-essential *MSB1* gene was previously identified as a multicopy suppressor of temperature-sensitive mutations in *CDC24* and *CDC42* genes, both of which are required for the establishment of cell polarity and bud formation during the yeast cell division cycle. These authors used a colony-sectoring screen to identify additional genes interacting functionally with *MSB1*. Starting with a *Δmsb1 ade2 ade3* strain, containing a *MSB1* allele on a multicopy *ADE3*-plasmid, mutants were isolated in which *MSB1* had become essential for viability. After mutagenesis, colonies that appeared completely red were studied further. After discarding all undesirable candidates (revertants of the *ade3* mutation, integrants of the *ADE3*-plasmid, candidates requiring the *ADE3* marker rather than *MSB1*, *cdc24* or *cdc42* mutants), mutations in two new genes, *BEM1* and *BEM2*, were identified implicating both genes in the control of bud formation. The *bem2* mutation requires at least one copy of *MSB1* at low temperature (synthetic lethality, see below) and multiple copies at normal growth temperature (multicopy suppressee). The *bem1* mutation requires at least one functional *MSB1* gene even at low temperatures. This example demonstrates that the multicopy suppressee method, with a null allele as the starting point, can lead to the isolation of multicopy suppressee or synthetic lethal mutations. More recently, using the *ade2 ade3/ADE3* colony-sectoring assay, Mack *et al.* (1996) isolated a multicopy suppressee mutation, *bem4*, requiring high dosage of *CDC42*. *BEM4* is also involved in bud formation. Further genetic screens, including the two-hybrid approach (see Chapter 15), synthetic enhancement and multicopy suppression, confirmed that Bem4p physically interacts with

Cdc42p. Both of these examples exemplify the validity of this screen for searching physically or functionally interacting partners of a gene of interest.

This method has been used to analyse the function of yeast eIF4B, which is encoded by the *STM1* gene. To do so, the *STM1* gene was cloned on a multicopy plasmid carrying the *ADE2* and *URA3* markers. An *ade2 ura3* strain harboring this plasmid was mutagenized and screened for a non-sectoring phenotype. One candidate showing non-sectoring colonies and unable to grow on 5-FOA containing medium, was defective in generalized translation owing to impaired pre-rRNA processing (de la Cruz *et al.*, 1998). Thus, the defect in ribosome biogenesis can now be exploited to further analyse the possible function of eIF4B in preparing ribosomal subunits for their association with the mRNA.

E. High-Dosage Phenotype and Dominant-Negative Mutations

For various types of proteins, such as components of polymeric structures or regulatory proteins, their exact stoichiometry within the cell can be critical to ensure their proper function and their overproduction may be harmful. Nevertheless, rather than an undesirable effect, gene overexpression can be used to identify the function of a gene product. Two alternative strategies that depend on the overexpression of the wild-type or a mutant allele must be considered.

In the first strategy, it is the increased amount of the wild-type gene product that is harmful and leads to a phenotype. For example, Liu *et al.* (1992) constructed a yeast cDNA library under the control of the *GAL1* promoter. After transformation of this library into a wild-type strain, 15 genes were found to confer galactose-induced lethality. Amongst these genes were *ACT1* (actin), *TUB2* (β-tubulin) and *ABP1* (actin binding protein 1), all three of which are components of polymeric structures, and *TPK1* (cAMP-dependent protein kinase) and *GLC7* (type 1 protein phosphatase), both involved in regulatory pathways. In a similar approach, Meeks-Wagner *et al.* (1986) screened a multicopy genomic library for genes that caused a high frequency of chromosomal loss. Two genes so identified, *MIF1* and *MIF2*, were found to interfere with chromosomal transmission during mitosis when overexpressed. In these examples, screening for a phenotype was carried out in a wild-type strain, but it is also possible to apply such a screen to mutant strains deficient in a particular process. In this case, overexpression of a negative regulator may have little effect on growth of wild-type cells but reveal a synthetic enhancement in mutants affected in a particular pathway (de la Cruz *et al.*, 1997).

In the second strategy, the consequence of overproduction of a mutant protein that is harmful to the cell and confers a phenotype even in presence of the wild-type gene product is exploited (Figure 4). One refers to this situation as high-dosage dominant-negative mutations. Such mutations have been found in proteins having independent functional domains. For example, a transcriptional activator such as Gcn4p that has

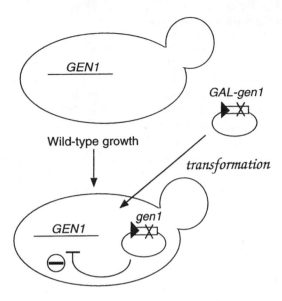

GEN1

GAL-gen1

Wild-type growth

transformation

GEN1   gen1

No growth on Galactose

Figure 4. Isolation of high-dosage dominant-negative mutations. A strain is transformed with a pool of plasmids carrying different mutations in the gene of interest which in turn is put under the control of a strong inducible promoter (e.g. *GAL-gen1*). Candidates are screened for non-growth under induced conditions.

been mutated within its activation domain, yet retains the capability of interacting with its target DNA sequence, can, when overexpressed, compete with its wild-type counterpart (Hope and Struhl, 1986). To a similar extent, a mutant subunit of a complex that retains the capability of interacting with its partners, but has lost its catalytic activity, can block the overall function of the complex. This has been observed for components of the splicing machinery such as Prp2p and Prp16p (Plumpton *et al.*, 1994; Schwer and Guthrie, 1992). Dominant-negative mutant forms of these two proteins bind to the spliceosome with apparently wild-type affinity but block its function. In some cases, overexpression of a truncated protein may also lead to a dominant-negative phenotype. Therefore mutants of interest may be also obtained by constructing a partial deletion of the gene, as has been described for *MATα2* and recombination/repair genes (Hall and Johnson, 1987; Milne and Weaver, 1993).

In general, to obtain dominant-negative mutations, a gene under the control of an inducible promoter is mutagenized *in vitro* and, after transformation into a wild-type strain, the colonies are screened for appearance of a phenotype under induced conditions. A dominant-negative mutation may be the starting point for the isolation of extragenic suppressors, thus allowing the characterization of interacting protein partners (Hosaka *et al.*, 1994; Powers *et al.*, 1991; Sugimoto *et al.*, 1995).

F. Synthetic Enhancement

The synthetic enhancement is the flip side of suppression (Rose, 1995) since the combination of two mutations results in a more severe phenotype than the one caused by the individual mutations (Rose, 1995). The molecular explanations of synthetic enhancement are diverse, and are usually not simply due to the combination of two detrimental mutations leading non-specifically to a non-growth phenotype. In many instances, synthetic enhancement strongly suggests that the two affected gene products are functionally related; for example, the two genes show functional redundancy, the two gene products may physically interact or they may act in the same pathway. Another explanation for a synthetic enhancement phenotype is the presence of mutations that affect parallel or crosstalking pathways. The synthetic enhancement approach can be carried out even in the absence of a phenotype caused by the primary mutation.

Plasmid segregation is the basis of synthetic enhancement screens (Figure 5). A mutant strain carrying its corresponding wild-type allele on a plasmid is mutagenized and synthetic mutations are isolated based on the fact that the provided wild-type allele becomes essential for viability and thus no plasmid segregation takes place. Numerous examples of synthetic phenotypes have been described. Here, we give three examples involving the isolation of redundant genes, genes whose products are members of related pathways, or genes whose products physically interact.

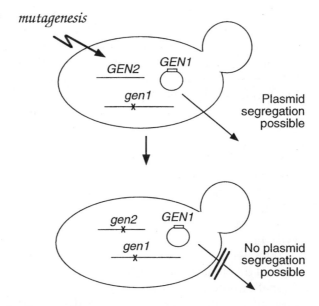

Figure 5. Synthetic enhancement. A strain, mutated in the gene of interest (*gen1*) and harboring its corresponding wild-type allele (*GEN1*) on a plasmid, is mutagenized and screened for mutations in other genes (*gen2*) strictly dependent on *GEN1*.

The sensitivity of wild-type $URA3^+$ cells to 5-FOA has been used to isolate $MYO5$, a gene highly homologous to $MYO3$, which encodes myosin (Goodson *et al.*, 1996). Because deletion of the $MYO3$ gene is not lethal, the presence of a possible redundant function in the genome was suggested. To isolate such a potentially redundant gene, the $MYO3$ gene was cloned on a $URA3$ plasmid and the genomic copy of $MYO3$ inactivated by gene disruption. After mutagenesis, colonies were screened for 5-FOA sensitivity, i.e. for the inability to segregate the $MYO3$ gene on the $URA3$ plasmid. The wild-type copy of the mutated $MYO5$ gene was cloned by complementation of the synthetic lethal phenotype.

The *ade2 ade3/ADE3* system has been used to isolate mutants affected in the pathway of the stepwise assembly of lipid-linked oligosaccharide (Zufferey *et al.*, 1995). Mutations in the *WBP1* gene, coding for a component of oligosaccharyl transferase, are synthetically lethal with mutations in the *ALG* genes, whose products are involved in oligosaccharide assembly. This observation was used to isolate new mutations synthetically lethal with a *wbp1* mutation. Starting with a *wbp1 ade2 ade3 ura3* strain carrying a *WBP1* gene on an *ADE3–URA3* plasmid, synthetic lethal mutations were first isolated by screening red, non-sectoring colonies. Furthermore, the *URA3* marker facilitated the assignment of the different mutations to complementation groups by using the 5-FOA sensitivity of the diploid cells. Because some mutations in the assembly pathway of oligosaccharides do not confer a conditional growth phenotype, the synthetic lethality facilitated the cloning of the corresponding wild-type genes by complementation of the synthetic phenotype (Stagljar *et al.*, 1994).

In the analysis of the multisubunit nuclear pore complex, the screen for synthetic lethal mutations has been very fruitful and permitted the isolation of many of its components (Doye and Hurt, 1995). The screens were also based on the red/white sectoring assay using the *ade2 ade3/ADE3* system.

Even though synthetic enhancement is a powerful approach to isolate components of a pathway or a complex, in some instances it leads to the identification of genes that are only distantly related to the process under investigation. It is important to emphasize, however, that some controls have to be performed before the putative gene candidates are analysed further. These controls are needed in order to demonstrate that the synthetic lethal phenotype is linked to the gene of interest rather than to the plasmid markers *URA3* or *ADE3*. The plasmid-shuffling technique is therefore adapted to eliminate those candidates in which the plasmid has been integrated or a gene conversion event has taken place at the marker locus and to exclude that the isolated mutation is a synthetic lethal with the genomic *ade3* or *ura3* mutations.

G. Non-allelic Non-complementation

Although classical complementation of a mutant phenotype is usually used to define the gene, allelic complementation and non-allelic non-complementation are also useful genetic tools. In the case of allelic complementation, two mutant alleles of the same gene can complement each other providing the mutations affect different functional domains of the gene product. In the case of non-allelic non-complementation (often called unlinked non-complementation), a diploid strain carrying recessive mutations in two different (i.e. non-allelic) genes reveals the phenotype of one of the parental haploid strains, despite the fact that the wild-type copies of the two mutated genes are present in its genome (i.e. non-complementation). The general idea behind this screen reflects the requirement for a certain level of activity of a heteromeric complex. A diploid cell may still survive because of the remaining active complexes even if the gene coding for one subunit of a complex is mutated. However, if two genes encoding different subunits of a complex are mutated, even less active complexes can be formed resulting in a growth defect.

To obtain non-allelic non-complementing mutations, a haploid wild-type strain is mutagenized and the resulting colonies are mated to a tester strain harboring a defined mutation (Figure 6). If the diploid cell still retains the mutant phenotype of the tester strain, the candidate will be further analysed. It either contains a mutation in the same gene, a dominant mutation or a non-allelic non-complementing mutation. Stearns and Botstein (1988) used the non-allelic non-complementation approach to isolate new mutations in the tubulin genes, *TUB1*, *TUB2* and *TUB3*, using a tester strain carrying a cold sensitive (cs) *tub2* mutation. To distinguish between allelic (*tub2*) and non-complementing mutations, the diploid strains retaining the cs *tub2* phenotype were sporulated and segregation of the markers followed. To exclude dominant mutations and non-mating mutants, the candidates were also mated to a wild-type strain. By this procedure they isolated a mutation in *TUB1*. In a subsequent screen, mutations in the *TUB3* gene have also been isolated. Because the Tub proteins were known to interact physically, these results validate the genetic approach by non-allelic non-complementation.

Although this genetic screen is potentially very powerful for studying protein–protein interactions in heteromeric complexes, its use in defining new genes has not been widely exploited. Nevertheless, several reports in the literature used non-allelic non-complementation to show interaction of two gene products. For example, mutations in the *SAC6* and *TPM1* genes, encoding fimbrin and tropomyosin respectively, have been identified as non-complementers of *act1* (actin) mutants (Vinh *et al.*, 1993). Also, proteins of the nuclear pore complex (Nup1p and Nup2p) have been shown, by non-allelic non-complementation and co-immuno-precipitation, to interact with Srp1p, a protein of unknown function (Belanger *et al.*, 1994). Mutations in the *FPR1*, *TOR1* and *TOR2* genes, responsible for a cell cycle arrest in the presence of rapamycin, have also been shown to exhibit non-allelic non-complementation, strongly suggesting an interaction between their gene products (Heitman *et al.*, 1991). Indeed, a direct interaction between Fpr1p and Tor1p or Tor2p has been confirmed.

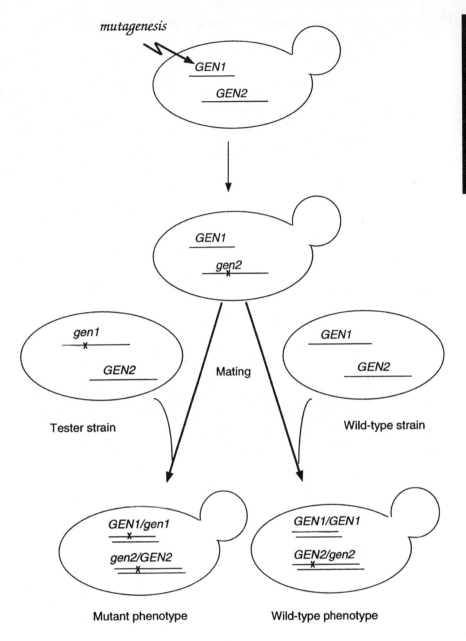

Figure 6. Isolation of non-allelic non-complementing mutations. A wild-type strain is mutagenized. The resulting colonies are mated with a tester strain bearing a mutation in *GEN1* (left) and with a wild-type strain (right). Non-allelic non-complementing mutations (*gen2*) will result in diploids having *gen1* (left) and wild-type (right) phenotypes.

H. Screening Conditional-Lethal Strain Collections

Because of its easy manipulation and the possibility it provides to analyse conditional phenotypes by biochemical and cellular methods, yeast has been used successfully to unravel key questions of cell cycle control. In a pioneering approach, Hartwell (1967) created a collection of ts strains to isolate mutants defective in cell cycle progression. At a first view, this approach seems very labour-intensive, but the screening of a collection of conditional-lethal mutations can be a powerful approach to characterize genes that are so far not amenable to genetic selection. Although almost 30 years old, the original collection of Hartwell is still being used with success. For instance, it has been recently used to identify mutants defective in replication control, leading to overreplication of DNA (Heichman and Roberts, 1996). In this study, the collection of ts mutants was analysed by flow cytometry to isolate those mutants with elevated DNA contents following a shift to the non-permissive temperature. Other collections of ts mutations and their application to different topics have been described (Lygerou *et al.*, 1994; Vijayraghavan *et al.*, 1989).

For some genes, it is very difficult or even impossible to obtain ts mutations. To circumvent this problem, collections of cs mutations may also be constructed. The target genes isolated from such cs mutant collections only partially overlap those from ts mutants, as shown for the analysis of cell cycle and pre-mRNA splicing mutants (Moir *et al.*, 1982; Noble and Guthrie, 1996). Indeed, whereas ts mutations affect the biochemical activity of a particular protein, it is generally assumed that cs mutants are most likely defective in the assembly of multiprotein complexes, as has been observed in bacteriophage assembly and ribosome biogenesis (Cox and Strack, 1971; Guthrie *et al.*, 1969).

I. The EUROFAN Screen

In contrast to the random conditional-lethal strain collections, the EURO-FAN project (**Euro**pean **f**unctional **a**nalysis **n**etwork), founded by the European Community (see Chapter 1), has as its principal aim the systematic analysis of new ORFs of unknown function that have been identified following the systematic sequencing of the complete *S. cerevisiae* genome. The project consists of disrupting all these ORFs in a diploid background and studying the phenotype of the haploid disruptants. In the case of non-essential genes, the morphology and behavior of the viable haploids under different growth conditions, as well as their mating efficiency, are analyzed. Diploid homozygotic disruptants are also analyzed under different growth conditions and for sporulation defects. All strains and plasmid constructs, together with the preliminary phenotypic data, will be made available. The collection of disrupted strains will therefore provide the scientific community with a collection of mutants that can be screened in the future for a multitude of phenotypes. Moreover, the systematic study of Eurofan will be followed by more detailed functional analyses. A similar and complementary approach has also been undertaken by the American yeast community.

J. *LacZ* Insertion and Fusion Libraries

The *lacZ* gene from *Escherichia coli* encoding β-galactosidase is a valuable tool for measuring gene expression in yeast (Guarente and Ptashne, 1981; Rose *et al.*, 1981). In general, the 5′ end of a gene of interest is fused in frame to the *lacZ*-coding region and the expression of the hybrid gene is followed by monitoring β-galactosidase activity. This "one-gene" approach has been extended to a collection of genes by using libraries of gene fusions. The approach described by Burns *et al.* (1994) used a bacterial mini-Tn3 to create random *lacZ* fusions in a library of cloned yeast fragments (see Chapter 10). In this study the cloned fragments were released from the vector and transformed into a diploid strain. The resulting strains were analysed for their phenotypes resulting from the gene disruptions by the inserted *lacZ* fusions in both diploid and haploid backgrounds. This approach allows both the identification and analysis of genes expressed at different times during the yeast life cycle and under different growth conditions. Other advantages are the easy analysis of the intracellular localization of the fusion proteins by immunofluorescence using anti-β-galactosidase antibodies. New multipurpose transposons for the creation of fusions to the green fluorescent protein and for the introduction of epitope tags have also recently been described (Ross-Macdonald *et al.*, 1997). A complementary approach has been developed by Daignan-Fornier and Bolotin-Fukuhara (see Chapter 11).

K. Genetic Screens with Heterologous Genes Expressed in Yeast

The power of yeast genetics is not restricted to the analysis of genes from *S. cerevisiae*, but it can also be used to either find functional homologs of yeast genes or yeast homologs of genes from other species (Elledge *et al.*, 1993; Holm, 1993; Thukral *et al.*, 1993; Whiteway *et al.*, 1993). In general, the screens using heterologous genes are based on those described in the previous sections. Here we describe three applications, using heterologous genes, as an illustration of the potential such screens offer.

The identification of functional homologs from other species can be achieved by transforming a mutant yeast strain with a heterologous cDNA expression library and by complementing the mutant phenotype. In the case of a conditional-lethal mutant, the screen for loss of the phenotype takes place under non-permissive conditions, for example at elevated temperature. However, in some cases the heterologous gene product is not active enough under such conditions. To circumvent this, it is possible to use the regulated expression of an essential gene under the control of an inducible promoter and to screen for the ability of transformants to grow under non-induced conditions. Alternatively, it is possible to use the plasmid-shuffling technique, by cloning the essential yeast gene on a counterselectable plasmid and testing growth on counterselective medium in presence of the heterologous gene. The successful isolation of genes complementing mutations affected in the cell cycle (Lew *et al.*, 1991), transcription factors (Becker *et al.*, 1991) and biosynthetic pathways (Schild *et al.*, 1990) are only a few examples of the potential of this method.

The heterologous gene isolated may be the strict homolog of the starting mutant gene or a heterologous suppressor (Colicelli *et al.*, 1989).

As in the multicopy suppressee approach (see section III.D), a yeast carrying a heterologous gene on a plasmid can be mutagenized and screened for mutants that absolutely require expression of the heterologous gene. This may result in the isolation of the yeast counterpart of the heterologous gene or in the characterization of functionally interacting proteins. This strategy has been described by Kranz and Holm (1990) for the isolation of the yeast topoisomerase gene, *TOP2*, by using the *Drosophila* homolog, *TOP2^D*.

Finally, the expression of certain heterologous genes may interfere with processes in yeast leading to dominant-negative phenotypes. This can be due to a partial activity of the heterologous protein that can, for example, assume the function of the yeast protein but escape its normal mode of regulation. It is also possible that the heterologous protein is incorporated into a complex, thereby reducing its activity drastically. In order to identify dominant-negative phenotypes caused by heterologous genes, a heterologous cDNA library is transformed into yeast and the appearance of a negative phenotype is scored. Because this screen is aimed at isolating genes exerting a dominant-negative phenotype, it is advantageous to use a cDNA library under the control of a strong inducible promoter. As an example, this approach has been used to identify *Candida albicans* genes interfering with the pheromone pathway of *S. cerevisiae* (Whiteway *et al.*, 1992). After transformation of a pheromone hypersensitive *S. cerevisiae* strain with a cDNA library from *C. albicans*, transformants still able to grow in the presence of α-factor were identified. This screen identified a set of cDNA clones, most of which turned out to possess sequence similarities with components of the *S. cerevisiae* pheromone pathway.

◆◆◆◆◆◆ IV. CONCLUSIONS

The ease of manipulation and the combination of genetic, molecular and biochemical methods have given yeast several advantages in its use as a valuable model organism for studying fundamental eukaryotic processes. The powerful genetic approaches that can be carried out in yeast have proven useful for unraveling complex cellular aspects that are inherently more difficult to address by strictly biochemical approaches. In this review, we have described the most common examples of classical and modern genetic screens and selections to study the function of a given gene or to identify its interacting partners. It is important to realize that such "smart" genetic screens can be used to isolate likely gene candidates involved in a particular process, but that further investigations using other genetic, biochemical or cell biological methods are required to confirm the molecular role of the candidate gene in the process studied.

Finally, success with such screens is often a question of chance and it requires a certain perseverance on the part of the investigator. The power of yeast genetics is almost unlimited, but whether a screen is smart or not depends on its elaboration, the definition of questions that may be answered and finally in the downstream analysis of the candidate genes so identified.

Acknowledgements

We are very grateful to Martine Collart, Costa Georgopoulos and Kyle Tanner for comments on the manuscript. J. de la Cruz is supported by fellowships from the Spanish Government, the Sandoz-Stiftung and the CIBA-GEIGY Jubiläums-Stiftung. Work in the authors' laboratory is supported by grants from the Swiss National Science Foundation (to P.L.). Jesús de la Cruz and Marie-Claire Daugeron contributed equally to this chapter.

References

Altmann, M., Müller, P. P., Wittmer, B., Ruchti, F., Lanker, S. and Trachsel, H. (1993). A *Saccharomyces cerevisiae* homologue of mammalian translation initiation factor 4B contributes to RNA helicase activity. *EMBO J.* **12**, 3997–4003.

Ausubel, F. M., Brent, R., Kingston, R. E., Moore, D. D., Seidman, J. G., Smith, J. A. and Struhl, A. (1994). In *Current Protocols in Molecular Biology*, Chapter 13. Wiley, Chichester.

Becker, D. M., Fikes, J. D. and Guarente, L. (1991). A cDNA encoding a human CCAAT-binding protein cloned by functional complementation in yeast. *Proc. Natl Acad. Sci. USA* **88**, 1968–1972.

Belanger, K. D., Kenna, M. A., Wei, S. and Davis, L. I. (1994). Genetic and physical interactions between Srp1p and nuclear pore complex proteins Nup1p and Nup2p. *J. Cell Biol.* **126**, 619–630.

Bender, A. and Pringle, J. R. (1991). Use of a screen for synthetic lethal and multicopy suppressee mutants to identify two new genes involved in morphogenesis in *Saccharomyces cerevisiae*. *Mol. Cell. Biol.* **11**, 1295–1305.

Bi, E. and Pringle, J. R. (1996). ZDS1 and ZDS2, genes whose products may regulate Cdc42p in *Saccharomyces cerevisiae*. *Mol. Cell. Biol.* **16**, 5264–5275.

Boeke, J. D., Lacroute, F. and Fink, G. R. (1984). A positive selection for mutants lacking orotidine-5'-phosphate decarboxylase activity in yeast: 5-fluoro-orotic acid resistance. *Mol. Gen. Genet.* **197**, 345–346.

Boeke, J. D., Trueheart, J., Natsoulis, G. and Fink, G. R. (1987). 5-Fluoroorotic acid as a selective agent in yeast molecular genetics. *Methods Enzymol.* **154**, 164–175.

Boles, E., Lehnert, W. and Zimmermann, F. K. (1993). The role of the NAD-dependent glutamate dehydrogenase in restoring growth on glucose of a *Saccharomyces cerevisiae* phosphoglucose isomerase mutant. *Eur. J. Biochem.* **217**, 469–477.

Burns, N., Grimwade, B., Ross-Macdonald, P. B., Choi, E.-Y., Finberg, K., Roeder, G. S. and Snyder, M. (1994). Large-scale analysis of gene expression, protein-localization, and gene disruption in *Saccharomyces cerevisiae*. *Genes Dev.* **8**, 1087–1105.

Chattoo, B. B., Sherman, F., Azubalis, D. A., Fjellstedt, T. A., Mehnert, D. and Ogur, M. (1979). Selection of *lys2* mutants of the yeast *Saccharomyces cerevisiae* by the utilization of α-aminoadipate. *Genetics* **93**, 51–65.

Colicelli, J., Birchmeier, C., Michaeli, T. O., Neill, K., Riggs, M. and Wigler, M. (1989). Isolation and characterization of a mammalian gene encoding a high-affinity cAMP phosphodiesterase. *Proc. Natl Acad. Sci. USA* **86**, 3599–3603.

Coppolecchia, R., Buser, P., Stotz, A. and Linder, P. (1993). A new yeast translation initiation factor suppresses a mutation in the eIF-4A RNA helicase. *EMBO J.* **12**, 4005–4011.

Cost, G. J. and Boeke, J. D. (1996). A useful colony colour phenotype associated with the yeast selectable/counter-selectable marker *MET15*. *Yeast* **12**, 939–941.

Cox, J. H. and Strack, H. B. (1971). Cold-sensitive mutants of bacteriophage lambda. *Genetics* **67**, 5–17.

de la Cruz, J., Iost, I., Kressler, D. and Linder, P. (1997). The p20 and Ded1 proteins have antagonistic roles in eIF4E-dependent translation in *Saccharomyces cerevisiae*. *Proc. Natl Acad. Sci. USA* **94**, 5201–5206.

de la Cruz, J., Kressler, D., Tollervey, D. and Linder, P. (1998). Dob1p (Mtr4p) is a putative ATP-dependent RNA helicase required for the 3' end formation of 5.8S rRNA in *Saccharomyces cerevisiae. EMBO J.* in press.

Doye, V. and Hurt, E. C. (1995). Genetic approaches to nuclear pore structure and function. *Trends Genet.* **11**, 235–241.

Elledge, S. J., Bai, C. and Edwards, M. C. (1993). Cloning mammalian genes using cDNA expression libraries in *Saccharomyces cerevisiae. Methods* **5**, 96–101.

Fisher, C. R. (1969). Enzymology of the pigmented adenine-requiring mutants of *Saccharomyces* and *Schizosaccharomyces. Biochem. Biophys. Res. Commun.* **34**, 306–310.

Goodson, H. V., Anderson, B. L., Warrick, H. M., Pon, L. A. and Spudich, J. A. (1996). Synthetic lethality screen identifies a novel yeast myosin I gene (*MYO5*): myosin I proteins are required for polarization of the actin cytoskeleton. *J. Cell Biol.* **133**, 1277–1291.

Greenberg, M. L., Myers, P. L., Skvirsky, R. C. and Greer, H. (1986). New positive and negative regulators for general control of amino acid biosynthesis in *Saccharomyces cerevisiae. Mol. Cell. Biol.* **6**, 1820–1829.

Guarente, L. (1993). Synthetic enhancement in gene interaction: a genetic tool comes of age. *Trends Genet.* **9**, 362–366.

Guarente, L. and Ptashne, M. (1981). Fusion of *Escherichia coli lacZ* to the cytochrome c gene of *Saccharomyces cerevisiae. Proc. Natl Acad. Sci. USA* **78**, 2199–2203.

Guthrie, C. and Fink, G. R. (1991). *Guide to Yeast Genetics and Molecular Biology*. Academic Press, London.

Guthrie, C., Nashimoto, H. and Nomura, M. (1969). Structure and function of *E. coli* ribosomes. 8. Cold-sensitive mutants defective in ribosome assembly. *Proc. Natl Acad. Sci. USA* **63**, 384–391.

Hall, M. N. and Johnson, A. D. (1987). Homeo domain of the yeast repressor alpha 2 is a sequence-specific DNA-binding domain but is not sufficient for repression. *Science* **237**, 1007–1012.

Hall, M. N. and Linder, P. (1993). *The Early Days of Yeast Genetics*. Cold Spring Harbor Laboratory, Cold Spring Harbor.

Harper, J. W., Adami, G. R., Wei, N., Keyomarsi, K. and Elledge, S. J. (1993). The p21 Cdk-interacting protein Cip1 is a potent inhibitor of G1 cyclin-dependent kinases. *Cell* **75**, 805–816.

Hartwell, L. H. (1967). Macromolecule synthesis in temperature-sensitive mutants of yeast. *J. Bacteriol.* **93**, 1662–1670.

Heichman, K. A. and Roberts, J. M. (1996). The yeast *CDC16* and *CDC27* genes restrict DNA replication to once per cell cycle. *Cell* **85**, 39–48.

Heitman, J., Movva, N. R. and Hall, M. N. (1991). Targets for cell cycle arrest by the immunosuppressant rapamycin in yeast. *Science* **253**, 905–909.

Hinnebusch, A. G. and Fink, G. R. (1983). Positive regulation in the general amino acid control of *Saccharomyces cerevisiae*. *Proc. Natl Acad. Sci. USA* **80**, 5374–5378.

Hinnebusch, A. G. and Liebman, S. W. (1991). Protein synthesis and translational control in *Saccharomyces cerevisiae*. In *The Molecular and Cellular Biology of the Yeast Saccharomyces* (J. R. Broach, J. R. Pringle and E. W. Jones, eds), pp. 627–735. Cold Spring Harbor Laboratory Press, Cold Spring Harbor.

Hohmann, S., Van Dijck, P., Luyten, K. and Thevelein, J. M. (1994). The *byp1-3* allele of the *Saccharomyces cerevisiae GGS1/TPS1* gene and its multi-copy suppressor tRNA(GLN) (CAG): Ggs1/Tps1 protein levels restraining growth on fermentable sugars and trehalose accumulation. *Curr. Genet.* **26**, 295–301.

Holm, C. (1993). A functional approach to identifying yeast homologs of genes from other species. *Methods* **5**, 102–109.

Hope, I. A. and Struhl, K. (1986). Functional dissection of a eukaryotic transcriptional activator protein, GCN4 of yeast. *Cell* **46**, 885–894.

Hosaka, K., Nikawa, J.-I., Kodaki, T. and Yamashita, S. (1994). Cloning and characterization of the *SCS1* gene required for the expression of genes in yeast phospholipid synthesis. *J. Biochem.* **115**, 131–136.

Huffaker, T. C., Hoyt, M. A. and Botstein, D. (1987). Genetic analysis of the yeast cytoskeleton. *Annu. Rev. Genet.* **21**, 259–284.

Kaiser, C., Michaelis, S. and Mitchell, A. (1994). *Methods in Yeast Genetics*. Cold Spring Harbor Laboratory, Cold Spring Harbor.

Koshland, D., Kent, J. C. and Hartwell, L. H. (1985). Genetic analysis of the mitotic transmission of minichromosomes. *Cell* **40**, 393–403.

Kranz, J. E. and Holm, C. (1990). Cloning by function: an alternative approach for identifying yeast homologs of genes from other organisms. *Proc. Natl Acad. Sci. USA* **87**, 6629–6633.

Leberer, E., Dignard, D., Harcus, D., Hougan, L., Whiteway, M. and Thomas, D. Y. (1993). Cloning of *Saccharomyces cerevisiae STE5* as a suppressor of a Ste20 protein kinase mutant: structural and functional similarity of Ste5 to Far1. *Mol. Gen. Genet.* **241**, 241–254.

Lew, D. J., Dulic, V. and Reed, S. I. (1991). Isolation of three novel human cyclins by rescue of G1 cyclin (Cln) function in yeast. *Cell* **66**, 1197–1206.

Liu, H., Krizek, J. and Bretscher, A. (1992). Construction of a *GAL1*-regulated yeast cDNA expression library and its application to the identification of genes whose overexpression causes lethality in yeast. *Genetics* **132**, 665–673.

Lygerou, Z., Mitchell, P., Petfalski, E., Seraphin, B. and Tollervey, D. (1994). The *POP1* gene encodes a protein component common to the RNase MRP and RNase P ribonucleoproteins. *Genes Dev.* **8**, 1423–1433.

Mack, D., Nishimura, K., Dennehey, B. K., Arbogast, T., Parkinson, J., Toh-e, A., Pringle, J. R. *et al.* (1996). Identification of the bud emergence gene *BEM4* and its interactions with Rho-type GTPases in *Saccharomyces cerevisiae*. *Mol. Cell. Biol.* **16**, 4387–4395.

Mann, C., Buhler, J. M., Treich, I. and Sentenac, A. (1987). *RPC40*, a unique gene for a subunit shared between yeast RNA polymerases A and C. *Cell* **48**, 627–637.

Meeks-Wagner, D., Wood, J. S., Garvik, B. and Hartwell, L. H. (1986). Isolation of two genes that affect mitotic chromosome transmission in *S. cerevisiae*. *Cell* **44**, 53–63.

Milne, G. T. and Weaver, D. T. (1993). Dominant negative alleles of *RAD52* reveal a DNA repair/recombination complex including Rad51 and Rad52. *Genes Dev.* **7**, 1755–1765.

Mitsuzawa, H., Uno, I., Oshima, T. and Ishikawa, T. (1989). Isolation and characterization of temperature-sensitive mutations in the *RAS2* and *CYR1* genes of *Saccharomyces cerevisiae*. *Genetics* **123**, 739–748.

Moir, D., Stewart, S. E., Osmond, B. C. and Botstein, D. (1982). Cold-sensitive cell-division-cycle mutants of yeast: isolation, properties, and pseudoreversion studies. *Genetics* **100**, 547–563.

Moreau, V., Madania, A., Martin, R. P. and Winsor, B. (1996). The *Saccharomyces cerevisiae* actin-related protein Arp2 is involved in the actin cytoskeleton. *J. Cell Biol.* **134**, 117–132.

Noble, S. M. and Guthrie, C. (1996). Identification of novel genes required for yeast pre-mRNA splicing by means of cold-sensitive mutations. *Genetics* **143**, 67–80.

Plumpton, M., McGarvey, M. and Beggs, J. D. (1994). A dominant negative mutation in the conserved RNA helicase motif 'SAT' causes splicing factor PRP2 to stall in spliceosomes. *EMBO J.* **13**, 879–887.

Powers, S., Gonzales, E., Christensen, Cubert, J. and Broek, D. (1991). Functional cloning of *BUD5*, a *CDC25*-related gene from *S. cerevisiae* that can suppress a dominant-negative *RAS2* mutant. *Cell* **65**, 1225–1231.

Printen, J. A. and Sprague, G. F., Jr (1994). Protein–protein interactions in the yeast pheromone response pathway: Ste5p interacts with all members of the MAP kinase cascade. *Genetics* **138**, 609–619.

Ramer, S. W., Elledge, S. J. and Davis, R. W. (1992). Dominant genetics using a yeast genomic library under the control of a strong inducible promoter. *Proc. Natl Acad. Sci. USA* **89**, 11589–11593.

Riles, L. and Olson, M. V. (1988). Nonsense mutations in essential genes of *Saccharomyces cerevisiae*. *Genetics* **118**, 601–607.

Rose, M. D. (1995). In *The Yeasts*, Vol 6. (A. H. Rose, A. E. Wheals and J. S. Harrison, eds), pp. 69–120. Academic Press, London.

Rose, M., Casadaban, M. J. and Botstein, D. (1981). Yeast genes fused to β-galactosidase in *Escherichia coli* can be expressed normally in yeast. *Proc. Natl Acad. Sci. USA* **78**, 2460–2464.

Ross-Macdonald, P., Sheehan, A., Roeder, G. S. and Snyder, M. (1997). A multipurpose transposon system for analyzing protein production, localization, and function in *Saccharomyces cerevisiae*. *Proc. Natl Acad. Sci. USA* **94**, 190-195.

Rotenberg, M. O., Moritz, M. and Woolford, J. L. J. (1988). Depletion of *Saccharomyces cerevisiae* ribosomal protein L16 causes a decrease in 60S ribosomal subunits and formation of half-mer polyribosomes. *Genes Dev.* **2**, 160–172.

Schild, D., Brake, A. J., Kiefer, M. C., Young, D. and Barr, P. J. (1990). Cloning of three human multifunctional *de novo* purine biosynthetic genes by functional complementation of yeast mutations. *Proc. Natl Acad. Sci. USA* **87**, 2916–2920.

Schwer, B. and Guthrie, C. (1992). A dominant negative mutation in a spliceosomal ATPase affects ATP hydrolysis but not binding to the spliceosome. *Mol. Cell. Biol.* **12**, 3540–3547.

Schwer, B., Linder, P. and Stewart, S. (1998). Effects of deletion mutations in the yeast Ces1 protein on cell growth and morphology and on high copy suppression of mutations in mRNA capping enzyme and translation initiation factor 4A. *Nucleic Acids Res.*, in press.

Singh, A. and Sherman, F. (1974). Association of methionine requirement with methyl mercury resistant mutants of yeast. *Nature* **247**, 227–229.

Song, J. M., Picologlou, S., Grant, C. M., Firoozan, M., Tuite, M. F. and Liebman, S. (1989). Elongation factor EF-1 alpha gene dosage alters translational fidelity in *Saccharomyces cerevisiae*. *Mol. Cell. Biol.* **9**, 4571–4575.

Stagljar, I., te Heesen, S. and Aebi, M. (1994). New phenotype of mutations deficient in glucosylation of the lipid-linked oligosaccharide: cloning of the *ALG8* locus. *Proc. Natl Acad. Sci. USA* **91**, 5977–5981.

Stearns, T. and Botstein, D. (1988). Unlinked noncomplementation: isolation of new conditional-lethal mutations in each of the tubulin genes of *Saccharomyces cerevisiae*. *Genetics* **119**, 249–260.

Sugimoto, K., Matsumoto, K., Kornberg, R. D., Reed, S. I. and Wittenberg, C. (1995). Dosage suppressors of the dominant G1 cyclin mutant *CLN3-2*: identification of a yeast gene encoding a putative RNA/ssDNA binding protein. *Mol. Gen. Genet.* **248**, 712–718.

Thukral, S. K., Chang, K. K. H. and Bitter, G. A. (1993). Functional expression of heterologous proteins in *Saccharomyces cerevisiae*. *Methods* **5**, 86–95.

Vazquez de Aldana, C. R. and Hinnebusch, A. G. (1994). Mutations in the GCD7 subunit of yeast guanine nucleotide exchange factor eIF-2B overcome the inhibitory effects of phosphorylated eIF-2 on translation initiation. *Mol. Cell. Biol.* **14**, 3208–3222.

Vijayraghavan, U., Company, M. and Abelson, J. (1989). Isolation and characterization of pre-mRNA splicing mutants of *Saccharomyces cerevisiae*. *Genes Dev.* **3**, 1206-1216.

Vinh, D. B., Welch, M. D., Corsi, A. K., Wertman, K. F. and Drubin, D. G. (1993). Genetic evidence for functional interactions between actin noncomplementing (Anc) gene products and actin cytoskeletal proteins in *Saccharomyces cerevisiae*. *Genetics* **135**, 275–286.

Whiteway, M., Dignard, D. and Thomas, D. Y. (1992). Dominant negative selection of heterologous genes: isolation of *Candida albicans* genes that interfere with *Saccharomyces cerevisiae* mating factor-induced cell cycle arrest. *Proc. Natl Acad. Sci. USA* **89**, 9410–9414.

Whiteway, M., Csank, C. and Thomas, D. Y. (1993). Dominant negative selection of heterologous genes in yeast. *Methods* **5**, 110–115.

Yu, Y., Jiang, Y. W., Wellinger, R. J., Carlson, K., Roberts, J. M. and Stillman, D. J. (1996). Mutations in the homologous *ZDS1* and *ZDS2* genes affect cell cycle progression. *Mol. Cell. Biol.* **16**, 5254–5263.

Zufferey, R., Knauer, R., Burda, P., Stagljar, I., te Heesen, S., Lehle, L. and Aebi, M. (1995). STT3, a highly conserved protein required for yeast oligosaccharyl transferase activity *in vivo*. *EMBO J.* **14**, 4949–4960.

17 Metabolic Control Analysis as a Tool in the Elucidation of the Function of Novel Genes

Bas Teusink[1,2], Frank Baganz[1], Hans V. Westerhoff[2,3] and Stephen G. Oliver[1]

[1]*Department of Biomolecular Sciences, UMIST, Manchester, UK,*
[2]*E.C. Slater Institute, BioCentrum, University of Amsterdam, Amsterdam, The Netherlands, and*
[3]*Department of Microbial Physiology, BioCentrum, Free University, Amsterdam, The Netherlands*

◆◆

CONTENTS

List of Abbreviations

F6P	Fructose 6-phosphate
F16bP	Fructose 1,6-*bis*phosphate
F26bP	Fructose 2,6-*bis*phosphate
G3P	Glycerol 3-phosphate
GPD	Glycerol-phosphate dehydrogenase
GPP	Glycerol-phosphate phosphatase
HXK	Hexokinase
IPTG	Isopropyl-β-D-thiogalactopyranoside
MCA	Metabolic control analysis
ORF	Open reading frame
PCA	Perchloric acid
PCR	Polymerase chain reaction
PDC	Pyruvate decarboxylase
PFK	Phosphofructokinase
PF-2-K	6-Phosphofructo-2-kinase
PYK	Pyruvate kinase
Pyr	Pyruvate

METHODS IN MICROBIOLOGY, VOLUME 26
ISBN 0–12–521526–6

◆◆◆◆◆◆ I. INTRODUCTION

The completion of the entire genome sequence of the brewers' and bakers' yeast, *Saccharomyces cerevisiae*, has defined some 6000 protein-encoding genes (Goffeau *et al.*, 1996) for which the function of less than half is known with any confidence. The search for the function of these unknown genes is called "functional analysis" here. The first, and very powerful, step in functional analysis is to compare the amino acid sequences of the predicted protein products of these genes with the sequences contained within the public data libraries to see if useful pointers to the likely functions of these novel yeast genes can be discovered by their similarity to genes or proteins from other organisms (for a good example of this approach, see Casari *et al.*, 1996, and Das *et al.*, 1997). Powerful as this approach is, its results need to be treated with caution and intelligence (Oliver, 1996). Moreover, it is notable that there is a growing fraction of novel yeast genes whose predicted products show sequence similarity to proteins from other species, but in none of these organisms is the function of the protein understood. We must ask ourselves why it is that the efforts of all the molecular geneticists, working on various systems all over the world, have failed to discover these genes by classical (or "function-first") approaches. One possibility is that our ability to uncover these genes has been constrained by the usual experimental paradigm of genetics in which experiments are designed (often with great ingenuity) to give qualitative (yes/no) results. If this is so, then it is important that we now adopt a quantitative approach to the analysis of phenotype. In this chapter, we describe how the concepts and mathematical formalisms of metabolic control analysis (MCA), as pioneered by Kacser and Heinrich and co-workers (Heinrich and Rapoport, 1974; Kacser and Burns, 1973), may be adapted to provide useful strategies for a systematic quantitative approach to the elucidation of gene function.

In this chapter, we will explain the basic concepts and theorems of MCA as it may be applied to the functional analysis of novel genes. A more general overview of MCA is given in some reviews (Cornish-Bowden, 1995; Fell, 1992; Kell and Westerhoff, 1986; Schuster and Heinrich, 1992), and recently two books on the topic of control of metabolism have been published (Fell, 1997; Heinrich and Schuster, 1996). Readers unfamiliar with this area are encouraged to read Henrik Kacser's update of the original paper on MCA (Kacser *et al.*, 1995), which is an exceptionally clear account that introduces the main concepts and terminology of the theory. There are also several useful internet sites dealing with MCA, including tutorials (P. Mendes: http://gepasi.dbs.aber.ac.uk/metab/mca_home.htm), a course (P. Butler: http://www.bi.umist.ac.uk/courses/2IRM/MCA/default.htm), a FAQ site (A. Cornish-Bowden: http://ir2lcb.cnrs-mrs.fr/lcbpage/athel/mcafaq.html; see also http://ir2lcb.cnrs-mrs.fr/lcbpage/athel/mcai.html), and a newsgroup (see, for example, http://www.bio.net:80/hypermail/BTK-MCA/).

Thus much of what is to come can be found elsewhere, but we will explain MCA with respect to its implications for the elucidation of the functions of novel genes. Many researchers in this area, most of them molecular biologists, are likely to be unfamiliar with the method. This account is therefore designed to be read by molecular biologists and the mathematics is kept to an absolute minimum. This is achieved by doing a functional analysis experiment, i.e. knocking out or overexpressing a novel gene and looking for the resultant phenotype. All the "experiments" in this paper will be done *in silico* using a simple kinetic model that mimics our functional analysis experiments (see Appendix 1 in this chapter, for details). Because such "virtual" or *gedanken* experiments give us the freedom to do things that will be difficult, or even impossible, to achieve with living cells, they provide helpful illustrations of the principles involved. However, we will, whenever possible, refer to real experimental data.

Throughout this section, we will use a simple example to illustrate the principles of MCA: the regulation of phosphofructokinase (PFK) by fructose, 2,6-*bis*phosphate (F26bP) (Bartrons *et al.*, 1982; Hers and Van Schaftingen, 1982). This metabolite is produced by the enzyme 6-phosphofructo-2-kinase (PF-2-K), encoded by the genes *PFK26* (Kretschmer and Fraenkel, 1991; Kretschmer *et al.*, 1991) and *PFK27* (Boles *et al.*, 1996). Fructose 2,6-bisphosphatase, encoded by the gene *FBP26*, dephosphorylates F26bP back to F6P (Paravicini and Kretschmer, 1992). In this chapter we pretend that we lack genetic information about the enzymes involved in F26bP metabolism, and that *PFK27* is just an open reading frame for which we need to determine the function. *PFK26* will turn out to be the isoenzyme that complicates matters. Such cases of apparent genetic redundancy are widespread in the yeast genome (Thomas, 1993; Wolfe and Shields, 1997) and represent a major obstacle to the systematic analysis of gene function (Oliver, 1996). It has been shown that deletion of *PFK27* affects the concentration of F26bP (Boles *et al.*, 1996). In Figure 1,

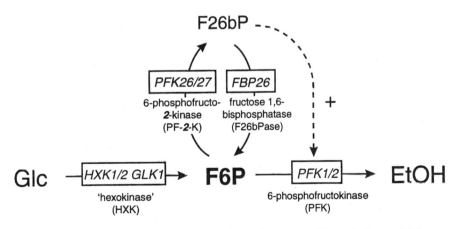

Figure 1. The system used as an example in this paper. Phosphofructo-2-kinase (PF-2-K) synthesizes fructose 2,6-*bis*phosphate (F26bP) from F6P. F26bP is a positive regulator of phosphofructokinase (PFK). In yeast, PF-2-K is encoded by the genes *PFK26* and *PFK27*. We shall take *PFK27* to be our novel gene of unknown function. In our functional analysis, we will be able to measure the concentration of F6P and the overall rate (flux) of the pathway.

this situation is shown as a simple glycolytic pathway, consisting of two enzymes – hexokinase (HXK) and PFK – that are connected via the metabolite fructose 6-phosphate (F6P).[1] The activity of PFK is regulated by the activator F26bP, which is produced by the enzyme that is encoded by our novel gene, *PFK27*, whose phenotype we are seeking.[2] Unfortunately, *PFK26* also produces F26bP, so deleting *PFK27* will decrease F26bP by only 10%, whereas overexpression of *PFK27* will give a 100% increase in F26bP.

In the experimental context, we envisage that the glucose concentration does not change significantly (or glucose may be continuously added to compensate for its consumption), and the ethanol concentration remains too low to be harmful. This situation is achieved in a real experiment by using a chemostat, although batch cultures at low cell density may also meet these conditions. Such a system will reach a *steady state* in which the growth rate (μ), the glycolytic flux (*J*), and the concentration of F6P are all constant (Fell, 1997).

◆◆◆◆◆◆ II. DEFINITIONS AND CONCEPTS OF METABOLIC CONTROL ANALYSIS

First, we must distinguish between *parameters* and *variables*. Parameters are set, either by nature (e.g. kinetic constants) or by the experimenter (e.g. pH and temperature, or the concentration of some inhibitor). They determine the behavior of the variables. The variables include the reaction rates and the concentrations of the metabolites internal to the pathway; they will change with time until a steady state is achieved. To illustrate the difference between parameters and variables, the steady state of our system was perturbed by deleting *PFK27*, which decreases the concentration of F26bP by 10%. In Figure 2A, the effect of a decrease in the parameter F26bP is shown; after inoculation of the wild type and *pfk27Δ* mutant in glucose medium, the rates of HXK and PFK converge to the same rate, the steady-state flux (equivalent to the "growth rate"). At this steady state, the concentration of F6P remains constant in time (Figure 2A). The deletion of *PFK27* caused (small) changes in the steady-state flux and F6P concentration compared with the wild type. In a qualitative or semiquantitative approach, such as comparison on plates, the small

---

[1]For simplicity, the rest of glycolysis is lumped into the HXK and PFK reactions. Thus, phosphoglucose isomerase is lumped into HXK, and the enzymes from aldolase to alcohol dehydrogenase are lumped into PFK. Lumping of reactions that are not important (for the purpose of the model) is good practice in mathematical modeling.

[2]F26bP is said to be an *external* regulator, as it does not take part in the metabolic reaction sequence. In the case of feedback regulation by the product of a pathway, e.g. in amino acid biosynthesis, the product would be called an *internal* regulator.

differences in growth rate shown in Figure 2A will not be detected. In a quantitative approach to functional analysis, however, we are trying to identify function on the basis of (small) quantitative differences between wild type and mutants.

In another experiment, F6P was injected into the wild type at *t*=25 so as to bring its concentration to the level of the *pfk27Δ* mutant (Figure 2B).

A

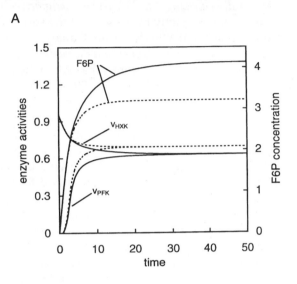

B

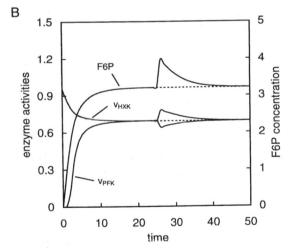

Figure 2. Simulated effect of a change in the parameter F26bP and the variable F6P on the steady-state characteristics of the system of Figure 1. In the figure, F6P means the concentration of F6P, V_{HXK} means the activity of HXK, and V_{PFK} stands for the activity of PFK. All units are arbitrary. (A) Relaxation to a steady state after inoculation in glucose medium. The transient of the wild-type (dotted lines) and that of the *pfk27Δ* mutant (solid lines) are shown. The 10% decrease in the concentration of F26bP in the mutant was set to be equal to a decrease of the activity of PFK by 20%. (B) The dotted line shows the same transient as in (A), for the wild type. The solid line shows a computer simulation "experiment" where, at *t* = 25, the F6P concentration was increased by 20% in these cells.

After a transition period, the new steady state at $t=50$ is identical to the steady state before the perturbation. The concentration of F6P is a variable of the system, and is unable to affect the steady state, whereas (in the $pfk27\Delta$ mutant) the change in the parameter, F26bP,[3] led to a new, *different*, steady state (Figure 2A).

The main question that is traditionally addressed in MCA is: *to what extent does a change in the system's parameters affect the behavior of the variables?* This is exactly the inverse of the sort of questions we ask in a quantitative approach to functional analysis: *how can we understand which parameters have been changed, on the basis of changes in the behavior of the metabolic variables?* Only recently has MCA been extended into this reverse direction (Westerhoff *et al.*, 1994). This complementarity of the two approaches means that it is essential to understand MCA in order to understand quantitative systematic functional analysis.

One can measure the effect of changes in the parameters (e.g. in the concentration of an enzyme) on the steady-state variables (e.g. growth rate or metabolite concentrations) without any understanding of the system. The *response coefficient* is then used to quantify the extent to which a parameter affects a variable. The response coefficient is defined as the *relative* change in a steady-state variable, such as the flux or a concentration, caused by a *relative* change in a parameter. Choosing relative changes rather than absolute ones, this coefficient becomes dimensionless: it has no units (see Hofmeyr, 1995). To measure a response coefficient, one would measure a steady-state variable in the wild type (in our example, the growth rate or the concentration of F6P) and compare it with the corresponding variable in the $pfk27\Delta$ mutant.[4]

Mathematically, the response coefficient of F6P to a change in F26bP would read:

$$R^{\text{F6P}}_{\text{F26bP}} \equiv \left(\frac{\Delta[\text{F6P}] / [\text{F6P}]}{\Delta[\text{F26bP}] / [\text{F26bP}]} \right)_{\Delta\text{F26bP}\to0} = \frac{d[\text{F6P}] / [\text{F6P}]}{d[\text{F26bP}] / [\text{F26bP}]} = \frac{d\ln[\text{F6P}]}{d\ln[\text{F26bP}]} \quad (1)$$

(using the mathematical relation: $1/x \cdot dx = d\ln x$ to derive at the right term). The square brackets indicate concentrations. The subscript of the response coefficient is used to indicate the parameter that is changed, the superscript indicates the variable that is evaluated.

[3]One can argue, and in fact one would be right, that the concentration of F26bP is also a variable, and the parameter that we change is the PF-2-K activity. To simplify matters, however, we take the concentration of F26bP to be set by the experimenter (albeit via the activity of PF-2-K), and in that sense consider F26bP as a parameter. We can do this without problems, because F26bP is an external regulator that does not take part in the biochemistry of the metabolic reactions, provided we can neglect its dependence on the concentration of F6P (Kahn and Westerhoff, 1991).
[4]Obviously, in a chemostat, one would not be able to see an increase in the steady-state growth rate because it is set by the experimenter via the dilution rate. Rather, the effect will be visible via the residual substrate concentration: for the theory see Snoep *et al.* (1994), and for an experimental example, see Yap *et al.* (1996).

The mathematics of the above formulation adds that if the change in [F26bP] is only small, then the response coefficient is the derivative of [F6P] with respect to [F26bP]. The advantage of such small modulations is that second-order effects remain negligible. Accordingly the use of derivatives enables the strict derivation of the laws of control of metabolic systems. Therefore, all coefficients within MCA are defined in terms of derivatives, and strictly speaking apply to small (infinitesimal) changes only. The use of derivatives also furnishes a graphical method to determine response coefficients, by plotting the logarithm of the steady-state variable against the logarithm of the parameter: the slope of the curve will give the response coefficient. The use of derivatives has been proposed a drawback as it is not always feasible to make a very small change in a parameter experimentally, or the resultant effect on the variable is not measurable with sufficient accuracy (Fell, 1992). Indeed, the standard experimental approach (Groen *et al.*, 1982b; Jensen *et al.*, 1993b; Ruyter *et al.*, 1991; Van der Vlag *et al.*, 1995) is to modulate the parameter over a fairly wide range around the steady-state value of interest. This then allows a fairly accurate determination of the slope at the interesting steady state (e.g. the *extent* of a change under a particular set of growth conditions). Moreover, response coefficients can be used as good indicators of what happens in case of larger changes (see Small and Kacser, 1993, for improvements on this point).

In functional analysis experiments, the magnitude of the change in the unknown enzyme is obviously not under the control of the experimenter. In our case, we had to live with a decrease in the concentration of F26bP of 10% when deleting *PFK27*, and the use of the concepts of MCA will not be strictly valid anymore (since theoretically, they deal with infinitesimal changes). Still, they should be very useful, as we will discuss at length later on.

Successful attempts have been made to extend MCA into the domain of large changes. The first approach included second-order derivatives, which enabled larger changes in parameters to be evaluated (Hoefer and Heinrich, 1993). However, the approach requires knowledge of more detailed kinetic information, and leads to mathematical equations that would make most molecular biologists, physiologists and biochemists feel miserable, and this will restrict its applications. Small and Kacser (1993) introduced the "deviation index", which can be used to predict the effect of a large change in a parameter on the flux. The same authors also developed a "universal method" to increase a particular flux (Kacser and Acerenza, 1993), and to increase a particular metabolite concentration (Small and Kacser, 1994) (see also Westerhoff and Kell, 1996, for discussion on large changes and a somewhat different strategy).

Having simply measured the response coefficient, we may want to understand which factors determine the response of a metabolic system to a parameter change. The first important determinant will be the extent to which the activity of an enzyme is affected by the parameter. In the *pfk27Δ* mutant the concentration of F26bP will be lower than in the wild type. The lower concentration of F26bP will at first only affect the activity of PFK. The change in this activity will cause changes in F6P. The latter change will then be sensed by HXK and, in the case of a larger (more realistic) pathway, the effect will spread further and further until a new steady state is reached. The magnitude of the *initial, local* change in the activity of PFK that was effected by F26bP can be described by the *elasticity* coefficient (Burns *et al.*, 1985; Kacser *et al.*, 1995).

In general, elasticity is defined as the relative change in the activity of an enzyme, divided by the relative change in a parameter or variable. At first glance, the elasticity may seem the same as the response coefficient, but conceptually it is very different. The response coefficient describes the (small) change in the variables before and after a new steady state has been achieved, and is therefore evaluating *global* (or *systemic*) properties: the system as a whole is allowed to respond to the change in F26bP. The elasticity coefficient, however, evaluates the *immediate* effect of a change in a parameter or variable on the rate of an enzymic reaction, without allowing the whole system to react. It quantifies the sensitivity of an enzyme as if *in isolation* and is therefore a *local* property of that enzyme. Importantly, one has to evaluate the elasticity of an enzyme towards a particular metabolite in the presence of the other metabolites and effectors at their steady-state levels. Elasticities are not constant properties of enzymes: other metabolites and effectors may alter the sensitivity of an enzyme. For example, the concentration of ATP may affect the impact of a change in F26bP (see, for example, Bigl *et al.*, 1991; Smolen, 1995). The presence of intracellular glucose will affect the sensitivity of the transport step to extracellular glucose (Teusink *et al.*, 1996b).

In mathematical terms, the elasticity coefficient of PFK with respect to F26bP reads:

$$\varepsilon_{F26bPX}^{v_{PFK}} \equiv \frac{\partial v_{PFK} / v_{PFK}}{\partial [F26bP] / [F26bP]} = \frac{\partial \ln v_{PFK}}{\partial \ln [F26bP]} \tag{2}$$

where v_{PFK} is the activity of PFK in our case. The derivatives are partial, as we evaluate only one of the many (possible) effectors of PFK, while keeping the others constant at their steady-state values.

The elasticity of HXK with respect to F26bP would be 0 (F26bP does not affect the activity of HXK), and the elasticity of PFK with respect to F26bP depends on the specific kinetics of PFK and, in a real cell, also on the concentration of the substrates and products of PFK in the steady state (F6P, F16bP, ATP, ADP and all the other effectors). When the parameter is not some (external) regulator such as F26bP, but the concentration of the enzyme itself, the elasticity will generally be 1, as the activity of an enzyme is usually proportional to its concentration. If such proportionality does not apply, more sophisticated forms of MCA must be implemented (Kholodenko and Westerhoff, 1995b).

In exactly the same way as with parameters such as F26bP, elasticity coefficients can also be used to describe the sensitivity of an enzyme towards its substrate or product (which are variables[5]). This type of

---

[5]It may occasionally be important to distinguish between sensitivity towards parameters and variables, in which case Kacser *et al.* (1995) offers different symbols.

elasticity is extremely important for the control laws of metabolic pathways that emerge through the theorems of MCA. Using the kinetics that were used in our computer simulations, the rate of HXK and the corresponding elasticity for glucose were calculated as functions of the glucose concentration (Figure 3A). At very high substrate concentrations, the enzyme is close to saturation and will be very insensitive to a change in the substrate concentration. In other words, its elasticity approaches zero when the substrate concentration increases. At very low substrate concentrations, the activity of HXK is proportional to its substrate concentration, and its elasticity approaches 1.

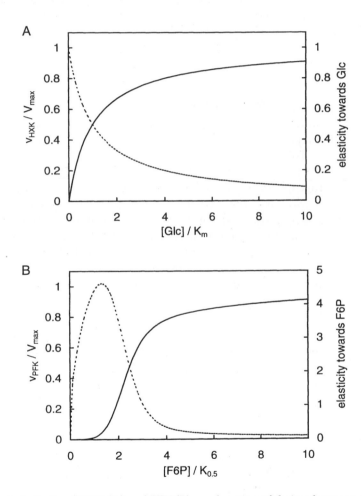

Figure 3. Activity of HXK (A) and PFK (B) as a function of their substrates, and the concomitant elasticity coefficients. The substrate concentration is normalized to the concentration that gives the half maximal rate. The rate of HXK followed simple Michaelis–Menten kinetics (Figure 3A, solid line); the rate of PFK follows Monod, Wyman and Changeux kinetics, which can be seen by the sigmoidal shape of the curve (Figure 3B, solid line). See Appendix I in this chapter for details. The dotted lines indicate the elasticities of HXK and PFK towards their substrates, Glc and F6P, respectively.

Figure 3B shows the rate and elasticity of PFK as a function of the concentration of its substrate, F6P. It illustrates that elasticities can also be much higher than 1, due (in this case) to substrate cooperativity. In the case of inhibitory effects, elasticities will be negative (e.g. the elasticity of an enzyme towards its product is usually negative, unless there is product activation).[6]

Thus, the immediate effect of a change in F26bP is described by the elasticity coefficient of PFK towards F26bP. The effect of that initial change in the activity of PFK on our system's variables, after a new steady state is reached, is described by yet another coefficient: the *control coefficient*. The control coefficient is defined as the relative change in a steady-state variable (such as the glycolytic flux, J) divided by the relative change in the local activity of an enzyme (v_{PFK}) (Schuster and Heinrich, 1992). Thus the control coefficient is also a dimensionless number.

Again, in mathematical terms:

$$C^J_{v_{PFK}} \equiv \left(\frac{\Delta J / J}{\Delta v_{PFK} / v_{PFK}} \right)_{\Delta v PFK \to 0} = \frac{dJ / J}{d v_{PFK} / v_{PFK}} = \frac{d\ln J}{d 1 n v_{PFK}} \tag{3}$$

Also with these derivatives one waits for the new steady state to settle and compares the two steady states, before and after the change in enzyme activity.

To carry out an experiment to determine the control coefficient of PFK on the flux, one should change the activity of PFK by changing a parameter that specifically affects PFK. One may change the concentration of PFK by modulating the expression of one of its genes (Jensen *et al.*, 1993b; Niederberger *et al.*, 1992). Alternatively, one may change the activity of PFK by specific inhibitors (Groen *et al.*, 1982a). Obviously, in our case we can change the activity of PFK specifically via the concentration of the effector F26bP. The resultant change in enzyme activity (the extent of which depends on the elasticity of PFK towards F26bP) will affect the glycolytic flux and the concentration of F6P. The magnitude of the changes in flux and F6P concentration will depend on the control of PFK on this flux and on the F6P concentration, respectively. In the end, one measures the response of the system to the change in the concentration of parameter F26bP:

$$R^J_{F26bP} \equiv \frac{d\ln J}{d 1 n[F26bP]} = \frac{d 1 n J}{d\ln v_{PFK}} \cdot \frac{\partial\ln v_{PFK}}{\partial 1 n[F26bP]} \tag{4}$$

$$= C^J_{v_{PFK}} . \varepsilon^{v_{PFK}}_{[F26bP]}$$

This equation shows the "combined response property" of metabolic systems (Hofmeyr, 1995; Kacser and Burns, 1973). It states mathematically that an agent affects the steady-state flux to an extent that equals the effect of the agent on its target enzyme multiplied by the control that target enzyme exerts on the flux. The control coefficient is calculated, from the measured response coefficient and elasticity, i.e. by rearranging equation

[6]Westerhoff and Van Dam (1987) and Hofmeyr (1995) give accounts of the contributions of thermodynamic and enzyme kinetic elements to the elasticity coefficient.

4 ($C = R/\varepsilon$). It should be noted that when a parameter is affecting more than one enzyme (e.g. temperature), the total response of the system will be the sum of the combined responses of the individual enzymes (Hofmeyr and Cornish-Bowden, 1991; Kahn and Westerhoff, 1993; Westerhoff and Chen, 1984). This will be the basis of the derivation of the so-called connectivity theorems in the next section.

Although the control coefficient is by far the most famous coefficient of MCA, it is conceptually the most difficult one (especially for experimentalists) because, as we have seen, one cannot measure a control coefficient *directly*. One can only change the activity of an enzyme by a parameter that (specifically) affects that activity, and we have defined such effects of parameters in terms of response coefficients. Why burden molecular biologists (and many biochemists and physiologists) with control coefficients, then? An example will justify this imposition.

Suppose one had constructed a mutant with altered PFK expression, such that the activity of PFK in this mutant (called *PFK20*) was exactly the same as that in our *pfk27Δ* mutant, but now with a wild-type level of F26bP. In order to do this, one should have to decrease the concentration (V_{max}) of PFK by 20%, because the 10% decrease in F26bP decreased the activity of PFK by 20% (see legend to Figure 2). As these two mutants have the same activity of PFK, the eventual flux and F6P concentration will be identical. However, the corresponding response coefficients of *PFK20* will be half that of the *pfk27Δ* mutant, because the change in the enzyme concentration needed to be double the change in [F26bP]. In other words, *response coefficients depend on the parameter that was used to affect the activity of an enzyme*. The control coefficient of an enzyme, however, is *independent of which parameter was used to change its activity*. Control coefficients really describe the contribution of each step in a pathway to the overall behavior of that pathway.

We have already noted that the elasticity of a reaction rate towards the enzyme catalyzing that reaction will normally be 1 (if the activity of an enzyme is proportional to its concentration). Under these circumstances, the response coefficient to that enzyme concentration will be the same as the control coefficient (see equation 4). Originally, "control coefficients" were defined in terms of enzyme concentrations by Kacser and co-workers (Kacser and Burns, 1973; Kacser *et al.*, 1995), and this definition is still in use (Fell, 1997). If necessary one may describe this "control coefficient" as an "enzyme concentration response coefficient" because, in principle, one measures the response of the system to a change in a parameter: the enzyme concentration.[7] In their original paper, the Berlin group has

---

[7]The control coefficient is equal to the enzyme concentration response coefficient when the activity of the enzyme is proportional to its concentration. There are cases, however, where the enzyme activity is *not* proportional to its enzyme concentration (e.g. in group transfer pathways and channeled pathways; Kholodenko *et al.*, 1994; Kholodenko and Westerhoff, 1995a; Van Dam *et al.*, 1993), and some essential theorems yet to come would appear to fail were the control coefficients to be defined in terms of enzyme concentrations rather than activities. This warning is therefore included for those people interested in, but still inexperienced with, the MCA jungle. For discussion about these issues, see Fell (1997), Kholodenko *et al.* (1995a) and Kholodenko and Westerhoff (1995b).

defined control coefficients in terms of enzyme *activities* rather than enzyme *concentrations*, as we have done (Heinrich and Rapoport, 1974).

◆◆◆◆◆◆ III. THE THEOREMS OF MCA: HOW ENZYMES BEHAVE IN A NETWORK

There are several laws that govern the control and regulation of metabolic pathways. These can be formulated precisely in terms of relationships between control coefficients, and between control coefficients and elasticities. Here we only discuss the relations and make them plausible by thought experiments. Mathematical proofs are in the literature to which we shall refer.

A. Summation Theorems

When, in our system, the activities of HXK and PFK are both increased by 1%, we expect to see a new steady state with a 1% increase in the flux J, the concentration of F6P remaining constant. In terms of control coefficients, this reads:

$$C^J_{v_{HXK}} + C^J_{v_{PFK}} = 1, \text{ and } C^{F6P}_{v_{HXK}} + C^{F6P}_{v_{PFK}} = 0 \tag{5a,b}$$

In a realistic pathway with n enzymes, these *summation theorems* will still hold: when the control coefficients of all n enzymes are summed, the sum will be 1 in the case of fluxes, and 0 in the case of metabolite concentrations (Heinrich and Rapoport, 1974; Kacser and Burns, 1973; Westerhoff and Van Dam, 1987).

The *flux* control coefficient summation theorem indicates that there is flux limitation, and it suggests that this limitation need not be conferred to a single *rate-limiting step*. In many both theoretical and experimental studies flux control was shown to be distributed among the enzymes that constituted the pathway (see, for example, Brand, 1996; Fell, 1992; Groen *et al.*, 1982a; Niederberger *et al.*, 1992; Van der Vlag *et al.*, 1995). Fell's book (Fell, 1997) describes numerous examples in detail. In the extreme case, each enzyme would have a control of about $1/n$, n being the number of enzymes in the system. In this situation, because n will be large in a real cell, each enzyme will exert (very) little flux control. Kacser and Burns (1980) reasoned along these lines to explain why most mutations are recessive, the rationale being that the reduction in enzyme level in a heterozygous diploid was insufficient to exert a significant metabolic phenotype. In practice, control may not always be equally distributed and the occurrence of negative control coefficients[8] causes the sum of absolute

---

[8]The control of an enzyme in one branch of a pathway over the flux in another branch of the pathway is generally negative.

values of the control coefficients to diverge from 1 substantially (Kahn and Westerhoff, 1991). Yet the conclusion should be that the average control coefficient of an enzyme tends to be small.

This conclusion has a significant implication for functional analysis. It can be expected that, in many cases where the expression of an enzyme is affected by the deletion of a novel regulatory gene, the control of this enzyme (i.e. the effect of its altered level upon flux) is too low to be measured. This problem is increased by the presence of isoenzymes, whose effect would be to decrease the relative change in enzyme activity[9] (Van Heeswijk *et al.*, 1995). Superficially, it might appear that many genes encoding enzymes or regulatory proteins exert little effect under particular growth conditions. This may explain why, in many cases, no obvious phenotype is observed at all. Appreciating this fundamental explanation for the lack of an obvious phenotype, one may be in a position to do something about it. With this in mind we will first explore the consequences of the flux control coefficient summation theorem a bit further.

Suppose we constructed several mutants with different concentrations of PFK around the wild-type level. In *E. coli*, modulation of the enzyme level of the plasma membrane ATPase and of the components of the phosphotransferase system was achieved by cloning an IPTG-inducible promoter in front of the corresponding genes (Jensen *et al.*, 1993b; Van der Vlag *et al.*, 1995). In yeast, there are similar systems available (Gari *et al.*, 1997; Korch *et al.*, 1991), but there remains a general need for promoters in yeast which can be externally regulated by non-metabolized inducers. In our simulation experiment, the glycolytic flux in the form of ethanol production is measured in each of the *PFK* mutants. This is shown in Figure 4. A decrease in the concentration of PFK compared with the wild-type level led to a decrease in the ethanol production rate (flux). The more PFK was decreased, the stronger became the dependence of the flux on the PFK concentration (Figure 4A). Eventually, PFK became a truly rate-limiting step: its flux control coefficient increased to 1 (Figure 4B). Of course this is not quite new. It is a well-known strategy to make a step rate-limiting by decreasing its activity, e.g. by inhibitory drugs.

Importantly, because of the summation theorem, the control of HXK must concomitantly decrease to 0, such that the sum of flux control coefficients remained 1 at all times (Figure 4B). This illustrates an important conclusion from MCA, namely that *the extent to which an enzyme controls a flux (or metabolite concentration) is not a fixed property of the enzyme but will depend on the conditions*. Changing the conditions (modulating the concentration of PFK in our case) will lead to a *redistribution of control*. This will

[9]If the control coefficient were to be defined in terms of the concentration of each individual gene product, the presence of isoenzymes would decrease that control coefficient. In the post-genome era, with many isoenzymes to be discovered by sequence similarities, we may have to reconsider how we want to define control, i.e. on the enzymic level in terms of the rate of the step in the pathway where this step is perhaps catalyzed by more than one gene product (as was done in this paper), or on the genomic level in terms of expression levels of each gene product (see also footnote 7).

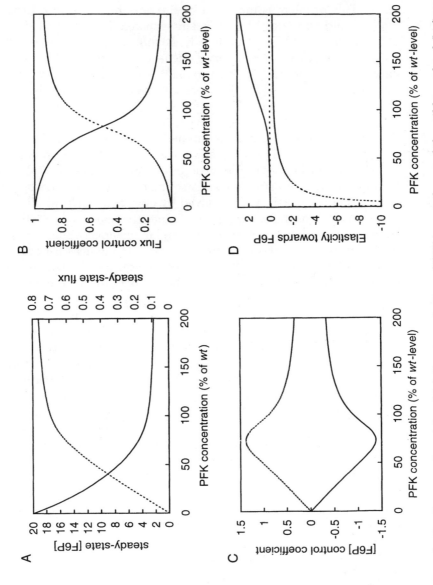

Figure 4. Simulation of a titration experiment, in which the concentration of PFK is varied around the wild-type level. In figures A–D, the effect of the PFK concentration is shown for: (A) the steady-state flux (dotted line) and F6P concentration (solid line); (B) the flux control coefficients of HXK (dotted line) and PFK (solid line); (C) the F6P-concentration control coefficients of HXK (dotted line) and PFK (solid line); and (D) the elasticities of HXK (dotted line) and PFK (solid line) with respect to the concentration of F6P.

happen whenever a non-infinitesimal change in one of the parameters is made. Therefore, each steady state has its own control distribution. Stating that "the glucose transporter controls the glycolytic flux" (Gancedo and Serrano, 1989) is therefore not useful, because the experimental conditions need to be specified since they will affect the control of the transporter (see Teusink *et al.*, 1996b).[10]

Apart from making life more complicated than it already was, we can use this insight to manipulate the control distribution. Decreasing the concentration of an enzyme will increase its control on the flux and will therefore enlarge the phenotypic effects of genes that regulate the expression of that enzyme. Conversely, increasing the activity of an enzyme with a large flux control coefficient will decrease the control coefficient of that enzyme and, owing to the summation theorem, that control will shift to other steps, again increasing the chances of detecting the effects of regulatory genes. In these and other ways one may implement understanding of metabolic control to create new, "sensitized" genetic backgrounds for functional analysis.

For the *concentration* control coefficients summation theorem (relating to the influence of an enzyme upon the concentration of a specific metabolite), no *a priori* statements can be made about the magnitude of the control coefficients. In our case, the concentration control of HXK for F6P is opposite to the concentration control of PFK, but the magnitudes can be (0.1 and –0.1), or (4 and –4), depending on the elasticities of HXK and PFK towards F6P. Figure 4C demonstrates that the concentration control coefficients also vary with the steady state conditions. Importantly, negative *concentration* control coefficients are expected to be much more common than negative *flux* control coefficients, and total controllability of metabolite concentrations (again defined as the sum of the absolute values of the control coefficients) should be expected to be often higher than that for fluxes. This suggests that monitoring metabolite concentrations in particular should be a good way to detect phenotypes. Metabolite concentrations are also less sensitive to the redundancy problem as the concentration control coefficient need not drop to zero at very high enzyme levels. (In this example, at the 200% level, another 100-fold increase in the V_{max} of PFK decreased the flux control coefficient of PFK to 0.02, whereas its F6P concentration control coefficient was still 0.24.)

[10]This shift in control has been stressed by critics of MCA, because what should be the use of determining control coefficients when they are likely to change in any other condition? In a biotechnological context, it may be quite frustrating to see that overexpression of the enzyme with the highest flux control coefficient does not significantly increase the flux because, by increasing its concentration, one decreased the extent to which this enzyme controlled the flux. We will not go into this debate here, except to point out that this is not a property of MCA itself, but of the metabolic systems that are described by MCA (cf. Fell, 1992; Hofmeyr and Cornish-Bowden, 1995; Kacser, 1995; Kell and Westerhoff, 1986; Westerhoff and Kell, 1996). Apart from redistribution of control, another reason for limited effects of overexpressing specific enzymes may be the "protein burden", i.e. the negative effect of overexpressing one gene on the expression of other genes (Snoep *et al.*, 1995).

Control on metabolite concentrations will be discussed further later in this section, and its use in functional analysis will be discussed extensively in the section on metabolic snapshots (section V).

B. Connectivity Theorems

Slightly more complicated and less familiar, but extremely important, are the connectivity laws of metabolic control. They allow control of fluxes and concentrations to be understood in terms of enzyme properties. To understand the relationships, we shall perform the following thought experiment: in our system, we increase the concentration of F6P by 20%. This will not affect the steady state in the long run, because no *parameter* is changed (see Figure 2B). Consider the flux through the system. The impact on flux of the change in F6P via enzyme HXK and PFK is described by $C^J_{v_{HXK}} \cdot \varepsilon^{v_{HXK}}_{F6P}$ and $C^J_{v_{PFK}} \cdot \varepsilon^{v_{PFK}}_{F6P}$, respectively. Note that this is the "*combined response*" that was discussed in the preceding section: F6P concentration affects the activities of HXK and PFK (described by the elasticities), and these affect the flux (via the flux control coefficients). In the end, once steady-state is reached, however, the total response of the system to a change in the concentration of F6P is zero:

$$C^J_{v_{HXK}} \cdot \varepsilon^{v_{HXK}}_{F6P} + C^J_{v_{PFK}} \cdot \varepsilon^{v_{PFK}}_{F6P} = 0 \tag{6}$$

The crux of the argument is that the total response of the system is the sum of all the responses of the individual enzymes and is zero for the steady-state condition when a variable is changed (see Westerhoff and Van Dam, 1987, for mathematical proof). This argument will be valid for any number of enzymes in a pathway. The equation demonstrates one of the connectivity theorems linking flux control coefficients to elasticities (Heinrich and Rapoport, 1974; Kacser and Burns, 1973).

There are similar connectivity theorems for concentration control coefficients (Westerhoff and Chen, 1984). In our case, there is only one free metabolite, F6P. Because, in the new steady state, the concentration of F6P will have to be the same as in the old steady state before the addition of F6P, the response of the system must act to reduce F6P to the original level. The response of the system towards an increase of F6P by 20% will be a 20% decrease in F6P to the original level (see Figure 2B). Hence,[11]

$$C^{F6P}_{v_{HXK}} \cdot \varepsilon^{v_{HXK}}_{F6P} + C^{F6P}_{v_{PFK}} \cdot \varepsilon^{v_{PFK}}_{F6P} = -1 \tag{7}$$

For larger metabolic systems with many metabolites, there will be many more equations, for each metabolite to enter the control coefficient (representing the metabolite for which to evaluate the response) and for each metabolite to enter the elasticity (representing the metabolite con-

[11]For details and proof, see Westerhoff and Chen (1984).

centration that is perturbed). In the case of two metabolites, there will be four such connectivity theorems linking concentration control coefficients with elasticities. When the response of the perturbed variable itself is evaluated, equation 7 applies. When the response of any other metabolite than the "perturber" is evaluated, the sum of responses should be zero following the same reasoning as above with respect to the flux (equation 6).

C. The Combination of the Theorems

Until now, on the one hand we have defined coefficients that quantify the sensitivity of *global, systemic* properties such as fluxes and steady-state metabolite concentrations towards enzyme activities (control coefficients), and the sensitivity of *local* enzyme activities towards parameters and metabolite concentrations (elasticities) on the other. Furthermore, we have found relationships between these control coefficients and elasticities in the form of summation and connectivity theorems. We are now ready to do the most important and most satisfactory trick of MCA: *combining the summation theorems with the connectivity theorems, we are able to express all control coefficients in terms of the elasticities of the enzymes.*

For our system (Figure 1), there are four equations (two summation and two connectivity theorems) and four unknowns (flux and concentration control coefficients of HXK and PFK). Such a set of equations can be solved, and the solutions for the four control coefficients read as follows:

$$C_{v_{HK}}^{J} = \frac{\varepsilon_{F6P}^{v_{PFK}}}{\varepsilon_{F6P}^{v_{PFK}} - \varepsilon_{F6P}^{v_{HK}}}, \quad C_{v_{PFK}}^{J} = \frac{\varepsilon_{F6P}^{v_{HK}}}{\varepsilon_{F6P}^{v_{PFK}} - \varepsilon_{F6P}^{v_{HK}}}$$

$$C_{v_{HK}}^{F6P} = \frac{1}{\varepsilon_{F6P}^{v_{PFK}} - \varepsilon_{F6P}^{v_{HK}}}, \quad \text{and} \quad C_{v_{PFK}}^{F6P} = \frac{-1}{\varepsilon_{F6P}^{v_{PFK}} - \varepsilon_{F6P}^{v_{HK}}}. \tag{7a–d}$$

This exercise shows the real power of MCA: one can understand the extent to which a certain enzyme controls the flux or a metabolite concentration on the basis of limited kinetic information concerning the enzymes in the pathway. Note that the elasticities of *both* enzymes appear in the control coefficients for this two-enzyme pathway. This clearly demonstrates that (control of) flux and concentration are determined by the whole system, and are therefore *systemic* properties.

In the expression of the *flux* control coefficients, the elasticity of the enzyme itself appears only in the denominator. The larger this term, the lower the control coefficient. Thus, the higher the elasticity of PFK towards F6P, the smaller its control on the flux. This is illustrated in Figure 4D, where the elasticities of HXK and PFK are plotted against the activity of PFK. When the elasticity of PFK towards F6P is very large (at high PFK activities), its flux control coefficient is low, and vice versa (cf. Figure 4B).

Highly regulated enzymes are expected to have high elasticities (this would be one way of defining regulatability; Westerhoff and Chen, 1984), and although they will be recognized as "key-enzymes" by many, they tend to have little control on the flux. PFK, again, is a particularly nice example. Biochemical textbooks tend to designate PFK as the key enzyme or "pacemaker" of glycolysis (e.g. Stryer, 1988), controlling the flux of that pathway. However, attempts to increase the glycolytic flux by over-expressing PFK have failed (Davies and Brindle, 1992; Schaaff et al., 1989). This may have been because the higher amount of enzyme was less active due to down-regulation of F26bP (Davies and Brindle, 1992). In a pfk26/pfk27 double deletion mutant, not able to synthesize F26bP, the flux was not affected because of increased levels of F6P that compensated for the absence of the activator (Boles et al., 1996). This should make clear that it will be impossible to decide whether PFK is a pacemaker enzyme based on the kinetic properties of that enzyme alone. If one were to make a prediction, however, the extensive regulation of the enzyme should make PFK a poor, rather than a good, candidate for having a high flux control coefficient.

In the case of *concentration* control coefficients, the elasticities of both enzymes towards F6P appear only in the denominator. Because $\varepsilon_{F6P}^{v_{HXK}}$ will be negative in most cases, HXK has a positive control on F6P, whereas PFK has a negative control on F6P (which is not at all surprising). We can see, however, that because the elasticities appear only in the denominator, a very high elasticity of any of the two enzymes will cause *both* concentration control coefficients to be small. This makes sense because, if an enzyme is very sensitive to a substrate, it effectively buffers the concentration of that metabolite. The concentration of metabolites that take part in a near-equilibrium reaction, which is characterized by extremely high elasticities of the participating enzymes (Hofmeyr, 1995; Westerhoff and Van Dam, 1987), are expected to be difficult to change (see also Stucki, 1980).

◆◆◆◆◆◆ IV. CONTROL OF FLUX: COMPETITION EXPERIMENTS IN CHEMOSTATS

A. The Need for Sensitive Flux Measurements: Hierarchical Control

Apart from the fact that "the number of enzymes n is large", there are other more sophisticated (and, biologically, highly relevant) reasons for low flux control coefficients, the most important of which is regulation of gene expression. The effect of a change in the activity of an enzyme will have some direct metabolic effect, but can also indirectly affect the expression of other enzymes, such that the cell compensates for the original change. This so-called hierarchical control will lead to lowered control by the altered enzyme. An experimental example is the work by Jensen and co-workers on the H$^+$-ATPase in *E. coli* (Jensen et al., 1995a,b). When they modulated the activity of the H$^+$-ATPase, they found a control coefficient for growth rate of 0.0 (i.e. no effect on growth rate). It turned out that H$^+$-ATPase had a negative control on the rate of respiration, so that an increase of H^+-ATPase activity was balanced by a decrease in respiration, a phenomenon they called "inverse respiratory control" (Jensen *et al.*, 1993a).

Hierarchical control may also be expected when isoenzymes are present that are expressed only under specific growth conditions. The family of hexose transporters, the *HXT*-gene family, may be a good example. Some enzymes are induced by glucose, whereas others are repressed (Ozcan and Johnston, 1995). When one of the isoenzymes is deleted, therefore, this may affect the expression of the other genes. Thus, *HXT1* encodes a low-affinity carrier (Reifenberger *et al.*, 1995) that is glucose induced (Ozcan and Johnston, 1995), but its deletion affected high affinity transport as well (Lewis and Bisson, 1991).

Hierarchical control tends to counterbalance the original change in enzyme activity (Kahn and Westerhoff, 1991) and, hence, minimize the phenotypic effect of deleting or overexpressing the novel gene. Therefore, sensitive methods need to be developed to measure the impact which such genetic manipulations have on the phenotype of the organism.

B. Accurate Measurement of Metabolic Fluxes: Competition Experiments in Chemostats

Under steady-state conditions, the growth rate is equivalent to a metabolic flux. Because the impact of single-gene deletions is likely to be small (see above), we have decided to perform our experiments in chemostat culture by competing deletant strains against one another and a standard strain. Such competition experiments are capable of detecting, and accurately quantifying, growth-rate differences that would be impossible to measure directly in either batch or single-strain continuous culture (Danhash *et al.*, 1991). In such competitions, it is important to have an appropriate standard strain against which to measure the growth-rate differences. It is also desirable that the replacement marker used to generate the deletion mutations has no phenotypic effect of its own, under the test conditions used. Although the use of a standard strain controls for any possible phenotypic effect of the replacement marker, such an effect might vary according to the site of insertion of the marker into the yeast genome. It is, therefore, best if the replacement marker has no effect on growth rate at all.

We have chosen to generate a standard strain in which the *HO* gene has been replaced by the same selectable marker as used to delete the test open reading frames (ORFs). *HO* has no known role, apart from mating-type switching, and it has been used as the site of insertion of heterologous genes in brewing yeasts without any perceptible effect on the fermentation characteristics of the organism or the quality of the product (Hammond *et al.*, 1994; Yocum, 1986). A diploid strain, homozygous for the *ho* deletion, is used both because *HO* is inactive in diploids and because the mating of independently derived haploid transformants allows any transformation-induced genetic lesions (Danhash *et al.*, 1991) to be nullified through complementation.

We have evaluated our choice of *HO* as a neutral site, and the phenotypic impact of two popular replacement markers (*kanMX*: Wach *et al.*, 1994; and *HIS3*: Baudin *et al.*, 1993), by competition of *ho* deletants produced by replacement with either of these two markers, against their wild-type parents (Baganz *et al.*, 1997). The *kanMX* marker was without significant phenotypic effect under either C-limited or N-limited conditions. In contrast, the *ho::HIS3* deletant was found to confer a 23% growth-rate advantage under C-limited conditions, but a 25% growth-rate penalty in an N-limited chemostat. In a more extensive study (Baganz *et al.*, 1997), a homozygous *ho:kanMX/ho:kanMX* deletant was competed against the wild-type EUROFAN diploid strain, FY1679. The competition experiments were carried out in chemostats under the following physiological conditions: glucose-limited/anaerobic, glucose-limited/aerobic, ethanol-limited, N-limited, P-limited, S-limited. It was found that the *ho:kanMX* deletion was without discernible effect on growth rate within a margin of error of ± 4%. This condones our choice of *HO* as a neutral site for replacement in standard strains and also indicates that resistance genes, such as *kanMX*, should be used in preference to nutritional markers in constructing standard strains.

Competition experiments between test deletants and the *ho:kanMX/ho:kanMX* standard strain have been performed and quantitative polymerase chain reaction (PCR) analyses employed to determine the proportions of test and standard strains in the culture at different times. In this approach, $N + 2$ oligonucleotide primers are required to evaluate a competition between N test strains and the standard strain: one common primer within the *kanMX* module, one primer complementary to a region flanking *HO*, and a primer flanking each of the target ORFs. PCR is performed on bulk DNA extracts from the competition culture and the products quantified. This can be done using GeneScan software and an ABI sequencing machine, provided fluorescently labeled, ORF-specific, primers are used. Alternatively the PCR products can be quantified densitometrically from negatives of agarose gel photographs. Both methods, and especially the second, are limited in the extent that they can be multiplexed. This means that repeated PCR reactions must be performed on a single DNA extract from a culture where several test strains are in competition with the standard. Of course, only two primers would suffice if one would first assess whether the wild type is outgrown at all by any of the competing mutants.

The recent development of methods to give deletion mutants specific oligonucleotide tags (so-called "molecular bar-codes"; Shoemaker *et al.*, 1996) offers the prospect of more efficient quantitation by employing hybridization array technology (Schena *et al.*, 1996). In this approach, a sequence is included in the replacement cassette, next to the *kanMX* marker, which consists of a 20 bp sequence that is unique to that particular mutant. This unique tag, or "bar-code", is flanked by priming sites that are common to all replacement cassettes used. Thus, in addition to the technical facility of the hybridization array method, costs are saved because just 2 (rather than $N + 2$) primers are required for the PCR amplification of DNA from the bulk culture. This would allow a wider range of selective conditions to be investigated in competition experiments.

A very attractive alternate approach to competition analysis has been used by Smith *et al.* (1995, 1996). They have used pools of Ty-generated yeast mutants and, again, have employed a sequencing machine with GeneScan software to analyse the proportions of mutant types in the population, thus a different oligonucleotide primer is required for each mutant ORF analysed. The drawback of this system for the MCA approach is that each mutant line carries mutations in addition to the one sampled by the PCR reaction. It is assumed that these mutations occur at random and that their effects are averaged out on a population basis. However, in contrast to the specific deletion mutants used in the EURO-FAN project, no individual mutant line can be carried forward for further analyses because it will have its own peculiar set of additional mutations. Moreover, the averaging-out process does not allow the accurate quantitation of the phenotypic impact of inactivating a single gene that we would wish to exploit using MCA.

◆◆◆◆◆◆ V. CONTROL OF METABOLITE CONCENTRATIONS: TAKING "METABOLIC SNAPSHOTS"

A. Why are Metabolite Concentrations Interesting to Measure?

Although the control of metabolite concentrations is more difficult to predict on the basis of "rules of thumb" compared with control of fluxes, measurements of changes in metabolite concentrations are likely to be very useful in the functional analysis of novel genes. This is mainly due to the expectation that concentration control coefficients are less sensitive to genetic redundancy (i.e. they tend to decrease less due to the presence of isoenzymes). There are two reasons to assume this.

(1) The redundancy problem, in the case of fluxes, was related in part to the summation theorem: there is a limit to the degree of control shared by all enzymes (the limit being 1), implying that when there is a large number of enzymes, the control of each enzyme will be low. In the case of metabolite concentrations, there is also a summation theorem, but it does not imply a limit to the extent to which an enzyme controls the concentration of a metabolite. Moreover, the extensive regulatory mechanisms that cells have evolved appear more likely to assess growth *rates* (fluxes) rather than changes in individual metabolite concentrations[12] (with a few

[12]This does not contradict the observations that in many cases large changes in flux are found with relatively small changes in metabolites (Fell, 1997; Fell and Thomas, 1995). To achieve such an effect, it has been proposed that multiple modulations of enzyme activities are needed (Fell and Thomas, 1995; Kacser and Acerenza, 1993; but see Westerhoff and Kell, 1996). This differs from an artificial (from the organism's point of view) change in the activity of only one enzyme.

possible exceptions, such as ATP and NADH; however, see, for example, Hohmann *et al.* (1996) and Richard *et al.* (1996) for studies where large changes in such "homeostatically controlled" metabolites were observed).

(2) The control on fluxes exerted by an enzyme normally varies between 1 at low enzyme concentration and 0 at high enzyme concentration (cf. Figure 4B). This is not the case for the control on metabolites. A factor here is that negative concentration control coefficients are more normal than negative flux control coefficients. Thus, effects on metabolites may still be seen, even in the presence of a high concentration of enzyme (where the control on the flux would be very small), e.g. when a gene is deleted in the presence of (many) isoenzymes. The obvious prerequisite is, still, that the relative change in the activity of the enzyme in question is large enough to see any effect. When the activity of enolase is changed by only 5%, owing to the presence of other isoenzymes (Chambers *et al.*, 1995; Entian *et al.*, 1987), it may be very difficult to see any change in a metabolite (which would also have a magnitude of only 5% if the control coefficient were as high as 1). This is a problem for any functional analysis (is the effect large enough to be observable?) and is not directly related to the MCA-inspired quantitative approach. In fact, the MCA approach recognizes the potentially small effects of deletion of novel genes, and seeks ways around the problem.

B. How to Take a Metabolic Snapshot

Whenever metabolites are to be measured accurately, there are three important conditions to be met. First, metabolism should be quenched as fast as possible, and metabolic processes should be completely inactive during subsequent treatment of the samples. Second, the extraction of metabolites should be quantitatively complete. Third, the amount of metabolites should be large enough for subsequent analysis.

In quantitative functional analysis, especially in the hierarchical approach (Oliver, 1996), there are a few more demands. A whole range of metabolites, throughout the metabolic map, should be assayed in a single sample. Moreover, because of the high throughput of mutants, the procedure should be as simple as possible. In the literature, there are reports of several methods of measuring intracellular metabolites. They involve shaking with glass beads, quenching and extraction with trichloracetic acid and subsequent diethylether treatment to remove the trichloracetic acid, and freeze–thaw cycles (see Gustafsson, 1979, for a comparison of some methods). These methods are too elaborate for high-throughput functional analysis. Two other methods have been reported: perchloric acid (PCA) quenching (which involves a simple neutralization step to remove the perchlorate) (Betz and Chance, 1965; Wallace *et al.*, 1994; Weibel *et al.*, 1974) and cold methanol quenching (De Koning and Van Dam, 1992; Richard *et al.*, 1996; Saez and Lagunas, 1976). We have, so far, focused our attention on the cold methanol method, for several reasons. The spraying of cells into methanol at −40°C is the fastest quenching method yet described for yeast. The crux of the method is that cells in cold methanol do not leak, which is an important advantage over the PCA extraction method: the sample can be concentrated by subsequent centrifugation. PCA extraction would yield quite low concentrations and a tedious lyophilization step might be necessary to concentrate the samples. Another important advantage over the PCA method is that the extraction can be done at neutral pH, and an acid-labile compound such as NADH can be measured in the same extract as NAD$^+$ (Richard *et al.*, 1993).

There are also some disadvantages attached to the cold methanol method, at least as originally practised (De Koning and Van Dam, 1992; Richard *et al.*, 1993, 1996). The equipment required is quite expensive and is not usually found in a molecular biology laboratory: a –20ºC centrifuge, a –40ºC cryostat, and an incubator at –35ºC, to vortex the samples in (although a freezer at –20ºC should also work). Moreover, the method is quite labour-intensive: the extraction of 24 samples would take at least 2 days, without any analysis. We are therefore adapting the methanol extraction method to a simple "1 hour" method using standard laboratory equipment. The outline of the method is as follows. A sample of a liquid culture is quenched 1:1 in –40ºC methanol. The sample volume and cell density used should suffice for the analysis of the required metabolites. For advanced analytical methods, such as mass spectrometry, 0.1 ml of sample with metabolite concentrations around 1–10 μM should be enough. Enzymatic assays usually require larger amounts, which depends on the number of assays performed, the sensitivity of the method (e.g. fluorometric detection of NADH would greatly enhance the sensitivity of the assay), and the size of the cuvettes. Automatic analyzers often employ cuvettes that need less sample than conventional spectrophotometers that only take 1 ml cuvettes.

In order to maintain the cells at the low temperature, a cryostat can be used, but a styrofoam ice bucket filled with dry-ice-cooled ethanol works equally well. Centrifugation at low temperature may be a problem, but depending on the type of centrifuge, the buckets may be precooled in a freezer. In our hands, we can centrifuge 40 ml of a methanol-quenched culture in a table centrifuge at room temperature without the temperature of the mixing rising above –15ºC, provided that precooled buckets are used and the total handling time is less than 5 min. The pellet is then resuspended in a suitable solvent that will open up the cells. For analysis by HPLC and capillary electrophoresis, possibly in combination with mass spectrometry, chloroform may be used as a solvent, although the solubility of metabolites in chloroform may be problematic. Moreover, chloroform is obviously not a suitable solvent for enzymatic analysis, and certain plastics are not resistant to chloroform either. One will therefore need to extract the metabolites from the suspension of chloroform and cells (as was done in the original protocol; De Koning and Van Dam, 1992). We are currently evaluating other solvent mixtures that do permeabilize the cell membrane, but do not have the disadvantages of chloroform. Mixtures of ethanol, water and 1-butanol or 1-octanol may be used (for a review on cell permeabilization, see Felix, 1982). It should be noted that some enzymes may partly survive chloroform and other solvents (e.g. adenylate kinase; De Koning and Van Dam, 1992) and appropriate action should be taken to inactivate them (low temperatures, addition of EDTA).

An extremely simple method for extraction of metabolites and inactivation of proteins has recently been developed by the group of Jean Marie Francois (Gonzalez *et al.*, 1997). We have adapted the method for analysis on a mass spectrometer. Methanol-quenched cells are centrifuged, and the pellet is resuspended in hot ethanol (75% v/v), and boiled for 3 min. The extract is cooled, transferred to Eppendorf tubes and spun at maximal speed to remove cell debris. We can quench 20 ml of a cell culture of about 0.2–0.5 g protein liter^{-1}, and resuspend it in only 2 ml of ethanol. In this way, the sample is concentrated by a factor of 10, which yields concentrations of metabolites that are in the right range for analysis by mass spectrometry.

C. How to Interpret a Metabolic Snapshot

The objective of taking a metabolic snapshot is to establish the possible area of metabolism in which the novel gene has an effect. What can we expect from metabolic snapshots? MCA gives us some clues as to what changes in metabolites can be expected, and whether they tend to spread throughout metabolism. Suppose we take the *PFK27* example again, and partly "unpack" the PFK reaction that lumped the lower part of glycolysis, to extend our metabolic pathway by two enzymes – pyruvate kinase (PYK) and pyruvate decarboxylase (PDC) – with two metabolites linking the enzymes: fructose 1,6-*bis*phosphate (F16bP) and pyruvate (Pyr), respectively (Figure 5). Again, some enzymes are lumped into others: the enzymes aldolase to enolase are now lumped into the PYK reaction, and alcohol dehydrogenase is lumped into the PDC reaction. The question addressed here is: what will happen to the concentrations of F6P, F16bP and Pyr when *PFK27* is deleted?

The answer is most obvious when the affected enzyme, PFK, has no control on the flux through the pathway (which is the case in real life; Boles *et al.*, 1996; Davies and Brindle, 1992; Schaaff *et al.*, 1989). This is the only case we will discuss here, but it is quite a likely one for many enzymes (Kacser and Burns, 1980). No flux control means that, in the new steady state (of the *pfk27Δ* mutant), the rate of PFK and of all the other enzymes in our glycolytic pathway is the same as in the wild type. However, PFK lacks 10% of the activator F26bP, and the only way of attaining the same rate would be to increase its substrate F6P and decrease its product F16bP. The magnitude of the changes in substrate and product concentration is not determined by PFK alone, but by all enzymes (remember that control on steady-state variables is a *systemic* property). We will illustrate some of the possibilities, the reality of which should depend on the kinetic properties of the enzymes of the system.

(1) PFK is very sensitive to its substrate F6P, and a small change in F6P will bring about a large change in the rate of PFK. Only a small change in F6P and no change in F16bP will be needed to compensate for the lack of F26bP. As the deletion of *PFK27* did not alter the flux, and F16bP did not

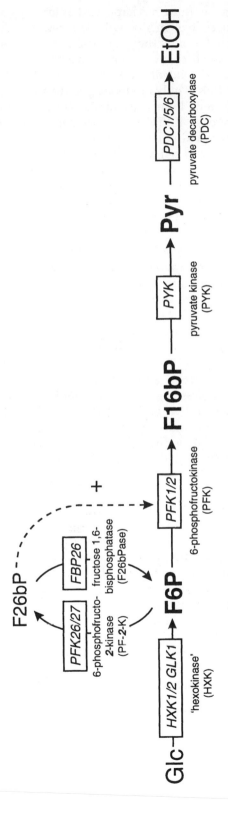

Figure 5. Modification of the system of Figure 1 to include pyruvate kinase (PYK) and pyruvate decarboxylase (PDC). Metabolites in bold are the variables to be measured in a metabolic snapshot approach.

change, nothing has happened as far as pyruvate kinase is concerned, and so no change in Pyr will be seen. No changes in metabolites will be detected, unless the analytical method is sensitive enough to detect the very small change in the concentration of F6P. *No noticeable differences between the mutant and the wild type should be expected.*

(2) PFK is only moderately and equally sensitive to F6P and F16bP. F6P would tend to increase in the *pfk27Δ* mutant, whereas F16bP would tend to decrease. If PFK were to be considered in isolation, the magnitudes of the changes in those metabolites would be the same. This would also happen in a real cell, when HXK and PYK are saturated with F6P and F16bP, respectively. Saturation means an elasticity of zero (see Figure 4).[13] In that case, F6P and F16bP can vary without affecting the rate of HXK and PYK. This means that metabolism beyond HXK and PYK does not see any change in metabolites, and no changes in metabolites apart from F6P and F16bP will be seen. *Only a few metabolites change, locally.* This could be the ideal case: the allocation of function should be easy. Enzymes with low flux control coefficients flanked by enzymes with low elasticities will be good candidates to observe only local changes.

(3) In most cases, however, the other enzymes will affect this distribution and will be affected themselves. To give an example of how this works, suppose in our system that PYK is sensitive to F16bP. Its rate would be significantly decreased by the decrease in [F16bP] as a result of the deletion of *PFK27*. A decrease in the rate of PYK then will tend to increase F16bP again, as PYK will consume the F16bP more slowly. Nevertheless, the increase in F16bP will have to be compensated for by an increase of Pyr, because the rate of PYK cannot change in the long term (again demanded by the zero flux control of PFK). So, the change in PFK activity can, indirectly, affect the concentration of Pyr. In a larger system, *changes at more than a single location on the metabolic map are detected.*

To summarize, when knocking out a gene, a battle over the metabolites will arise that, upon reaching steady state, will end up in a "truce" depending on the elasticities of the participating enzymes. In this battle, as a rule of thumb, high elasticities tend to buffer changes in concentrations. It is important to realize there are (many) exceptions to this rule of thumb. Moreover, we analyzed a simple linear pathway with a perturbation in an enzyme with no flux control. There are obviously more complicated systems, and the analysis would be more tedious (significant flux control, branches, moiety conserved sums (such as ATP, ADP and AMP, or NADH and NAD$^+$), long-distance regulatory loops, "multipurpose" co-enzymes such as ATP). The latter three features of metabolism, in particular, can cause "jumps" over long distances in the metabolic map, and one will quickly end up in situation 3. Nevertheless, our analysis should

---

[13]Alternatively (and, in fact, more likely) the F6P concentration is much lower than its Michaelis–Menten constant of HXK, i.e. F6P/K_m ≪ 1. In that case, the elasticity of HXK towards F6P is also (very close to) zero. Note that when the *substrate* is low compared with its K_m, the elasticity would be close to 1 (Figure 4).

give a feeling for what to expect when a deletion of a novel gene affects the activity of an enzyme somewhere in the metabolic map.

In situation 3, the obvious problem will be to localize the origin of all those changes. In some cases, we may use our biochemical intuition and discover a pattern that provides a clue as to the function of the novel gene (e.g. a low ATP concentration and a high concentration of hexose phosphates may point at a "*ggs1/tps1*-type" of function; Hohmann *et al.*, 1996). This will probably be successful for only a few genes. In the case that the many changes stem from pleiotropic effects of the gene deletion, rather than from the networking properties of metabolic pathways, one should hope that the metabolic snapshot is comparable to a metabolic snapshot of a known gene with pleiotropic effects, e.g. protein kinases operating at different points in the same cascade, or two signal transduction pathways that exhibit significant crosstalk. If not, taking a metabolic snapshot may be of little help, and other methods may be more successful. One could, for instance, look for synthetic phenotypes with genes of known function (although a metabolic snapshot may still be a useful way of revealing such a phenotype).

In MCA, there is a systematic approach to the situation where many metabolite concentrations have been changed, as pointed out by Westerhoff *et al.* (1994). If one were to change all enzymes in the system by different small amounts, then the effects on concentrations and fluxes can be predicted by the product $\mathbf{C} \cdot \delta\ln[e]$, where $\mathbf{C}$ is a matrix of control coefficients and $\delta\ln[e]$ a column of all changes in enzymes. This procedure can be reversed by multiplying the changes in fluxes and metabolite concentrations with the inverse of $\mathbf{C}$. However, for this method a reasonable estimate of individual control coefficients is required, making this approach difficult in practice.

When the novel gene acts on the activity of a single enzyme, we suggest here an approach that may be very helpful in finding the origin of the perturbation in the maze of metabolite changes. We have called the method FANCY: Functional ANalysis by Co-responses in Yeast.

D. The FANCY Method

Once again, we change our pathway to extend it with two enzymes (GPD, glycerol-phosphate dehydrogenase; and GPP, glycerol-phosphate phosphatase) to include a branch from F16bP to glycerol formation (Figure 6). One additional intermediate can be measured: glycerol 3-phosphate (G3P). Four metabolic snapshots are taken:

1. Two snapshots of the wild type, with 5 and 20 mM extracellular glucose (for simplicity we ignore any glucose signaling effects in this *in silico* experiment).
2. One snapshot in the strain *PFK27+*, in which *PFK27* is overexpressed, which gives a 100% increase in F26bP, equivalent to a two-fold increase in the V_{max} of PFK.
3. One in a strain in which *GPD1* has been deleted, which leaves the GPD activity at only 30% of its wild-type level (due to the presence of an isoenzyme *GPD2*; Eriksson *et al.*, 1995).

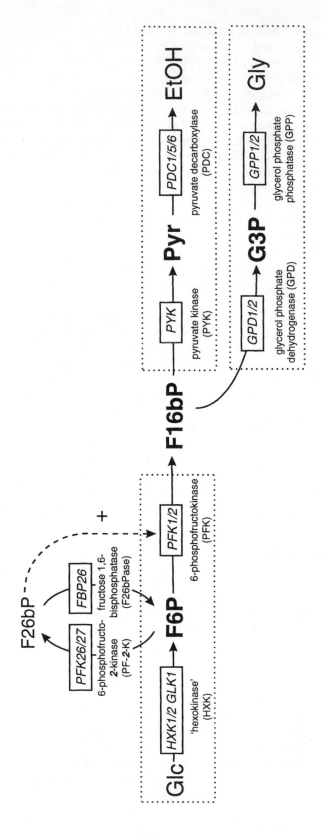

Figure 6. Extension of the system in Figure 5 with glycerol 3-phosphate dehydrogenase (GPD) and glycerol 3-phosphate phosphatase (GPP). The dotted boxes indicate monofunctional units in this system.

Table 1 gives the results of our "experiments" (again, see Appendix 1, this chapter, for details on the kinetics used in the simulation). It can be seen that all the metabolite concentrations have changed in all the three experiments. Can we allocate the origin of the change? From the fact that when *PFK27* was overexpressed, all metabolites before PFK decreased, and all metabolites after PFK increased (the so-called crossover theorem; Chance and Williams, 1955) one may identify the point of change as PFK. For more complicated systems, such as a living cell, this would not be as easy (Heinrich and Rapoport, 1974; Westerhoff and Chen, 1984). We therefore need a fancier method.

In Table 2, we have calculated the so-called *co-responses* of pairs of metabolites to the parameter changes made in our experiments. A co-response (symbol Ω) quantifies how two metabolites are affected by a parameter change relative to one another. It is the ratio of the two individual response coefficients (Hofmeyr *et al.*, 1993). For the concentrations of Pyr and G3P, the co-response coefficient would read (in the case of the *PFK27+* mutant):

$$^{[F26bP]}\Omega_{[G3P]}^{[Pyr]} = R_{[F26bP]}^{[Pyr]} / R_{[F26bP]}^{[G3P]} \approx \frac{\ln[Pyr]_{PFK27+} - \ln[Pyr]_{wt}}{\ln[G3P]_{PFK27+} - \ln[G3P]_{wt}} \tag{9}$$

The last term describes the way in which co-responses can be simply estimated in functional analysis experiments by comparing the steady-state concentrations of two metabolites in the wild type and a mutant. This equation was used to calculate the co-responses in Table 2, for

Table 1. Steady state metabolite concentrations for the model depicted in Figure 6. See the text for explanations.

Strain	$[F6P]_{ss}$	$[F16bP]_{ss}$	$[Pyr]_{ss}$	$[G3P]_{ss}$
Wild type (20 mM Glc)	3.22	1.98	0.45	0.61
Wild type (5 mM Glc)	2.17	0.61	0.17	0.25
PFK27+ (20 mM Glc)	2.25	2.45	0.53	0.71
GPD 30% (20 mM Glc)	3.22	4.98	0.88	0.29

Glc, glucose.

Table 2. Co-response analysis of the steady state metabolite concentrations shown in Table 1. Calculation of the co-responses is explained in the text.

X^a	$^x\Omega_{[F16bP]}^{[F6P]}$	$^x\Omega_{[Pyr]}^{[F6P]}$	$^x\Omega_{[G3P]}^{[F6P]}$	$^x\Omega_{[Pyr]}^{[F16bP]}$	$^x\Omega_{[G3P]}^{[F16bP]}$	$^x\Omega_{[G3P]}^{[Pyr]}$
HXK	0.33	0.41	0.43	**1.22**	**1.28**	**1.05**
PFK	−1.72	−2.28	−2.40	**1.33**	**1.40**	**1.05**
GPD	0.00	0.00	0.00	**1.39**	−1.22	−0.88

[a]The enzyme that was changed compared with the wild type (with 20 mM glucose) in Table 1. Thus, HXK was affected by the decrease in its substrate glucose, PFK by the increase in F26bP caused by overexpressing PFK27, and GPD by deleting one of the isoenzymes.

several pairs of metabolites. It is clear that some of these co-responses are very similar (printed in bold), whereas others are very different. This will be the crux of the FANCY method. We need a little bit more theory first.

Theoretical work of Rohwer *et al.* (1996) has shown that some enzymes can be grouped into so-called "monofunctional units" (see also Kholodenko *et al.*, 1995b). The useful property of enzymes within such a unit is that analogous perturbations in any of these enzymes will always produce exactly the same co-responses of variables outside the unit, *irrespective of which enzyme in the unit was perturbed, and irrespective of the magnitude of that perturbation.*[14] In our system, HXK and PFK can be grouped within such a unit and, indeed, the co-responses of metabolites outside the unit (F16bP, Pyr and G3P) are the same. To the extent that monofunctional units abound in cell physiology, there will be groups of enzymes (of many different sizes) that will produce identical co-responses outside that group. We can use this property for functional analysis. *When two deletion mutants exhibit the same co-responses, they will affect the same monofunctional unit. Knowing the unit in which one of the genes causes an effect, we can infer that of the other.* By this device, we have located the origin of the changes in metabolites – the aim of the FANCY method.

As not all enzymes produce the same co-responses, there must be some rules to this game. Indeed, Rohwer *et al.* (1996) have defined the conditions under which a group of enzymes behaves as a monofunctional unit:

1. There is only one independent flux linking the unit and the rest of metabolism.
2. There are no metabolites inside the unit that directly affect the rate of any enzyme outside the unit.
3. There are no metabolites in the unit that are part of a moiety-conserved sum (e.g. [NADH + NAD$^+$]) outside the unit.

In short, the unit should behave as if it were an enzyme catalyzing a (completely coupled) reaction. Two simple enzymes in a chain of reactions will usually fulfil these conditions (e.g. phosphoglycerate mutase and enolase most likely behave as a monofunctional unit) but larger groups of enzymes may also be grouped (e.g. glycolysis as a whole, leaving out the adenine nucleotides and the redox couple NADH/NAD$^+$; see Rohwer, 1997; Rohwer *et al.*, 1996 for discussion).

The FANCY method should work as follows (Figure 7). First, one has to take metabolic snapshots of mutants in the novel genes and, importantly, of many regulatory mutants that cause alterations to known enzymes spread throughout the metabolic map. Then the metabolic map should be divided into monofunctional units. A comparison of the co-

---

[14]Strictly speaking, the co-response of variables X and Y is only *exactly* the same for large (non-infinitesimal) changes, when the changes in enzyme activities v_1 and v_2 (in mutant mut1 and mut2, respectively) affect X to the same extent, i.e. $(\ln X_{wt} - \ln X_{mut1}) = (\ln X_{wt} - \ln X_{mut2})$. As the changes in v_1 and v_2 are not in the hands of the experimenter, the co-responses will not be *exactly* the same (see Table 2), but similar enough for functional analysis.

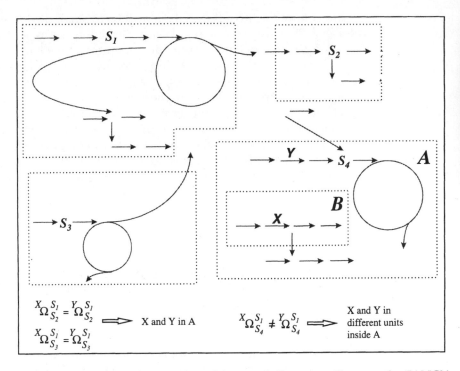

Figure 7. Schematic representation of the metabolic map to illustrate the FANCY method. The dotted boxes indicate monofunctional units within the metabolic map. Genes X and Y are novel genes; S_1, S_2, S_3 and S_4 are measured metabolites.

responses of the unknown with the known genes should give clues as to which unit the product of a novel gene belongs or regulates.

In the interests of experimental efficiency, these units obviously have to be fairly large in the early phases of the analysis. Thus, in common with other systematic approaches to functional analysis, a hierarchical strategy is required. (Oliver, 1996). Suppose gene X was identified to have an effect in unit A in Figure 7. The function of gene X is then investigated by more detailed tests, until its exact function is revealed. It turns out that gene X acts on unit B, a subunit of A. Once the function of gene X is known, its metabolic snapshot can be used to check whether gene Y, whose co-responses also pointed to an effect somewhere in unit A, acts inside or outside unit B. The information on gene X therefore speeds up the functional analysis of gene Y. Our FANCY method is therefore autocatalytic, a very desirable property in view of the number of genes to be analyzed.

However, some problems and uncertainties remain that may limit the usefulness of the proposed method. First, the rules for the delineation of monofunctional units may be too stringent to enable the construction of usefully large units. The metabolic map is a network with numerous interactions, many of which we do not yet know. The stringent rules would appear to restrict the monofunctional units to rather small stretches of metabolism. This appears to be a serious problem, but we do not know how sensitive the analysis will be to small deviations from the rules, e.g. if there were a regulatory metabolite from within the unit directly acting on an enzyme outside the unit, and we had not taken this into account. Nevertheless, it should be realized that groups of enzymes *will* behave as units even though we have not been clever enough to place them in a box ourselves. One can therefore expect to find positive matching of mutants of known and novel genes without drawing any boxes. FANCY just tells us how to look for the similarities. Drawing the boxes is a useful way of deciding which mutants of known genes should be used to cover the metabolic map, but is not essential for the method itself.

Second, the accuracy of metabolic snapshots may not suffice to determine the differences between co-responses accurately enough. This may be true, but the number of independent co-responses, n, is one less than the number of metabolites one can measure, N, i.e. $n = N - 1$.[15] If we were able to measure as many as 50 metabolites in a single snapshot, we would have 49 co-responses. This is quite a large number, which may compensate for the experimental error. We expect that, with increasing numbers of metabolites (and thus, co-responses), the sign rather than the magnitude of the co-response may be enough to allocate the potential function of a novel gene.

Third, the method cannot account for pleiotropic effects such as multi-site modulation. This would indeed be a serious limitation on the method, but it is not a universal one. Whenever the pleiotropic effects occur within a single monofunctional unit, the same co-responses will be achieved as if only one enzyme was affected. It very much depends on the scale at which the novel gene affects metabolism, and on the scale at which mono-functional behavior is being observed (i.e. how large the boxes we can draw are, and how many metabolites we measure).

There are other (so-called chemometric) methods to compare metabolic snapshots of mutants defective in a gene of known function, with those defective in novel genes. Statistical methods such as principle component analysis may be used to discriminate between snapshots that are similar, and those that are different. Artificial neural networks (Goodacre *et al.*, 1992) may even be trained to predict possible areas of function of novel genes, based on snapshots of mutants defective in genes of known function. These tools need large multivariate data sets, provided by techniques such as Fourier-transform infrared spectroscopy (Goodacre *et al.*, 1996b) or (pyrolysis) mass spectrometry (Goodacre *et al.*, 1996a). They also allow direct measurement on whole cell suspensions. We are currently evaluating their use in quantitative functional analysis, and it appears that these chemometric techniques can be a powerful complementation to the outlined FANCY method, in that they may give a first clue as to what area of functionality might be affected (also called the cladistic phase; Oliver, 1997). This information may be used to identify which metabolites should be analyzed using the metabolic snapshots (analytical phase).

[15]When there are N metabolites, the set of $N - 1$ co-responses (R_1/R_2), (R_1/R_3), ... (R_1/R_N) can be used to derive the other co-responses, e.g. $(R_{20}/R_{43}) = (R_1/R_{20})^{-1} \cdot (R_1/R_{43})$.

◆◆◆◆◆◆ VI. CONCLUSIONS

The ultimate objective of the life sciences is the understanding of the behavior of organisms on the basis of their molecular components and their interactions. Component identification is one of the steps towards such a goal. Consequently, a vast number of researchers are looking for and characterizing genes responsible for a particular phenotype in which they are interested. In fact, one might compare today's hunt for genes with the hunt in the first half of this century for metabolic intermediates and enzymes that constitute the biochemical pathways. In essence, either hunt is component identification. The EUROFAN project aims at the elucidation of the function of some 1000 novel components. This is an immense challenge, and the identification of many new components in yeast is bound to boost our understanding of living cells in general.

However, it is unthinkable that we will completely understand how a living cell functions – which is (or should be) our ultimate objective – even when we have identified all the components. Jensen *et al.* (1995b) have compared the living cell with a television set: even if one were to have all the basic components, such as resistors, transistors and capacitors, and if one also were to have a circuit diagram (i.e. the "stoichiometry"), one would still miss the World Cup final when the magnitudes of the resistances and capacitances were not indicated on the diagram. *"The proper voltages must arise at all positions and must respond correctly to changes in input signals so as to produce the response that is optimal in functional terms"* (Jensen *et al.*, 1995b). Therefore, even when you have identified all the essential components, true understanding of the function of the component within the system requires *quantitative* knowledge of how the components interact.

MCA is an attempt to understand *quantitatively* the global behavior of complex systems on the basis of the local properties of the components, the enzymes. In this paper, we have surveyed the basic definitions and theorems of MCA. This may leave the impression that MCA can only deal with idealized pathways that do not exist in real life. This is not so: in the last decade, substantial progress has been made to extend the theory to more and more complex systems, such as large systems (Brand, 1996; Brown *et al.*, 1990; Schuster *et al.*, 1993; Westerhoff and Van Dam, 1987), group transfer pathways (Kholodenko and Westerhoff, 1995a; Van Dam *et al.*, 1993), channeling (Kholodenko *et al.*, 1994), regulatory cascades (Kahn and Westerhoff, 1991), dynamic and oscillatory systems (Acerenza, 1990; Acerenza *et al.*, 1989; Kholodenko *et al.*, 1997; Teusink *et al.*, 1996a; Westerhoff *et al.*, 1990), and many more (see, for example, Westerhoff and Kell, 1996, and references therein).

In this chapter, we have perhaps proceeded one step further. We have suggested that in many cases only a *quantitative* approach will lead to the identification of the function of the components themselves. We have shown that in many cases only small, quantitative effects are to be expected when deleting novel genes, and that only with careful measurement and thoughtful interpretation can we find the function of such genes. This should not be a surprise once the complexity of the network of enzymatic activity that constitutes a living cell is appreciated. Inspired by MCA, and working within its framework of mathematically and conceptually clearly defined concepts such as control and elasticity, we have outlined some strategies to do a kind of *inverse* metabolic control analysis: how to understand which components have been changed on the basis of changes in the behavior of the whole metabolic network.

Acknowledgements

This work has been supported by the European Commission (in both the EUROFAN network and in a scholarship for B.T.), the BBSRC, the Wellcome Trust, and by Pfizer Central Research, Applied Biosystems, Amersham International and Zeneca. We thank Barbara Bakker, Johann Rohwer, Mark Sefton and Douglas Kell for discussions and critical reading of the manuscript.

◆◆◆◆◆◆ APPENDIX 1: DESCRIPTION OF THE KINETIC MODEL

A kinetic model was built to illustrate the principles of MCA, and the behavior was calculated using the metabolic modeling software package Gepasi 3.01 (Mendes, 1996). The model consists of a series of enzymes, each with kinetic properties described below. The stoichiometry of the model can be seen in Figures 1, 5 and 6 (see corresponding sections in the main text for explanation). When all parameters, such as the kinetic parameters and the extracellular glucose concentration, were specified the program calculated either a time simulation (such as in Figure 2) or a steady state (such as in Table 1). The following kinetics were used to obtain the results described in the main text.

PFK

Standard Monod, Wyman and Changeux kinetics for allosteric enzymes were used (Monod *et al.*, 1965; see also Hofmeyr, 1995), with $K_{0.5, F6P} = 1$; $n = 8$, $L_0 = 10^4$; $V_{max} = 1$:

$$v = V_{max} \cdot \frac{1}{1+[F6P]/K_{0.5}} \cdot \frac{1}{1+L} \qquad (A1)$$

$$\text{with } L = L_0 \cdot \frac{1}{(1+[F6P]/K_{0.5})^n} \qquad (A2)$$

where $K_{0.5}$ is the concentration of F6P at half the maximal velocity, n is the number of subunits and L is the equilibrium constant for the transition between the active (relaxed) and the less active or inactive (tense) state of the enzyme. This equilibrium is influenced by the binding of substrate.

HXK, PYK, ADH, GPD and GPP

We used Michaelis–Menten kinetics for reversible reactions with one substrate and one product (see, for example, Cornish-Bowden, 1996):

$$v = V_f \cdot \frac{\dfrac{S}{K_s} \cdot \left(1 - \dfrac{P/S}{K_{eq}}\right)}{1 + \dfrac{S}{K_s} + \dfrac{P}{K_p}} \tag{A3}$$

with V_f the maximal forward rate, S the substrate concentration, P the product concentration, K_s the Michaelis–Menten constant for the substrate, K_p the Michaelis–Menten constant for the product, and K_{eq} the equilibrium constant of the reaction. For all reactions, the Michaelis–Menten constants for substrate and product, and the maximal forward velocity were set to 1. For simplicity, co-factors such as ATP, NADH and phosphate have not been included. The following values for the equilibrium constants were used: $K_{eq,HXK} = 1$; $K_{eq,PYK} = 0.5$; $K_{eq,ADH} = 0.1$; $K_{eq,GPD} = 1$; $K_{eq,GPP} = 1$.

REFERENCES

Acerenza, L. (1990). Temporal aspects of the control of metabolic processes. In *Control of Metabolic Processes*, A. Cornish-Bowden and M. L. Cardenas, eds, pp. 297–302. Plenum Press.

Acerenza, L., Sauro, H. M. and Kacser, H. (1989). Control analysis of time-dependent metabolic systems. *J. Theor. Biol.* **137**, 423–444.

Baganz, F., Hayes, A., Marren, D., Gardner, D. C. J. and Oliver, S. G. (1997). Suitability of replacement markers for functional analysis studies in *Saccharomyces cerevisiae*. *Yeast* **13**, 1563–1573.

Baudin, A., Ozier-Kalogeropoulos, O., Denouel, A. F. L. and Cullin, C. (1993). A simple and efficient method for direct gene deletion in *Saccharomyces cerevisiae*. *Nucl. Acids Res.* **21**, 3329–3330.

Bartrons, R., Van Schaftingen, E. and Hers, H. G. (1982). The stimulation of yeast phosphofructokinase by fructose 2,6-bisphosphate. *FEBS Lett.* **143**, 137–140.

Betz, A. and Chance, B. (1965). Phase relationship of glycolytic intermediates in yeast cells with oscillatory metabolic control. *Arch. Biochem. Biophys.* **109**, 585–594.

Bigl, M., Eschrich, K. and Hofmann, E. (1991). Kinetics of phosphofructokinase from a yeast mutant. *Biomed. Biochim. Acta* **50**, 239–250.

Boles, E., Goehlmann, W. H. and Zimmermann, F. K. (1996). Cloning of a second gene encoding 6-phosphofructo-2-kinase in yeast, and its characterization of mutant strains without fructose-2,6-bisphosphate. *Mol. Microbiol.* **20**, 65–76.

Brand, G. C. (1996). Top-down metabolic control analysis. *J. Theor. Biol.* **182**, 351–360.

Brown, G. C., Hafner, R. P. and Brand, M. D. (1990). A 'top-down' approach to the determination of control coefficients in metabolic control theory. *Eur. J. Biochem.* **188**, 321–325.

Burns, J. A., Cornish-Bowden, A., Groen, A. K., Heinrich, R., Kacser, H., Porteous, J. W., Rapoport, S. M. *et al.* (1985). Control analysis of metabolic systems. *Trends Biochem. Sci.* **10**, 16.

Casari, G., De Daruvar, A., Sander, C. and Schneider, R. (1996). Bioinformatics and the discovery of gene-function. *Trends Genet.* **12**, 244–245.

Chambers, A., Packham, E. A. and Graham, I. R. (1995). Control of glycolytic gene expression in the budding yeast (*Saccharomyces cerevisiae*). *Curr. Genet.* **29**, 1–9.

Chance, B. and Williams, G. R. (1955). Respiratory enzymes in oxidative phosphorylation III. The steady state. *J. Biol. Chem.* **217**, 409–427.

Cornish-Bowden, A. (1995). Metabolic control analysis in theory and practice. *Adv. Mol. Cell. Biol.* **11**, 21–64.

Cornish-Bowden, A. (1996). *Fundamentals of Enzyme Kinetics*. Portland Press, London.

Danhash, N., Gardner, D. C. J. and Oliver, S. G. (1991). Heritable damage to yeast caused by transformation. *Bio/Technology* **9**, 179–182.

Das, S., Yu, L., Galtatzes, C., Rogers, R., Freeman, J., Blenkowska, J., Adams, R. M. *et al.* (1997). Biology's new Rosetta stone. *Nature* **385**, 29–30.

Davies, S. E. and Brindle, K. M. (1992). Effects of overexpression of phospho-fructokinase on glycolysis in the yeast *Saccharomyces cerevisiae*. *Biochemistry* **31**, 4729–4735.

De Koning, W. and Van Dam, K. (1992). A method for the determination of changes of glycolytic metabolites in yeast on a subsecond time scale using extraction at neutral pH. *Anal. Biochem.* **204**, 118–123.

Entian, K.-D., Meurer, B., Koehler, H., Mann, K.-H. and Mecke, D. (1987). Studies on the regulation of enolases and compartmentation of cytosolic enzymes in *Saccharomyces cerevisiae*. *Biochim. Biophys. Acta* **923**, 214–221.

Eriksson, P., Andre, L., Ansell, R., Blomberg, A. and Adler, L. (1995). Cloning and characterization of *GPD2*, a second gene encoding *sn*-G3PDH (NAD⁺) in *Saccharomyces cerevisiae*, and its comparison with *GPD1*. *Mol. Microbiol.* **17**, 95–107.

Felix, H. (1982). Permeabilized cells. *Anal. Biochem.* **120**, 211–234.

Fell, D. A. (1992). Metabolic control analysis: a survey of its theoretical and experimental development. *Biochem. J.* **286**, 313–330.

Fell, D. A. (1997). *Understanding the Control of Metabolism*. Portland Press, London.

Fell, D. A. and Thomas, S. (1995). Physiological control of metabolic flux: the requirement for multisite modulation. *Biochem. J.* **311**, 35–39.

Gancedo, C. and Serrano, R. (1989). Energy-yielding metabolism. In *The Yeast*, Vol. 3 (A. H. Rose and J. S. Harrison, eds), pp. 205–209. Academic Press, London.

Gari, E., Piedrafita, L., Aldea, M. and Herero, E. (1997). A set of vectors with a tetracycline-regulatable promoter system for modulating gene expression in *Saccharomyces cerevisiae*. *Yeast* **13**, 837–848.

Goffeau, A., Barrell, B. G., Bussey, H., Davis, R. W., Dujon, B., Feldmann, H., Galibert, F. *et al.* (1996). Life with 6000 genes. *Science* **274**, 546.

Gonzalez, B., François, J. and Renaud, M. (1997). A rapid and reliable method for metabolite extraction in yeast using boiling ethanol. *Yeast* **13**, 1347–1356.

Goodacre, R., Kell, D. B. and Bioanchi, G. (1992). Neural networks and olive oil. *Nature* **539**, 594.

Goodacre, R., Neal, M. J. and Kell, D. B. (1996a). Quantitative analysis of multivariate data using artificial neural networks – a tutorial review and applications to the deconvolution of Pyrolysis mass-spectra. *Zentrallblatt Bakteriologie* **284**, 516–549.

Goodacre, R., Timmins, E. M., Rooney, P. J., Rowland, J. J. and Kell, D. B. (1996b). Rapid identification of streptococcus and enterococcus species using diffuse reflectance-absorbency Fourier-transform infrared-spectroscopy and artificial neural networks. *FEMS Microbiol. Lett.* **140**, 233–239.

Groen, A. K., Van der Meer, R., Westerhoff, H. V., Wanders, R. J. A., Akerboom, T. P. M. and Tager, J. M. (1982a). Control of metabolic fluxes. In *Metabolic Compartmentation* (H. Sies, ed.), pp. 9–37. Academic Press, London.

Groen, A. K., Wanders, R. J. A., Westerhoff, H. V., Van der Meer, R. and Tager, J. M. (1982b). Quantification of the contribution of various steps to the control of mitochondrial respiration. *J. Biol. Chem.* **257**, 2754–2757.

Gustafsson, L. (1979). The ATP pool in relation to the production of glycerol and heat during growth of the halotolerant yeast *Debaryomyces hansenii*. *Arch. Microbiol.* **120**, 15–23.

Hammond, J. R. M., Lancashire, W. D., Meaden, P. G., Oliver, S. G. and Smith, N. A. (1994). *Stability of Genetically Modified Yeasts in Relation to Beer of Good and Consistent Quality*, Report 07/63M. MAFF (Ministry of Agriculture, Fisheries and Food, UK).

Heinrich, R. and Rapoport, T. (1974). A linear steady-state treatment of enzymatic chains. *Eur. J. Biochem.* **42**, 89–95.

Heinrich, R. and Schuster, S. (1996). *The Regulation of Cellular Systems.* Chapman & Hall, New York.

Hers, H. G. and Van Schaftingen, E. (1982). Fructose 2,6-bisphosphate 2 years after its discovery. *Biochem. J.* **206**, 1–12.

Hoefer, T. and Heinrich, R. (1993). A second-order approach to metabolic control analysis. *J. Theor. Biol.* **164**, 85–102.

Hofmeyr, J.-H. S. (1995). Metabolic regulation: a control analytic perspective. *J. Bioenerget. Biomembr.* **27**, 479–490.

Hofmeyr, J.-H. S. and Cornish-Bowden, A. (1991). Quantitative assessment of regulation in metabolic systems. *Eur. J. Biochem.* **200**, 223–236.

Hofmeyr, J. H. S. and Cornish-Bowden, A. (1995). Strategies for manipulating metabolic fluxes in biotechnology. *Bioorg. Chem.* **23**, 439–449.

Hofmeyr, J.-H. S., Cornish-Bowden, A. and Rohwer, J. M. (1993). Taking enzyme kinetics out of control; putting control into regulation. *Eur. J. Biochem.* **212**, 833–837.

Hohmann, S., Bell, W., Neves, M. J., Valckx, D. and Thevelein, J. M. (1996). Evidence for trehalose-6-phosphate-dependent and trehalose-6-phosphate-independent mechanisms in the control of sugar influx into yeast glycolysis. *Mol. Microbiol.* **20**, 981–991.

Jensen, P. R., Westerhoff, H. V. and Michelsen, O. (1993a). Excess capacity of H^+-ATPase and inverse respiratory control in *Escherichia coli*. *EMBO J.* **12**, 1277–1282.

Jensen, P. R., Westerhoff, H. V. and Michelsen, O. (1993b). The use of *lac*-type promoters in control analysis. *Eur. J. Biochem.* **211**, 181–191.

Jensen, P. R., Michelsen, O. and Westerhoff, H. V. (1995a). Experimental determination of control by the H^+-ATPase in *Escherichia coli*. *J. Bioenerget. Biomem.* **27**, 543–554.

Jensen, P. R., Snoep, J. L., Molenaar, D., Van Heeswijk, W. C., Kholodenko, B. N., Van der Gugten, A. A. and Westerhoff, H. V. (1995b). Molecular biology for flux control. *Biochem. Soc. Trans.* **23**, 367–370.

Kacser, H. (1995). Recent developments beyond metabolic control analysis. *Biochem. Soc. Trans.* **23**, 387–391.

Kacser, H. and Acerenza, L. (1993). A universal method for achieving increases in metabolite production. *Eur. J. Biochem.* **216**, 361–367.

Kacser, H. and Burns, J. A. (1973). The control of flux. *Symp. Soc. Exp. Biol.* **27**, 65–104.

Kacser, H. and Burns, J. A. (1980). The molecular basis of dominance. *Genetics* **97**, 639–666.

Kacser, H., Burns, J. A. and Fell, D. A. (1995). The control of flux. *Biochem. Soc. Trans.* **23**, 341–366.

Kahn, D. and Westerhoff, H. V. (1991). Control theory of regulatory cascades. *J. Theor. Biol.* **153**, 255–285.

Kahn, D. and Westerhoff, H. V. (1993). The regulatory strength: how to be precise about regulation and homeostasis. *Biotheor. Acta* **41**, 85–96.

Kell, D. B. and Westerhoff, H. V. (1986). Metabolic control theory: its role in microbiology and biotechnology. *FEMS Microbiol. Rev.* **39**, 305–320.

Kholodenko, B. N. and Westerhoff, H. V. (1995a). Control theory of group transfer pathways. *Biochim. Biophys. Acta* **1229**, 256–274.

Kholodenko, B. N. and Westerhoff, H. V. (1995b). The macroworld versus the microworld of biochemical regulation and control. *Trends Biochem. Sci.* **20**, 52–54.

Kholodenko, B. N., Cascante, M. and Westerhoff, H. V. (1994). Control theory of metabolic channeling. *Mol. Cell. Biochem.* **133/134**, 313–331.

Kholodenko, B. N., Molenaar, D., Schuster, S., Heinrich, R. and Westerhoff, H. V. (1995a). Defining control coefficients in non-ideal metabolic pathways. *Biophys. Chem.* **56**, 215–226.

Kholodenko, B. N., Schuster, S., Rohwer, J. M., Cascante, M. and Westerhoff, H. V. (1995b). Composite control of cell function: metabolic pathways behaving as single control units. *FEBS Lett.* **368**, 1–4.

Kholodenko, B. N., Denim, O. V. and Westerhoff, H. V. (1997). Control analysis of periodic phenomena in biological systems. *J. Phys. Chem.* **101**, 2070–2081.

Korch, C., Mountain, H. A. and Bystrom, A. S. (1991). Cloning, nucleotide sequence and regulation of *MET14*, the gene encoding the APS kinase of *Saccharomyces cerevisiae. Mol. Gen. Genet.* **229**, 96–108.

Kretschmer, M. and Fraenkel, D. G. (1991). Yeast 6-phosphofructo-2-kinase: sequence and mutant. *Biochemistry* **30**, 10663–10672.

Kretschmer, M., Tempst, P. and Fraenkel, D. G. (1991). Identification and cloning of yeast phosphofructokinase 2. *Eur. J. Biochem.* **197**, 367–372.

Lewis, D. A. and Bisson, L. F. (1991). The HXT1 gene product of *Saccharomyces cerevisiae* is a new member of hexose transporters. *Mol. Cell. Biol.* **11**, 3804–3813.

Mendes, P. R. (1996). Gepasi 3. A 32 bit microsoft windows computer program for simulating biochemical dynamics. In *BioThermoKinetics of the Living Cell* (H. V. Westerhoff, J. L. Snoep, J. E. Wijker, F. E. Sluse and B. N. Kholodenko, eds), pp. 258–261. BioThermoKinetics Press, Amsterdam.

Monod, J., Wyman, J. and Changeux, J.-P. (1965). On the nature of allosteric transitions. *J. Mol. Biol.* **12**, 88–118.

Niederberger, P., Prasad, R., Miozzari, G. and Kacser, H. (1992). A strategy for increasing an *in vivo* flux by genetic manipulations – the tryptophan system of yeast. *Biochem. J.* **287**, 473–479.

Oliver, S. G. (1996). From DNA-sequence to biological function. *Nature* **379**, 597–600.

Oliver, S. G. (1997). Yeast as a navigational aid in genome analysis. The 1996 Kathleen Barton-Wright memorial lecture. *Microbiology* **143**, 1483–1487.

Özcan, S. and Johnston, M. (1995). Three different regulatory mechanisms enable yeast hexose transporter (*HXT*) genes to be induced by different levels of glucose. *Mol. Cell. Biol.* **15**, 1564–1572.

Paravicini, G. and Kretschmer, M. (1992). The yeast *FBP26* gene codes for a fructose-2,6-bisphosphatase. *Biochemistry* **31**, 7126–7133.

Reifenberger, E., Freidel, K. and Ciriacy, M. (1995). Identification of novel *HXT* genes in *Saccharomyces cerevisiae* reveals the impact of individual hexose transporters on glycolytic flux. *Mol. Microbiol.* **16**, 157–167.

Richard, P., Teusink, B., Westerhoff, H. V. and Van Dam, K. (1993). Around the growth phase transition *S. cerevisiae*'s make-up favours sustained oscillations of intracellular metabolites. *FEBS Lett.* **318**, 80–82.

Richard, P., Teusink, B., Hemker, M. B., Van Dam, K. and Westerhoff, H. V. (1996). Sustained oscillations in free energy state and hexose phosphates in yeast. *Yeast* **12**, 731–740.

Rohwer, J. M. (1997). Interaction of functional units in metabolism. PhD, University of Amsterdam.

Rohwer, J. M., Schuster, S. and Westerhoff, H. V. (1996). How to recognize monofunctional units in a metabolic system. *J. Theor. Biol.* **179**, 213–228.

Ruyter, G. J. G., Postma, P. W. and Van Dam, K. (1991). Control of glucose metabolism by enzyme-IIGlc of the phosphoenolpyruvate-dependent phosphotransferase system in *Escherichia coli*. *J. Bacteriol.* **173**, 6184–6191.

Saez, M. J. and Lagunas, R. (1976). Determination of intermediary metabolites in yeast. Critical examination of the effect of sampling condition and recommendations for obtaining true levels. *Mol. Cell. Biochem.* **13**, 73–78.

Schaaff, I., Heinisch, J. and Zimmermann, F. K. (1989). Overproduction of glycolytic enzymes in yeast. *Yeast* **5**, 285–290.

Schena, M., Shalon, D., Heller, R., Chai, A., Brown, P. O. and Davis, R. W. (1996). Parallel human genome analysis–microarray-based expression monitoring of 1000 genes. *Proc. Natl Acad. Sci. USA* **93**, 10614–10619.

Schuster, S. and Heinrich, R. (1992). The definitions of metabolic control analysis revisited. *BioSystems* **27**, 1–15.

Schuster, S., Kahn, D. and Westerhoff, H. V. (1993). Modular analysis of the control of complex metabolic pathways. *Biophys. Chem.* **48**, 1–17.

Shoemaker, D. D., Lashkari, D. A., Morris, D., Mittmann, M. and Davis, R. W. (1996). Quantitative phenotypic analysis of yeast deletion mutants using a highly parallel molecular bar-coding strategy. *Nature Genetics* **14**, 450–456.

Small, J. R. and Kacser, H. (1993). Responses of metabolic systems to large changes in enzyme activities and effectors. 1. The linear treatment of unbranched systems. *Eur. J. Biol.* **226**, 649–657.

Small, J. R. and Kacser, H. (1994). A method for increasing the concentration of a specific internal metabolite in steady-state systems. *Eur. J. Biochem.* **226**, 649–657.

Smith, V., Botstein, D. and Brown, P. O. (1995). Genetic footprinting – a genomic strategy for determining a gene's function given its sequence. *Proc. Natl Acad. Sci. USA* **92**, 6479–6483.

Smith, V., Chou, K. N., Lashkari, D., Botstein, D. and Brown, P. O. (1996). Functional analysis of the genes of yeast chromosome V by genetic footprinting. *Science* **274**, 2069–2074.

Smolen, P. (1995). A model for glycolytic oscillations based on skeletal muscle phosphofructokinase kinetics. *J. Theor. Biol.* **174**, 137–148.

Snoep, J. L., Jensen, P. R., Groeneveld, P., Molenaar, D., Kholodenko, B. N. and Westerhoff, H. V. (1994). How to determine control of growth rate in a chemostat. Using metabolic control analysis to resolve the paradox. *Biochem. Mol. Biol. Int.* **33**, 1023–1032.

Snoep, J. L., Yomano, L. P., Westerhoff, H. V. and Ingram, L. O. (1995). Protein burden in *Zymomonas mobilis* – negative flux and growth control due to overexpression of glycolytic enzymes. *Microbiology* **141**, 2329–2337.

Stryer, L. (1988). *Biochemistry*. Freeman, New York.

Stucki, J. W. (1980). The thermodynamic-buffer enzymes. *Eur. J. Biochem.* **109**, 257–267.

Teusink, B., Bakker, B. M. and Westerhoff, H. V. (1996a). Control of frequency and amplitude is shared by all enzymes in three models for yeast glycolytic oscillations. *Biochim. Biophys. Acta* **1275**, 204–212.

Teusink, B., Walsh, M. C., Van Dam, K., Gustafsson, L. and Westerhoff, H. V. (1996b). The extent to which the glycolytic flux in *Saccharomyces cerevisiae* is controlled by the glucose transport system varies with the extracellular glucose concentration. In *BioThermoKinetics of the Living Cell* (H. V. Westerhoff, J. L. Snoep, J. E. Wijker, F. E. Sluse and B. N. Kholodenko, eds), pp. 417–421. BioThermoKinetics Press, Amsterdam.

Thomas, J. H. (1993). Thinking about genetic redundancy. *Trends Genet.* **9**, 395–399.

Van Dam, K., Van der Vlag, J., Kholodenko, B. N. and Westerhoff, H. V. (1993). The sum of the flux control coefficients of all enzymes on the flux through a group-transfer pathway can be as high as two. *Eur. J. Biochem.* **212**, 791–799.

Van der Vlag, J., Van't Hof, R., Van Dam, K. and Postma, P. W. (1995). Control of glucose metabolism by the enzymes of the glucose phosphotransferase system in *Salmonella typhimurium*. *Eur. J. Biochem.* **230**, 170–182.

Van Heeswijk, W. C., Stegeman, B., Hoving, S., Molenaar, D., Kahn, D. and Westerhoff, H. V. (1995). An additional P-II in *Escherichia coli* – a new regulatory protein in the glutamine-synthase pathway. *FEMS Microbiol. Lett.* **132**, 153–157.

Wach, A., Brachat, A., Pohlmann, R. and Philippsen, P. (1994). New heterologous modules for classical or PCR-based gene disruptions in *Saccharomyces cerevisiae*. *Yeast* **10**, 1793–1808.

Wallace, P. G., Pedler, S. M., Wallace, J. C. and Berry, M. N. (1994). A method for the detection of the cellular phosphorylation potential and glycolytic intermediates in yeast. *Anal. Biochem.* **222**, 404–408.

Weibel, K. E., Mor, J. R. and Fiechter, A. (1974). Rapid sampling of yeast cells and automated assays of adenylate, citrate, pyruvate and glucose-6-phosphate pools. *Anal. Biochem.* **58**, 208–216.

Westerhoff, H. V. and Chen, Y. (1984). How do enzyme activities control metabolite concentrations? *Eur. J. Biochem.* **142**, 425–430.

Westerhoff, H. V. and Kell, D. B. (1996). What biotechnologists knew all along . . . ? *J. Theor. Biol.* **182**, 411–420.

Westerhoff, H. V. and Van Dam, K. (1987). *Thermodynamics and Control of Biological Free-Energy Transduction*. Elsevier, Amsterdam.

Westerhoff, H. V., Aon, M. A., Van Dam, K., Cortassa, S., Kahn, D. and Van Workum, M. (1990). Dynamical and hierarchical coupling. *Biochim. Biophys. Acta* **1018**, 142–146.

Westerhoff, H. V., Hofmeyr, J. H. and Kholodenko, B. N. (1994). Getting to the inside using metabolic control analysis. *Biophys. Chem.* **50**, 273–283.

Wolfe, K. and Shields, D. (1997). Yeast Gene Duplications on World Wide Web (http:\\acer.gen.tcd.ie/~khwolfe/yeast/nova/index.html).

Yap, W. M. G. J., Van Verseveld, H., Snoep, J. L., Postma, P. W. and Van Dam, K. (1996). Enzyme IICB_Glc of the phosphoenolpyruvate: glucose phosphotransferase system controls the growth rate of *Escherichia coli* at fixed, low glucose concentrations as determined using glucose-limited chemostats. In *BioThermoKinetics of the Living Cell* (H. V. Westerhoff, J. L. Snoep, J. E. Wijker, F. E. Sluse and B. N. Kholodenko, eds), pp. 428–432. BioThermoKinetics Press, Amsterdam.

Yocum, R. (1986). Genetic engineering of industrial yeasts. *Proc. Bio. Expo.* **86**, 17.

18 Identifying Stress Genes

Willem H. Mager[1], Kick Maurer[1] and Peter W. Piper[2]

[1]*Department of Biochemistry and Molecular Biology, IMBW, Biocentrum Amsterdam, Vrije Universiteit, Amsterdam, The Netherlands and* [2]*Department of Biochemistry and Molecular Biology, University College London, London, UK*

◆◆◆

CONTENTS

List of Abbreviations

ER	Endoplasmic reticulum
FGM	Fermentable growth medium
HSE	Heat shock element
Hsps	Heat shock proteins
ORF	Open reading frame
PKA	Protein kinase A
STRE	Stress-responsive element
UPR	Unfolded protein response
yAREs	Yeast API-responsive elements

◆◆◆◆◆◆ I. INTRODUCTION

A. Yeast Stress Conditions

Yeasts exposed to adverse conditions display rapid adaptive responses (Mager and Moradas-Ferreira, 1993; Hohmann and Mager, 1997). These responses seem to be acting to increase the capacity of the cells for either *growth* under moderately stressful conditions, or *survival* under conditions of even more extreme stress. Stress responses thus increase stress tolerances, the response elicited by a mild exposure to the inducing stress that frequently leads to tolerance to a subsequent, much more severe exposure to the same stress. Intriguingly, multiple stress resistances can often be acquired by challenging with a single stress. This phenomenon is called cross-protection. It may be linked to what, in the discussion below, is termed the *general* stress response (Hohmann and Mager, 1997).

METHODS IN MICROBIOLOGY, VOLUME 26
ISBN 0–12–521526–6

Any unfavorable circumstance that adversely affects growth can be designated a stress. Such circumstances differ widely. They include not only environmental challenges such as heat or cold shock, osmotic dehydration, salt stress, ethanol stress, extremes of pH, heavy metal challenges, or oxidative stress, but also conditions of nutrient limitation. The molecular responses evoked to cope with such a diverse range of stress conditions are often not identical and are characteristic of a specific stress response. For example, osmostress induces massive glycerol accumulation, thus enabling cells to maintain their turgor pressure. Heat stress, in contrast, does not cause such glycerol synthesis, but instead mainly induces systems that protect cells from the effects of protein denaturation and aggregation. In general the induced response seems to lead to the induction of systems that protect against, or allow the repair of, the major cellular damages caused by the inducing stress. However, there are certain cellular changes (e.g. trehalose accumulation and arrest in the G1 phase of the cell cycle) that are triggered by multiple stress responses. These latter events are assumed to be part of the so-called *general* stress response (Hohmann and Mager, 1997).

With its amenability to genetic analysis, yeast has gained increasing importance in recent years as a model experimental system in which to unravel the stress responses of a eukaryotic cell. There is also the increasing appreciation that rapidly proliferating microbial cells need efficient responses to worsening physiological circumstances, both to survive and to subsequently resume growth and division. Therefore, many recent studies have investigated the responses of yeast to, for example, salt and other osmolytes (Varela and Mager, 1996), high ethanol concentrations (Piper, 1995) and compounds that, like hydrogen peroxide, generate reactive oxygen species (Moradas-Ferreira *et al.*, 1996).

In this chapter we will first discuss the *trans*-acting factors known to mediate stress-responsive transcriptional activation, and then the stress signaling pathways elucidated so far that influence the activity of these transcription factors. We will apply the term "stress-responsive" only to those genes whose transcription is elevated after stress exposure. We will next present results recently obtained with a grid filter of RNAs from cells exposed to different stresses. This promises to be a valuable experimental tool, both for examining the pattern of expression of well-known stress-responsive genes and for rapidly revealing the stress-induction patterns of novel stress-responsive genes. Apart from stress-responsive genes (i.e. those whose transcription is increased upon stress), many other genes also play a part in the stress response without themselves being stress-inducible. We will end this chapter by briefly discussing possible experimental strategies whereby these latter stress genes might be detected.

B. Stress-Responsive Genes

I. HSE/Hsf1p-responsive genes

The heat shock response was the first stress response to be discovered. Displayed by all cell types, the details of this response have been particularly well studied in yeast (Lindquist, 1986; Piper, 1997). Heat shock genes can be considered the prototypical stress-responsive genes, but they are in fact just one subfamily of stress genes. In *Saccharomyces cerevisiae* their transcription is strongly induced by either temperature upshift to the maximum temperatures of growth (approximately 37–39°C for cultures growing on glucose, 35–37°C for respiratory cultures) or exposure to ethanol concentrations of at least 4–6% (v/v) (Piper, 1995). Two-dimensional gel analysis of the proteins synthesized *de novo* before and after such a temperature shock has revealed that this stress leads to a decreased rate of synthesis for most proteins and a dramatically increased synthesis for a subset (the heat shock proteins, Hsps). It is still not known why most RNA polymerase II-catalyzed transcription is inhibited upon exposure to stress (*stress exposure*) while transcription of stress-responsive genes can escape such arrest.

Many Hsps play an essential role even under normal physiological conditions. As molecular chaperones, they are involved in pivotal cellular processes, for example ensuring the correct folding and membrane translocation of newly synthesized proteins. Apart from the genes for evolutionarily conserved Hsps (Hsp104, Hsp82, Hsp70s, Hsp60, Hsp26 and Hsp12), some genes encoding glycolytic enzymes (*PGK1*, *ENO1* and *TDH1*) have been identified as heat-shock inducible in yeast (Mager and Moradas-Ferreira, 1993). In addition, *HSP30* codes for a plasma membrane protein and *HSP150* for a secreted protein of unknown function. Hsp70 helps to prevent the thermal aggregation of proteins, while Hsp104 assists in the resolubilization of protein aggregates once these have formed. Both proteins assist in the recovery from thermal damage (Piper, 1997).

Heat-induced transcription of *HSP* genes is mediated through a specific promoter element, the heat shock element (HSE; Mager and De Kruijff, 1995). This element comprises at least three alternating repeats of the 5 bp sequence nGAAn and is the binding site for heat shock transcription factor, Hsf1p. Hsf1p shows strong similarity with the heat shock factors in higher eukaryotic cells (Mager and De Kruijff, 1995; Piper, 1997). It contains several distinct functional domains involved in specific DNA binding, and constitutive as well as stress-induced transcriptional activation. Yeast Hsf1p is bound to the HSEs in the promoters of target genes even in the absence of stress, providing a basal activity to these promoters. Hsf1p is also an essential protein in yeast, possibly because it has both basal and heat-induced activities as a transcription factor, its basal activity in unstressed cells providing essential Hsp chaperones during normal growth. The mechanism of heat activation of Hsf1p is still partly elusive. The heat-stimulated component of the activity is kept inactive in unstressed cells, possibly through interaction with Hsp70. In heat-shocked cells, increased association of Hsp70 with damaged protein may deplete the "free" Hsp70 pool, possibly leading to release of Hsp70 from Hsf1p, thereby stimulating Hsf1p activity and increasing transcription of HSE-regulated genes (Piper, 1997).

2. STRE-responsive genes

A few years ago it became clear that Hsf1p–HSE is not the only *cis–trans* regulatory combination mediating stress-induced transcription. Through studies of the *CTT1* gene encoding catalase, the *DDR2* gene involved in DNA damage repair, and *HSP12* encoding a small Hsp of unknown function, another *cis*-element was identified. This is the STRE element (general STress-Responsive Element; consensus CCCCT or AGGGG; Mager and De Kruijff, 1995; Ruis and Schüller, 1995). *CTT1*, *DDR2* and *HSP12* are all induced by heat shock, yet this induction was unexpectedly found to be Hsf1p- and HSE-independent. Moreover STREs, unlike HSEs, are induced by several other diverse forms of stress, including osmotic stress, oxidative stress, and nitrogen starvation. These different stresses appear to induce many of the same genes by activating (generally multiple) STREs in the promoters of these genes. Many such genes subject to activation by multiple stresses have since been identified, although in most cases, the STRE consensus sequences in their promoters have yet to be proven to be functional STREs. A recent computer search for STRE-consensus-containing promoters yielded about a hundred candidate stress-responsive genes (C. Schüller and H. Ruis, personal communication), suggesting that this is a fundamental gene control element in yeast. It has been proposed that STRE-directed gene expression may serve to reprogram cells for the survival of many forms of extreme stress, whereas HSE-directed expression may have the more defined purpose of allowing growth at high temperatures (Ruis and Schüller, 1995).

Recently, two zinc finger proteins, Msn2p and Msn4p, have been shown to bind STRE sequences *in vitro* (Martinez-Pastor *et al.*, 1996). Notably, mutant strains carrying disruptions of the *MSN2* and *MSN4* genes are only sensitive to *severe* stress conditions. The activation of certain STRE-containing promoters by many stresses is strongly diminished in these strains, although osmoshock conditions do induce a transcriptional response in the *msn2,msn4* mutant, albeit at a reduced level. Disruption of both *MSN2* and *MSN4* has no effect on transcription under non-stress conditions, *HSP12* gene expression still being repressed in unstressed *msn2,msn4* cells. If Msn2p and Msn4p fulfill a role in transcription activation, it is clear that this is dependent on other signals (see section I.C) and/or other factors. It is also possible that yet more factors can bind to the STRE and compete with Msn2p and Msn4p under certain conditions. It remains to be investigated if the strong negative effects of protein kinase A on STRE activity (see section I.C) are mediated through these factors.

3. Genes responsive to oxidative stress and to cadmium

Several yeast genes are induced by oxidative stress, an induction observed following the exposure of cells to prooxidants (Moradas-Ferreira *et al.*, 1996). Either hydrogen peroxide or compounds generating superoxide free radicals (e.g. menadione, paraquat) are generally used for these studies. Many of these genes are controlled by the Yap1p factor (also possibly the closely related Yap2p). Genes under the control of Yap1p include *TRX2* and *GSH1*, genes that are important in maintaining cellular thioredoxin and glutathione levels, respectively. Another example is *YCF1*, encoding a member of the ATP-binding cassette transporter proteins.

Yap1p and Yap2p belong to the family of AP1-like transcription factors, of which Gcn4p is also a member (Santoro and Thiele, 1997). These proteins contain a B-zipper motif specifying their mode of DNA binding. A MAP kinase phosphorylation site has been suggested to play a part in controlling their activity. In addition, major regulation of these factors can occur at the level of translation of the corresponding mRNAs. For Gcn4p this translational control has been examined extensively (Hinnebusch, 1994), but for the Yap proteins the data so far are extremely limited. Yap1p binds to promoter sites called yAREs (yeast AP1-responsive elements; consensus TTAC/GTAA). These show similarity to, but are distinct from, the binding site for the general amino acid control factor Gcn4p. The yAREs identified so far show quite considerable sequence variation.

Yap1p has also been implicated in the response to the toxic metal ion Cd^{2+} (Moradas-Ferreira *et al.*, 1996), although the induction of several genes such as *HSP12* by Cd^{2+} (as well as peroxide) occurs in a Yap1[P], Yap2[P]-independent manner (Varela *et al.*, 1995). There is increasing evidence that certain gene inductions observed with superoxide free radicals are Yap1p- and Yap2p-independent (Stephen *et al.*, 1995). In addition, cells lacking the two-component signaling protein Pos9p/Skn7p/Bry1p are very sensitive to oxidative stress (Krems *et al.*, 1995). This protein has recently been shown to bind to the promoter sequences of the oxidative stress-responsive *TRX2*, a gene regulated by both Yap1p and Bry1p (Morgan *et al.*, 1997). Thus, besides the Yap proteins, other more poorly characterized transcription factors also seem to participate in the response to oxidative stress.

4. The unfolded protein response element

The accumulation of unfolded proteins in the lumen of the endoplasmic reticulum (ER) triggers increased production of several ER-resident proteins. This unfolded protein response (UPR) is induced in yeast by tunicamycin, 2-deoxyglucose and 2-mercaptoethanol. A single 22 bp UPR-response element in the *KAR2* promoter is sufficient to activate transcription in response to these treatments (Shamu *et al.*, 1994). However, it is thought that only a small number of genes, specifically those encoding ER-resident proteins, may be under the control of this Ire1p-kinase-regulated stress response system (Shamu *et al.*, 1994).

C. Stress Signaling Pathways

Whereas the mechanisms whereby cells sense stress conditions remain largely elusive, several pathways or components have been identified

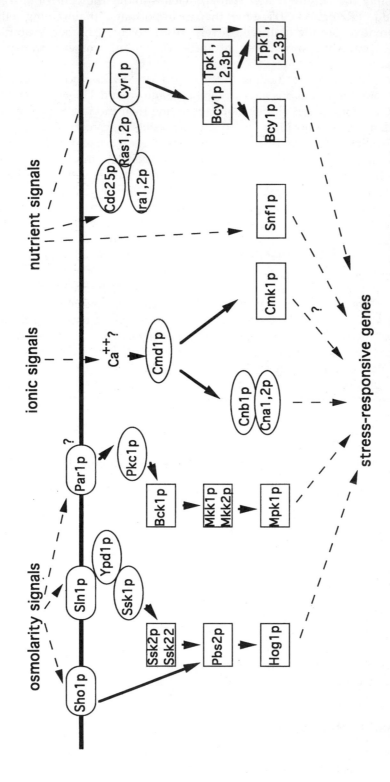

Figure 1. Schematic overview of signal transduction components so far identified as playing a part in the response of *S. cerevisiae* to stress conditions. Sho1p is a seven-transmembrane protein; Sln1p, Ypd1p and Ssk1p are constituents of a three-component sensing and signaling system; Ssk2p/Ssk22p, Pbs2p and Hog1p form the HOG MAP-kinase cascade (consisting of MAP kinase kinase kinase, MAP kinase kinase and MAP kinase, respectively). Par1p is a membrane protein, putatively regulating Pkc1p (protein kinase C) activity; Bck1p, Mkk1p/Mkk2p and Mpk1p form another MAP-kinase cascade. Cmd1p is calmodulin; Cnb1p and Cna1p/Cna2p represent the regulatory and catalytic subunits of calmodulin-dependent protein phosphatase; Cmk1p is calmodulin-dependent kinase. Snf1p is an AMP-kinase homolog. Cyr1p is adenylate cyclase whose activity is regulated by the action of the G-proteins Ras1p/Ras2p, which are in turn positively and negatively controlled by Cdc25p and Ira1p/Ira2p, respectively. Bcy1p and Tpk1p, Tpk2p and Tpk3p represent the cAMP-regulatory and catalytic subunits of protein kinase A.

through which stress signals are transmitted (Figure 1). In this respect, protein kinase A (PKA) should be mentioned first. PKA activity is predominantly regulated by the nutrient availability of cells. When the nutrient status is favorable, PKA activity is high, growth rate is maximal, and stress-responsive gene transcription is repressed (Thevelein, 1994; De Winde *et al.*, 1997). On the other hand, when PKA activity decreases, a general derepression of the transcription of many stress genes occurs. Indeed growth-related genes (e.g. those for ribosomal proteins) and stress genes of the "general" stress response display mirror image patterns of regulation, this being largely due to opposing responses to PKA activity (see, for instance, the patterns of *RPL25* and *HSP12* gene expression in Figures 2 and 3). cAMP plays a major role in the control of PKA activity. For instance, during the transition of cells from the nonfermentative to the fermentative growth mode (by glucose addition), the RAS adenylate cyclase pathway is activated transiently causing a peak in cAMP levels

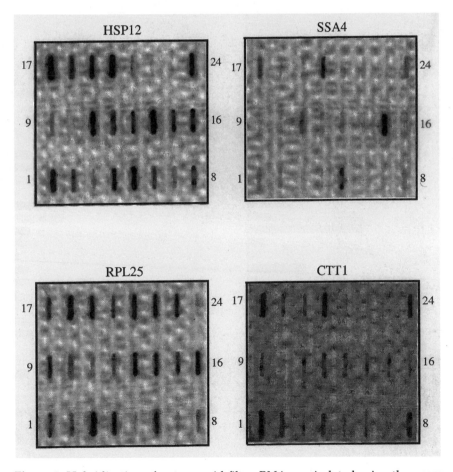

Figure 2. Hybridization of a stress grid filter. RNA was isolated using the procedure detailed in Protocol 1. The filter contained 24 samples as listed in Table 1. In this example *HSP12*, *SSA4*, *RPL25* and *CTT1* were used as the gene-specific probes.

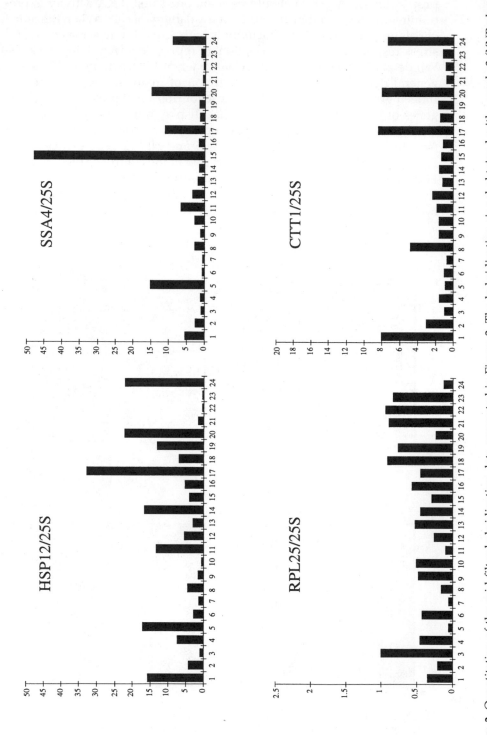

Figure 3. Quantitation of the grid filter hybridization data presented in Figure 2. The hybridization signal obtained with sample 3 (YNB glucose 24°C) was arbitrarily set at 1.0. Correction for loading artifacts was performed by calculating signal intensities relative to those obtained with a 25S ribosomal RNA-specific probe. Note the different scales used in the *y*-axis to demonstrate the alterations in the respective pattern of gene expression.

and a concomitant boost in PKA activity. Part of the regulation of PKA activity is, however, RAS-independent. The FGM (fermentable growth medium induced) pathway has been postulated to mediate the latter control (Thevelein, 1994). It is well documented that the stress resistance of cells is high under conditions of very low PKA activity, for example when cells are starved for an essential nutrient or when they approach stationary phase. In contrast, rapidly proliferating cells have a higher PKA content and are stress sensitive. Though it is tempting to speculate that a stress challenge may result in a decrease in PKA activity, experimental evidence for this is still lacking (Siderius and Mager, 1997).

PKA does not influence HSE/Hsf1p-mediated gene induction by heat shock (Ruis and Schüller, 1995). Hsf1p is phosphorylated after the stress, but this probably assists restoration of the less transcriptionally active state of the factor (Mager and De Kruijff, 1995). In contrast, the general STRE-mediated response is subject to strong negative regulation by cAMP (or PKA activity) (Siderius and Mager, 1997). Indeed, transcription of *HSP12*, a good monitor of general stress conditions, is inversely correlated with the PKA-linked nutrient status of yeast cells. Yap1p activity has been suggested to be regulated by cAMP-dependent PKA (Gounalaki and Thireos, 1994), but so far the actual evidence for this is rather weak. A strong handicap to establishing the functional role of PKA in the stress response is that the targets of this response at the level of transcription have still not been identified.

For some specific stress conditions the components mediating a stress response are much better understood. For example, an increase in external osmolarity immediately triggers a MAP-kinase signaling cascade, the HOG pathway (Hohmann, 1997). The components of this signal transduction pathway are depicted in Figure 1. The reverse condition, a downshock in external osmolarity, stimulates yet another MAP-kinase pathway, the PKC pathway (Levin and Errede, 1995). There is circumstantial evidence that the HOG pathway and the PKC pathway are interrelated and may "crosstalk" to each other. The PKC pathway is also stimulated by heat, though it is not the pathway signaling heat stress to Hsf1p (Piper, 1997). The HOG and PKC pathways are discussed further in Chapter 20. Another important signaling component has been found to be calcineurin, the Ca^{2+}-dependent protein phosphatase (see Figure 1; Serrano *et al.*, 1997). Calcineurin has been implicated in the response to salt stress, as a regulator of the sodium pump Ena1p. However, the actual targets of all of these signaling cascades at the level of transcription factors have yet to be firmly identified.

II. STRESS GENE ANALYSIS

A. Analysis of Stress-Responsive Genes

At present, any versatile method for the analysis of yeast stress genes has to allow for the possibility that important stress gene activation elements may still remain to be discovered. The presence of the HSE, STREs or

yARE consensus elements in upstream sequences can be used in preliminary screening for putative stress-responsive genes. The expression
patterns of various STRE-regulated genes show, however, that the mere
presence of STRE consensus sequences is not sufficient to predict the
actual stress regulation. Rather, it is very likely that the context of these
STREs in a promoter plays a major part in defining the level and kinetics
of the transcriptional response (Siderius and Mager, 1997).

We have recently developed a grid filter approach to aid in the identification of stress-responsive genes and to identify rapidly their major
inducing stresses. To this end we have exposed cells to a variety of different stress conditions, as well as a few normal growth conditions as a reference. Total RNA was then isolated from these cultures to give several
RNA samples, these being used subsequently to prepare a standard grid
filter of 24 slot-blotted RNAs. In parallel, the sequences of certain genes
whose stress induction patterns have been well characterized in earlier
work were prepared by polymerase chain reaction. Probes were then
made from the latter sequences and hybridized to the RNA filters;
hybridization was quantitated using a phosphoimager and corrected for
loading artifacts by using a 25S rDNA probe as a control. Full details of
the experimental protocol are given in Protocol 1, while Table 1 indicates
the different stress exposures of the cultures from which the RNAs of the
standard grid filter were prepared.

Figures 2 and 3 show sample filter hybridization results obtained with
probes comprising the sequences of three stress genes, *HSP12*, *CTT1* and
the Hsp70-encoding *SSA4*, and sequences of the ribosomal protein-

---

Protocol 1. Protocol for samples 4–24 of the stress response grid filter

1. Preculture strain FY73 in yeast nitrogen base medium containing 2%
   glucose, histidine and uracil.
2. Make a 1:25 dilution of the culture in the same medium and grow aer-
   obically at 24°C to an OD_{660} of 0.4.
3. Expose the cells to stress for either 60 min (samples 5, 6, 9–24) or
   overnight (samples 7, 8).
4. Harvest the cells and freeze in liquid nitrogen.
5. Isolate RNA by the glass bead method of Bromley *et al.* (1982).
6. Slot-blot the RNA samples, with denaturation in a large volume: 5 µg
   RNA in 18.75 µl, 7.5 µl 10 × NBC (0.5 M boric acid, 10 mM sodium
   citrate, 50 mM NaOH, pH 7.5), 11.25 µl 37% formaldehyde, and 37.5 µl
   formamide; total 75 µl. Incubate for 5 min at 65°C, then keep the
   samples on ice.
7. Transfer the samples to Whatman positively charged membrane,
   prewetted in 1 × NBC, using slot-blot apparatus. First, add 100 µl 1 ×
   NBC (test), then the RNA sample; wash with 100 µl 20 × SSC buffer
   and crosslink for 5 min on an ultraviolet box.
8. Label probes with ^{32}P using the Prime-It II random primer labeling kit
   (Stratagene) and hybridize to the RNA filters at 65°C overnight in 7%
   SDS, 1 mM EDTA and 0.5 M phosphate buffer (pH 7.0).
9. Wash 2 × in SSC containing 0.1% SDS and incubate for 15 min at 65°C.
10. Quantitate the RNA signal using a phosphorimager.

Table 1. Composition of stress response grid filter

Sample	Composition
1	YNB 3% lactate, 0.1% bactocasaminoacids 24°C growth
2	YNB 3% lactate, 0.1% bactocasaminoacids 37°C growth
3	YNB glucose 24°C growth
4	YNB glucose 37°C growth
5	YNB glucose 24°C to 37°C
6	YNB glucose 37°C to 16°C
7	C-starvation overnight
8	N-starvation overnight
9	0.4 mM hydrogen peroxide
10	1.0 mM menadione
11	1.0 mM CdSO$_4$
12	1.0 mg ml^{-1} tunicamycin
13	50 mM LiCl
14	6% ethanol
15	12% ethanol
16	pH 4.5 to pH 2.5
17	Hyperosmotic shock NaCl 0.1 M to 1.4 M
18	Hyperosmotic shock NaCl 0.1 M to 0.7 M
19	Hyperosmotic shock KCl 0.1 M to 0.7 M
20	Hyperosmotic shock KCl 0.1 M to 1.4 M
21	Hyperosmotic shock sucrose 0.1 M to 1.2 M
22	Hypo osmotic shock NaCl 0.7 M to 0.1 M
23	Hypo osmotic shock sucrose 1.2 M to 0.1 M
24	Hyperosmotic shock maltose 0.1 M to 1.2 M

encoding *RPL25* gene. The relative transcript levels, expressed in numbers normalized to the value found in sample 3 (YNB-glucose, 24°C growth), are depicted in Figure 2. *HSP12* and *CTT1* are both STRE-regulated genes (Varela *et al.*, 1995). However, comparison of the expression patterns of *HSP12* and *CTT1* confirms the above statement that STRE-regulated genes can display markedly dissimilar stress responses. *SSA4* is a gene under HSE control (Boorstein and Craig, 1990) that lacks STRE elements. Comparison of the *SSA4* and *HSP12* expression patterns (Figures 2 and 3) reveals the marked differences between these two heat shock genes. It is tempting to suggest that this reflects *SSA4* being HSE-regulated while *HSP12* is STRE-regulated, but the full explanation for these differences is probably much more complicated. Furthermore, comparison of the relative *HSP12* and *RPL25* transcription patterns provides a clear demonstration of the finding, discussed in section I.C, that stress-responsive genes (such as *HSP12*) are regulated inversely to growth-controlled genes such as those for ribosomal proteins. It is evident from the results presented in Figures 1 and 2 that this approach will be a useful initial assay to analyze the expression of candidate stress-responsive genes. Such an analysis is currently in progress using a set of open reading frames (ORFs) containing multiple upstream STREs selected by a computer search of the *S. cerevisiae* genome sequence.

B. Analysis of Genes of Stress Sensing and Stress Signaling

It is clear that the summary of stress signaling components in Figure 1 is only the tip of the iceberg. It will be a challenge of future work to identify the other constituents of the different stress response pathways and to unravel their functions. This work will need to distinguish the true stress-*sensing* pathways from those stress-induced alterations to activities of other signal transduction pathways that are not a part of stress sensing or stress adaptation. A discussion of methods for the dissection of signaling pathways can be found in Chapter 20. Unlike the stress-responsive genes discussed above (which are stress-inducible because they, for the most part, increase stress tolerances), genes of stress sensing and signaling may not be strongly regulated at the level of transcription. For example, we have recently analysed *HOG1* gene expression using the grid filter approach and observed only minor alterations to the levels of the *HOG1* transcript. To screen for mutants defective in this type of stress gene it would be attractive to use transient expression assays of a suitable stress-sensitive monitor gene. We propose to make use of the *HSP12* promoter (Varela *et al.*, 1995), fused to a reporter gene encoding a readily assayable product. In this way altered stress signaling to this promoter following different stress exposures could be readily detected. Using this assay disruptants of candidate ORFs or transformants overexpressing these ORFs could be screened for altered stress responses.

Acknowledgements

The described stress filter was developed with support specifically provided for the Stress Node of the EC EUROFAN project.

References

Boorstein, W. R. and Craig, E. A. (1990). Structure and regulation of the *SSA4* HSP70 gene of *Saccharomyces cerevisiae. J. Biol. Chem.* **265**, 18912–18921.

Bromley, S., Hereford, L. and Rosbash, M. (1982). Further evidence that the *rna2* mutation of *Saccharomyces cerevisiae* affects mRNA processing. *Mol. Cell. Biol.* **2**, 1205–1211.

De Winde, J. H., Thevelein, J. M. and Winderickx, J. (1997). From feast to famine: adaptation to nutrient depletion in yeast. In *Yeast Stress Responses* (S. Hohmann and W. H. Mager, eds), pp. 7–52. Molecular Biology Intelligence Unit, R.G. Landes Co., Austin, TX.

Gounalaki, N. and Thireos, G. (1994). Yap1, a yeast transcriptional activator that mediates multidrug resistance, regulates the metabolic stress response. *EMBO J.* **13**, 4036–4041.

Hinnebusch, A. (1994). Translational control of GCN4: an in vivo barometer of initiation-factor activity. *Trends Biochem. Sci.* **19**, 409–414.

Hohmann, S. (1997). Shaping up: the response of yeast to osmotic stress. In *Yeast Stress Responses* (S. Hohmann and W. H. Mager, eds), pp. 101–145. Molecular Biology Intelligence Unit, R.G. Landes Co., Austin, TX.

Hohmann, S. and Mager, W. H. (eds) (1997). *Yeast Stress Responses*. Molecular Biology Intelligence Unit, R.G. Landes Co., Austin, TX.

Krems, B., Charizanis, C. and Entian, K.-D. (1995). Mutants of *Saccharomyces cerevisiae* sensitive to oxidative and osmotic stress. *Curr. Genet.* **27**, 427–434.

Levin, D. E. and Errede, B. (1995). The proliferation of MAP kinase signalling pathways in yeast. *Curr. Opin. Cell. Biol.* **7**, 197–202.

Lindquist, S. (1986). The heat shock response. *Annu. Rev. Biochem.* **55**, 1151–1191.

Mager, W. H. and De Kruijff, A. J. J. (1995). Stress-induced transcriptional activation. *Microbiol. Rev.* **59**, 506–531.

Mager, W. H. and Moradas-Ferreira, P. M. (1993). Stress response of yeast. *Biochem. J.* **290**, 1–13.

Martinez-Pastor, M. T., Marchler, G., Schüller, C., Marchler-Bauer, A., Ruis, H. and Estruch, F. (1996). The *Saccharomyces cerevisiae* zinc finger proteins Msn2p and Msn4p are required for transcriptional induction through the stress-response element (STRE). *EMBO J.* **15**, 101–109.

Moradas-Ferreira, P., Costa, V., Piper, P. and Mager, W. H. (1996). The molecular defences against reactive oxygen species in yeast. *Mol. Microbiol.* **19**, 651–658.

Morgan, B. A., Banks, G. R., Toone, W. M., Raitt, D., Kuge, S. and Johnston, L. H. (1997). The Skn7 response regulator controls gene expression in the oxidative stress response of the budding yeast *Saccharomyces cerevisiae*. *EMBO J.* **16**, 1035–1044.

Piper, P. W. (1995). The heat shock and ethanol stress responses of yeast exhibit extensive similarity and functional overlap. *FEMS Microbiol. Lett.* **134**, 121–127.

Piper, P. (1997). The heat shock response. In *Yeast Stress Responses* (S. Hohmann and W. H. Mager, eds), pp. 75–99. Molecular Biology Intelligence Unit, R.G. Landes Co., Austin, TX.

Ruis, H. and Schüller, C. (1995). Stress signalling in yeast. *Bioessays* **17**, 959–965.

Santoro, N. and Thiele, D. J. (1997). Oxidative stress responses in the yeast *Saccharomyces cerevisiae*. In *Yeast Stress Responses* (S. Hohmann and W. H. Mager, eds), pp. 171–211. Molecular Biology Intelligence Unit, R.G. Landes Co., Austin, TX.

Serrano, R., Marquez, J. A. and Rios, G. (1997). Crucial factors in salt stress tolerance. In *Yeast Stress Responses* (S. Hohmann and W. H. Mager, eds), pp. 147–169. Molecular Biology Intelligence Unit, R.G. Landes Co., Austin, TX.

Shamu, C. E., Cox, J. S. and Walter, P. (1994). The unfolded protein response pathway in yeast. *Trends Cell. Biol.* **4**, 56–60.

Siderius, M. H. and Mager, W. H. (1997). General stress response: in search of a common denominator. In *Yeast Stress Responses* (S. Hohmann and W. H. Mager, eds), pp. 213–230. Molecular Biology Intelligence Unit, R.G. Landes Co., Austin, TX.

Stephen, D. W. S., Rivers, S. L. and Jamieson, D. J. (1995). The role of the *YAP1* and *YAP2* genes in the regulation of the adaptive oxidative stress response of *Saccharomyces cerevisiae*. *Mol. Microbiol.* **16**, 415–423.

Thevelein, J. M. (1994). Stress signalling in yeast. *Yeast* **10**, 1753–1790.

Varela, J. C. S. and Mager, W. H. (1996). Response of *Saccharomyces cerevisiae* to changes in external osmolarities. *Microbiology* **142**, 721–731.

Varela, J. C. S., Praekelt, U. M., Meacock, P. A., Planta, R. J. and Mager, W. H. (1995). The *Saccharomyces cerevisiae HSP12* gene is activated by the high-osmolarity glycerol pathway and negatively regulated by the protein kinase A. *Mol. Cell. Biol.* **15**, 6232–6245.

19 Identifying Genes Encoding Components of the Protein Synthesis Machinery of the Yeast *Saccharomyces cerevisiae*

Mick F. Tuite[1], Ian Stansfield[2] and Rudi J. Planta[3]

[1]*Department of Biosciences, University of Kent, Canterbury, Kent,* [2]*Department of Molecular and Cell Biology, Institute of Medical Sciences, University of Aberdeen, Aberdeen, UK and* [3]*Department of Biochemistry and Molecular Biology, Vrije Universiteit, Amsterdam, The Netherlands*

◆◆

CONTENTS

Introduction
Studying protein synthesis defects *in vivo*
Studying protein synthesis defects *in vitro*
Genetic screens for protein synthesis mutants
Detecting translational errors
Identifying genes for ribosomal proteins

List of Abbreviations

ID	One-dimensional
2D	Two-dimensional
NEPHGE	Non-equilibrium pH gradient
ORF	Open reading frame
PABP	Poly(A)-binding protein
PAGE	Polyacrylamide gel electrophoresis
PMSF	Phenylmethylsulfonyl fluoride
poly(A)	Polyadenylate
rp	Ribosomal protein
UTR	Untranslated region
YNB	Yeast nitrogen base

◆◆◆◆◆◆ I. INTRODUCTION

We have learned a great deal about the process of translation (i.e. protein synthesis) from both genetic and biochemical studies with the yeast

METHODS IN MICROBIOLOGY, VOLUME 26
ISBN 0–12–521526–6

Saccharomyces cerevisiae. These studies have led to the general conclusion that, with a few notable exceptions, there is significant structural and functional conservation of the translational machinery between yeast and higher eukaryotic cells.

A. The Translation Cycle

The process starts with the small ribosomal 40S subunit binding the initiator Met-tRNA$_i$ and this 43S complex in turn binding to the 5′ methylated guanosine (the cap) of the mRNA. The 43S complex then scans along the 5′ untranslated region (5′ UTR) of the mRNA until it locates the AUG initiation codon *via* an interaction between the Met-tRNA$_i$ bound to the 40S subunit, and the AUG codon. The larger 60S ribosomal subunit then joins the 40S initiation complex to form the 80S initiation complex to complete the initiation phase of the process. The ribosome then decodes the mRNA leading to the sequential addition of amino acid residues to the carboxy-terminus of the growing nascent polypeptide chain. Amino acids are brought to the ribosome by aminoacyl-tRNAs as dictated by the codon being translated until one or other of the three stop codons – UAA, UAG or UGA – are reached. This signals the end of the elongation phase of protein synthesis and, in the absence of tRNAs able to decode the stop codon, the final termination phase begins. Termination results in the release of the completed polypeptide chain from the ribosome and, subsequently, the ribosomal subunits from the mRNA. It is beyond the scope of this article to discuss in full detail the molecular events underlying each of the three phases; the reader is referred to an excellent recent volume for such information (Hershey *et al.*, 1996).

B. Translation Factors

The translation cycle (Figure 1) requires the interplay of a large number of protein translation factors and various RNAs in addition to the ribosomes and mRNA. The translation factors, many of which are multisubunit in composition, are required for each of the three phases: eIFs for initiation, eEFs for elongation and eRFs for termination. Many of these factors are shared in common with mammalian cells, the notable exception being the elongation factor eEF3, which appears to be a soluble factor unique to the fungal translational machinery (Belfield and Tuite, 1993). Many of the translation factors bind to and hydrolyse GTP (and, albeit less commonly, ATP) and so there must also exist one or more recycling factors that facilitate nucleotide exchange. For example, the five-subunit recycling factor eIF2B is the guanine nucleotide exchange factor for eIF2. Furthermore, the function of many of the translation factors is regulated by phosphorylation and therefore both kinases and corresponding phosphatases must play important roles in protein synthesis. While the consequences of phosphorylation are perhaps less influential in yeast than in mammalian cells, many of the yeast factors are also phosphorylated.

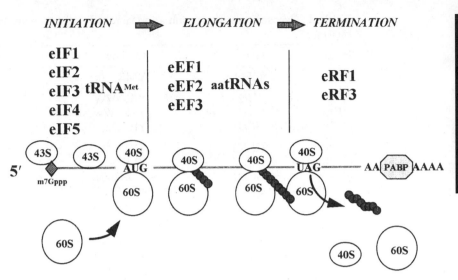

Figure 1. The stages and components of the protein synthesis cycle in yeast.

Genetic studies in yeast have also identified several other protein factors that are implicated as playing either a direct or indirect role in protein synthesis; for example, a protein bound to the polyadenylate (poly(A)) tail called PABP (poly(A)-binding protein encoded by the *PAB1* gene) has been implicated as playing a role in the initiation phase possibly by facilitating 60S subunit joining (Sachs and Davis, 1989). Such observations lead to this question: How does one identify a gene whose product plays either a direct role or an indirect role in yeast protein synthesis? In this chapter, we will consider both *in vitro* and *in vivo* approaches to this question. Given that relatively few of the translation factor genes are duplicated (unlike ribosomal protein genes, see below) and that most, if not all, translation factors are essential for viability, such studies are greatly aided by the use of conditional-lethal alleles, which can be readily constructed and screened for using the techniques described in Chapter 6.

Many of the translation factors identified biochemically in higher eukaryotic cells have now been identified in *S. cerevisiae* primarily through genetic screens (see below). Table 1 summarizes the yeast eIF genes so far identified and their gene products. The four unassigned eIF3 subunits have been identified as open reading frames (ORFs) but have not yet been assigned a gene name. The genes encoding the five subunits of the eIF2 recycling factor eIF2B have also been identified from the *GCN4*-based screen (see below). The number of eEF-encoding genes is less than for eIFs, although, in contrast to the eIF-encoding genes, there are generally two copies of each gene (Table 2). The yeast gene encoding the eEF1B α-subunit remains to be identified. The genes encoding the two essential subunits of the yeast termination release factor (eRF) have also been identified (Table 2).

Table 1. Cytosolic translation initiation factors (eIFs) in the yeast *Saccharomyces cerevisiae*

Factor	Gene	ORF no.	Mass (kD)
eIFlA	*TIF11*	YMR260c	17.4
eIF2α	*SUI2*	YJR007W	34.7
eIF2β	*SUI3*	YPL237w	31.6
eIF2γ	*GCD11*	YERO25w	57.9
eIF2Bα	*GCN3*	YKR026c	34.0
eIF2Bβ	*GCD7*	YLR291c	42.6
eIF2Bγ	*GCD1*	YOR260w	65.7
eIF2Bδ	*GDC2*	YGR083c	70.9
eIF2Bε	*GDC6*	YDR211w	81.2
eIF3 p16	*SUI1*	YNL244c	12.3
eIF3 p39	*TIF34*	YMR146c	38.7
eIF3 p62	*GCD10*	YNL062c	54.3
eIF3 p90	*PRT1*	YOR361c	88.1
eIF3 p110[a]	nd	YBR079c	110.3
eIF4AI	*TIF1*	YKR059w	45.0
eIF4AII	*TIF2*	YJL138c	44.6
eIF4B	*TIF3/STM1*	YPR163c	48.5
eIF4E	*TIF45/CDC33*	YOL139c	24.3
eIF4G	*TIF4632*	YGR162w	107.1
	TIF4632	YGL049c	103.9
p20	*CAF20*	YOR276w	18.2
eIF5	*TIF5*	YPR041w	45.2
eIF5A	*TIF51/HYP2*	YEL034w/YJR047c	17.1

[a] The gene for three other eIF3 subunits remains to be defined.
nd, gene nomenclature not yet defined.
Reference: see Http://www.mips.biochem.mpg.de/mips/yeast/funcat/fc05_04.htm.

Table 2. Cytosolic translation elongation (eEF) and termination (eRF) factors in the yeast *Saccharomyces cerevisiae*

Factor	Gene	ORF no.	Mass (kD)
Elongation factors			
eEFlA	*TEFl/TEF2*	YPR080w/YBRll8w	50.1
eEFlBα	ni	ni	
eEFlBβ	*TEF5*	YAL003w	22.6
eEFlBγ	*TEF3*	YPL048w	47.1
	TEF4	YKL081w	46.7
eEF2	*EFT1*	YOR133w	93.3
	EFT2	YDR385w	93.3
eEF3	*YEF3*	YLR249w	115.8
eEF4	nd	YNL163c	124.4
Termination factors			
eRFl	*SUP45*	YBR143c	49
eRF3	*SUP35*	YDR172w	76.6

ni, not yet identified in the yeast genome; nd, gene nomenclature not yet defined.

◆◆◆◆◆◆ II. STUDYING PROTEIN SYNTHESIS DEFECTS *IN VIVO*

Protein synthesis can be studied *in vivo* in a variety of ways, both qualitatively and quantitatively. In looking for defects in protein synthesis, one can address several different aspects experimentally:

- Is the rate of protein synthesis (i.e. amino acids polymerized per unit of time) impaired?
- Is there any change in the pattern (either qualitatively or quantitatively) of the proteins being synthesized?
- If there is a defect in protein synthesis, what stage of the process is impaired: initiation, elongation or termination?

A. Radiolabeling Proteins *In Vivo*

To monitor the rate of protein synthesis *in vivo*, yeast cells can be pulse-labeled with either a [³H]- or [¹⁴C]-labeled amino acid. The choice of which amino acid to use is important with amino acids such as leucine or phenylalanine being preferred because they have low endogenous pool sizes. Following the labeling period – usually between 10 and 15 min – incorporation of the radiolabeled amino acid into protein is rapidly terminated by the addition of a potent yeast protein synthesis inhibitor such as cycloheximide. Care must be taken in choosing the growth medium and is usually one based on yeast nitrogen base (YNB) and buffered to pH 5.2 with a sodium succinate–sodium hydroxide buffer (see Appendix II).

[³⁵S]-methionine or a mixture of [³⁵S]-methionine/cysteine are widely used to label proteins synthesized *in vivo*, particularly if polyacrylamide gel electrophoresis (PAGE)/autoradiography is to be used to analyse the pattern of proteins synthesized. Although the use of the sulfur-containing amino acids in this context has several advantages – wide availability, high specific activities, and rapid isotopic equilibration – the problem in using Met/Cys is that they are comparatively rare in yeast proteins. An alternative option is to use a high specific activity [¹⁴C]-amino acid mixture because this will provide a truer reflection of the total number and relative abundance of proteins upon PAGE/autoradiography.

It is also possible to specifically label mitochondrial proteins *in vivo* by terminating protein synthesis with an antibiotic that inhibits protein synthesis on cytoplasmic, but not mitochondrial, ribosomes. Particularly effective in this context is cycloheximide (Douglas *et al.*, 1979).

B. Electrophoretic Analysis of Yeast Proteins

PAGE provides an effective means of studying yeast protein synthesis both qualitatively and quantitatively (Grant *et al.*, 1996). Radiolabeled proteins can be separated in one dimension (1D) either on the basis of their different molecular weights (SDS-PAGE), on their isoelectric points (isoelectric focusing) or a combination of both in two-dimensional (2D)

SDS-PAGE or non-equilibrium pH gradient (NEPHGE) gels. Any of these systems can be used in combination with western blotting to detect specific proteins.

The choice of which gel system to use very much depends on what one is trying to assess. 1D SDS-PAGE can give good resolution of proteins over a wide molecular weight range, although only the most abundant proteins will be detected and co-migration of proteins can be problematical. 1D isolectric focusing gels, in combination with western blotting, can be used to analyse the phosphorylation status of proteins. For example, such a system has been used to assess the phosphorylation status of the yeast translation initiation factor eIF2α in a variety of kinase-defective strains (Dever *et al.*, 1992).

2D gel systems are much better at resolving complex mixtures of proteins with high resolution. Current methods can be used to resolve several hundred different proteins from a total yeast cell extract and, by using computer-assisted analysis, one can investigate both qualitative and quantitative differences in protein synthesis. There have been recent significant efforts to establish a gene-protein index using 2D gel analysis (e.g. Boucherie *et al.*, 1996) with over 250 proteins so far identified and mapped.

C. 2D Gel Analysis of Ribosomal Proteins

Ribosomal proteins of yeast have been analyzed in different 2D PAGE systems, all of which are essentially modifications of the procedures developed by Kaltschmidt and Wittmann (1970; basic urea/acidic urea) and Mets and Bogorad (1974; acidic urea/SDS), respectively, for *Escherichia coli* ribosomal proteins. Any one of four different systems can be used to study yeast ribosomal proteins: acidic–SDS, basic–SDS, basic–acidic, and acidic–acidic. The latter was also used by Michel *et al.* (1983) and previously described for *E. coli* ribosomal proteins (Madjar *et al.*, 1979).

These electrophoretic separations can be carried out conveniently and rapidly in a minigel system developed by Bollen *et al.* (1981). Protocols for the preparation of 60S and 40S ribosomal subunits from yeast cells, and the extraction and subsequent separation of the ribosomal proteins have been previously published (Raué *et al.*, 1991).

D. Polysome Analysis in Yeast

The separation of the various components of a yeast cell's ribosome population by sucrose gradient centrifugation has proved a powerful tool for analysing yeast mutants defective in some aspect of the translational process (Petersen and McLaughlin, 1974). In some cases identification of the specific biochemical defect responsible for causing the inhibition of protein synthesis is possible.

Methods for polysome analysis vary slightly, but all essentially involve the preparation of a post-mitochondrial supernatant from a yeast lysate using a lysis buffer which includes concentrations of $MgCl_2$ and KCl sufficient to preserve ribosomal integrity. This lysate is then loaded onto a sucrose gradient and ultracentrifuged to resolve the 40S and 60S subunits, 80S ribosomes, and polysomal ribosomes (Protocol 1; Stansfield et al., 1992). Gradients can be centrifuged for increasing lengths of time if greater resolution of the subunits is required. Ordinarily, during cell lysis and lysate processing, ribosomes continue to elongate until they reach the natural termination codon and subsequently leave the mRNA, thus producing a polysome profile predominantly consisting of 80S ribosomes and lacking polysomes (Figure 2A). Such a profile is often referred to as a "run-off" profile. The translation elongation inhibitor cycloheximide can, however, be added to a culture before cell harvesting, in order to "freeze"

Protocol 1. Polysome profile analysis in yeast (after Stansfield et al., 1992)

1. A yeast strain is grown to a cell density of approximately 1×10^7 cells ml^{-1}. If required, 15 min before harvesting the cells, cycloheximide can be added to a final concentration of 200 µg ml^{-1} to produce an elongation-inhibited profile (see text and Figure 2b for details). 100 ml of cell culture is harvested, washed, and resuspended in 1 ml buffer A. At this point phenylmethylsulfonyl fluoride (PMSF) is added to 1 mM final concentration to inhibit serine protease activity following lysis.

2. Cells are broken by vortexing for 3×30 s with an equal volume of glass beads (0.4 mm diameter).

3. Centrifuge the yeast lysate at 3000g for 5 min. Retain the supernatant, to which is added Triton X-100 detergent to 1% v/v final concentration, and centrifuge at 12 000g for 15 min.

4. Load the supernatant from step 3 onto the surface of a 15–42.5% w/v sucrose gradient made up in an ultracentrifuge tube (tubes to suit a Beckman SW41 rotor or equivalent would be appropriate). An amount of lysate equivalent to 400 µl of a solution of 40 A_{280} units should be loaded. Sucrose gradients can be made using 15 sucrose solutions of increasing concentration (15%, 17.5%, 20%, etc.) dissolved in buffer A. Centrifuge tubes are placed in dry ice, and an appropriate volume of the most concentrated sucrose solution added. Successive layers of equal volume and decreasing concentration of sucrose are added as the preceding one freezes until the tube is filled. Gradients can be stored for up to 3 months at –20°C, and thawed at room temperature for 2–3 h before use.

5. Centrifuge the loaded gradient for 3 h at 170 000 g, and unload using a gradient fractionator apparatus (e.g. Isco 184). To obtain the polysome profile, the sucrose gradient solution withdrawn using the gradient fractionator is passed through an ultraviolet flow-through monitor (280 nm) connected to a suitable paper chart recorder.

Solutions
Buffer A: 25 mM Tris, pH 7.2, 50 mM KCl, 5 mM $MgCl_2$, 5 mM 2-mercaptoethanol (add this last reagent immediately before use)
PMSF: 100 mM stock solution in 100% ethanol (store at 4°C, and add to buffer A immediately before cell lysis)
Cycloheximide: 10 mg ml^{-1} stock solution in sterile water. Stocks can be frozen at –20°C

---

the ribosomes on the mRNA, inhibiting run-off and thereby producing a "snapshot" of an *in vivo* polysome profile. A characteristic wild-type yeast polysome gradient prepared from cells treated with cycloheximide (Figure 2B) will usually give approximately 80% of the ribosomes present in the polysome fraction. The individual peaks identified indicate one extra ribosome per mRNA: dimer, trimer, tetramer, etc.

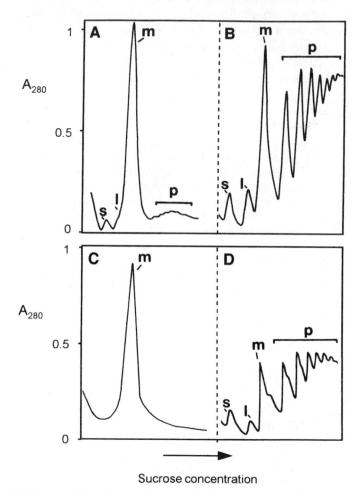

Sucrose concentration

Figure 2. The use of sucrose density gradient analysis to study the relative proportions of ribosomal subunits, monosomes and polysomes in yeast cells carrying various defects in protein synthesis. (A) A run-off polysome profile, where ribosomes were prepared as described (Protocol 1) from a wild-type yeast strain in the absence of cycloheximide. (B) A translation elongation-inhibited polysome profile, where ribosomes were prepared from a wild-type yeast strain treated with 200 µg ml^{-1} cycloheximide 15 min before harvest. (C) The polysome profile exhibited by the conditional-lethal *prt1* initiation block mutant shifted to the nonpermissive 36°C for 5 min, followed by cycloheximide addition to "freeze" translation elongation before cell lysis (Petersen and McLaughlin, 1974). (D) A polysome profile exhibiting the presence of halfmers, appearing as shoulders on the trailing edges of the monosome and smaller polysome peaks. The positions of the polysome (p) and 80S monosome/ribosome (m) peaks are indicated, as are the small 40S (s) and large 60S (l) subunits.

A defect in protein synthesis may be suspected when gene disruption or mutation results in a phenotype such as antibiotic sensitivity. In such cases polysome analysis can be used to probe the biochemical defect. As an analytical tool it has been particularly successful in the investigation of protein synthesis defects resulting from conditional-lethal mutations, particularly temperature-sensitive mutants. Polysome profiles have, however, also been employed to analyse the consequences of disruptions of genes encoding translation factors (e.g. Coppolecchia *et al.*, 1993).

1. Polysome analysis of translation initiation

A stringent block of the initiation phase of protein synthesis is characterized by a marked decrease in the proportion of polysomes with a concomitant increase in the size of the 80S ribosome peak (Figure 2C). Most initiation blocks act before the joining of the 60S subunit to the 43S initiation complex; often the increased proportion of 80S ribosomes present in these cases represents inactive aggregations of 40S and 60S subunits, rather than functional 80S monosomes on an mRNA. These two species of 80S particle can be distinguished by their resistance to dissociation by 0.8 M KCl. 80S ribosomes will readily dissociate into 40S and 60S subunits under these conditions while monosomes will not (Martin and Hartwell, 1970). The archetypal initiation block is represented by the temperature-sensitive *prt1* mutation, which causes an almost complete conversion from polysomes to 80S ribosomes within 5 min of a shift from the permissive (25°C) to the non-permissive temperature (37°C) (Petersen and McLaughlin, 1974). *PRT1* encodes a subunit of the yeast eIF3 initiation factor complex (Table 1; Naranda *et al.*, 1994).

Some initiation blocks produce a more subtle change in polysome profile, such as the appearance of "halfmers" (Helser *et al.*, 1981; Figure 2D). Halfmers are defined as polyribosomes that have an extra 40S or 43S ribosomal subunit attached to the mRNA, and were first visualized in polysome profiles derived from yeast treated with low (10 μg ml⁻¹) concentrations of cycloheximide. At this low concentration, cycloheximide inhibits the initiation step of translation, rather than the elongation block which results from treatment with a higher cycloheximide concentration (200 μg ml⁻¹). Examples of halfmers on polysome profiles have been observed in yeast strains depleted for ribosomal protein L16 (Rotenberg *et al.*, 1988), which may reflect a reduced ability of the 60S subunit to join the 43S complex on the mRNA.

2. Polysome analysis of translation elongation

The polysome profile of a mutant blocked at the elongation step of translation should mimic a wild-type cycloheximide-treated culture, i.e. inhibition of ribosomal run-off with the preservation of polysomal ribosomes after cell lysis in the absence of cycloheximide treatment (Hartwell and McLaughlin, 1968). However, partial elongation block phenotypes may allow slow run-off in the absence of cycloheximide; in this instance, cycloheximide treatment of a disruptant strain before lysis might be expected to produce a larger proportion of ribosomes in polysomes than the parent strain.

3. Polysome analysis of translation termination

Mutations in genes encoding termination factors can also be analysed using polysome gradients. Theoretically, such profiles would be expected to exhibit blocked polysome run-off in the absence of cycloheximide, in much the same way as for elongation mutants. Treating the termination mutant strain with cycloheximide *in vivo* before harvest should produce a profile shifted towards larger polysomes in comparison to the parent strain, as stalled terminating ribosomes cause ribosomal queueing on the mRNA. We are, however, unaware of any such examples in yeast having been described in the literature. Known mutations with a nonsense suppressor phenotype in the gene encoding the translation release factor eRF1 (*SUP45*), including a conditional-lethal mutation, display almost wild-type polysome profiles following treatment with cycloheximide (Stansfield *et al.*, 1996; I. Stansfield and M. F. Tuite, unpublished). This is thought to occur because in the absence of the binding of the release factor complex to a stop codon, endogenous tRNAs able to decode a termination codon (albeit at a low efficiency) catalyse an alternative elongation event, thus alleviating a block.

4. Polysome analysis of ribosome integrity and biogenesis

Polysome profiles can also be used to examine the effects of mutations that might affect the structure of the ribosome, or its biogenesis, because such defects often produce altered subunit ratios. For example, mutations in the gene encoding in 60S ribosomal protein L46, which suppresses a PABP temperature-sensitive defect, produce decreased ratios of 60S:40S subunits (Sachs and Davis, 1989). Similarly, a mutation in the *DRS2* gene, involved in ribosome biogenesis, causes increased levels of 60S compared with 40S subunits (Ripmaster *et al.*, 1992).

◆◆◆◆◆◆ III. STUDYING PROTEIN SYNTHESIS DEFECTS *IN VITRO*

Studying protein synthesis mutants *in vivo* can be problematic because one is not always readily able to separate transcriptional defects from translational defects. One way around this is to study protein synthesis *in vitro* using a cell-free lysate prepared from the mutant strain. Several methods have been described to prepare active cell-free translation systems from yeast that are capable of efficiently translating either synthetic mRNA templates such as polyuridylic acid (polyU) or natural mRNAs (Hartley *et al.*, 1996).

A. Poly(U)-Directed Translation Systems

In this relatively easy-to-prepare system, yeast ribosomes are able to translate the poly(U) template into polyphenylalanine at high efficiency. Providing supraoptimal Mg^{2+} concentrations are used (>12 mM) rather than the physiologically relevant 2 mM, the only translation factors that are required are eEF1, eEF2 and eEF3 (see Figure 1, Table 2). Protein synthesis is monitored by use of [^{3}H]- or [^{14}C]-labeled phenylalanine. This system can therefore be used to assess defects in the elongation factors but can also be used to study other translational defects.

1. Studying mistranslation *in vitro*

If a mutant shows a high level of general mistranslation (i.e. amino acid misincorporation), this can be detected by assaying for the (mis)incorporation of [^{3}H]- or [^{14}C]-labeled leucine (codons UUG/UUC) into polypeptides in the poly(U)-directed system. This can only arise by mistranslation of the UUU codon by the leu-tRNA that decodes UUG/UUC, i.e. by third base misreading. Using this assay, Eustice *et al.* (1986) showed that mutations in some ribosomal proteins led to higher endogenous levels of mistranslation, while Tuite and McLaughlin (1984) showed, using this assay, that the antibiotic paromomycin caused a high level of mistranslation at the UUU codon.

2. Studying ribosome-mediated antibiotic resistance *in vitro*

Poly(U)-directed systems have been widely used to demonstrate that resistance to a specific antibiotic is due to a change in the ribosome. Furthermore, the subunit can be identified by using ultracentrifugation to remove the ribosomes from both a wild type and an antibiotic resistant strain cell lysate, dissociating the ribosome into its two subunits, mixing and matching the subunits and then reassaying in a poly(U)-directed system. In this way, for example, Schindler *et al.* (1974) showed that resistance to trichodermin was mediated at the level of the 60S subunit.

Defects in either aminoacylation of tRNAs (e.g. mutations in an aminoacyl-tRNA synthetase gene) or in elongation factor recycling may also be detected by the poly(U)-directed assay.

B. Natural mRNA-Directed Translation Systems

Both homologous (yeast) and heterologous (various) mRNAs can be translated both efficiently and accurately in yeast lysates (Tuite *et al.*, 1980). These RNAs can be purified directly from their natural source or they can be generated by *in vitro* transcription using the widely available T7 or SP6-based vectors. In the case of *in vitro* transcripts, if one wishes to achieve a maximal level of translation, then the transcript needs to have a poly(A) tail of at least 25 As added either co-transcriptionally or post-transcriptionally (Gerstel *et al.*, 1992). The addition of a 5′ m⁷Gppp cap at the 5′ end of the transcript is less important for translation.

Translation of a natural mRNA can be readily assayed by radiolabeling the *in vitro* translation product with [³⁵S]-methionine, for example, and analysing by SDS-PAGE and autoradiography. For an authentic primary amino acid sequence, the cell-free lysate must carry out all three phases of translation accurately. Defects in specific steps of the initiation process (e.g. Feinberg *et al.*, 1982) and the elongation process (e.g. Hutchinson *et al.*, 1984) can be monitored in such systems.

Defects in translation termination in yeast mutants can also be identified using a natural mRNA-directed cell-free translation system (Tuite *et al.*, 1981). In these assays, the cell-free system is programmed with an mRNA with a known termination codon that, when translated, will allow the ribosomes to continue translation into the 3′UTR. This will generate a new "readthrough" protein, which can be differentiated from the correctly terminated protein by SDS-PAGE and autoradiography. Adding a known nonsense suppressor tRNA to this assay enhances its sensitivity significantly. Such an assay has been used to confirm translation termination defects in prion-containing [*PSI*⁺] strains (Tuite *et al.*, 1983) and in a conditional-lethal mutant encoding a defective eRF1 subunit (Stansfield *et al.*, 1997).

C. Transient Expression Assays as Alternatives to Cell-Free Assays

Cell-free extracts are difficult to prepare. Recently, however, an alternative means of assaying the translation of an mRNA by yeast has become available. Essentially, the target mRNA is introduced into spheroplasts prepared from the mutant strain, by electroporation (Everett and Gallie, 1992). Translation of the introduced mRNA peaks 2–3 h after electroporation and can be best monitored by use of a reporter mRNA whose gene product is readily assayable, e.g. luciferase (see Chapter 9). Such "transient expression" systems show both a 5′ cap and a poly(A)-tail dependency suggesting that they use authentic translational mechanisms.

◆◆◆◆◆◆ IV. GENETIC SCREENS FOR PROTEIN SYNTHESIS MUTANTS

Several genetic screens have been developed for specifically identifying genes encoding components of the protein synthesis machinery in yeast. In many cases these screens have led to the identification of translation factors that have not been revealed by *in vivo* biochemical studies. Several screens have also been developed to find components of the protein synthesis regulatory machinery. It is beyond the scope of this article to review all the available screens. Rather, specific examples will be outlined to illustrate the type of screens and their outcomes.

A. Screen for Ribosomal Protein Mutants

Mutations leading to resistance to 80S ribosome-active antibiotics is a particularly powerful screen for identifying ribosomal protein (rp) genes. For example, resistance to cryptopleurine, an inhibitor of the EF2-GTP-dependent step in ribosomal translocation, maps to two unlinked genes: *CRY1* and *CRY2* (Larkin and Woolford, 1983). These two genes are identical and encode ribosomal proteins S14A and S14B (formerly called rp59; see Appendix I). Mutations that confer resistance to high concentrations of cycloheximide, an inhibitor of the peptidyl-transferase center, have identified the *CYH2* gene, which encodes rp L28 (formerly called L29/rp44; Fried and Warner, 1982), while mutations conferring resistance to trichodermin, another inhibitor of translation elongation, map to the *TCM1* gene, which encodes rp L3 (Fried and Warner, 1981). Although many of the rp genes are duplicated (see Appendix I), mutations leading to antibiotic resistance mediated by a mutation in an rp gene are generally semi-dominant with respect to this phenotype and are not therefore masked by the second, wild-type allele. Gene disruption experiments have suggested that the majority of ribosomal proteins (but by no means all) are essential for cell viability. For duplicated rp genes, deletion of one copy will usually have minimal effect on cell phenotype.

Increased sensitivity to antibiotics can also be used as a phenotype to screen for rp mutants. For example, mutation in either the *SUP44* or *SUP46* genes (encoding rp S2, formerly called S4, and S9B, formerly called S13, respectively) can lead to increased sensitivity to the aminoglycoside antibiotic paromomycin (Wakem and Sherman, 1990). This antibiotic acts, via the ribosome, to induce mistranslation and the enhanced sensitivity phenotype of the *SUP44* and *SUP46* mutants would suggest that these two ribosomal proteins play an important role in maintaining translational accuracy during mRNA decoding.

B. Screen for Translation Initiation Mutants

Several different genetic screens have been established in yeast to identify components of the translation initiation machinery. Two will be briefly described here.

1. Selection of mutants defective in start codon recognition

This genetic screen selects for mutants that allow the scanning 43S complex to initiate on a non-AUG codon and made use of a strain carrying an *in vitro* engineered *HIS4* gene in which the initiation codon had been mutated to AUU. By selecting for His⁺ revertants, three unlinked genes (*SUI1-3*) were identified which, when mutated, allowed translation to initiate from a UUG codon located two codons 3′ of the mutated AUG codon (Castilho-Valavicius *et al.*, 1990). Replacing the mutated AUU codon with an UUG codon resulted in translation also being able to initiate at the 5′-most UUG codon. Each of the three genes identified by this screen (*SUI1–3*) encode a component of the translation initiation machinery (Table 1).

2. Selection of mutants defective in eIF2 recycling factor eIF2B

eIF2 is a key component of the translation initiation phase of protein synthesis and eIF2-bound GTP is hydolyzed to GDP during this phase. For eIF2 to reenter the translation cycle the GDP must be exchanged for GTP and this is achieved by the five-subunit eIF2 recycling factor eIF2B (Table 1). The genes encoding the five subunits of eIF2B were identified in a genetic screen initially developed to identify how the expression of the gene encoding the Gcn4p transcription factor (*GCN4*) was regulated. Gcn4p is required to activate the transcription of a large number of genes involved in amino acid biosynthesis. During periods of amino acid starvation, the *GCN4* mRNA is translated whereas, in amino acid rich conditions, although the *GCN4* mRNA is synthesized, it is not translated. Unusually, the 5′ UTR of the *GCN4* mRNA is long (591 nucleotides) and contains four short ORFs, which play a central role in the posttranscriptional regulation of this gene. [For full details of this novel gene-specific posttranscriptional regulatory system, see the recent review by Hinnebusch (1996).]

In this screen two different classes of *GCN4* regulatory mutant were identified:

- *gcd* mutants − which failed to repress translation of the *GCN4* mRNA in amino-acid-rich conditions; and
- *gcn* mutants − which failed to derepress translation of *GCN4* mRNA under amino-acid-starvation conditions.

In addition to identifying the γ subunit of eIF2 (*GCD11*) and the p62 subunit of eIF3 (*GCD10*), this screen has identified the protein kinase that phosphorylates the eIF2α subunit (*GCN2*) and the five subunits of eIF2B, the eIF2 recycling factor (*GCN3, GCD7, GCD1, GCD2* and *GCD6*). Although not originally designed to identify translation initiation factors, it has without question been a very fruitful screen for genes encoding these yeast eIFs.

To assess whether a newly identified mutant or gene disruptant has an effect on *GCN4* regulation (and may therefore define another component of the translation initiation regulatory mechanism of yeast) the most straightforward assay is to introduce a plasmid-borne copy of a *GCN4–lacZ* fusion gene by transformation and then to determine the levels of expression of the encoded β-galactosidase in the transformant under either non-amino-acid-starvation conditions (i.e. repressing conditions) or under conditions of histidine starvation (i.e. derepressing conditions). *gcn*-type mutants would result in no elevation of β-galactosidase levels under derepressing conditions while *gcd*-type mutants would show high levels of enzyme activity under both repressing and derepressing conditions.

C. Genetic Screens for Termination (eRF) Factors

Genetic studies in yeast have identified several genes that, when mutated, give rise to a defect in translation termination (Stansfield and Tuite, 1994). One genetic screen used to identify such mutants exploits a weak nonsense (ochre) suppressor tRNA encoded by the *SUQ5* gene (Cox, 1965). The *SUQ5*-encoded tRNA^{Ser}, although having an anticodon mutation that allows it to translate the UAA codon by a cognate interaction, does so very inefficiently because it is unable to compete effectively with the endogenous eRF-driven termination mechanism. In an *SUQ5* strain carrying a defined ochre mutation in a readily scorable gene (e.g. *ade2-1*) no suppression of the mutation can be detected (red, Ade⁻). By screening, postmutagenesis, for mutants that allow the *SUQ5* suppressor to suppress the *ade2-1* mutation, five so-called allosuppressor genes (*SAL1–SAL5*) were identified (Cox, 1977). Two of these genes, the *SAL3 (SUP35)* and *SAL4 (SUP45)* genes, define the two essential subunits of the yeast eRF (Stansfield *et al.*, 1995b). These two genes were also identified in a screen for mutations that led to the suppression of a range of nonsense mutations (i.e. as codon non-specific suppressors) in the absence of a defined suppressor tRNA, so-called omnipotent suppressor mutants (Stansfield and Tuite, 1994). The *SUQ5*-based screen also led to the identification of the extrachromosomal determinant called [*PSI*] that has subsequently been shown to be due to the prion-like behavior of the product of the *SUP35* gene (Tuite and Lindquist, 1996).

◆◆◆◆◆◆ V. DETECTING TRANSLATIONAL ERRORS

The disruption of, or mutations in, genes encoding translation factors, ribosome-associated proteins and integral ribosome proteins will in many cases have direct functional consequences for the accuracy of protein synthesis. Various assays for nonsense, frameshift and missense translation errors can be used to indicate a direct or peripheral role for a gene product in protein synthesis.

A. Nonsense Error Assays

During translation of an mRNA, errors in the recognition of stop codons by the ribosomal machinery can result in a suppression event, in which a tRNA binds the stop codon causing translation elongation, rather than termination, to occur. Such events can be assayed by placing a reporter gene downstream of, and in-frame with, the stop codon; nonsense suppression will result in translation of the downstream reporter, with reporter protein activity proportional to the efficiency of nonsense suppression. A shuttle vector-based assay system has been developed for yeast using this principle and consists of the first 33 codons of the yeast

PGK1 ORF, followed by an in-frame stop codon and then the *E. coli lacZ* gene (Figure 3A; Firoozan *et al.*, 1991; Stansfield *et al.*, 1995a). Three vectors are available with either UAA, UAG or UGA as the premature stop codon (plasmids pUKC817, pUKC818 and pUKC819 respectively), together with a fourth *lacZ* expression control vector lacking a premature stop codon (plasmid pUKC815). β-Galactosidase levels determined in a

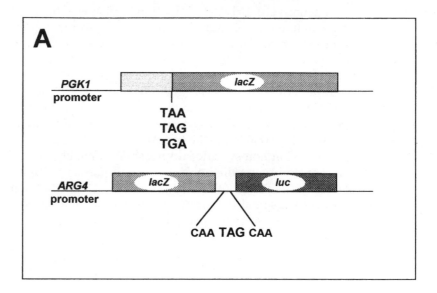

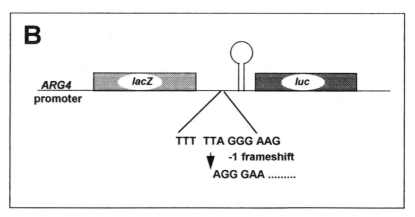

Figure 3. Plasmid-based reporters for assaying translation errors in *Saccharomyces cerevisiae*. (A) Termination errors: two different reporter constructs have been developed. *Upper:* A *PGK1–lacZ* fusion in which the two reading frames are contiguous apart from an in-frame stop codon (Firoozan *et al.*, 1991; Stansfield *et al.*, 1995a). *Lower:* A *lacZ–luc* fusion in which both cistrons are intact with translation of the linking termination codon (TAG in this case) giving rise to the synthesis of functional luciferase (Stahl *et al.*, 1995). (B) Ribosomal frameshifts: construct is essentially the same as the *lacZ–luc* fusion in (A) above except the *lacZ* and *luc* genes are linked by a sequence that shifts the ribosomes into the –1 reading frame and is based on a sequence from the retrovirus HIV-1 (Stahl *et al.*, 1995).

strain transformed with either pUKC817, 818 or 819 are expressed as a percentage of reporter levels in the same strain transformed instead with the control plasmid pUKC815. The single-copy vectors carry the *URA3* selectable marker and standard β-galactosidase assays methods are used to assay reporter activity in the transformed strains (Finkelstein and Strausberg, 1983).

In an otherwise wild-type strain, nonsense suppression levels resulting from disruption of a gene encoding a component of the termination machinery may be low. In such cases it may be advantageous to assay the disruption event in a weak nonsense suppressor tRNA genetic background, and look for enhancement of the low, but detectable suppressor activity of the tRNA (i.e. allosuppression). For example, certain mutations in either of the yeast release factor genes (*SUP45* and *SUP35*) exhibit an allosuppressor phenotype in a weak ochre suppressor tRNA[ser] background (the *SUQ5* allele; Cox, 1977). This strategy relies on the strain carrying one or more nonsense mutations in genes whose loss of function and restoration of function can be readily assayed *in vivo*.

The limitation of this approach, other than having to construct strains carrying these mutations, is that the identity of the amino acid inserted by the suppressor tRNA at the "reporter" codon may not be compatible with the ability of the corresponding gene product to remain functional. In contrast, the identity of the amino acid inserted at the PGK-β-galactosidase junction in the *PGK1–lacZ* reporter system has no effect on β-galactosidase activity *in vivo*.

One caveat to add to the use of these various reporter systems is that apparent increases in nonsense suppression levels may result from gene disruptions causing defects not in termination, but in the nonsense-mediated mRNA decay pathway (e.g. the *UPF1* gene; Ruiz-Echevarria *et al.*, 1996). The observed increased "nonsense suppression" in this case is due to stabilization of the premature stop codon-containing reporter mRNA, which increases its abundance. The use of vectors expressing two tandem in-frame reporter genes (*lacZ* and *luc*) can be used to circumvent this problem (Stahl *et al.*, 1995; Figure 3B). For example, the reporter genes are separated by an in-frame stop codon in vector pACTMV or by an ORF in the control vector pACTQ (Stahl *et al.*, 1995). By using relative levels of β-galactosidase activities expressed from the upstream *lacZ* gene in a pACTMV and a pACTQ transformant, respectively, measured luciferase activities arising from nonsense suppression can be normalized. This acts as a control for differences in reporter mRNA levels and stabilities. The premature UAG stop codon separating the *lacZ* and *luc* reporters in this assay system is flanked by two CAA codons giving rise to a nucleotide context which, in tobacco mosaic virus, stimulates nonsense suppression. This particular context, i.e. CAA XXX CAA, results in 20% UAG readthrough frequency (with respect to termination) in a wild type yeast, thus making the system suitable for analysing genetic lesions that result in either decreased or increased nonsense suppression frequencies (Stahl *et al.*, 1995).

B. Frameshift Error Assays

Once initiated on the authentic AUG (start) codon, the ribosomes elongate three bases (i.e. one codon) at a time, maintaining this reading frame until the termination codon is reached. The ability of ribosomes to move out of frame during elongation into either the +1 or −1 reading frame is rare although the random spontaneous frameshift error rate has never been accurately estimated for *S. cerevisiae*. In order to assay frameshift errors, it is necessary to provide a nucleotide sequence in the mRNA coding region at which such errors are more likely to occur, i.e. shifty sequences. For frameshift events occurring during retroviral gene expression, the shifty nucleotide sequences that stimulate −1 frameshifts take the form X XXY YYZ, where the triplets denote the 0 frame. There are also examples of several "shifty sequences" which promote relatively high efficiency at specific sites within mRNAs. One endogenous "shifty sequence" is the CUUAGGC sequence in the retrotransposon-Ty1-encoded mRNA which promotes +1 frameshifting (Belcourt and Farabaugh, 1990). These slippery sites usually precede RNA secondary structural motifs such as hairpin loops or pseudoknots which, by pausing the ribosome, stimulate frameshifting.

To assay frameshifts in yeast, the HIV-1 −1 frameshift sequence together with the stimulatory downstream hairpin has been placed between and in frame with an upstream *lacZ* gene, but in the +1 frame relative to the downstream *luc* gene (Figure 3; Stahl *et al.*, 1995). A −1 frameshift event is required in order to express functional luciferase. The β-galactosidase levels in frameshift and control vector transformants are used to normalize luciferase expression levels.

The frequency of −1 frameshift errors at the HIV-1 frameshift site in yeast is 6%, indicating that the mammalian recoding signals are also functional in yeast (Stahl *et al.*, 1995). Using such a basal level of −1 frameshifting, the stimulatory or inhibitory effect of gene disruptions on frameshifting can be measured. The yeast Ty1 frameshift sequence can also be exploited into the same double reporter system, allowing assay of +1 frameshift events in the same way (Stahl *et al.*, 1995).

An alternative frameshift assay, based on the *CUP1* gene reporter encoding copper resistance, has also been developed (Lee *et al.*, 1995). This frameshift reporter, incorporating the mouse mammary tumour virus −1 frameshift site, allows direct selection on solid medium for gene mutations or disruptions which increase frameshift frequency containing copper. However, the single reporter *CUP1*-based system cannot control for changes in the stability of the mRNA and therefore defects in the nonsense-mediated mRNA decay pathway (see above) can result in apparent increases in frameshift frequency owing to stabilization of the reporter mRNA.

C. Missense Error Assays

Perhaps the most common translation error is that of misincorporation of an incorrect amino acid into a growing polypeptide chain. The levels of such non-cognate decoding at a given codon has been variably estimated

---

Protocol 2. Assaying missense translation in yeast (Santos *et al.*, 1996)

1. A yeast strain expressing β-galactosidase is grown to a cell density of approximately 1×10^7 cells ml^{-1}; 5 ml of cell culture is harvested, washed, and resuspended in 1 ml Z-buffer.
2. Permeabilize the cells by adding 20 μl SDS (0.1% w/v) and 50 μl chloroform to the cell suspension in Z-buffer and vortexing for 15 s.
3. Incubate the yeast suspension at 47°C for varying lengths of time (0–15 min) to inactivate the β-galactosidase before transferring to ice for 30 min.
4. Transfer the suspensions to 37°C for 5 min.
5. Add 200 μl *o*-nitrophenolgalactoside (4 mg ml^{-1} in deionized water) prewarmed to 37°C and allow yellow color to develop at 37°C before stopping the reaction with the addition of 0.8 ml 1 M sodium carbonate.
6. Spin the reaction mixture to remove the cells and measure the absorbance of the supernatant at 420 nm.

Solutions

Z-buffer; 60 mM Na$_2$HPO$_4$, 40 mM Na$_2$H$_2$PO$_4$, 10 mM KCl, 1 mM MgSO$_4$, 50 mM 2-mercaptoethanol (add this last reagent immediately before use)

---

in eukaryotes at between 10^{-3} and 10^{-5} although an accurate determination of the levels of such errors in *S. cerevisiae* is lacking. However, an assay that can be used to determine global translation error rates has been developed for *S. cerevisiae* based on a method originally developed for *E. coli*. In this system, the thermal lability of β-galactosidase is used as an indicator of global levels of missense translation, because missense errors make the enzyme less resistant to prolonged incubations at high temperature (Branscomb and Galas, 1975). In *E. coli*, β-galactosidase expressed in cells grown in the presence of streptomycin (an inducer of translational errors) exhibits a more rapid loss of activity during a period of incubation at 58°C than the enzyme expressed in a control culture (Branscomb and Galas, 1975). This assay has been adapted for use in *S. cerevisiae* (Santos *et al.*, 1996) transformed with a suitable β-galactosidase expression vector (e.g. pUKC815; Stansfield *et al.*, 1995a). In this assay (Protocol 2) the β-galactosidase-expressing yeast culture is heated to 47°C for increasing lengths of time, followed by a period of 30 min on ice. This incubation on ice seems to reduce variability between replicates, as well as making the assay more manageable, allowing the physical separation of the heat-denaturation and assay parts of the experiment. Reporter activity (β-galactosidase) is then assayed according to standard protocols (Finkelstein and Strausberg, 1983).

◆◆◆◆◆◆ VI. IDENTIFYING GENES FOR RIBOSOMAL PROTEINS

In order to make a complete inventory of the rp genes on the genome of *S. cerevisiae*, several criteria may be used:

- The promoters of the various rp genes in yeast have a very characteristic architecture. The promoters of the rp genes from S. cerevisiae studied in detail so far contain either (usually two) binding sites for the global gene regulator Rap1p (the majority of the rp genes) or a single binding site for another factor, Abf1p. In a few cases, e.g. the *RPS2* gene (new nomenclature; see Appendix 1), the promoter carries a binding site for both Rap1p and Abf1p. In addition, downstream of these strong protein-binding sites, a characteristic T-rich region is present (two T-stretches of 6–10 nucleotides, interrupted by a short purine-rich sequence), which has been shown to play an important role in the transcriptional regulation of the rp genes (Gonçalves *et al.*, 1995).
- The codon adaptation index of the candidate rp genes should be around 0.6–0.9, but at least 0.4.
- The expression of the candidate rp genes should display a typical response to a nutritional shift-up (a three- to four-fold increase in transcription level by about 60 min after adding glucose to a yeast culture growing on a non-fermentable carbon source), and to a heat shock (a decline of the transcript levels to about 25% within 10 min after shifting a culture from 23°C to 36°C).
- In most cases the products of the candidate rp genes, identified by applying the criteria indicated above, show a clear similarity with ribosomal proteins from other species, in particular from the rat.

Appendix I contains the complete list of yeast ribosomal proteins and their genes. As can be inferred from these data, the small ribosomal subunit contains 32 and the large ribosomal subunit 48 different ribosomal proteins.

Acknowledgements

Work on yeast protein synthesis in the authors' laboratories has been supported by the BBSRC (M.F.T.), the Wellcome Trust (M.F.T., I.S.). and the European Community Framework IV (M.F.T., R.J.P.).

References

Belcourt, M. F. and Farabaugh, P. J. (1990). Ribosomal frameshifting in the yeast retrotransposon Ty: tRNAs induce slippage on a 7 nucleotide minimal site. *Cell* **62**, 339–352.

Belfield, G. P. and Tuite, M. F. (1993). Translation elongation factor 3: a fungus-specific translation factor? *Mol. Microbiol.* **9**, 411–418.

Bollen, G. H. P. M., Mager, W. H. and Planta, R. J. (1981). High resolution mini-two-dimensional gel electrophoresis of yeast ribosomal proteins. *Mol. Biol. Rep.* **8**, 37–44.

Boucherie, H., Sagliocco, F., Joubert, R., Maillet, I., Labarre, J. and Perrot, M. (1996). Two-dimensional gel protein database of *Saccharomyces cerevisiae*. *Electrophoresis* **17**, 1683–1699.

Branscomb, E. W. and Galas, D. J. (1975). Progressive decrease in protein synthesis accuracy induced by streptomycin in *Escherichia coli*. *Nature* **254**, 161–163.

Castilho-Valavicius, B., Yoon, H. and Donahue T. F. (1990). Genetic characterisation of the *Saccharomyces cerevisiae* translational initiation suppressors *sui1*, *sui2* and *SUI3* and their effects on *HIS4* expression. *Genetics* **124**, 483–495.

Coppolechia, R., Buser, P., Stotz, A. and Linder, P. (1993). A new yeast translation initiation factor suppresses a mutation in the eIF-4A RNA helicase. *EMBO J.* **12**, 4005–4011.

Cox, B. S. (1965). A cytoplasmic suppressor of super-suppressors in yeast. *Heredity* **20**, 505–521.

Cox, B. S. (1977) Allosuppressors in yeast. *Genet. Res.*, **30**, 187–205.

Dever, T. E., Feng, L., Wek, R. C., Cigan, A. M., Donahue, T. F. and Hinnebusch, A. G. (1992). Phosphorylation of initiation factor 2α by protein kinase GCN2 mediates gene-specific translation control of *GCN4* in yeast. *Cell* **68**, 585–596.

Douglas, M., Finkelstein, D. and Burow, R. A. (1979). Analysis of products of mitochondrial protein synthesis in yeast: genetic and biochemical aspects. *Methods Enzymol.* **56**, 58–66.

Eustice, D. C., Walkem, L. P., Wilhelm, J. M. and Sherman, F. (1986). Altered 40S ribosomal subunits in omnipotent suppressors of yeast. *J. Mol. Biol.* **188**, 207–214.

Everett, J. G., and Gallie, D. R. (1992). RNA delivery in *Saccharomyces cerevisiae* using electroporation. *Yeast* **8**, 1007–1014.

Feinberg, B., McLaughlin, C. S. and Moldave, K. (1982). Analysis of temperature-sensitive mutant ts187 of *Saccharomyces cerevisiae* altered in a component required for the initiation of protein synthesis. *J. Biol. Chem.* **257**, 10846–10851.

Finkelstein, D. B. and Strausberg, S. (1983). Heat shock regulated production of *E. coli* β-galactosidase in *Saccharomyces cerevisiae. Mol. Cell. Biol.* **3**, 1625–1633.

Firoozan, M., Grant, C. M., Duarte, J. and Tuite, M. F. (1991). Quantitation of readthrough of termination codons in yeast using a novel gene fusion assay. *Yeast* **7**, 173–183.

Fried, H. M. and Warner, J. R. (1981). Cloning of the yeast gene for trichodermin resistance and ribosomal protein L3. *Proc. Natl. Acad. Sci. USA* **78**, 238–242.

Fried, H. M. and Warner, J. R. (1982). Molecular cloning and analysis of the yeast gene for cycloheximide resistance and ribosomal protein L29. *Nucleic Acids Res.* **10**, 3133–3138.

Gerstel, B., Tuite, M. F. and McCarthy, J. E. G. (1992). The effects of 5'-capping, 3'-polyadenylation and leader composition upon the translation and stability of mRNA in a cell-free extract derived from the yeast *Saccharomyces cerevisiae. Mol. Microbiol.* **6**, 2339–2348.

Gonçalves, P. M., Griffioen, G., Minnee, R., Bosma, M., Kraakman, L. S., Mager, W. H. and Planta, R. J. (1995). Transcription activation of yeast ribosomal protein genes requires additional elements apart from binding sites for Abf1p and Rap1p. *Nucl. Acids Res.* **23**, 1475–1480.

Grant, C. M., Fitch, I. T. and Tuite, M. F. (1996). Electrophoretic analysis of yeast proteins. In *Methods in Molecular Biology, Vol. 53: Yeast Protocols* (I. Evans, ed.), pp. 259–267. Humana Press, New Jersey.

Hartley, A. D., Santos, M. A. S., Colthurst, D. R. and Tuite, M. F. (1996). Preparation and use of yeast cell-free translation lysates. In *Methods in Molecular Biology, Vol. 53: Yeast Protocols* (I. Evans, ed.), pp 249–257. Humana Press, New Jersey.

Hartwell, L. H. and McLaughlin, C. S. (1968). Temperature-sensitive mutants of yeast exhibiting a rapid inhibition of protein synthesis. *J. Bacteriol.* **96**, 1664–1671.

Helser, T. L., Baan, R. A. and Dahlberg, A. E. (1981). Characterisation of a 40S ribosomal subunit complex in polyribosomes of *Saccharomyces cerevisiae* treated with cycloheximide. *Mol. Cell. Biol.* **1**, 51–57.

Hershey, J. W. B., Matthews, M. B. and Sonerberg, N. (eds) (1996). *Translational Control.* Cold Spring Harbor Laboratory Press, New York.

Hinnebusch, A. G. (1996). Translation control of *GCN4*: gene-specific regulation by phosphorylation of eIF2. In *Translational Control* (J. W. B. Hersey, M. B. Matthews and N. Sonenberg, eds), pp 199–244. Cold Spring Harbor Laboratory Press, New York.

Hutchinson, J. S., Feinberg, B., Rothwell, T. C. and Moldave, K. (1984). Monoclonal antibody specific for yeast elongation factor 3. *Biochemistry* **23**, 3055–3060.

Kaltschmidt, E. and Wittmann, H. G. (1970). Ribosomal proteins. VII. Two-dimensional polyacrylamide gel electrophoresis for fingerprinting of ribosomal proteins. *Anal. Biochem.* **46**, 401–412.

Larkin, J. C. and Woolford, J. L. (1983). Molecular cloning and analysis of the *CRY1* gene: a yeast ribosomal protein gene. *Nucleic Acids Res.* **11**, 403–408.

Lee, S. I., Umen, J. G. and Varmus, H. E. (1995). A genetic screen identifies cellular factors involved in retroviral −1 frameshifting. *Proc. Natl. Acad. Sci. USA* **92**, 6587–6591.

Madjar, J.-J., Michel, S., Cozzone, A. J. and Reboud, J. P. (1979). A method to identify individual proteins in four different two-dimensional gel electrophoresis systems: application to *Escherichia coli* ribosomal proteins. *Anal. Biochem.* **92**, 174–182.

Martin, T. E. and Hartwell, L. H. (1970). Resistance of active yeast ribosomes to dissociation by KCl. *J. Biol. Chem.* **245**, 1504–1508.

Mets, L. J. and Bogorad, L. (1974). Two-dimensional polyacrylamide gel electrophoresis: an improved method for ribosomal proteins. *Anal. Biochem.* **57**, 200–210.

Michel, S., Traut, R. R. and Lee, J. C. (1983). Yeast ribosomal proteins: electrophoretic analysis in four two-dimensional gel systems. Correlation of nomenclatures. *Mol. Gen. Genet.* **191**, 251–256.

Naranda, T., Macmillan, S. E. and Hershey, J. W. B. (1994). Purified yeast translation initiation factor eIF3 is an RNA binding protein complex that contains the PRT1 protein. *J. Biol. Chem.* **269**, 32286–32292.

Petersen, N. S. and McLaughlin, C. S. (1974). Polysome metabolism in protein synthesis of yeast. *Mol. Gen. Genet.* **129**, 189–200.

Raué, H. A., Mager, W. H. and Planta, R. J. (1991). Structural and functional analysis of yeast ribosomal proteins. In *Methods in Enzymology* (C. Guthrie and G. R. Fink, eds), vol. 194, pp. 453–477. Academic Press, San Diego.

Ripmaster, T. L., Vaughn, G. P. and Woolford, J. C. (1992). A putative ATP-dependent RNA helicase involved in *Saccharomyces cerevisiae* ribosome assembly. *Proc. Natl. Acad. Sci. USA* **89**, 11131–11135.

Rotenberg, M. O., Moritz, M. and Woolford, J. L. (1988). Depletion of *Saccharomyces cerevisiae* ribosomal protein L16 causes a decrease in 60S ribosomal subunits and formation of half-mer polyribosomes. *Genes Devel.* **2**, 160–172.

Ruiz-Echevarria, M. J., Czaplinski, K. and Peltz, S. W. (1996). Making sense of nonsense in yeast. *Trends Biochem. Sci.* **21**, 433–437.

Sachs, A. B. and Davis, R. W. (1989). The polyA-binding protein is required for poly A shortening and 60S ribosomal subunit-dependent translation initiation. *Cell* **58**, 857–867.

Santos, M. A. S., Perreau, V. M. and Tuite, M. F. (1996). Transfer RNA structural change is a key element in the reassignment of the CUG codon in *Candida albicans*. *EMBO J.* **15**, 5060–5068.

Schindler, D., Grant, P. and Davies, J. (1974). Trichodermin resistance-mutation affecting eukaryotic ribosomes. *Nature* **248**, 535–538.

Stahl, G., Bidou, L., Rousset, J-P. and Cassan, M. (1995). Versatile vectors to study recoding: conservation of rules between yeast and mammalian cells. *Nucleic Acids Res.* **23**, 1557–1560.

Stansfield, I. and Tuite, M. F. (1994). Polypeptide chain termination in *Saccharomyces cerevisiae. Curr. Genet.* **25**, 385–395.

Stansfield, I., Grant, C. M., Akhmaloka and Tuite, M. F. (1992). Ribosomal association of the yeast *SAL4 (SUP45)* gene product: implications for its role in translation fidelity and termination. *Mol. Microbiol.* **6**, 3469–3478.

Stansfield, I., Akhmaloka and Tuite, M. F. (1995a). A mutant allele of the *SUP45 (SAL4)* gene of *Saccharomyces cerevisiae* shows temperature-dependent allosuppressor and omnipotent suppressor phenotypes. *Curr. Genet.* **27**, 417–426.

Stansfield, I., Jones, K. M., Kushnirov, V. V., Dagkesamanskaya, A. R., Poznyakovski, A. I., Paushkin, S. V., Nierras, C. R. *et al.* (1995b). The products of the *SUP45* (eRF1) and *SUP35* genes interact to mediate translation termination in *Saccharomyces cerevisiae. EMBO J.* **14**, 4365–4373.

Stansfield, I., Eurwilaichitr, L., Akhmaloka and Tuite, M. F. (1996). Depletion in the levels of the release factor eRF1 causes a reduction in the efficiency of translation termination in yeast. *Mol. Microbiol.* **20**, 1135–1143.

Stansfield, I., Kushnirov, V. V., Jones, K. M. and Tuite, M. F. (1997). A conditional-lethal translation termination defect in a *sup45* mutant of the yeast *Saccharomyces cerevisiae. Eur. J. Biochem.* **245**, 557–563.

Tuite, M. F. and Lindquist, S. L. (1996). The maintenance and inheritance of yeast prions. *Trends Genet.* **12**, 467–471.

Tuite, M. F. and McLaughlin, C. S. (1984). The effects of paromomycin on the fidelity of translation in a yeast cell-free system. *Biochim. Biophys. Acta* **783**, 166–170.

Tuite, M. F., Plesset, J., Moldave, K. and McLaughlin, C. S. (1980). Faithful and efficient translation of homologous and heterologous mRNA in a mRNA-dependent system from *Saccharomyces cerevisiae. J. Biol. Chem.* **255**, 8761–8766.

Tuite, M. F., Cox, B. S. and McLaughlin, C. S. (1981). A homologous *in vitro* assay for yeast nonsense suppressors. *J. Biol. Chem.* **256**, 7298–7304.

Tuite, M. F., Cox, B. S. and McLaughlin, C. S. (1983). *In vitro* nonsense suppression in [psi⁺] and [psi⁻] cell-free lysates of *Saccharomyces cerevisiae. Proc. Natl. Acad. Sci. USA* **80**, 2824–2828.

Wakem, L. P. and Sherman, F. (1990). Isolation and characterisation of omnipotent suppressors in the yeast *Saccharomyces cerevisiae. Genetics* **124**, 515–522.

20 MAP Kinase-Mediated Signal Transduction Pathways

María Molina, Humberto Martín, Miguel Sánchez and César Nombela

Departamento de Microbiología II, Facultad de Farmacia, Universidad Complutense, Madrid, Spain

◆◆

CONTENTS

List of Abbreviations

BCIP	5-Bromo-4-chloro-3-indolylphosphate
ERK	Extracellular signal-regulated kinase
GAP	GTPase-activating protein
HOG	High osmolarity glycerol
MAPK	Mitotic-activated protein kinase
PI	Propidium iodide
PNPP	p-Nitrophenylphosphate
SLAD	Synthetic low-ammonia dextrose
STRE	Stress response element
WWW	World Wide Web
YNB	Yeast nitrogen base
YPD	Yeast Protein Database/yeast extract peptone dextrose (*see Appendix II*)

◆◆◆◆◆◆ I. INTRODUCTION

All organisms with a cellular structure require sensory systems that allow them to monitor the changes occurring in their environment so that they can respond properly to external stimuli. Once the sensory system has been stimulated, a series of reactions is triggered that enables transduction of the signal from the cell surface to the nucleus, where certain transcription factors are responsible for activating the appropriate response.

METHODS IN MICROBIOLOGY, VOLUME 26
ISBN 0–12–521526–6

Organisms as distant on the phylogenetic scale as yeasts and mammalian cells not only share similar response capacities but in many cases also share a similar molecular basis of signal transduction.

In mammalian cells, numerous external signals initiate a mitogenic response mediated by a protein kinase cascade. The final components of these cascades have been designated ERKs (extracellular signal-regulated kinases) or MAPKs (mitotic activated protein kinases). These kinases form a group of enzymes involved in the activation of other proteins by phosphorylation of their serine and threonine residues. They themselves are regulated by phosphorylation on tyrosine and threonine residues.

In yeast, several MAPK-mediated pathways have been identified controlling different cellular processes (Figure 1). The Slt2/Mpk1-mediated pathway regulates the generation of a rigid cell wall responsible for cellular shape and for maintenance of cellular integrity. The Hog1 pathway is involved in osmoregulation and is essential for adaptation to hyperosmotic media. MAPK pathways have also been implicated in haploid–diploid transitions, both in sporulation and mating. In addition, certain haploid strains are able to grow invasively into the agar, and diploid cells try to spread around solid media during nutrient starvation by developing a polarized pseudohyphal form of growth. Some elements that act in the pheromone-response mating pathway are also involved in these differential growth patterns.

Because a detailed description of each of these pathways is beyond the scope of this chapter and because they are reviewed elsewhere (Herskowitz, 1995), we shall briefly outline the cell integrity pathway only, pointing out the most relevant similarities and peculiarities with the other pathways. The stimuli responsible for activating this pathway are not clearly understood, although responses to high temperature, exposure to pheromones and hypotonic shock seem to be involved in MAPK activation. Signal transduction requires the action of protein kinase C (Pkc1p), which is regulated by the small GTPase, Rho1. After Pkc1p activation, the signal is transmitted to a protein phosphorylation cascade composed of the MAPKKK (MAP kinase kinase kinase) Bck1p, the MAPKKs Mkk1p and Mkk2p, and the MAPK Slt2p/Mpk1p. Although an additional branch activated by Pkc1p has been proposed, the nature of its components remains obscure (Levin et al., 1994). The MAPK cascade is a conserved module among the different pathways. While in most of them a complete MAPK module has been identified, in the sporulation and the pseudohyphal/invasive pathways some components are still unknown. Downstream from these modules are usually transcription factors, whose activation is controlled by the MAPKs through phosphorylation. There is greater heterogeneity in the components that function upstream from these pathways. These components range from receptors with seven transmembrane domains coupled to a heterotrimeric G protein, to sensor molecules similar to the prokaryotic two-component system. Small G proteins of the Ras and Rho subfamilies and members of the Ste20p/PAK-like protein kinase family are also involved in regulating the MAPK cascade. The protein Ste5p, which is the only component that lacks obvious homology to proteins of known function, has been implicated in the

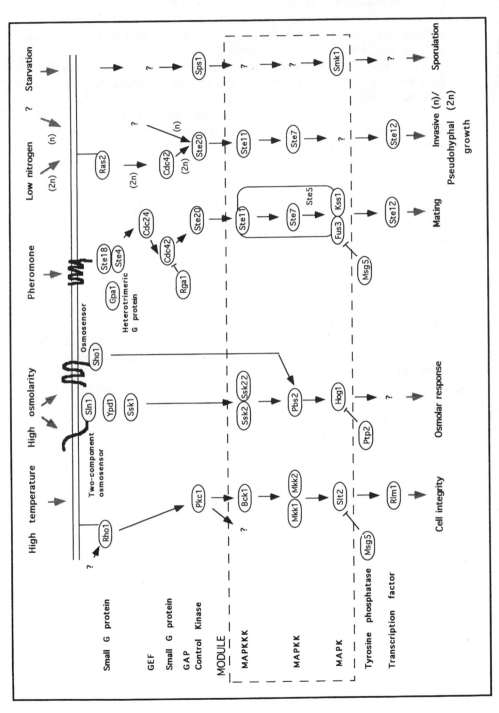

Figure 1. Components of the MAPK-mediated signal transduction pathways in *S. cerevisiae*. GEF, GTPase-exchange factor; GAP, GTPase-activating protein; MAPK, MAP kinase; MAPKK, MAPK kinase; MAPKKK, MAPKK kinase.

formation of a scaffold that holds together the components involved in the mating MAPK cascade. The formation of such complexes may prevent crosstalk between protein kinases from different signaling modules. However, it remains to be seen how the fidelity of the other pathways is maintained and to what extent a certain degree of promiscuity is tolerated in some pathways under some conditions.

In this chapter we focus on the approaches that can be used to investigate a possible role for a protein in yeast cell signaling, in particular in MAPK-mediated transduction pathways. Analysis of the protein sequence to identify any significant domains (for example, a protein kinase domain) could be considered the first step. This could indicate whether the protein of interest is a member of a known protein family or simply whether the protein might be associated with signal transduction. The development of genetic studies might be a second step in elucidating the role of the gene or protein under study. Over the past years, genetic analysis has been widely and successfully used to dissect the function of many genes in yeast. This analysis is extraordinarily useful in the case of cell signaling studies. The use of different strategies allows researchers to test not only the participation of a given gene in known pathways, but also the site of action of this gene in the pathway. The results of this analysis can be interpreted easily because simple standardized assays have been developed to detect alterations in transmission across the different pathways. Although almost every tool (genetic, molecular and biochemical) has been used in the dissection of these pathways, the biochemical approach is usually undertaken once structural and genetic studies have afforded an insight into the role of the protein. Therefore, in this chapter we discuss the different strategies that can be used to identify proteins involved in signaling pathways and, finally, the methods for assaying the processes regulated by the currently known pathways.

◆◆◆◆◆◆ II. STRATEGIES FOR IDENTIFYING PROTEINS INVOLVED IN SIGNALING PATHWAYS

A. Structural Analysis

Most signaling proteins possess characteristic modules or domains that are essential for their activity through their ability to mediate protein–protein, protein–DNA or protein–lipid interactions or to perform catalytic functions. The presence of such modules in a protein often indicates that it belongs to a general class or family of proteins, suggesting its involvement in specific functions. These modules include protein kinase catalytic domains, nucleotide binding domains or GAP (GTPase-activating protein) domains. These domains can be identified by scanning the amino acid sequence under study against PROSITE, a database of biologically significant protein sites, patterns and profiles (http://expasy.hcuge.ch/sprot/prosite.html). However, some other well-characterized motifs are exhibited by signaling proteins of different classes. Therefore, the presence of any of these modules would suggest a potential role in signal transduction for the protein of interest. Among these modules, the SH3 and PH domains are the most relevant ones. SH3 domains are modules of 50 to 100 amino acids in length found in many signaling proteins and cytoskeletal components. These domains bind proline-rich peptides of approximately 10 amino acids. PH domains are protein modules of around 100 amino acids that are also exhibited by several proteins involved in signal transduction. They have been proposed to bind βγ subunits of G proteins, and phospholipids. In addition, other motifs indicate a more specific interaction in signaling. This is the case for the CRIB (Cdc42/Rac-interactive-binding sequence) domain found in several protein kinases, which binds the Rho GTPases Cdc42 and Rac in different organisms.

The complete yeast genome sequence is now available. Therefore, an exhaustive structural analysis can be carried out to obtain information about potential protein function. Indeed, there are different World Wide Web (WWW) servers offering data regarding the yeast genome (see Chapter 3; Appendix III). For example, information about protein localization, function, molecular environment or genetic properties can be obtained from the YPD (Yeast Protein Database) (http://quest7.proteome.com/YPDhome.html) and MIPS (http://www.mips.biochem.mpg.de/mips/yeast/index.htmlx).

B. Genetic Analysis

As described in the Introduction, the signaling pathways are regulatory hierarchies that are activated by some sort of signal to elicit specific cellular responses. Mutations in their components can cause the pathway to fail to respond properly to the signal. Therefore, alterations in the normal amount or activity of the protein under study are the simplest way to address the elucidation of the role of the protein. Subsequent analysis of signaling in mutant cells would indicate the relevance of the protein in transduction processes.

I. Alteration in protein activity

Gene disruption to abolish the function of the gene product is the obvious first experiment. Currently, in addition to standard methods (Rothstein, 1991), gene disruptions can be achieved by faster and easier ways (see Chapter 5). Because signal transduction pathways modulate events such as mating and sporulation, haploid null mutants of both mating types (**a** and α) and homozygous and heterozygous mutant diploid strains should be obtained. In addition, because the genetic background can influence the phenotype of signaling mutants, the gene of interest should be disrupted in different backgrounds.

To identify a potential cell-signaling-related function, another possibility is to manipulate the gene of interest to increase the activity of the encoded product. The classical approach to enhancing protein activity is to overexpress the gene. A wide array of techniques is available for the overproduction of the protein at different levels (see Chapters 5 and 6). An additional approach involves the construction and expression of a constitutively activated dominant allele of the gene, leading to a hyper-activated mutant protein. Because modulation of the activity of signal transduction proteins is often brought about by posttranslational modifications, this latter strategy frequently proves to be more effective than the former. In fact, the expression of gain-of-function alleles of different genes has been the most powerful way to hyperactivate signaling pathways. Table 1 illustrates various mutations that increase the activity of the specific signaling proteins. Constructing analogous mutations in the homologous domains of our target protein would be expected to activate the protein.

The signaling assays described in section III can be used to study transduction processes in strains carrying mutations in the gene of interest. In addition, biochemical assays for protein phosphorylation (Hunter and Sefton, 1991) can be achieved to detect transmission defects. When a mutation causes abnormal signaling, epistasis experiments should be performed to study the relationship between the gene and its pathway, and its position within the pathway. A set of mutant strains affected in the various genes of the pathway, and plasmids carrying the wild-type genes are needed for these experiments. We will summarize the different possibilities that can arise when these genetic interactions are studied.

Table I. Representative hyperactivated signaling proteins

Protein family	Hyperactivated protein	Reference
Small G proteins	$Cdc42^{V12}$/$Cdc42^{L61}$/$Cdc42^{A118}$ (1)	Ziman *et al.* (1991)
Pkc1	$Pkc1^{P398}$	Nonaka *et al.* (1995)
	$Pkc1^{A398}$	Watanabe *et al.* (1994)
Ste20	$Ste20\Delta N$	Ramer and Davis (1993)
	$hPAK1^{F107}$ (2)	Brown *et al.* (1996)
Ssk1	$Ssk1^{N554}$	Maeda *et al.* (1994)
MAPKKK	$Ste11\Delta N$	Cairns *et al.* (1992)
	$Ste11^{S279}$/$Ste11^{I596}$	Stevenson *et al.* (1992)
	$Bck1^{P1174}$/Bck^{V1146}/Bck^{L1120}/Bck^{T1120}/ Bck^{P1119}	Lee and Levin (1992)
	$Ssk2\Delta N$	Maeda *et al.* (1995)
	$Ssk22\Delta N$	Maeda *et al.* (1995)
MAPKK	$Mkk1^{P386}$	Watanabe *et al.* (1995)
	$Ste7^{P368}$	Errede *et al.* (1995)
MAPK	$Fus3^{L161}$/$Fus3^{N48}$/$Fus3^{N227}$/$Fus3^{Y28}$/ $Fus3^{K9}$/$Fus3^{H7}$/$Fus3^{P363}$	Hall *et al.* (1996); Brill *et al.* (1994)
	$Rolled^{N334}$ (3)	Brunner *et al.* (1994)

(1) Substitutions of homologous residues in other Rho- and Ras-like GTPases lead to similar dominant activated proteins.
(2) hPAK1 is a human Ste20 homolog.
(3) Rolled is a *Drosophila melanogaster* MAP kinase.

2. Epistasis analysis

The simplest case would occur when null mutants display similar phenotypic traits as cells defective for a known signaling pathway. This would indicate that the gene of interest might be involved in the transduction of the signal. Then, genetic epistasis might clearly uncover the functional position of the gene in the pathway. A set of double mutants must be constructed, each of which carries disruptions in the gene of interest and in one of the known genes on the pathway. A similar phenotype for the single and double mutants would suggest that the gene of interest functions within the pathway. In contrast, if the double mutants display more complex phenotypes than the single mutants, this would indicate that the gene of interest does not act in the same linear pathway. One should also test whether overexpression of the wild-type gene or an activated allele of the gene is able to suppress the phenotypic defects of cells lacking individual components of the pathway. Reciprocally, one should test whether overexpression or hyperactivation of other genes on the pathway can suppress the phenotypic effects of disrupting the gene of interest. If gene A acts upstream of B, overexpression of B might suppress inactivation of A. In contrast, overexpression of A would fail to suppress inactivation of B. Several downstream elements in different MAPK-mediated pathways have been identified by screening for genes that, when overexpressed, are able to suppress the defects of mutations in an earlier component in the pathway. In other cases, hyperactivation of the pathway leads to a detectable phenotype. In this case, analysing the effects of inactivating the gene of interest in strains carrying hyperactived alleles of other genes can help to define its position in the pathway. In a linear pathway, inactivation of a downstream element will eliminate the phenotype caused by hyperactivation of an upstream element.

What are we dealing with if loss of the gene function leads to increased signaling, or gene activation produces the same phenotype as a block in the transduction pathway? Both of these possibilities could occur if the gene of interest encodes a negative regulator of the pathway. For example, the *SLN1* gene encodes a protein kinase sensor which negatively controls the osmoregulatory Hog1 pathway (Figure 1). Therefore, disruption of *SLN1* leads to a constitutive activation of this pathway. A search for mutations that suppress the phenotype of an *sln1* mutation yielded recessive mutations in different downstream elements of the pathway (Maeda *et al.*, 1994). This example indicates the types of epistasis experiments that could be carried out. Inactivation of downstream elements would suppress the phenotype produced by disrupting a negative regulator. In contrast, constitutive activation of these downstream elements would suppress the negative consequences of hyperactivating a negative regulator. Other negative regulators that specifically modulate the activity of a component on the pathway can be identified through their ability to suppress the phenotypic effects of constitutively activating an upstream component in the pathway. For example, overexpression of the tyrosine phosphatase gene *MSG5* suppresses a constitutively activated mating pathway, whereas loss of *MSG5* function causes an increase in Fus3p activity (Doi *et al.*, 1994).

Another possible result is the absence of an apparent phenotype or the lack of altered signal transduction in null mutants. This does not exclude the possibility that the protein of interest is involved in signaling processes. This could be a consequence of the existence of another gene(s) that performs a similar overlapping function, i.e. functional redundancy. For example, Irie *et al.* (1993) isolated *MKK1* and *MKK2* as homologs functioning at the same level in the pathway regulated by *PKC1*. Deletion of either *MKK* gene alone does not cause any apparent phenotypic effect, but deletion of both *MKK1* and *MKK2* results in the defects associated with a block in the pathway. Redundancy analysis can be performed to uncover the presence of proteins with overlapping functions. One approach is to screen for genes whose inactivation produces a synthetic phenotypic effect (Chapter 16). Alternatively, the availability of the entire *S. cerevisiae* genome sequence can be exploited to search for homologous genes or proteins. Double or multiple mutants of structural homologs can be generated and their phenotypes compared with those displayed by single mutants. Again, epistasis experiments should allow the functional position of the genes in the pathway to be determined (see above). The need to use the double mutant, in which both redundant genes are defective, is the only caution to be exercised for such studies. For example, overexpression of *SLT2/MPK1* suppresses the *mkk1 mkk2* double deletion. However, overexpression of *MKK1* or *MKK2* fails to suppress the *slt2/mpk1* deletion mutant, whereas overexpression of *MKK1* can suppress the *bck1* inactivation. These results suggest that *MKK1* and *MKK2* act upstream of *SLT2/MPK1* and downstream of *BCK1* in the *PKC1*-regulated pathway (Figure 1). In this way the point of action of a gene in a linear pathway can be determined.

The problem becomes more complicated if complex pathways are considered. Signaling pathways are often highly adaptable, with branches feeding in and out at different levels. Two basic models can be considered: a model in which the pathway bifurcates from a point, and an integration model, in which several inputs are integrated at a point. To see how the position of a protein within these transduction pathways can be defined, two examples will be explored:

(a) The lytic phenotype of *pkc1* mutants was used to screen for downstream components on the pathway. *BCK1* was identified because, when this gene was activated mutationally, the cell lysis defect was suppressed. However, the inactivation of *BCK1* or other downstream components results in temperature-dependent cell lysis whereas *pkc1* mutants display an unconditional cell lysis defect. For this reason, it has been proposed that the pathway bifurcates after *PKC1*. In general, differences in the severity of the phenotype caused by mutations in genes functioning on the same pathway are attributed to the involvement of different branches in the pathway.

(b) Maeda *et al.* (1995) showed that the signal generated by the Sln1p–Ssk1p two-component osmosensor is transduced to a MAPK cascade composed of Ssk2p/Ssk22p, Pbs2p and Hog1p (see Figure 1). However, while signal transduction down this pathway was completely blocked in *pbs2* mutants, Hog1p was still activated in *ssk2 ssk22* double mutants in response to high external osmolarity. Therefore, an alternative mechanism to activate Pbs2p in a Ssk2p/Ssk22p-independent manner should exist. A screen based on this assumption yielded *SHO1*. A combination of the three mutations (*ssk2*, *ssk22* and *sho1*) on two pathways completely abolishes the activation of Hog1p. This is a typical example of signal integration. In general, if there is signal integration at one point, inactivation of elements downstream of this point block hyperactivation by mutations in upstream components in any branch, but inactivation of some of these upstream components will not completely block the pathway. The phenotype of a strain that combines inactivating mutations in all of the branches that feed into the common pathway must be the same as the phenotype caused by complete inactivation of the common pathway.

◆◆◆◆◆◆ III. METHODS FOR MONITORING MAPK-MEDIATED SIGNALING

A. Cell Integrity Pathway

Alterations in the signaling pathway that control cell integrity lead to defects in cell wall assembly or maintenance, resulting in cell lysis. As shown in Figure 1, this pathway is thought to be bifurcated with a MAPK cascade acting on one of the branches. While inactivation of the initial common components of the pathway results in cell lysis at any temperature, loss of function of any of the components of the MAPK branch results in cell lysis only at high temperatures or in the presence of low concentrations of caffeine (3–12 mM, depending on the genetic background). In all cases, the lytic defect is characterized by increased cell permeability and release of intracellular contents to the external medium. Therefore, cell lysis can be detected either by assaying for an intracellular marker on the outside of the cells (e.g. alkaline phosphatase) or by the uptake of a marker molecule that can only enter cells when the plasma membrane has lost its selective permeability (e.g. propidium iodide). Because the cellular defect is confined to the cell wall structure and does not affect the plasma membrane directly, the lytic phenotype can be remedied by providing osmotic support to the growth medium (e.g. 1 M sorbitol). For this reason, phenotypic assays for cell lysis should be carried out both in the absence and presence of an osmotic stabilizer.

1. Assay for alkaline phosphatase release

(a) Qualitative test on solid medium

Qualitative assessment of cell lysis can be achieved in cells grown on solid medium by detecting the activity of the intracellular enzyme alkaline phosphatase, without previous permeabilization, using a chromogenic substrate. PNPP (*p*-nitrophenylphosphate) hydrolysis can be detected by the traditional overlay assay described by Cabib and Duran (1975). However, we have developed a simpler method that incorporates the substrate BCIP (5-bromo-4-chloro-3-indolylphosphate) into YPD solid medium (Appendix II, Table 1; Sherman, 1991) at a concentration of 40 µg ml^{-1} (YPD-BCIP). To analyze cell lysis, strains are patched onto YPD-BCIP plates and YPD-BCIP plates containing 1 M sorbitol (Martín *et al.*, 1993). Alternatively, 10 µl of a 10^7 cells ml^{-1} suspension of each strain can be spotted onto the appropriate plates in order to achieve similar inocula (10^5 cells per spot). Duplicate plates are incubated for 2 days at 24ºC and 37ºC, respectively. When cell lysis occurs, the patches progressively turn blue, whereas they remain white if the cells do not lyse. The intensity of the blue color correlates with the level of cell lysis.

Because of differences in genetic background, some thermosensitive lytic strains are unable to proliferate when incubated directly at 37ºC. In this case, patches or spots grown at 24ºC are replica-plated onto the appropriate plates and incubated at 37ºC.

(b) Quantitative test on liquid medium

To quantify the release of alkaline phosphatase to the external medium, PNPP hydrolysis is measured in supernatants from liquid cultures by spectrophotometric determination at 420 nm (OD_{420}) (Protocol 1). The PNPP-hydrolyzing activity of the culture medium can be normalized to the OD_{600} of the culture and to the assay time in minutes (t).

Relative alkaline phosphatase activity (arbitrary units) = $OD_{420} \times 1000 / OD_{600} \times t \times (V_s / V_t)$

where V_s = sample volume and V_t = total volume.

2. Propidium iodide staining and flow cytometry

The fluorescent marker propidium iodide (PI) allows quantification of cell lysis by flow cytometry (Protocol 2). Intact yeast cells are not permeable to PI and thus give only background staining. However, in permeable cells, the nucleic acids are accessible to the dye and the resulting fluorescent

---

Protocol 1. Qualitative assay for alkaline phosphatase on liquid cultures

1. Inoculate the strain to be tested into YPD liquid medium. Incubate overnight at 24ºC, shaking at 200 r.p.m.
2. Add an appropriate volume to YPD alone and YPD plus 1 M sorbitol liquid medium to give an OD_{600} of 0.1–0.2. Incubate at 37ºC, shaking at 200 r.p.m. for 48 h.
3. Harvest cells by low-speed centrifugation. The resulting supernatants are collected to be used as samples for the enzymatic assay.
4. 0.5 ml of 20 mM Na$_2$PNPP (dissolved in 0.1 M glycine/NaOH buffer at pH 9.8) is added to 0.5 ml of sample. The reaction mix is incubated at 37ºC until a yellow color appears, and then the OD_{420} is measured.

> **Protocol 2.** Flow cytometric determination of cell lysis
>
> 1. Inoculate the strain to be tested into YPD liquid medium. Incubate overnight at 24°C, shaking at 200 r.p.m.
> 2. Inoculate an appropriate volume of overnight culture into YPD alone and YPD plus 1 M sorbitol liquid medium to give an OD_{600} of 0.1–0.2. Incubate at 37°C, shaking at 200 r.p.m. for 4–8 h.
> 3. Harvest 100 µl of each culture in a microcentrifuge tube (approximately $1–2 \times 10^6$ cells). Carefully discard supernatant.
> 4. Resuspend pellet in 100 µl of PI solution (0.005% in phosphate-buffered saline) to be analyzed in a flow cytometer. If required, stained samples may be left for several hours at 4°C in the dark before flow cytometry analysis.

signal can be observed by fluorescence microscopy or analyzed by flow cytometry. It should be stressed that the determination of cell viability by colony-forming units after cell sorting clearly demonstrates that only PI-negative cells are viable. Also, comparative observations under fluorescence and phase-contrast microscopy show that loss of birefringence, a criterion traditionally used for cell lysis detection, is not accurate because uptake of fluorescent dye occurs in cells that are still birefringent (de la Fuente *et al.*, 1992). In conclusion, flow cytometric determination of fluorescence is not only a simple, rapid, accurate and reliable method, but also permits quantitative analysis of individual cells in a given population, which is not possible by the enzymatic assay described above. Flow cytometric results are expressed as percentage of lysed cells (fluorescent cells). Quantitation is essential for the measurement of lysis kinetics and to establish the degree of heterogeneity in the expression of a lytic mutation.

3. Assay for caffeine sensitivity

To assay caffeine sensitivity, strains are patched or spotted onto YPD plates containing different concentrations of caffeine (ranging from 3 to 12 mM), incubated for 3 days at 24°C, and then scored for growth. Caffeine sensitivity is expressed as the lowest caffeine concentration that inhibits cell growth. Strains unable to grow in less than 12 mM caffeine are considered sensitive, and are then assayed for cell lysis to confirm that the inhibitory effect of this compound can be attributed to a failure to maintain cell integrity.

Cell lysis can be assessed qualitatively as described above by replica-plating patches or spots grown in solid YPD at 24°C onto solid YPD-BCIP containing caffeine at the inhibitory concentration and onto the same medium osmotically stabilized with 1 M sorbitol, and incubating at 24°C.

Quantitative analysis of cell lysis can be carried out by the methods described in Protocols 1 and 2, except that cells pregrown at 24°C should be transferred to liquid YPD and YPD containing 1 M sorbitol supplemented with caffeine at the inhibitory concentration.

B. Mating Pathway

The mating process is initiated by extracellular pheromones which activate a signaling pathway that leads to a variety of physiological changes which prepare cells for mating. This cellular response includes the transcriptional induction of genes required for cell and nuclear fusion (e.g. *FUS1*), the arrest of the cell cycle in the G1 phase, and the morphological change to pear-shaped cells (shmoos). Because a functional pheromone response pathway is required for mating, it is possible to assay for the ability of a strain to mate (mating assays), to produce pheromones (halo assay for pheromone production), and to respond to pheromones (halo assay for pheromone sensitivity; *FUS1* transcriptional induction; microscopic examination of shmoo formation). In this way it has been possible to identify genes involved in this signaling process. The methods currently used to carry out the above-mentioned assays have been thoroughly described by Sprague (1991) and hence they are not described in this chapter. However, some modifications of the assay for *FUS1* induction are discussed.

I. Assay for FUSI induction

This assay has proved to be very useful for quantifying the activation of the mating pathway. The assay can be performed either by Northern analysis to measure *FUS1* transcript levels or by measuring β-galactosidase levels in strains harboring a plasmid-borne *FUS1–lacZ* gene fusion (Trueheart *et al.*, 1987) before and after pheromone treatment (Sprague, 1991). The latter method is faster and less expensive, but requires cells to be grown on selective medium for maintenance of the plasmid. The use of strains harboring the *FUS1–lacZ* construct in a centromeric plasmid (Cairns *et al.*, 1992) or integrated in a chromosome (for example at the *URA3* locus; Rhodes *et al.*, 1990) reduces plasmid loss in rich media and increases reproducibility by limiting copy-number variation. A multicopy plasmid is not needed for *FUS1* because its expression is increased several hundred-fold by mating factors.

An alternative hybrid reporter gene (*FUS1::HIS3*) can be used to detect a response to pheromones (Stevenson *et al.*, 1992). Induced cells are selected by screening for their capacity to grow in the absence of histidine. Moreover, a quantitative measure of the expression level can be achieved by determining the level of resistance to aminotriazole, a competitive inhibitor of the HIS3 enzyme. Strains carrying both reporter gene fusions (*FUS1–lacZ* and *FUS1::HIS3*) have also been constructed (Boone *et al.*, 1993).

C. Sporulation Pathway

This is a developmentally regulated transduction pathway required for the completion of the sporulation program (Figure 1). Cells defective in the two known members of the pathway (*SPS1* and *SMK1*) are blocked after meiosis but before spore wall assembly. Consistent with this defect, abnormal asci are obtained after transferring homozygous diploid mutants to sporulation medium. The defective asci do not exhibit natural fluorescence, ether resistance or birefringence under phase-contrast microscopy characteristic of mature spore walls. They also display enhanced sensitivity to heat shock and cell wall lytic enzymes.

Procedures for sporulation, for detecting the natural fluorescence of dityrosine in spore walls and for spore viability (ether test) are extensively described in various chapters of the book by Guthrie and Fink (1991).

Assays for viability after heat shock and after treatment with glusulase can be performed as described by Briza et al. (1990).

D. Pseudohyphal Growth Pathway

The MAPK signal transduction pathway controlling filamentous growth is activated in response to nitrogen starvation and is specific for diploid cells (Figure 1). Diploids with a Σ1278 background undergo the most extensive, uniform and easily controlled transition to pseudohyphal growth in low-ammonia solid medium (Gimeno et al., 1992). The morphological transition results from a reiterated pattern of unipolar cell division to form the chain of elongated cells that constitutes the pseudohypha. Consequently, there is a transition from non-polarized colonial growth to polarized colonies, with the pseudohypha growing away from the center of the colony in all directions, resembling the colonies formed by filamentous fungi. Pseudohyphal cells also penetrate the surface of the agar plate and grow invasively into the medium.

I. Assay for filamentous growth

First, diploid strains in a Σ1278 background must be obtained. These strains are streaked to obtain single cells on synthetic low-ammonia dextrose (SLAD) medium (Gimeno and Fink, 1994) and incubated for 3 days at 30ºC in order to examine the characteristic rough and filamentous morphology of microcolonies using an inverted microscope. Cell shape can be analysed by phase-contrast or Nomarski optics after removing cells from the agar surface (noninvasive) or by scraping them off the plate to reveal invasive growth.

2. Assay for FG(TyA) induction

Like the use of the *FUS1* induction to assay activation of the mating pathway, a *lacZ* fusion derived from transposon Ty1 can be used as a transcriptional reporter to assay the activity of the filamentous growth pathway (Mösch et al., 1996). Conditions that stimulate pseudohyphal development also induce the *FG(TyA)::lacZ* reporter in an *STE12*-dependent manner. This induction can be examined using a strain harboring a plasmid-borne *FG(TyA)::lacZ* gene fusion to compare β-galactosidase levels after growth in rich medium (control culture) and under nitrogen starvation conditions (Protocol 3). The previous comments about *FUS1*

Protocol 3. Assay for *FG(TyA)* induction

1. Grow the *FG(TyA)::lacZ*-bearing strain overnight in the appropriate medium (selective medium for strains carrying plasmids or complete medium for strains carrying an integrative construct).
2. Harvest cells in exponential growth phase by centrifugation and wash them with 2% glucose.
3. Spread on an SLAD plate and on a YPD plate (control culture). Incubate at 30°C for 3 days.
4. Harvest cells from the plates by resuspension in water. Wash once with water.
5. Resuspend cells in an eppendorf tube using Z-buffer. Add acid-washed glass beads (0.45 mm diameter) to the height of the meniscus of the cell suspension.
6. Break cells by vortexing at high speed for 30 s and cool on ice for 30 s. Repeat vortex and cool for four to six times. Check the amount of cell breakage by visual inspection under the microscope.
7. Make a hole in the bottom of the tube; place this tube into an empty eppendorf tube and collect the lysate by centrifugation at 2000 r.p.m. for 2 min.
8. Collect the supernatant after a 15-min top-speed centrifugation as the soluble cell extract and assay for β-galactosidase activity using *o*-nitrophenyl-β-D-galactoside as substrate (Guarente, 1983).

Media and solutions
- Z-buffer: per liter, 16.1 g of $Na_2HPO_4 . 7H_2O$, 5.5 g of $NaH_2PO_4 . H_2O$, 0.75 g of KCl, 0.246 g of $MgSO_4 . 7H_2O$, and 2.7 ml of 2-mercaptoethanol; pH adjusted to 7.0
- SLAD medium: 0.67% of yeast nitrogen base (YNB) without amino acids and ammonium sulfate (Difco laboratories), 0.05 mM ammonium sulfate as sole nitrogen source, 2% of anhydrous D-glucose, and 2% of washed agar. The nitrogen source and the concentrated 4× stock solution of the YNB are filter-sterilized. Other components are autoclaved as separated stock solutions (40% glucose and 4% agar; agar is washed a few times with water before autoclaving). Stock solutions and sterile water are mixed to make a 2 × solution to which an equal volume of molten 4% agar is added. Petri dishes are filled with 25 ml of medium. These plates yielded uniform results during 2–3 weeks following preparation (Gimeno and Fink, 1994)

induction concerning the use of centromeric or integrative plasmids can also be taken into account for this assay (section III.B.1).

E. Invasive Growth Pathway

Haploid cells of both mating types of the Σ1278 background can exhibit invasive growth behavior with many similarities to pseudohyphal development, including filament formation and agar penetration, and dependence on the same components of the MAPK pathway (Figure 1; Roberts and Fink, 1994). However, the haploid agar penetration does not occur on low-ammonia medium but on rich medium. Moreover, diploid filaments extend beyond the colony, whereas haploid filaments (which are also less elongated) extend only beneath the colony.

I. Assay for invasive growth

As with the assay for diploid filamentous growth (section III.D.1), haploid strains with a Σ1278 background must be used to obtain an extensive manifestation of this developmental pathway. Strains are patched carefully onto YPD plates to avoid scratching of the agar surface, incubated at 30°C for 3 days, and then at room temperature for an additional 2 days. A gentle stream of deionized water is then used to rinse all the cells from the agar surface in order to observe the existence of cells below the surface to confirm invasiveness. These invasive cells can be examined microscopically by removing a piece of agar from washed plates (the cells that had invaded the agar remained as visible patches inside the agar after washing) with a toothpick and crushing the agar piece between a coverslip and microscope slide.

F. Osmosensing Pathway

This MAPK pathway, also named the HOG (high osmolarity glycerol response) pathway, plays a central role in mediating cellular responses to an increase in the osmolarity of the extracellular environment (Figure 1). Increasing the external osmolarity of the medium by addition of NaCl or sorbitol induces yeast to accumulate glycerol and thereby restore the osmotic gradient across the cell membrane. This response, together with the induction of expression of other genes involved in protection against osmotic stress (e.g. *CTT1*), is blocked in mutants lacking components of the osmosensing signaling pathway, which are hypersensitive to high osmolarity.

I. Assay of osmotic sensitivity

To analyse osmotic sensitivity, the strains to be tested are patched onto the surface of solid YPD plates and YPD plates supplemented with 0.9 M NaCl or 1.5 M sorbitol. Alternatively, 10 μl of a cell suspension of each strain may be spotted onto the appropriate plates, in order to achieve similar inocula (10^5 cells per spot). Plates are incubated at 30°C for 48–72 h. Strains that grow well on YPD but not on high-osmolarity medium are considered sensitive to osmotic stress and are then assayed for high-osmolarity-induced intracellular glycerol accumulation.

2. Analysis of intracellular glycerol

The intracellular glycerol level can be used as an indicator of the activity of the signaling cascade. Glycerol concentration can be easily determined enzymatically in cell extracts with a commercial glycerol determination kit (Boehringer Mannheim) (Protocol 4).

Protocol 4. Intracellular glycerol analysis under osmotic stress

1. Inoculate the strain to be tested into YPD liquid medium. Incubate overnight at 30°C, shaking at 200 r.p.m.
2. Dilute cells from overnight culture in YPD to $OD_{600} = 0.1$.
3. Divide exponentially growing culture at $OD_{600} = 0.5$ into two aliquots. Add a volume of concentrated salt solution to one of them to give a 0.4 M NaCl final concentration, and add the same volume of YPD to the other aliquot (control culture). Incubate at 30°C, shaking at 200 r.p.m. for 1 h.
4. Harvest cells by low-speed centrifugation. Resuspend cells in water and treat cell suspension at 95°C for 10 min. Centrifuge for 1 min at 10 000g and use the supernatant for the intracellular glycerol assay (Blomberg *et al.*, 1988).

---

3. Assay for STRE-driven gene expression

The STRE DNA element has been named "stress response element" because it mediates the activation of transcription in response to various types of stress, including an increase in osmolarity, and plays an important role in the induction of stress resistance. The osmosensing MAPK pathway regulates the osmotic induction of *CTT1* (catalase T encoding gene) and other genes via STRE. Therefore, strains bearing an STRE–*LEU2*–*lacZ* reporter gene construct (Marchler *et al.*, 1993) can be used to assay HOG-pathway-dependent STRE-induced expression. It is important to keep in mind that STRE sequences are activated by a broad variety of stress signals in a HOG-independent manner, but the induction controlled by the HOG pathway is limited to osmotic stress. Therefore, epistatic analysis should be carried out to confirm activation by the HOG pathway.

The procedure is similar to that of *FUS1::lacZ* induction (Sprague, 1991) and *FG(TyA)::lacZ* induction (Protocol 3), except that cells from the overnight cell culture are exposed to the osmotic stress as described in Protocol 4 (steps 2 and 3).

◆◆◆◆◆◆ IV. CONCLUDING REMARKS AND FUTURE PROSPECTS

The genetic approaches based on the strategies and methods compiled in this chapter are the most straightforward ways to test the role of a protein in a MAPK-mediated signaling pathway. Further analysis is usually needed in order to confirm its interacting partners. Different methods to identify protein–protein interactions *in vivo* and *in vitro* and to assess the strengths of these interactions have emerged in recent years (Chapter 15). Additionally, the activity of a given enzyme should be assayed *in vitro* to characterize accurately its catalytic function. The techniques to perform biochemical analysis with typical signaling proteins have been described in several manuals (e.g. Milligan, 1992; Hunter and Sefton, 1991), including methods for expression, purification, and assay of different protein activities.

In future years, genetic and phenotypical studies will be used widely to analyze the functional role of a great array of genes that have arisen from the yeast sequencing project (see Chapter 1). The continuing productive genetic analysis will surely be extremely useful to fill both unanticipated and expected gaps in the cell signaling machinery. The use of promoter–reporter fusions has been proved to be the most convenient of the assays to monitor signaling via individual pathways. Therefore, the development of new promoter–reporter fusions would improve signal transduction studies, especially in the cell integrity, sporulation and invasive pathways for which no reporter systems have been used yet. These studies might also benefit from the use of other reporter genes. For example, the green fluorescent protein would serve to quantify expression levels by flow cytometry in intact living yeast and to follow the kinetics of induction *in vivo*. Because signal transduction mechanisms are often conserved among different organisms, the analysis of signaling in *S. cerevisiae* strains expressing heterologous proteins is expected to be used more frequently to provide clues about the functional role of these proteins in signaling processes of their respective organisms.

Acknowledgements

This study was supported by grants BIO95-0303 from CICYT (Ministerio de Educación y Ciencia, Spain) and 95/0073-01 from FIS, Spain.

References

Blomberg, A., Larsson, C. and Gustafsson, L. (1988). Microcalorimetric monitoring of growth of *Saccharomyces cerevisiae*: osmotolerance in relation to physiological state. *J. Bacteriol.* **170**, 4562–4568.

Boone, C., Davis, N. G. and Sprague, Jr G. F. (1993). Mutations that alter the third cytoplasmic loop of the a-factor receptor lead to a constitutive and hypersensitive phenotype. *Proc. Natl. Acad. Sci. USA* **90**, 9921–9925.

Brill, J. A., Elion, E. A. and Fink, G. R. (1994). A role for autophosphorylation revealed by activated alleles of *FUS3*, the yeast MAP kinase homolog. *Mol. Biol. Cell* **5**, 297–312.

Briza, P., Breitenbach, M., Ellinger, A. and Segall, J. (1990). Isolation of two developmentally regulated genes involved in spore wall maturation in *Saccharomyces cerevisiae*. *Genes Dev.* **4**, 1775–1789.

Brown, J. L., Stowers, L., Baer, M., Trejo, J., Coughlin, S. and Chant, J. (1996). Human Ste20 homologue hPAK1 links GTPases to the JNK MAP kinase pathway. *Curr. Biol.* **6**, 598–605.

Brunner, D., Oellers, N., Szabad, J., Biggs III, W. H., Zipursky, S. L. and Hafen, E. (1994). A gain-of-function mutation in Drosophila MAP kinase activates multiple receptor tyrosine kinase signaling pathways. *Cell* **76**, 875–888.

Cabib, E. and Duran, A. (1975). Simple and sensitive procedure for screening yeast mutants that lyse at non-permissive temperatures. *J. Bacteriol.* **124**, 1604–1606.

Cairns, B. R., Ramer, S. W. and Kornberg, R. D. (1992). Order of action of components in the yeast pheromone response pathway revealed with a dominant allele of the STE11 kinase and the multiple phosphorylation of the Ste7 kinase. *Genes Dev.* **6**, 1305–1318.

de la Fuente, J. M., Alvarez, A. M., Nombela, C. and Sánchez, M. (1992). Flow cytometry analysis of *Saccharomyces cerevisiae* autolytic mutants and protoplasts. *Yeast* **8**, 39–45.

Doi, K., Gartner, A., Ammerer, G., Errede, B., Shinkawa, H., Sugimoto, K. and Matsumoto, K. (1994). MSG5, a novel protein phosphatase promotes adaptation to pheromone response in *S. cerevisiae. EMBO J.* **13**, 61–70.

Errede, B., Cade, R. M., Yashar, B. M., Kamada, Y., Levin, D. E., Irie, K., Matsumoto, K. (1995). Dynamics and organization of MAP kinase signal pathways. *Mol. Reprod. Dev.* **42**, 477–485.

Gimeno, C. J. and Fink, G. R. (1994). Induction of pseudohyphal growth by overexpression of *PHD1*, a *Saccharomyces cerevisiae* gene related to transcriptional regulators of fungal development. *Mol. Cell. Biol.* **14**, 2100–2112.

Gimeno, C. J., Ljungdahl, P. O., Styles, C. A. and Fink, G. R. (1992). Unipolar cell divisions in the yeast *S. cerevisiae* lead to filamentous growth: regulation by starvation and RAS. *Cell* **68**, 1077–1090.

Guarente, L. (1983). Yeast promoters and *lacZ* fusions designed to study expression of cloned genes in yeast. In *Methods in Enzymology, vol. 101* (M. Rose and D. Botstein, eds), pp. 181–191. Academic Press, London.

Guthrie, C. and Fink, G. R. (1991). *Methods in Enzymology: Guide to Yeast Genetics and Molecular Biology*, vol. 194. Academic Press, London.

Hall, J. P., Cherkasova, V., Elion, E., Gustin, M. C. and Winter, E. (1996). The osmoregulatory pathway represses mating pathway activity in *Saccharomyces cerevisiae*: isolation of a *FUS3* mutant that is insensitive to the repression mechanism. *Mol. Cell. Biol.* **16**, 6715–6723.

Herskowitz, I. (1995). MAP kinase pathways in yeast: for mating and more. *Cell* **80**, 187–197.

Hunter, T. and Sefton, B. M. (1991). *Methods in Enzymology: Protein Phosphorylation*, vol. 200. Academic Press, London.

Irie, K., Takase, M., Lee, K. S., Levin, D. E., Araki, H., Matsumoto, K. and Oshima, Y. (1993). *MKK1* and *MKK2*, which encode *Saccharomyces cerevisiae* mitogen-activated protein kinase–kinase homologs, function in the pathway mediated by protein kinase C. *Mol. Cell. Biol.* **13**, 3076–3083.

Lee, K. S. and Levin, D. E. (1992). Dominant mutations in a gene encoding a putative protein kinase (*BCK1*) bypass the requirement for a *Saccharomyces cerevisiae* protein kinase C homolog. *Mol. Cell. Biol.* **12**, 172–182.

Levin, D. E., Bowers, B., Chen, C.-Y., Kamada, Y. and Watanabe, M. (1994). Dissecting the protein kinase C/MAP kinase signalling pathway of *Saccharomyces cerevisiae. Cell. Mol. Biol. Res.* **40**, 229–239.

Maeda, T., Wurgler-Murphy, S. M. and Saito, H. (1994). A two-component system that regulates an osmosensing MAP kinase cascade in yeast. *Nature* **369**, 242–245.

Maeda, T., Takekawa, M. and Saito, H. (1995). Activation of yeast PBS2 MAPKK by MAPKKKs or by binding of an SH3-containing osmosensor. *Science* **269**, 554–558.

Marchler, G., Schuller, C., Adam, G. and Ruis, H. (1993). A *Saccharomyces cerevisiae* UAS element controlled by protein kinase A activates transcription in response to a variety of stress conditions. *EMBO J.* **12**, 1997–2003.

Martín, H., Arroyo, J., Sánchez, M., Molina, M. and Nombela, C. (1993). Activity of the yeast MAP kinase homologue Slt2 is critically required for cell integrity at 37ºC. *Mol. Gen. Genet.* **241**, 177–184.

Milligan, G. (1992). *Signal Transduction*. IRL Press, Oxford.

Mösch, H. V., Roberts, R. Y. and Fink, G. R. (1996). Ras2 signals via the Cdc42/Ste20/mitogen-activated protein kinase module to induce filamentous growth in *Saccharomyces cerevisiae. Proc. Natl. Acad. Sci. USA* **93**, 5352–5356.

Nonaka, H., Tanaka, K., Fujiwara, T., Kohno, H., Umikawa, M., Mino, A. and Takai, Y. (1995). A downstream target of *RHO1* small GTP-binding protein is *PKC1*, a homolog of protein kinase C, which leads to activation of the MAP kinase cascade in *Saccharomyces cerevisiae. EMBO J.* **14**, 5931–5938.

Ramer, S. W. and Davis, R. W. (1993). A dominant truncation allele identifies a gene, *STE20*, that encodes a putative protein kinase necessary for mating in *Saccharomyces cerevisiae. Proc. Natl. Acad. Sci. USA* **90**, 452–456.

Rhodes, N., Connell, L. and Errede, B. (1990). STE11 is a protein kinase required for cell-type-specific transcription and signal transduction in yeast. *Genes Dev.* **4**, 1862–1874.

Roberts, R. L. and Fink, G. R. (1994). Elements of a single MAP kinase cascade in *Saccharomyces cerevisiae* mediate two developmental programs in the same cell type: mating and invasive growth. *Genes Dev.* **8**, 2974–2985.

Rothstein, R. (1991). Targeting disruption, replacement, and allele rescue: integrative DNA transformation in yeast. In *Methods in Enzymology, vol. 194: Guide to Yeast Genetics and Molecular Biology* (C. Guthrie and G. R. Fink, eds), pp. 281–301. Academic Press, London.

Sherman, F. (1991). Getting started with yeast. In *Methods in Enzymology, vol. 194: Guide to Yeast Genetics and Molecular Biology* (C. Guthrie and G. R. Fink, eds), pp. 3–21. Academic Press, London.

Sprague, Jr G. F. (1991). Assay of yeast mating reaction. In *Methods in Enzymology, vol. 194: Guide to Yeast Genetics and Molecular Biology* (C. Guthrie and G. R. Fink, eds), pp. 77–93. Academic Press, London.

Stevenson, B. J., Rhodes, N., Errede, B. and Sprague, Jr G. F. (1992). Constitutive mutants of the protein kinase STE11 activate the yeast pheromone response pathway in the absence of the G protein. *Genes Dev.* **6**, 1293–1304.

Truehcart, J., Boeke, J. D. and Fink, G. R. (1987). Two genes required for cell fusion during yeast conjugation: evidence for a pheromone-induced surface protein. *Mol. Cell. Biol.* **7**, 2316–2328.

Watanabe, M., Chen, C. Y. and Levin, D. E. (1994). *Saccharomyces cerevisiae PKC1* encodes a protein kinase C (PKC) homolog with a substrate specificity similar to that of mammalian PKC. *J. Biol. Chem.* **269**, 16829–16836.

Watanabe, Y., Irie, K. and Matsumoto, K. (1995). Yeast *RLM1* encodes a serum response factor-like protein that may function downstream of the mpk1 (slt2) mitogen-activated protein-kinase pathway. *Mol. Cell. Biol.* **15**, 5740–5749.

Ziman, M., O'Brien, J. M., Ouellette, L. A., Church, W. R. and Johnson, D. Y. (1991). Mutational analysis of *CDC42Sc*, a *Saccharomyces cerevisiae* gene that encodes a putative GTP-binding protein involved in the control of cell polarity. *Mol. Cell. Biol.* **11**, 3537–3544.

21 Analysis of the *Candida albicans* Genome

P. T. Magee

Department of Genetics and Cell Biology, University of Minnesota, St Paul, Minnesota, USA

◆◆

CONTENTS

List of Abbreviations

ARS	Autonomously replicating sequence
5-FOA	5-Fluoro-orotic acid
FIGE	Field inversion gel electrophoresis
LTR	Long terminal repeat
MRS	Major repeat sequence
OFAGE	Orthogonal field agarose gel
PCR	Polymerase chain reaction
STS	Sequence tagged site
TAFE	Transverse alternating field electrophoresis

◆◆◆◆◆◆ I. INTRODUCTION

A. Medical Importance of *Candida albicans*

The most commonly isolated fungal agent of human disease, *Candida albicans*, is a diploid fungus that is capable of growing in a variety of morphologies, including yeast, pseudohyphae (chains of cells with constrictions at the septae), or true hyphae. Although the yeast form (called blastospores, historically) is the one most commonly found in the laboratory, pseudohyphae and hyphae form under a variety of conditions that may reflect the environment found in infections. *In situ* studies of infection usually reveal cells of all morphologies: yeast, pseudohyphae, and hyphae. *Candida* infections are becoming an increasing problem and a recent survey indicated that *Candida* now ranks third among the isolates found in clinical laboratories studying infectious disease, being more frequent than all Gram-negative bacteria combined. Because the organism is an opportunistic pathogen, its frequency of isolation reflects the increasing number of immunosuppressed patients occurring as a result of modern medical practices. There seems no doubt that as the frequency of AIDS patients increases, more transplants are carried out, and chemotherapy becomes more sophisticated, *Candida* infections will increase as a medical problem.

B. The Diploid Nature of the Organism

Because *C. albicans* is diploid as usually isolated and has no known sexual cycle, classical genetic approaches to the analysis of pathogenesis have not been very fruitful. The approach of isolating a mutant with a particular phenotype, isolating the gene, disrupting it, and examining the null phenotype is problematic because mutant isolation is difficult. Molecular genetics has been used more successfully as a way to study the biology of the organism, especially the yeast-to-hyphal transition. An excellent recent review gives many details of the molecular genetics of the organism (Pla *et al.*, 1996). However, the obligate diploid nature of the organism means that gene knockouts must be carried out twice to inactivate both alleles of a given gene, and this makes even molecular genetics cumbersome.

C. Implications of Asexuality

The lack of a sexual cycle in *C. albicans* has important implications for the structure of the genome. First, in the absence of chromosome pairing during meiosis, there is no obvious reason for homologs to remain similar. However, although there are minor variations in the size of homologs in various strains and several translocations have been identified in specific isolates, the overall structure of the genome is remarkably constant from isolate to isolate. This suggests that not much time (in an evolutionary sense) has elapsed since *C. albicans* lost its sexual cycle.

Second, there is no reason that this organism might not have accumulated several recessive lethal mutations, so long as these are balanced by the dominant allele on the other homolog. Indeed, some early work suggested that such lethal alleles were relatively common in clinical isolates (Whelan and Soll, 1982). However, it has now been shown that several strains can grow (albeit slowly) while monosomic for chromosome 3 (Barton and Gull, 1992) or for chromosome 7 and parts of chromosomes 5 and 6 (Magee and Magee, 1997). Once again, this suggests that the organism has only recently become an obligate diploid.

D. The Deviation from the Standard Genetic Code

An early observation of molecular biologists studying *C. albicans* was that this organism rarely expresses foreign genes, although its genes are often expressed in *Saccharomyces cerevisiae* and sometimes in *Escherichia coli*. This paradox was explained when it was discovered that in *C. albicans* (and in several other *Candida* species) the codon CUG encodes serine, rather than leucine as is the case with most organisms (Santos and Tuite, 1995). The coding difference exists at the level of the tRNA, which has the structure of a tRNA[ser] but with the GAC anticodon. This codon is rarely used in *Candida*, so that most of the genes isolated do not contain it and can be expressed using the standard genetic code. The coding deviation has greatly hampered the development of reporter genes for this organism.

◆◆◆◆◆◆ II. ANALYSIS OF GENOMIC STRUCTURE IN *C. ALBICANS*

A. History

The exact number of chromosomes in *C. albicans* was not determined until pulsed-field electrophoresis able to separate DNA molecules in the 1–4 Mb range was developed. Use of probes for specific chromosomes gave evidence for eight separate sets of homologs (Wickes *et al.*, 1991), although the details of the karyotype vary from strain to strain because of translocations. Figure 1 shows a pulsed-field gel separation of the chromosomes of a typical set of *C. albicans* strains. The chromosomes range in size from 4.3 to 1 Mb, and they are numbered from the largest to the smallest. The ribosomal-DNA-bearing chromosome varies greatly in the size of its homologs, most likely because of sister chromatid exchange at the rDNA repeat. The variation in homolog size in this chromosome takes place frequently enough so that differences can sometimes be seen in serial clones of a particular laboratory strain. *C. albicans* has about 100 rDNA repeats per haploid genome, and the size of the repeat is 12 kb. Homologs of this chromosome, labeled R, can vary in size from >4 to <3 Mb and can have from 50 to 150 repeats per homolog (Iwaguchi *et al.*, 1992b; Wickes *et al.*, 1991). Chromosome R has been given a letter rather than a number because it has no consistent place in the electrophoretic karyotype, running sometimes ahead of (smaller than) the next largest chromosome and sometimes behind (larger than) it. Chromosomes 1 to 7 range in size from 3 to 1 Mb, and except for translocations their sizes and hence their order on the karyotype are highly conserved among isolates despite the asexual life style of the organism.

B. Electrophoretic Separation of *C. albicans* Chromosomes

The technique for electrophoretic separation of DNA the size of *C. albicans* chromosomes is called pulsed-field electrophoresis. In this technique the

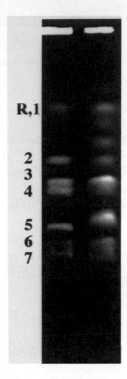

Figure 1. CHEF separation of chromosomes from strain 1006 and strain A81-Pu. The numbers on the left of 1006 indicate the numbering of the chromosomes for that lane. A81-Pu has several translocations which account for the extra bands.

DNA is subjected to electric fields at oblique angles to the direction of migration; these fields are reversed in polarity with a particular frequency. The principle of pulsed-field separation requires an alternating electric field so that the DNA, which is alternately rigidly aligned with the instantaneous electric field and relaxed as the direction of the field changes, can reptate (wiggle like a snake) through the agarose. The determination of appropriate separation conditions remains empirical, and the choice of regimens is dictated by the size of the molecules in the region where maximum separation is desired. Because the size of the chromosomes of *C. albicans* varies over a four-fold range, one has two choices: one can use conditions that give the best resolution overall, sacrificing the maximum separation of the larger and/or smaller chromosomes, or one can use conditions that separate one extreme or the other most effectively.

I. Preparation of DNA

The DNA for electrophoretic separation is prepared essentially as described for *S. cerevisiae* (see Chapter 2). Either agarose plugs or beads may be used for electrophoresis.

2. Pulsed-field instruments

There are three instruments commercially available for the pulsed-field separation of molecules the size of *C. albicans* chromosomes: they are the BioRad CHEF, the LKB Pulsaphor, and the Beckmann TAFE (transverse alternating field electrophoresis). The first two instruments mentioned above are contour-clamped machines, meaning that electric fields at angles to the direction of migration are maintained during the run, resulting in electropherograms with parallel lanes (Vollrath and Davis, 1987). In TAFE the field alternation is perpendicular to the surface of the gel. OFAGE (orthogonal field agarose gel electrophoresis), without the contour clamping, results in hour-glass-shaped electropherograms; there are some claims that better resolution is achieved with this technique, but it is not widely used at this time. A completely different approach is field inversion gel electrophoresis (FIGE) (Carle *et al.*, 1986). In this method the field, parallel to the direction of migration, is run in one direction, then reversed for a short period of time. Interestingly, under some conditions, FIGE results in the opposite order of separation of *Candida* chromosomes compared with OFAGE or CHEF, with the smaller ones migrating more slowly.

Three sets of conditions are given in Table 1: one for the largest chromosomes, one that gives the best resolution of the karyotype as a whole, and one for separating the *Sfi*I fragments (section III) The protocols outlined here for *C. albicans* refer to the CHEF DRIII instrument. Different regimes are necessary for the other machines.

The gels can be blotted and hybridized with probes just like any gel electrophoresis separation. Probing the blots is essential for preparing a physical map and for identifying the source of unusual bands, which are for the most part translocations. Chromosomes can be isolated from the gel, labeled, and used as probes to examine relationships among clones.

Table I. CHEF separation of *Candida albicans* chromosomes

1. For the largest chromosomes	The gel[a] is 0.6% Amersco PFGE agarose in 0.5 × TBE. Switching time is ramped from 120 s to 300 s at 2.5 Vcm^{-1}, field angle 106°, followed by 40 h with switching time from 420 s to 900 s, using the same field strength and angle.
2. For the karyotype as a whole	The gel[a] is 0.9% regular agarose. Switching time is ramped from 60 s to 300 s over 24 h with the field strength at 4.5 Vcm^{-1} and field angle at 120°, followed by 24 h with switching time ramped from 720 s to 900 s, with field strength at 2.0 Vcm^{-1} and angle at 106°.
3. For the smallest chromosomes and the *Sfi*I fragments	The gel[a] is 0.9% regular agarose. Switching time is ramped from 60 s to 300 s over 36 h. The field is 4.0 Vcm^{-1} at a 120° angle.

[a] All gels are run in 0.5 × TBE (0.045 M Tris borate, 0.001 M EDTA, pH 8.0) at 14°C on a BioRad CHEF DRIII instrument.

However, as will be discussed below, the presence of several common repeats means that one cannot use an isolated chromosome as a probe to identify a band on a gel as a homolog. However, one can use a chromosome as a probe to determine the provenance of a particular clone (unless the clone contains a repeated sequence).

C. Identifying Chromosomes

Because of translocations, band position does not always indicate the identity of chromosomes. For example, in the well-studied strain WO-1, a band with the mobility of chromosome 5 in a normal *Candida* strain is actually made up of a translocation product of chromosomes 4 and 7 (Chu *et al.*, 1993). Thus, to determine chromosome identity, specific probes must be used. A set of such probes, containing 16 sequences of telomere-proximal genes, one from each arm of each chromosome, is available (B. B. Magee, personal communication). This set allows one to identify translocations and to characterize them using a minimal number of Southern blots of chromosome separations.

D. The Structure of Chromosomes

I. Repeated sequences

Although *C. albicans* chromosomes vary four-fold in size, they share some common features. Chief among these are the repeated sequences RPS (Iwaguchi *et al.*, 1992a), CARE2 (Lasker *et al.*, 1992), and Ca7. Each is related to the structure of the chromosome. Ca7 is a probe isolated by Sadhu *et al.* (1991) and contains the *C. albicans* telomere repeat. This probe can be used on chromosome digests to identify the fragments derived from the chromosome ends. We have recently discovered that CARE2, another *Candida* repeat, is also located quite near the telomeres of several chromosomes. Finally, the repeated sequence RPS, which has been well characterized at Nagoya University Medical School in Japan, is a part of a large, complex repeat which we have called the major repeat sequence (MRS) (Chindamporn *et al.*, 1995). The MRS includes the fragment called HOK, from one to 10 repeats of RPS, and the fragment called RB2. RPS, in turn, is composed of several repeats of a 29 base-pair sequence called COM29; COM29 contains a site for the restriction enzyme *Sfi*I. The MRS is found on all *C. albicans* chromosomes except chromosome 3. On chromosomes 4 and 7 (and possibly 1) it occurs twice, separated by several hundred kilobases of unique DNA. It is not found in most other *Candida* species, the exception being *Candida stellatoidea*. A homologous sequence exists in *Candida dubliniensis*, a recently discovered species, but it is not identical to the MRS. Hence, the MRS is a very reliable way of identifying *C. albicans* and its subspecies *C. stellatoidea*.

The MRS includes the sequences represented in the probes Ca3 (Sadhu *et al.*, 1991) and 27a (Scherer and Stevens, 1988), which have been extensively used for epidemiological studies. In restriction enzyme digests of genomic DNA of clinical isolates, the size of the fragments containing these sequences varies extensively, providing a way of identifying a particular strain and its close relatives. The molecular basis of the variation found in the genome using these probes is still not known, but it may be related to a difference in the number of repeats of RPS in the MRS.

The MRS also seems to be a favored site for translocations of *C. albicans* chromosomes; of more than 15 translocations we have analyzed, only one seems to have taken place away from the MRS. It seems most likely that the MRS serves as a region of homology common to non-homologous chromosomes, and that ectopic pairing at this region, followed by homologous recombination, leads to translocation. However, there are only a few data to support this model. Some supportive evidence comes from characterization of the molecular orientation of the MRS. The mapping we have carried out (see section IV) has allowed us to determine the direction of the repeats, and in every translocation we have characterized, the repeats are so arranged that homologous recombination would yield the translocation found. For example, Figure 2 shows that a translocation yielding a 5I–6C product is allowed, while homologous recombination could not yield a 5M–6C product, and, indeed, the former product is found in strain WO-1. Whatever the molecular explanation for the high frequency of translocation at the MRS, the effect is that large-scale restriction mapping with the restriction enzyme *Sfi*I (see below), which cuts within the MRS, gives a characteristic pattern of fragments whether translocations exist or not.

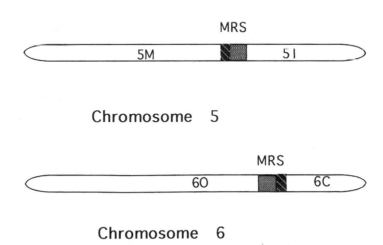

Figure 2. Orientation of the MRS in chromosomes 5 and 6 in strain 1006. The numbers on the chromosomes refer to the fragments generated when the restriction enzyme *Sfi*I cuts in the MRS. The translocation products 6O–5M and 6C–5I have been observed, but not 6O–5I or 6C–5M. Cross-hatched boxes (HOK) and stippled boxes (RPS) indicate the orientation of sequences within the MRSs on chromosomes 5 and 6.

2. Telomere structure

The telomere structure of *C. albicans* is different from that of *S. cerevisiae*, having 23 bp complex repeats that are very uniform both from chromosome to chromosome and from strain to strain (McEachern and Hicks, 1993). This uniformity means that use of the Ca7 probe to identify telomeres works in every strain so far examined. On at least one chromosome in strain 1006, there is a reverse transcriptase gene near the telomere; this RT is homologous to the enzymes from non-(LTR) type retrotransposons. Elements of non-LTR retrotransposons have been found in *Drosophila* telomeres, so the location of this sequence may be related to telomere function.

3. Centromere structure

The centromere of *C. albicans* has not been cloned. Circumstantial evidence, based on the pattern of translocations and the stability of small chromosomal fragments, suggests that centromeres are located on *Sfi*I fragments 7F, 6C, 5I, 4, and 2U.

◆◆◆◆◆◆ III. LARGE-SCALE RESTRICTION MAPPING USING *Sfi*I

A. Choice of Enzyme

To carry out a large-scale restriction map of a genome one needs a restriction enzyme that fits several criteria: it must cut relatively infrequently, it must give a relatively wide spectrum of band size, and, ideally, it should not have a site in the ribosomal DNA repeat. For *C. albicans*, *Sfi*I fits these criteria (Chu *et al.*, 1993). *Not*I fits the first criterion, but its digestion products are either quite small or quite large, so that it is difficult to separate and characterize them. However, *Sfi*I (which recognizes the sequence GGCCNNNNNGGCC) gives 50 large fragments ranging in size from about 50 kb to about 2.2 Mb. (Each of the duplicate fragments from homologs is counted in this reckoning, because polymorphisms in *C. albicans* are frequent enough so that each *Sfi*I fragment is unique.) Many of the digestion points are in the MRS. This block has at its core the 2 kb RPS sequence which contains from three to five *Sfi*I sites; RPS itself is often repeated (up to five times in the MRS of chromosome 5 of strain 1006 and as many as 30 times in some translocations examined in clinical strains). For the purposes of counting *Sfi*I sites, each MRS is considered one site, because the products of its digestion are small repeated fragments which are not important in large-scale mapping. Of the 34 *Sfi*I sites identified by the map, at least 16 are in MRS sequences. The remaining sites are in unique DNA. The ability to digest the chromosomes of *C. albicans* into smaller, easily separable fragments allows one to localize genes to regions of chromosomes by hybridization, providing a finer scale physical map.

◆◆◆◆◆◆ **IV. PREPARING A PHYSICAL MAP OF THE**
     ***C. ALBICANS* GENOME**

Because of the lack of a sexual cycle, molecular methods have come to predominate in *C. albicans* genetics. A major problem has been that there is no genetic map to use as a reference. We have therefore undertaken to generate a physical map. The strain we have chosen is 1006, a well-characterized strain that has been of major use in genetic studies. It must be emphasized that because of strain differences in chromosome structure, not all of the details of the map in this strain will be identical to those of all other strains. However, the overall map should be highly transferable to other *C. albicans* strains.

A. Preparation of a Sequence-Tagged Site Map

1. Assigning genes to the chromosomes and to large restriction fragments

Because pulsed-field electrophoresis separations can be blotted like other DNA gels, assignment of genes to chromosomes is straightforward. One caveat is that although the amount of DNA per band is of the order of magnitude of a normal DNA gel, because the DNA is the highly complex chromosome, any given gene is represented at a low frequency. Hence, the sensitivity is greatly reduced, and small probes (less than 1 kb) can be difficult to map. A second problem with this approach is that with chromosomes of the order of magnitude of several megabases, such assignment does not yield a very informative map. Nevertheless, matching genes to chromosomes is the essential first step in mapping the genome.

The *Sfi*I fragments allow a second level of resolution in mapping isolated genes. These fragments effectively divide the chromosomes up into domains which can be as small as 50 kb. Because the telomeric fragments are known, one can get a rough idea of which genes are telomere-proximal and which are on internal fragments. Such mapping also allows one to identify fragments generated by translocations and has led to tentative assignment of the centromeres to particular *Sfi*I fragments.

B. Correlating an Ordered Library with the Electrophoretic Karyotype

The next step in constructing a sequence-tagged site (STS) map is prepare a "contig map", essentially an alignment of overlapping genomic clones along the chromosomes. This is accomplished by hybridizing all the clones in the library with each probe. The clones that hybridize with a given sequence by definition constitute an overlapping set, and those that hybridize with two sequences will in many cases bridge two such sets. The result, after many repetitions, is a map like the one shown in Figure 3 for part of chromosome 6 in *C. albicans*.

The *C. albicans* library used in this mapping project consists of 40–50 kb pieces of genomic DNA in a vector called a fosmid (Kim *et al.*, 1992). This

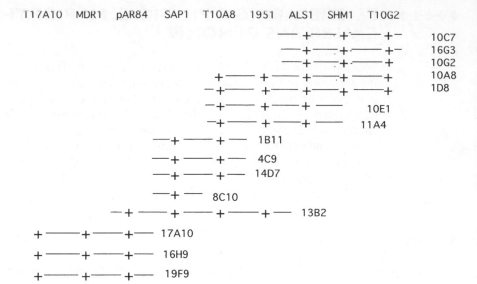

Figure 3. A contig map of an interior fragment of chromosome 6. The probes used to identify the various fosmids are at the top. The fosmid numbers, corresponding to their position in the gridded-out library, are listed next to the horizontal lines. (+) Indicates that the probe hybridizes with the fosmid listed. *ALS1* is estimated to be 270 kb from the telomere of chromosome 6. The markers are evenly spaced, but this is only a convention. Their actual position must be determined by a method that measures physical distance. The approximate length of this contig is 150 kb.

vector is similar to a cosmid but contains the F-episome origin. It thus exists in one or two copies per cell; as a result, repeated DNA is much more stable than in a classical cosmid. The library consists of approximately 40 000 clones, comprising about 10-fold coverage of the genome. Most screening is done in half the library for the sake of efficiency; the second half is screened only for sequences that do not appear in the first half.

Presently, about 65% of the fosmids in the first half of the library are positive for one or more probes. At this level of assignment, screening with random probes becomes inefficient, because only about 35% of new probes find new fosmids. At this stage it is necessary to use the technique of "sampling without replacement", in which new probes are made from fosmids that are not positive for any of the assigned probes. This immediately locates the new fosmid on the map and has the same chance as a random clone of hybridizing with other unassigned fosmids. Hence, the map construction becomes much more efficient.

C. Determination of Physical Distance on the Chromosome

The contig map does not give actual distances between mapped sequences because the location of the probes on the fosmids is not determined. Therefore, a method is needed to measure the distance between the probes. Random breakage mapping is ideal for this (Game *et al.*, 1990). In this technique, pulsed-field DNA preparations are subjected to gamma irradiation for various periods of time, then each sample is run on a pulsed-field gel. Gamma rays cause double-strand breaks in chromosomes, and at longer irradiation times the size of the fragment carrying a particular gene approaches as a limit the distance of that gene from the telomere. There will, of course, be two such fragments, one corresponding to each arm, and together they should sum to the size of the chromosome. Using this method, the size of the two homologs of chromosome 7 in strain 1006 has been determined as 1030 and 1010 kb, and genes have been localized to within 20 kb (H. Chibana, unpublished results).

◆◆◆◆◆◆ V. THE *C. ALBICANS* WEB PAGE

The mapping data, together with a large amount of information on *C. albicans* methods, strains, and cloned genes are available on a World Wide Web Page at http://alces.med.umn.edu/Candida.html

The results of the physical mapping effort and any new genes either identified during the mapping or submitted to the database are updated on a weekly basis. There are also links to the Candida Sequencing Project (see below). This Web Page is maintained by Dr Stewart Scherer.

◆◆◆◆◆◆ VI. FINDING NEW *C. ALBICANS* GENES

A. One-pass Sequencing

An STS map of the genome is most informative when the sequences are known genes. Many of the sequences assigned to the *C. albicans* genome are genes that have been isolated by workers in the scientific community for reasons related to specific research projects. However, these do not constitute a sufficient number to complete the STS map. For a map with markers spaced about 25 kb apart, about 650 probes are required for the 16 Mb haploid genome, and of course because markers isolated on a random basis will not be distributed evenly, more will be needed. We therefore carry out "one-pass" sequencing of all probes we use. This gives us between 400 and 700 base pairs of sequence. This sequence is compared with the database using blastx, and the genes are assigned names if the homology with a previously isolated gene (from any organism) is such that the blastx score is above 100. Identification of the gene is of course not definitive, because the sequence may be a highly conserved one, and the function of the gene in *Candida* may be different from the one in the database it most closely resembles. Publication of the complete sequence of the *S. cerevisiae* genome has greatly aided the process of gene identification, and at the present moment we find that about one in four sequences yields a match.

B. The *C. albicans* Genome Sequencing Project

The *C. albicans* genome sequencing project, the Burroughs Wellcome Fund and the National Institute of Dental Research have provided funds for one-pass sequencing of the genome of *Candida albicans* strain SC5314. The initial plan for the project will involve sequencing of each end of the inserts of a random shear library to generate about 25 Mb of sequence, or a little more than 1.5 genome's worth. For the first stage of the project, no effort will be made to complete the sequence. However, comparison of the sequence generated with the databases, especially the *S. cerevisiae* genome, is expected to identify 95% of the genes in the *Candida* genome that can be identified on the basis of their similarity to genes in other species. These partial sequences can then be used to generate polymerase chain reaction (PCR) primers to isolate the gene from any of the existing libraries. Thus, within 3 years, almost all *Candida albicans* genes should be accessible to investigators in a straightforward manner. Selected single-copy sequences will be placed on the physical map concurrently. These genes will provide additional markers to make the physical map more dense and will be important for subsequent efforts to assemble more complete genomic sequences.

◆◆◆◆◆◆ VII. CLONING *C. ALBICANS* GENES

A. Libraries

1. Libraries in *E. coli–S. cerevisiae* shuttle vectors

There are several genomic libraries constructed in vectors that will replicate in both *E. coli* and *S. cerevisiae*. One is available from the AIDS Research and Reference Reagent Program, sponsored by the NIH (E-mail: obcaids@ix.netcom.com); it consists of fragments of a *Bam*HI–*Hin*dIII partial digest of the strain WO-1 in the vector pEMBLy23. The inserts are from 5 to 10 kb in size, and it has the advantage that neither of the enzymes has a site in the rDNA repeat, so rDNA is not represented in the library. This library has been used to identify a variety of *C. albicans* genes by complementation of mutations isolated in *S. cerevisiae*. Several groups have constructed similar libraries.

2. Libraries in vectors that will replicate in *C. albicans* as well as in *E. coli* and *S. cerevisiae*

A library in the vector 1041, which contains origins of replication for *E. coli*, *S. cerevisiae* and *C. albicans*, the Amp^R gene for selection in bacteria, and the *C. albicans URA3* gene, and thus functions in *E. coli* and *Saccharomyces* as well as *Candida*, is also available from the AIDS Research and Reference Program. This library contains *Sau*3A fragments from 4 to 8 kb in size derived from the genome of the highly pathogenic strain C9. A similar library made from the well-characterized strain 1006 is available from the Magee laboratory. The vector in which these libraries were made contains the autonomously replicating sequence (ARS) originally isolated at Squibb Pharmaceuticals; this ARS is inefficient and leads to oligimerization of the vector in Candida (Kurtz *et al.*, 1987). Several other ARS sequences have been isolated since and libraries in vectors containing them have been prepared. It is not clear whether these vectors oligimerize or not, but the problem of oligimerization can be avoided by use of a vector with two of the ARS sequences on it. One library has been constructed in such a vector (Pla *et al.*, 1995).

3. Construction of a chromosome-specific library

It is possible to prepare chromosome-specific libraries from bands isolated from CHEF separations, but there are problems associated with this approach. The largest two chromosomes, R and 1, are very difficult to separate, and it is impossible with present techniques to purify one free of the other. The other chromosomes, when eluted from the gel, tend to be contaminated with fragments of the larger chromosomes. This contamination is due to the fact that some random breakage of the larger chromosomes is unavoidable, and the resultant fragments migrate down the gel and are found in all size ranges. For example, in preparation of a chromosome 5-specific library, half of the clones were from other chromosomes. In such a library, therefore, the origin of the clones must be verified. It is possible to do so by probing the library on filters with the isolated chromosome. Although the isolated chromosome preparation will contain random fragments from the rest of the genome, the great majority of the probe will come from the selected chromosome, and the clones from that chromosome will hybridize to a much greater extent than the clones from other chromosomes.

B. Identifying Specific Genes

I. Cloning directly in *C. albicans*

Cloning by complementation in *Candida* is in principle no more difficult than cloning in *Saccharomyces*. However, the difficulty in making mutants limits the usefulness of this technique. In addition, the fact that vectors oligimerize makes it difficult to isolate the complementing plasmid from *Candida* and transform *E. coli* directly, because the oligimers may consist of more than one plasmid and the bacteria are difficult to transform with large molecules. Goshorn *et al.* (1992) devised a way of getting around the problem by using vectors with the *S. cerevisiae* 2μ origin of replication and transforming directly into *cir*⁺ yeast. Recombination in *Saccharomyces* resolves the oligimers which can then be retransformed into *E. coli*.

2. Cloning by complementation of mutations in *S. cerevisiae*

Most *Candida* genes are expressed in *Saccharomyces* (see section I.D) and several interesting genes have been isolated this way. It is important to note that the fact that a *Candida* gene complements a mutation in *Saccharomyces* does not necessarily prove that it carries out the same function in *Candida*. The prevalence of gene families and differences in regulation between the two organisms will certainly lead to suppression by non-homologous genes in some cases.

3. Cloning by sequence homology

The best way to clone a *Candida* gene based on sequence homology is to use PCR. This technique, using a highly conserved region of a gene, is quite effective, because the amplified fragment can be used to probe a library to isolate the gene of interest. The differences in sequence between *Candida* genes and *Saccharomyces* genes are significant. At the amino acid level, the average *Candida* biosynthetic gene is 65% identical to the homolog in *Saccharomyces*. At the DNA level, the difference is such that hybridization even at moderate stringency rarely works to identify homologous genes.

◆◆◆◆◆◆ VIII. DISRUPTION OF GENES IN *C. ALBICANS*

A. The Problem of Diploidy

Because the organism is diploid, gene disruption in *Candida* is problematic. Each disruption must be carried out twice, to disrupt both copies of any gene. The most commonly used technique exploits the "URA-blaster" (Figure 4). Modified by Fonzi (Fonzi and Irwin, 1993) after the technique of Alani *et al.* (1987), it uses a *ura3* strain, usually CAI4, derived from the clinical isolate SC5314. Strain 1006 can also be used (Goshorn and Scherer, 1989).The strain CAI4 has both copies of *URA3* deleted and replaced by a small piece of lambda DNA. A copy of *URA3* flanked by direct repeats of the *Salmonella typhimurium HisG* gene (the URA-blaster cassette) is used to replace a part of the target gene to be disrupted and the resulting construct transformed into CAI4. Integration occurs by homologous recombination at the target locus, and Ura⁺ transformants are selected. From such a transformant a Ura⁻ derivative is selected on 5-fluoroorotic acid. (A large fraction of such derivatives are due to loss of the *URA3* gene by homologous recombination between the two adjacent repeats.) Then the transformation is repeated. The resulting strain is *URA3/ura3* with both copies of the target gene interrupted by an insertion. A similar approach, developed by Gorman *et al.* (1991), uses the *GAL1* gene as a selectable marker. Selection for Gal⁺ cells is performed by growth on galactose as a sole carbon source and selection for Gal⁻ cells can be carried out with 2-deoxygalactose. The Gal⁻ cells can be isolated from many clinical isolates after mild ultraviolet irradiation.

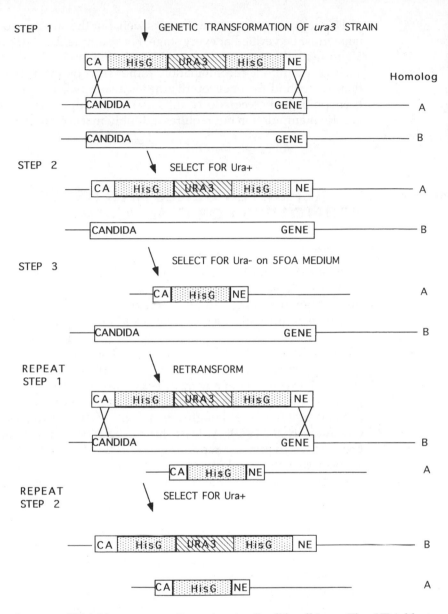

Figure 4. URA-blaster gene disruption in *Candida albicans*. The URA-blaster cassette, consisting of the intact *URA3* flanked by repeats of the *Salmonella typhimurium HisG* gene, is inserted into a cloned *Candida* gene by recombinant DNA techniques. The resulting construct is linearized, leaving regions of homology to the *Candida* genome at both ends, and transformed into a *ura3* strain (usually CAI4 or 1006) (step 1). Selection for the Ura⁺ phenotype yields some transformants that have replaced the chromosomal copy of the *Candida* gene with the construct (step 2). One of these transformants is subjected to selection for Ura⁻ with 5-fluoroorotic acid (5-FOA) (step 3). Among the Ura⁻ colonies are some that have lost *URA3* via a recombination event between the repeated HisG sequences, leaving a copy of HisG in the chromosome. Steps 1 and 2 are then repeated to disrupt the allele on the other homolog. The resulting strain is Ura⁺ and homozygous for the disruption of the *Candida* gene. If a Ura⁻ strain is desired, the 5-FOA selection can be repeated (not shown).

This technique is rapid and works well, but the structure of the disruptants must be verified at every stage. For example, Ura⁻ strains may arise after the first transformation step by gene conversion by the homologous allele or by mitotic recombination, rather than by intragenic recombination. A method for one-step disruption using two markers has recently been published (Negredo *et al.*, 1997). This approach avoids repeated genetic manipulation but requires a doubly marked strain.

◆◆◆◆◆◆ IX. REPORTER GENES AND REGULATABLE PROMOTERS FOR C. *ALBICANS*

A. Reporter Genes

Early efforts to develop reporter genes in *C. albicans* foundered, probably because of the specialized codon usage found in this group of yeasts. Thus many laboratories tried to develop expression systems using the *E. coli lacZ* gene product, β-galactosidase, all without success. The *Candida URA3* gene has been proposed as a reporter, but its use is limited by the lack of an easy assay and by the need to use a *ura3* homozygote to avoid endogenous expression (Myers *et al.*, 1995). Attempts to use firefly luciferase to study the expression of a gene differentially expressed during the white phase of the white-opaque phenotypic transition failed at the level of protein expression, although the mRNA of the fusion was measured (Srikantha *et al.*, 1995). At least three exogenous proteins have recently been shown to serve as reporter genes in *C. albicans*. Leuker *et al.* (1992) showed that the β-galactosidase of *Kluyveromyces lactis* can serve as a reporter gene for both *C. albicans* and *C. tropicalis* (despite the presence of two CTG codons). The luciferase of the sea pansy, *Renilla reniformis*, contains no CTG codons and is expressed as a fusion protein (Srikantha *et al.*, 1996). This enzyme has been used to study expression using several *Candida* promoters. Unfortunately, the substrate for this luciferase is quite expensive. The green fluorescent protein of the jellyfish *Aequorea victoria* has been re-engineered to remove CTG codons, to modify its codon usage for high expression in *Candida*, and to incorporate two mutations that enhance its fluorescence. The resulting product should be of great general use (Cormack *et al.*, 1997).

B. Regulatable Promoters

Several inducible promoters have been reported for *C. albicans*, although none seems likely to be as useful as the *GAL* promoter in *S. cerevisiae*. The *GAL1* promoter from *C. albicans* was used to show the usefulness of the luciferase from *R. reniformis* as a reporter gene (Srikantha *et al.*, 1996), but unlike its counterpart in *Saccharomyces*, it has a detectable basal level of transcription and thus cannot be used in cases where an absolute absence of gene product is needed. The *MAL2* promoter, which is repressed on glucose and expressed on sucrose (Geber *et al.*, 1992), has been used by Brown *et al.* (1996) to regulate the expression of the *URA3* gene. Regulation by this promoter also does not seem to be absolute, because cells whose only copy of the *URA3* gene is under the control of the *MAL2* promoter still grow somewhat on minimal-glucose medium lacking uridine. The promoter for *ADH1* (alcohol dehydrogenase) is repressed by ethanol and derepressed in cells grown on galactose, while glucose as a carbon source gives intermediate levels of activity (Bertram *et al.*, 1996). The *ADH1* promoter has been used for overexpression of the *Candida* *HYR1* (hyphally regulated) gene in both *C. albicans* and *S. cerevisiae* (Bailey *et al.*, 1996). The *PCK1* (phosphoenolpyruvate carboxykinase) promoter is repressed on glucose but active during growth on succinate, ethanol, or sucrose. It has been used to examine the effect of down-regulating *EFG1* (enhanced fungal growth), an essential gene that is involved in regulation of pseudohypha formation (Stoldt *et al.*, 1997). The *MET3* promoter from *C. albicans* has been isolated by Sudbury and his colleagues (P. Sudbury, personal communication). This promoter is repressed by methionine in the medium and its basal level of transcription is very low. The promoter is on in medium lacking methionine. This may be the most tightly regulated of the promoters presently available.

◆◆◆◆◆◆ X. THE USEFULNESS OF THE *S. CEREVISIAE* GENOME PROJECT FOR *C. ALBICANS* GENOMIC RESEARCH

A. Genetic Similarity Between *S. cerevisiae* and *C. albicans*

I. Frequency and processing of introns

As in *Saccharomyces*, many *Candida* genes lack introns. However, introns have been found in the following genes: *ACT1*, *TUB1*, *TUB2*, *CMD1*, *RPL39*, *PTR1*, *RSR1*, *DLH1*, *YPR55*, and *IMH3*. Although *S. cerevisiae* is much less closely related to *C. albicans* than to several other yeasts (Kurtzman and Robnet, 1997), the two fungi share mRNA processing consensus sequences; *Saccharomyces* will process *Candida* introns. *Candida* β-tubulin has been shown to be expressed in *Saccharomyces* (Smith *et al.*, 1990), and an artificial construct with the actin intron inserted into *URA3* will complement a *ura3* mutation in both *Candida* and *Saccharomyces* (B. Corner, unpublished data).

2. Comparative synteny of *Saccharomyces* and *Candida*

The arrangement of genes on that part of the *Candida* genome which has been mapped shows no relationship to the *Saccharomyces* map. The map positions of the 22 genes identified on chromosome 7 in *Candida* include all but three chromosomes in *Saccharomyces*.

3. Genome comparisons

The availability of the complete sequence of the *S. cerevisiae* genome has been, and will continue to be, of great use to research on *C. albicans*. The high degree of homology between the two organisms will provide information in two main ways. The genes which are found in both organisms will, for the most part, be characterized first in *Saccharomyces*, giving researchers a rapid and predictable way of generating the null phenotype of these genes in *Candida*. This will greatly speed the analysis of the possible roles of such genes in pathogenesis. As more becomes known about the relationship between the genes of the two organisms, finding the *Candida* homolog of a gene known in *Saccharomyces* will become much easier.

A second and equally important contribution to our understanding of the biology of *Candida* will come from those genes that do not occur in *Saccharomyces*. Not all of these genes will, of course, be related to virulence, but some of them undoubtedly will. Based on present information, we can expect about 3% of the *Candida* genes to be absent from *Saccharomyces*; this amounts to about 180 genes. As we expand our catalog of these genes, they may give us significant insight into the particular characteristics of *Candida* important for its commensal lifestyle.

A third category of genes of significant interest includes those that are clearly homologous but have significant sequence differences. An example of this is the *CDC4* gene from *Candida*, which contains a region of high similarity to the *CDC4* gene from *Saccharomyces* for much of the sequence, but which lacks the N-terminal region of the *Saccharomyces* protein. This gene complements a *cdc4* mutant of *Saccharomyces* for its mitotic but not its meiotic functions (J. Rosamond, personal communication). Such differences will provide the same kind of insights as will genes unique to *Candida*.

Several general insights about yeast genome organization have arisen from the *Saccharomyces* project. Among these are the number of gene families (an important consideration when considering particular genes as drug targets), the frequency of introns, and the density of genes. The latter two characteristics of genome organization will not be forthcoming in the first round of the *Candida* genome project, because many genes will not be completely sequenced and the amount of intergenic DNA will not be known (although it can be estimated). The role of the special biology of *C. albicans* in the significant differences in telomere and centromere structure, repeat organization, and chromosome size remains to be elucidated, but it seems very likely that some or all of these differences will be related to pathogenicity.

◆◆◆◆◆◆ XI. CONCLUSION

Genomic studies of *C. albicans* are well underway, and in the next few years 95% of the genes from this important human pathogen will have been identified and a physical map will be available. These tools will allow researchers to examine not only the gene but also its context in the genome. Comparison of these data with the *S. cerevisiae* genome sequence will greatly facilitate genetic studies on both the peculiar biology and the pathogenesis of the organism. We can look forward to a replacement of the current step-by-step attack on particular problems with a broader approach to the complex of properties that make this organism such an important medical problem.

Acknowledgements

I would like to thank Bebe Magee and Brian Corner for helpful criticisms and suggestions on the manuscript and Stew Scherer for providing much of the information, especially about the Candida Sequencing Project. I am grateful to Hiroji Chibana, Brian Corner, Peter Sudbery and John Rosamond for allowing me to cite their unpublished results. The work from my laboratory was supported by USPHS grants AI 16567 and AI 35109.

References

Alani, E., Cao, L. and Kleckner, N. (1987). A method for gene disruption that allows repeated use of *URA3* selection in the construction of multiply disrupted yeast strains. *Genetics* **116**, 541–545.

Bailey, D. A., Feldmann, P. J. F., Bovey, M., Gow, N. A. R. and Brown, A. J. P. (1996). The *Candida albicans HYR1* gene, which is activated in responses to hyphal development, belongs to a gene family encoding yeast cell wall proteins. *J. Bacteriol.* **178**, 5353–5360.

Barton, R. C. and Gull, K. (1992). Isolation, characterization, and genetic analysis of monosomic, aneuploid mutants of *Candida albicans. Mol. Microbiol.* **6**, 171–177.

Bertram, G., Swoboda, R. K., Gooday, G. W., Gow, N. A. and Brown, A. J. P. (1996). Structure and regulation of the *Candida albicans ADH1* gene encoding an immunogenic alcohol dehydrogenase. *Yeast* **12**, 115–127.

Brown, D. H., Slobodkin, I. V. and Kumamoto, C. A. (1996). Stable transformation and regulated expression of an inducible reporter construct in *Candida albicans* using restriction enzyme-mediated integration. *Mol. Gen. Genet.* **251**, 75–80.

Carle, G. F., Frank, M. and Olson M. V. (1986). Electrophoretic separations of large DNA molecules by periodic inversion of the electric field. *Science* **232**, 65–68.

Chindamporn, A., Nakagawa, Y., Homma, M., Chibana, H., Doi, M. and Tanaka, K. (1995). Analysis of the chromosomal localization of the repetitive sequences (RPSs) in *Candida albicans. Microbiology* **141**, 469–476.

Chu, W. S., Magee, B. B. and Magee, P. T. (1993). Construction of an *Sfi*I macro-restriction map of the *Candida albicans* genome. *J. Bacteriol.* **175**, 6637–6651.

Cormack, B., Bertram, G., Egerton, M., Gow, N. A. R. and Brown, A. J. P. (1997). Yeast-enhanced green fluorescent protein (yEGFP): a reporter of gene expression in *Candida albicans. Microbiology* **143**, 303–311.

Fonzi, W. A. and Irwin, M. Y. (1993). Isogenic strain construction and gene mapping in *Candida albicans. Genetics* **134**, 717–728.

Game, J. C., Bell, M., King, J. S. and Mortimer, R. K. (1990). Random-breakage mapping, a rapid method for physically locating an internal sequence with respect to the ends of a DNA molecule. *Nucl. Acids Res.* **18**, 4453–4461.

Geber, A., Williamson, P. R., Rex, J. H., Sweeney, E. C. and Bennet, J. E. (1992).

Cloning and characterization of a *Candida albicans* maltase gene involved in sucrose utilization. *J. Bacteriol.* **174**, 6992–6996.

Gorman, J. A., Chan, W. and Gorman, J. W. (1991). Repeated use of *GAL1* for gene disruption in *Candida albicans*. *Genetics* **129**, 19–24.

Goshorn, A. K. and Scherer, S. (1989). Genetic analysis of prototrophic natural variants of *Candida albicans*. *Genetics* **123**, 667–673.

Goshorn, A. K., Grindle, S. and Scherer, S. (1992). Gene isolation by complementation in *Candida albicans* and applications to physical and genetic mapping. *Infect. Immun.* **60**, 876–884.

Iwaguchi, S., Homma, M., Chibana, H. and Tanaka, K. (1992a). Isolation and characterization of a repeated sequence (RPS1) of *Candida albicans*. *J. Gen. Microbiol.* **138**, 1893–1900.

Iwaguchi, S., Homma, M. and Tanaka, K. (1992b). Clonal variation of chromosome size derived from the rDNA cluster region in *Candida albicans*. *J. Gen. Microbiol.* **138**, 1177–1184.

Kim, U. J., Shizuya, H., de Jong, P. J., Birren, B. and Simon, M. I. (1992). Stable propagation of cosmid sized human DNA inserts in an F factor based vector. *Nucl. Acids Res.* **20**, 1083–1085.

Kurtz, M. B., Cortelyou, M. W., Miller, S. M., Lai, M. and Kirsch, D. R. (1987). Development of autonomously replicating plasmids for *Candida albicans*. *Mol. Cell. Biol.* **7**, 209–217.

Kurtzman, C. P. and Robnet, C. J. (1997). Identification of clinically important ascomycetous yeasts based on nucleotide divergence in the 5′ end of the large-subunit (26S) ribosomal DNA gene. *J. Clin. Microbiol.* **35**, 1216–1223.

Lasker, B. A., Page, L. S., Lot, T. J. and Kobayashi, G. S. (1992). Isolation, characterization, and sequencing of *Candida albicans* repetitive sequence element 2. *Gene* **116**, 51–57.

Leuker, C. E., Hahn, A. M. and Ernst, J. F. (1992). Beta-galactosidase of *Kluyveromyces lactis* (Lac4p) as reporter of gene expression in *Candida albicans* and *C. tropicalis*. *Mol. Gen. Genet.* **235**, 235–241.

Magee, B. B. and Magee, P. T. (1997). WO-2, a stable aneuploid derivative of *Candida albicans* strain WO-1, can switch from white to opaque and form hyphae. *Microbiology* **143**, 289–295.

McEachern, M. J. and Hicks, J. B. (1993). Unusually large telomeric repeats in the yeast *Candida albicans*. *Mol. Cell. Biol.* **13**, 551–560.

Myers, K. K., Sypherd, P. S. and Fonzi, W. A. (1995). Use of *URA3* as a reporter of gene expression in *C. albicans*. *Curr. Genet.* **27**, 243–248.

Negredo, A., Monteoliva, L., Gil, C., Pla, J. and Nombela, C. (1997). Cloning, analysis, and one-step disruption of the *ARG5,6* gene in *Candida albicans*. *Microbiology* **143**, 297–302.

Pla, J., Perez-Diaz, R. M., Navarro-Garcia, F., Sanchez, M. and Nombela, C. (1995). Cloning of the *Candida albicans HIS1* gene by direct complementation of a *C. albicans* histidine auxotroph using an improved double-ARS shuttle vector. *Gene* **165**, 115–120.

Pla, J., Monteoliva, L., Navarro-Garcia, F., Sanchez, M. and Nombela, C. (1996). Understanding *Candida albicans* at the molecular level. *Yeast* **12**, 1677–1702.

Sadhu, C., McEachern, M. J., Rustchenko-Bulgac, E. P., Schmid, J., Soll, D. R. and Hicks, J. B. (1991). Telomeric and dispersed repeat sequences in *Candida* yeasts and their use in strain identification. *J. Bacteriol.* **173**, 842–850.

Santos, M. A. and Tuite, M. F. (1995). The CUG codon is decoded *in vivo* as serine and not leucine in *Candida albicans*. *Nucl. Acids Res.* **23**, 1481–1486.

Scherer, S. and Stevens, D. A. (1988). A *Candida albicans* dispersed, repeated gene family and its epidemiologic applications. *Proc. Natl Acad. Sci. USA* **85**, 1452–1456.

414

Smith, H. A., Gorman, J. W., Koltin, Y. and Gorman, J. A. (1990). Functional expression of the *Candida albicans* beta-tubulin gene in *Saccharomyces cerevisiae*. *Gene* **90**, 115–123.

Srikantha, T., Chandrasekhar, A. and Soll, D. R. (1995). Functional analysis of the promoter of the phase-specific *WH11* gene of *Candida albicans*. *Mol. Cell. Biol.* **15**, 1797–1805.

Srikantha, T., Klapach, A., Lorenz, W. W., Tsai, L. K., Laughlin, L. A., Gorman, J. A. and Soll, D. R. (1996). The sea pansy *Renilla reniformis* luciferase serves as a sensitive bioluminescent reporter for differential gene expression in *Candida albicans*. *J. Bacteriol.* **178**, 121–129.

Stoldt, V. R., Sonneborn, A., Leuker, C. E. and Ernst, J. F. (1997). Efg1p, an essential regulator of morphogenesis of the human pathogen *Candida albicans*, is a member of a conserved class of bHLH proteins regulating morphogenetic processes in fungi. *EMBO J.* **16**, 1982–1991.

Vollrath, D., and Davis, R. W. (1987). Resolution of DNA molecules greater than 5 megabases by contour-clamped homogeneous electric fields. *Nucl. Acids Res.* **15**, 7865–7876.

Whelan, W. L. and Soll, D. R. (1982). Mitotic recombination in *Candida albicans*: recessive lethal alleles linked to a gene required for methionine biosynthesis. *Mol. Gen. Genet.* **187**, 477–485.

Wickes, B., Staudinger, J., Magee, B. B., Kwon-Chung, K. J., Magee, P. T. and Scherer, S. (1991). Physical and genetic mapping of *Candida albicans*: several genes previously assigned to chromosome 1 map to chromosome R, the rDNA-containing linkage group. *Infect. Immun.* **59**, 2480–2484.

22 Applications of Yeast in the Discovery of Human Gene Function

Paul A. Moore

Human Genome Sciences, Department of Molecular Biology, Rockville, Maryland, USA

◆◆◆

CONTENTS

List of Abbreviations

BLAST	Basic local alignment search tool
EST	Expressed sequence tag
MIPS	Martinsried Institute for Protein Sequences
NCBI	National Center for Biotechnology Information
OMIM	Online Mendelian Inheritance in Man
PCR	Polymerase chain reaction
PRIMES	Peptide/receptor interactions in microbial expression systems

◆◆◆◆◆◆ **I. INTRODUCTION**

Its convenience of handling and ease of both classical and recombinant genetic manipulation have conspired to make the yeast *Saccharomyces cerevisae* an invaluable model system in which to dissect both metabolic pathways and fundamental cellular processes such as the cell cycle, DNA repair, transcription, translation and protein secretion. Subsequent analysis of these pathways in higher eukaryotes has demonstrated that, despite a large evolutionary distance, yeast and humans exhibit a remarkable conservation of gene function. A consequence of this conservation is the ability to identify the functional homolog (ortholog) of a yeast gene from another eukaryote. Once identified, the comparison of ortholog gene sequences isolated from species separated by evolution allows the identification of the domain(s) within proteins that are of the greatest structural and/or functional importance. In addition to improving our understanding of fundamental cellular processes, the identification of human orthologs is particularly attractive considering the ill-defined genetic predisposition of many human disease states. For example, the discovery of the human orthologs of a series of yeast genes involved in DNA repair was instrumental in their identification as genes mutated in some colon cancers (Papadopoulos *et al.*, 1994). Considering the developmental complexity associated with multicellular organisms, one would expect that only a subset of human genes will have a counterpart in yeast. Indeed, this is reflected in the 10–20-fold difference in the number of genes encoded by the yeast genome and the predicted number in the partially sequenced human genome. Despite this, various elegant yeast screening procedures have been developed that allow the identification of nuclear, cytoplasmic, membrane-bound and secreted protein-encoding genes for which no homolog necessarily exists in yeast.

This chapter describes approaches used to identify human homologs of a yeast gene by database homology searching, and yeast genetic screens suitable for the functional isolation of human genes that lack a counterpart in yeast.

◆◆◆◆◆◆ II. ANALYSIS OF HUMAN GENE FUNCTION BY COMPARATIVE YEAST GENOMICS

The isolation of a homolog of a yeast gene from another species can be achieved by a variety of methods including functional complementation, degenerate polymerase chain reaction (PCR), and low stringency hybridization of a cDNA library with either a DNA or an antibody probe. While these approaches have been rewarding in the past, they are relatively labor-intensive and are now being superseded by computer-based searches for gene homologs, which exploit the availability of ever-expanding genomic and cDNA sequence databases from a range of organisms. Particularly useful for human gene identification are expressed sequence tags (ESTs), which are sequences of approximately 300–500 bp, often derived from a longer cDNA fragment. Essentially, ESTs capture a snap-shot of an mRNA sequence. Considering that only a small percentage of the human genome is actually transcribed into mRNA, EST sequencing has provided a more efficient approach towards identifying human protein encoding genes than genomic sequencing.

Equipped with Internet links through a browser such as Netscape, the identification of the human ortholog of a specific yeast gene can be performed via multiple avenues. For example, the Martinsried Institute for Protein Sequences (MIPS) database (http://www.mips.biochem. mpg.de/) provides access to systematically performed comparisons of the complete yeast genome against human cDNA sequences (detailed in

Chapter 3). To perform a search for a human homolog of a specific yeast gene, the services provided by the National Center for Biotechnology Information (NCBI) (http://www.ncbi.nlm.nih.gov/) provide an alternative approach. Before performing a search for homologs, one must retrieve the nucleotide and/or amino acid sequence of the target yeast gene. These sequences can be obtained from a variety of sources, including the Entrez browser site provided by NCBI (http://www.ncbi.nlm.nih.gov/Entrez/). To perform a database search, the query sequence can be copied and pasted directly into the window provided by NCBI for performing either basic or advanced basic local alignment search tool (BLAST) homology searches (http://www.ncbi.nlm.nih.gov/BLAST/). Variations of the BLAST algorithm, designed by Altschul *et al.* (1990) are provided, and depending on the input query sequence and the database to be searched, the appropriate algorithm should be selected. A range of databases are available to compare the query sequence against, including a database of EST sequences (dbEST) which, at the time of writing, includes over 700 000 human EST sequences. The default parameters that control the stringency of the search and the data output are sufficient for initial analysis but can be refined if necessary. The results of the homology search can either be delivered by e-mail or via the Web, and they should be examined carefully to determine if a homolog has been identified. A review by Altschul *et al.* (1994) provides a detailed account of the various parameters to be considered when interpreting BLAST results, such as similarity scores and their statistical significance.

The failure to identify a human homolog of a particular yeast gene could indicate either the absence of such a homolog in humans, or that no EST corresponding to a conserved domain has been sequenced thus far. In cases where an EST displaying significant homology to the query sequence is observed, it is useful to determine whether it overlaps with other EST sequences, including those that were not observed in the initial BLAST search, possibly because they correspond to a non-conserved domain. This can be determined by subjecting the EST sequence to a BLAST analysis against dbEST and forming a contig of overlapping EST sequences using DNA sequence analysis software. By repeating this process of "dry" walking it is possible to generate a contig that could potentially represent the entire open reading frame. Alternatively, overlapping ESTs can be identified using the Unigene database (http://www.ncbi.nlm.nih.gov/UniGene/index.html), which has grouped together ESTs that correspond to the same transcription unit. The Unigene database also provides valuable information regarding ESTs that have been grouped in a single transcription unit, for example, the cDNA libraries from which they were derived, and in certain cases their chromosomal location. Thus, without physically handling a human EST sequence, it is possible to get an indication of both its tissue distribution and its chromosomal location. The chromosomal location of a human gene is particularly relevant to determining whether the gene is associated with a genetically inherited human disease state. Genetic disorders associated with a particular area of the human genome can be accessed using the Online Mendelian Inheritance in Man (OMIM) server (http://www.ncbi.nlm.nih.gov/Omim/).

To confirm and, if necessary, complete the cDNA sequencing of a putative homolog, the EST cDNA should be obtained from a source designated by the creator of the EST (a list of distributors is provided at http://www-bio.llnl.gov/bbrp/image/idist_add.html). In cases where the EST(s) identified does not correspond to full-length mRNA, a full-length cDNA can be obtained by screening a cDNA library (generated using the same tissue from which the EST was derived) using the EST as a probe. Once the full-length open reading frame is identified, the extent of sequence similarity between the identified human gene and yeast gene can be further established using algorithms such as BEST-FIT and FASTA, which introduce alignment gaps and hence can identify homologies potentially missed by BLAST searches.

A major attraction of identifying a human counterpart of a yeast gene (or *vice versa*) is the relative ease with which the human sequence can be analysed in yeast. The first step involves the construction of a yeast mutant in which the target gene has been inactivated. Such a strain might already be available through the European or American yeast genome analysis programs (Oliver, 1996; Shoemaker *et al.*, 1996). Alternatively, straightforward strategies for yeast gene disruption are described in Chapter 5 in this volume. The next step is to test whether the human sequence can functionally complement the yeast mutation. This would confirm that a genuine ortholog has been identified. A prerequisite for this is that the yeast mutant displays a scorable phenotype such as a certain growth defect or conditional lethality, which can be rescued by complementation. If this is the case, the human cDNA should be subcloned into a yeast expression vector. A set of yeast expression vectors created by Mumberg *et al.* (1995) are ideally suited for this purpose. In cases where the mutated yeast strain has a specific growth defect, the plasmid expressing the human cDNA is introduced under conditions in which the yeast mutant grows normally and is then tested for its ability to rescue the yeast mutant strain under the restrictive condition. Using this approach, Hartzog *et al.* (1996) demonstrated complementation of the yeast *spt4* mutant with a human ortholog *SUPT4H*, discovered by dbEST homology searching.

In situations where the yeast mutation is lethal, it is necessary to perform a plasmid shuffle, such as that performed by McKune *et al.* (1995), who demonstrated that specific human RNA polymerase subunits are capable of substituting for their essential yeast counterpart. An extensive list of human genes capable of complementing yeast mutants is provided at http://www.ncbi.nlm.nih.gov/XREFdb/. It should be borne in mind, however, that in certain cases human genes are only capable of partial complementation of a yeast mutant *in vivo*, either at selective temperatures or when the human gene is expressed from a multicopy plasmid as demonstrated by McKune *et al.* (1995). Furthermore, there are other situations in which a human gene is highly similar in sequence to a yeast gene and can be clearly demonstrated *in vitro* to perform the functions attributed to its yeast counterpart, but yet is incapable of complementing that gene *in vivo* (Cormack *et al.*, 1991). The lack of complementation in these cases might reflect an evolutionary divergence not of the core function of the protein itself, but rather of protein–protein interactions involving the target protein and additional components of a multi-subunit complex.

The identity of key functional amino acid residues can be gleaned from a closer inspection of amino acid sequence alignments of ortholog proteins from eukaryotic species positioned at various stages of evolution. To facilitate this, BLAST servers for *Caenorhabditis elegans* and *Schizosaccharomyces pombe* are provided by The Sanger Centre (http://www.sanger.ac.uk/) while NCBI (http://www.ncbi.nlm.nih.gov/BLAST/) provides a capability for the analysis of the non-redundant database, which contains sequences from all organisms.

Finally, in certain cases it is desirable to identify the yeast counterpart of a specific human gene. Subsequent characterization of the yeast ortholog using the various genetic and recombinant technologies available in yeast permits functional analyses that are impossible to perform in humans (e.g. gene knock-outs, structure/function analyses, and extragenic suppression). The identification of a yeast homolog of a specific human gene or EST sequence can be performed via the BLAST or FASTA server provided at the Stanford Saccharomyces Genome Database (http://genome/www.stanford.edu/Saccharomyces/). The characterization of yeast genes homologous to human disease genes is particularly attractive as it can potentially accelerate our understanding of the function of the gene and the molecular basis of the disease. A database of yeast sequences homologous to known human disease genes is available at http://www.ncbi.nlm.nih.gov/Bassett/Yeast/.

◆◆◆◆◆◆ III. IDENTIFICATION OF HUMAN GENES BY GENETIC SELECTION IN YEAST

A. The Two Hybrid Assay: Variations on a Theme

Since its inception by Fields and Song (1989), numerous variations of the yeast two hybrid assay have been used to extend the range of protein-encoding genes it can identify. While the theory and methodology of the yeast two hybrid assay are detailed in Chapter 15, extensions to the assay pertinent to the isolation of human genes are discussed here.

I. DNA-binding proteins

While the expression of eukaryotic genes can be controlled at multiple levels, most genes are regulated at the level of transcriptional initiation. As a result, there is great interest in identifying the transcriptional regulatory protein(s) that bind the promoter regions of specific genes and regulate their transcription. The yeast one hybrid assay (Figure 1) has been developed from the two hybrid assay to specifically identify DNA-binding proteins capable of binding a specific DNA sequence. In this approach, a DNA sequence that has been demonstrated previously to contain a target sequence for the regulatory protein under investigation (e.g. by gel shift or promoter-reporter analyses) is subcloned in the context of a basal promoter upstream of a yeast reporter gene such as *lacZ* or *HIS3* and stably integrated into a suitable yeast strain (Figure 1A). It is worth nothing that the neomycin resistance gene has also been successfully used as a reporter gene for one hybrid screening by Chan *et al.* (1993). Pivotal to the applicability of the one hybrid assay is that the DNA bind-

A.

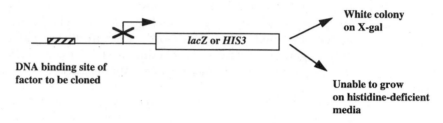

B.

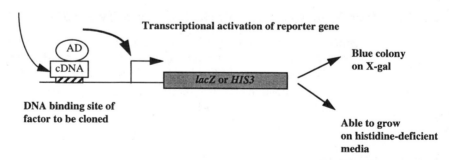

Figure 1. The yeast one hybrid assay. (A) The minimal DNA recognition site of the human DNA-binding protein encoding gene to be cloned is placed in the context of a minimal yeast promoter, upstream of a selectable reporter gene such as *lacZ* or *HIS3*. Assuming no endogenous yeast transcription factor is capable of binding the DNA recognition site and activating transcription, the reporter gene will not be activated. (B) A library of cDNAs fused to a transcriptional activation domain is introduced into the yeast reporter strain and an interaction between a cDNA-encoded protein with the target DNA recognition sequence will result in activation of the *lacZ* or *HIS3* reporter gene.

ing sequence is not itself capable of directing transcriptional activation of the reporter gene in yeast. This therefore precludes the use of most yeast promoter regions, but is ideal for the analysis of human promoter regions for which no endogenous yeast protein is capable of binding and activating transcription. To decrease possible activation by endogenous yeast factors, a well-defined DNA binding site for the regulatory factor of interest should be used. The sensitivity of the yeast reporter strain can then be increased through the use of concatemerized minimal binding sites.

The yeast reporter strain is then transformed with a plasmid library expressing human cDNA sequences fused to a transcriptional activation domain (such as the libraries used in two hybrid screens; Chapter 15), and transformants screened for activation of the reporter gene (Figure 1B). Plasmids are isolated from transformants that reproducibly demonstrate activation of the reporter gene, and reintroduced into yeast to confirm they are capable of activating transcription specifically through the promoter element. As a partial length cDNA fragment present in the activation domain fusion may correspond to the DNA-binding region only and may not contain complete sequence, it may be necessary to obtain a full-length cDNA. This can be achieved either by hybridization screening of cDNA libraries, or more rapidly by searching the public databases for sequences that overlap the cDNA fragment cloned in the activation domain fusion, as described in section II.

The feasibility of the one hybrid approach was initially demonstrated by Wang and Reed (1993) who devised the assay to isolate the rat olfactory neuronal transcription factor, Olf1. Subsequently, Chan *et al.* (1993) identified the human NF-E2-related protein (Nrf1) gene in a screen for proteins capable of binding the NF-E2 DNA recognition sequence. In addition to the cloning of transcription factors, this powerful assay should find applications in the discovery of other DNA-binding proteins including transcriptional repressors, proteins involved in DNA replication, and proteins that play a role in the structural organization of DNA.

2. RNA-binding proteins

Proteins capable of binding specific RNA sequences play integral roles in numerous cellular processes including RNA processing, translation, mRNA stability and RNA viral infection. To aid in the identification of RNA-binding proteins, SenGupta *et al.* (1996) have developed a yeast three hybrid assay system (Figure 2). In this system, a yeast reporter strain containing the *lacZ* and *HIS3* reporter genes under the control of a basal promoter bearing multiple copies of the lexA binding site was constructed. This strain was then stably transformed with a constitutively expressed lexA–MS2 fusion protein. LexA is an *Escherichia coli* DNA-binding protein that binds the lexA binding sites (now incorporated into the promoters of the reporter genes), while the bacteriophage MS2 coat protein is an RNA-binding protein that recognizes a 21-nucleotide RNA stem-loop structure sequence in the MS2 genome with high affinity. Binding of the lexA–MS2 fusion protein to the lexA binding site brings the

MS2 RNA-binding protein into the proximity of the promoter–reporter gene. The resulting strain can be used universally for the identification of any specific RNA-binding protein. The specificity of the screen is provided by an RNA fusion comprising the RNA target sequence fused to the MS2 stem loop sequence which binds the MS2 protein. To facilitate co-operative binding of the MS2 coat binding protein, two copies of this

A.

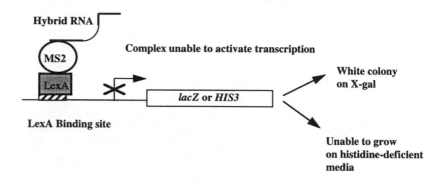

B.

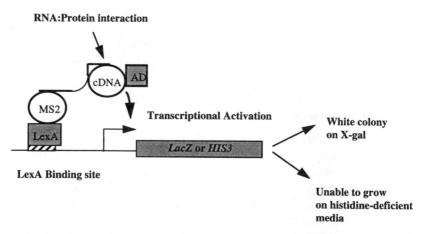

Figure 2. The yeast three hybrid assay. (A) A yeast strain containing a stably integrated selectable reporter gene (e.g. *lacZ* and/or *HIS3*) under the control of a minimal promoter bearing multiple lexA binding sites is transformed with a lexA–MS2 fusion protein together with a hybrid RNA containing MS2 protein binding sites fused to the target RNA sequence for which a binding protein is sought. (B) A cDNA library fused to a transcriptional activation domain is then introduced. An interaction between a cDNA-encoded protein and the RNA target sequence will result in activation of the selectable reporter gene (SenGupta *et al.*, 1996).

high-affinity MS2 RNA sequence are employed. The hybrid RNA encoding gene is placed under the control of polymerase III promoter and terminator sequences derived from the yeast *RPR1* gene, and transformed on a plasmid into the yeast reporter strain. Despite being in an artificial context, the resulting hybrid RNA is still capable of interacting with the MS2 domain of the lexA–MS2 fusion protein, bringing the hybrid RNA into the proximity of the promoter–reporter gene (Figure 2A). The RNA target sequence is then free to bind proteins expressed, for example, by a library of human cDNA sequences fused to a transcriptional activation domain. In this way an RNA–protein association can be monitored through activation of the reporter gene, as described for the two hybrid assay (Figure 2B). Plasmids are isolated from transformants demonstrating reproducible activation of both reporter genes, and reintroduced into yeast to confirm that the human cDNA sequence activates the reporters by interacting specifically with the target RNA sequence.

The usefulness of the three hybrid assay in identifying novel RNA-binding proteins was recently demonstrated by the isolation of the human gene encoding a histone mRNA hairpin binding protein by Martin *et al.* (1997). The histone mRNA hairpin binding protein had previously been identified in nuclear and polysomal extracts as a factor that binds the hairpin structure at the 3′ end of histone mRNAs, which plays an important role in nucleocytoplasmic translation and stability. However, the protein had proven difficult to purify for amino acid sequencing and its identity remained unknown until the cloning of the gene using the three hybrid assay. With its validity now established, it is anticipated that the three hybrid assay will be used for the identification of other specific RNA-binding proteins including those important in the regulation of human retroviruses such as HIV-1.

3. Receptors and ligands

In humans, many key processes during development, differentiation and survival are controlled by extracellular signals that bind to and activate membrane-bound receptors. In response to engagement of an extracellular ligand, the receptor directs signaling to the nucleus via an ill-defined set of intracellular ligands, culminating in coordinated nuclear gene activation. Identification of the various extracellular ligands, receptors and intracellular ligand associations is therefore critical towards gaining a complete understanding of the intricacies of human development. With regards to this challenge, the yeast two hybrid assay has proven and potential applications.

The cytoplasmic domains of various humans receptors, including members of the tyrosine kinase (Liu and Roth, 1995), serine/threonine kinase (Kawabata *et al.*, 1995) and tumor necrosis factor (Hsu *et al.*, 1995) receptor families, have been used as baits in two hybrid screens. This has led to the identification of both novel receptors and receptor-associated proteins which can form a heterodimer with the bait receptor (Kawabata *et al.*, 1995) and novel internal signaling molecules (Liu and Roth, 1995;

Hsu *et al.*, 1995). However, the inability of yeast to perform the accurate posttranslational modifications of human receptors critical for their ability to form protein–protein interactions limits the range of receptors that can be analysed in the standard yeast two hybrid assay. To overcome this limitation, Osbourne *et al.* (1995) have developed the "tribrid" system in which a third plasmid expressing a human gene capable of compensating for a specific defect in yeast posttranslational modification is introduced. Specifically, the assay was developed to identify proteins that interact with the cytoplasmic domain of the high affinity IgE receptor (FcεRI),

A.

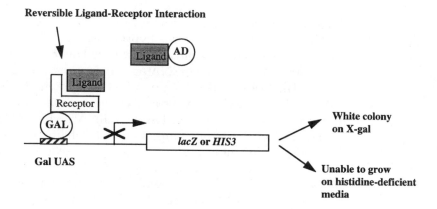

B.

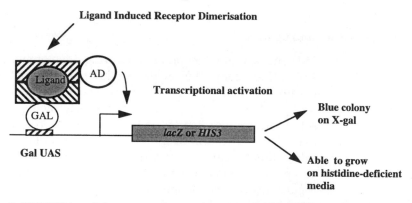

Figure 3. PRIMES (peptide receptor interactions in microbial expression systems). (A) The two hybrid association between the extracellular domain of a specific single transmembrane receptor fused to a DNA binding domain with its cognate ligand fused to a transcriptional activation domain can be disrupted by overexpression of free ligand, resulting in the blocking of reporter gene transcription. (B) Expression of free ligand can in specific cases mediate the homodimerization of a DNA binding domain–receptor fusion with an activation domain–receptor fusion resulting in transcriptional activation of a reporter gene. See Chapter 15 for details of the two hybrid system upon which this assay is based.

which requires phosphorylation by a tyrosine kinase to be capable of associating with internal signaling molecules such as the protein tyrosine kinase Syk. Because yeast lacks significant tyrosine kinase activity, FcεRI is not phosphorylated in yeast and is unable to associate with the SH2 domain of Syk. However, the co-expression in yeast of either the human Lck or Lyn protein tryosine kinase phosphorylates FcεRI, resulting in its ability to associate with Syk. The usefulness of the tribrid system to identify novel ligands was demonstrated in a screen for proteins capable of interacting with FcεRI in a yeast strain co-expressing Lck. This screen resulted in the isolation of a novel SH2-domain-containing protein (Osbourne *et al.*, 1995).

The two hybrid assay also has applications in the characterization of interactions involving single transmembrane receptors and their extracellular polypeptide ligands. Specifically, two members of the cytokine superfamily, growth hormone and prolactin, were demonstrated by Ozenberger and Young (1995) to be capable of interacting specifically with the extracellular domains of their cognate receptor by two hybrid analysis. Concurrent expression of free ligand (growth hormone or prolactin) was then shown to competitively inhibit association between the ligand–DNA binding domain fusion and the receptor–activation domain fusion. This illustrates that the yeast system is capable of mimicking the reversibility of the ligand–receptor interaction observed in mammalian systems (Figure 3A). In addition, the PRIMES (peptide/receptor interactions in microbial expression systems) technology was used to demonstrate ligand-induced receptor homodimerization (Figure 3B) for both the growth hormone receptor and the vascular endothelial growth factor receptor (Ozenberger and Young, 1995). Now that the validity of the yeast two hybrid assay and PRIMES for the analysis of receptor–ligand interactions has been established, it can be anticipated that this will facilitate both structure/function analyses of existing human ligand–receptor interactions, and lead to the identification of novel human ligands and receptors via cDNA library screening.

B. Signal Peptide Selection

Central to multicellular development and differentiation are the intracellular signals mediated by extracellular signaling molecules and membrane-bound receptors. The entry of both extracellular proteins and membrane-bound proteins into the secretory pathway is programmed by the presence of an amino-terminal signal peptide sequence which is cleaved off while the protein is translocated through the membrane. The signal peptide, which immediately follows the initiating methionine, generally consists of a basic amino acid followed by a hydrophobic core of 4–15 amino acids and a pair of small uncharged amino acids separated by a single intervening amino acid that defines the cleavage site. A complete understanding of the mechanisms of cellular signaling will require the identification of a large repertoire of secreted and membrane-bound proteins. A genetic screen in yeast recently described by Klein *et al.* (1996), which selects for signal peptide sequences, should greatly facilitate this process.

The signal peptide selection assay devised by Klein and co-workers (1996) revolves around the properties of the secreted yeast protein invertase, encoded by the *SUC2* gene. Invertase is required in yeast for the extracellular breakdown of sucrose to fructose and glucose. Deletion of the invertase signal peptide renders yeast incapable of growing on sucrose as a sole carbon source. However, Kaiser *et al.* (1987) have demonstrated previously that this signal sequence can be replaced by a heterologous signal sequence, resulting in the secretion of a protein capable of breaking down sucrose. Klein *et al.* (1996) took advantage of this observation and fused short rat cDNA fragments to an invertase deletion that lacks both its signal peptide and an initiating methionine. The library was then transformed into a *suc2* null mutant and transformants selected on the basis of their ability to grow on sucrose in the presence of antimycin (an inhibitor of any leaky growth of the *suc2* mutant strain on sucrose). Plasmid DNA was recovered from sucrose-utilizing yeast transformants, and sequenced. The usefulness of this screen for signal peptides for the isolation of secreted proteins is reflected in the fact that out of the 161 unique genes identified (including 49 novel cDNAs), at least 138 appeared to encode extracellular proteins (based on similarity to a known secreted protein or, in the case of a novel cDNA, on the fact that they carry a putative signal peptide sequence).

Now that the usefulness of this assay has been proven, it should now be possible to perform a yeast signal peptide screen on cDNA libraries from any human tissue type. It is worth noting that a similar signal peptide selection system has also been described for use in mammalian cells by Tashiro *et al.* (1993). While this system has certain advantages over the yeast system (for example, yeast may not recognize all mammalian secretory signals), it has the disadvantage of requiring multiple rounds of selection to isolate a pool of cDNA clones containing putative signal peptide sequences. A disadvantage common to both systems is the potential to isolate false positives that frequently correspond to cDNA fragments containing internal hydrophobic sequences such as transmembrane domains preceded by an in-frame methionine, and this should be borne in mind when interpreting the output of such a screen. Furthermore, because the screen is based on the isolation of amino-terminal sequences, it may be necessary to isolate cDNAs encoding the remainder of the protein before performing further biological studies. This can be achieved either by traditional screening of cDNA libraries with the isolated cDNA fragment, or by searching the databases for sequences that overlap with the isolated cDNA fragment, as described in section II.

◆◆◆◆◆◆ IV. CONCLUSIONS

The yeast *Saccharomyces cerevisiae* has provided a powerful tool for the isolation and functional analysis of human genes. Using resources available on the internet, it is relatively straightforward to identify human homologs of yeast genes of known function. Complementation studies using well-defined yeast mutants provide a convenient tool for testing whether yeast and human proteins represent functional homologs. Undoubtedly, as the functions of more yeast genes become defined, and as human DNA sequence databases expand, comparative genomics will continue to provide an invaluable approach. In addition, elegant yeast genetic screens expand the usefulness of yeast into the functional identification of human genes for which no homolog necessarily exists in yeast. These screens include variations of the two hybrid assay, the signal peptide selection assay, and the reconstituted human seven transmembrane assay described by Broach and Thorner (1996). Such screens are capable of uncovering a wide range of human proteins, including DNA-binding proteins, RNA-binding proteins, cytoplasmic proteins, membrane-bound proteins and extracellular proteins. These screens and other newly developed assays will continue to keep yeast at the forefront of the discovery of human genes and their function. Furthermore, it is clear that analogous assays are applicable to the functional analysis of genes from other multicellular organisms.

References

Altschul, S. F., Gish, W., Miller, W., Myers, E. W. and Lipman, D. J. (1990). Basic local alignment search tool. *J. Mol. Biol.* **215**, 403–410.

Altschul, S. F., Boguski, M. S., Gish, W. and Wootton, J. C. (1994). Issues in searching molecular sequence databases. *Nature Genetics* **6**, 119–129.

Broach, J. R. and Thorner, J. (1996). High-throughput screening for drug discovery. *Nature* **384** (Suppl), 14–16.

Chan, J. Y., Han, X-L. and Kan, Y. W. (1993). Cloning of Nrf1, an NF-E2-related transcription factor, by genetic selection in yeast. *Proc. Natl Acad. Sci. USA* **90**, 11371–11375.

Cormack, B. P., Strubin, M., Ponticelli, A. S. and Struhl, K. (1991). Functional differences between yeast and human TFIID are localized to the highly conserved region. *Cell* **65**, 341–348.

Fields, S. and Song, O. (1989). A novel genetic system to detect protein–protein interactions. *Nature* **340**, 245–246.

Hartzog, G. A., Basrai, M. A., Ricupero-Hovasse, S. L., Hieter, P. and Winston, F. (1996). Identification and analysis of a functional human homolog of the *SPT4* gene of *Saccharomyces cerevisiae*. *Mol. Cell. Biol.* **16**, 2848–2856.

Hsu, H., Xiong, J. and Goeddel, D. V. (1995). The TNF receptor 1-associated protein TRADD signals cell death and NF-κB activation. *Cell* **81**, 495–504.

Kaiser, C. A., Preuss, D., Grisafi, P. and Botstein, D. (1987). Many random sequences functionally replace the secretion signal sequence of yeast invertase. *Science* **235**, 312–317.

Kawabata, M., Chytil, A. and Moses, H. L. (1995). Coning of a novel type II serine/threonine kinase receptor through interaction with the type I transforming growth factor-beta receptor. *J. Biol. Chem.* **270**, 5625–5630.

Klein, R. D., Gu, Q., Goddard, A. and Rosenthal, A. (1996). Selection for genes encoding secreted proteins and receptors. *Proc. Natl Acad. Sci. USA* **93**, 7108–7113.

Liu, F. and Roth, R. A. (1995). Grb-IR: a SH2-domain-containing protein that binds to the insulin receptor and inhibits its function. *Proc. Natl Acad. Sci. USA* **92**, 10287–10291.

Martin, F., Schaller, A., Eglite, S., Schumperli, D. and Muller, B. (1997). The gene for histone RNA hairpin binding protein is located on human chromosome 4 and encodes a novel type of RNA binding protein. *EMBO J.* **16**, 769–778.

McKune, K., Moore, P. A., Hull, M. W. and Woychik, N. (1995). Six human RNA polymerase subunits functionally substitute for their yeast counterpart. *Mol. Cell. Biol.* **15**, 6895–6900.

Mumberg, D., Muller, R. and Funk, M. (1995). Yeast vectors for the controlled expression of heterologous proteins in different genetic backgrounds. *Gene* **156**, 119–122.

Oliver, S. G. (1996). From DNA sequence to biological function. *Nature* **379**, 579–600.

Osbourne, M. A., Dalton, S. and Kochan, J. P. (1995). The yeast tribrid system – genetic detection of trans-phosphorylated ITAM-SH2 interactions. *Biotechnology* **13**, 1474–1478.

Ozenberger, B. A. and Young, K. H. (1995). Functional interaction of ligands and receptors of the hematopoietic superfamily in yeast. *Mol. Endocrinol.* **9**, 1321–1329.

Papadopoulos, N., Nicolaides, N. C., Wei, Y. F., Ruben, S. M., Carter, K. C., Rosen, C. A., Haseltine, W. A. *et al.* (1994). Mutation of a mutL homolog in hereditary colon cancer. *Science* **263**, 1625–1629.

SenGupta, D. J., Zhang, B., Kraemer, B., Pochart, P., Fields, S. and Wickens, M. (1996). A three-hybrid system to detect RNA–protein interactions *in vivo*. *Proc. Natl Acad. Sci USA* **93**, 8496–8501.

Shoemaker, D. D., Lashkari, D. A., Morris, D., Mittmann, M. and Davis, R. (1996). Quantitative phenotypic analysis of yeast deletion mutants using a highly parallel molecular bar-coding strategy. *Nature Genetics* **14**, 450–456.

Tashiro, K., Tada, H., Heilker, R., Shirozu, M., Nakono, T. and Honjo, T. (1993). Signal sequence trap: a cloning strategy for secreted proteins and type I membrane proteins. *Science* **261**, 600–603.

Wang, M. M. and Reed, R. R. (1993). Molecular cloning of the olfactory neuronal transcription factor Olf-1 by genetic selection in yeast. *Nature* **364**, 121–126.

23 Yeast Mutant and Plasmid Collections

Karl-Dieter Entian and Peter Kötter

Institute for Microbiology, J.-W. Goethe-Universität, Frankfurt, Germany

◆◆

CONTENTS

Introduction
EUROSCARF
Other yeast mutant and plasmid collections
Conclusions and future developments

List of Abbreviations

ATTC	American Type Culture Collection
NPD	Non-parental ditype
ORF	Open reading frame
PCR	Polymerase chain reaction
PD	Parental ditype
TT	Tetrad type
YGSC	Yeast Genetic Stock Center

◆◆◆◆◆◆ I. INTRODUCTION

A. General Introduction

The yeast *Saccharomyces cerevisiae* is a very suitable organism for genetic analysis. Now that its genome has been completely sequenced, deletion mutants for each gene can easily be established by reverse genetics. Using the very precise recombination apparatus of *S. cerevisiae*, each gene locus can be replaced by selection markers such as amino acid auxotrophies, nucleoside auxotrophies or dominant resistance markers. This allows the generation of a collection of mutants for each of the 6200 open reading frames (ORFs) within the *S. cerevisiae* genome. Even deletions within essential genes can be collected in the form of heterozygous diploid strains. Additionally, deletion cassettes containing long flanking regions

METHODS IN MICROBIOLOGY, VOLUME 26
ISBN 0–12–521526–6

are also becoming available, so that the respective deletion can be introduced into any *S. cerevisiae* strain of interest. Furthermore, a complete set of genes is being collected on centromeric plasmids, and these cognate clones can be used for *trans*-complementation of the respective deletion mutations. At present, about 1200 deletion mutants, 1100 deletion cassettes and 900 cognate clones are available. The present collection (EUROSCARF, standing for European Saccharomyces cerevisiae archive for functional analysis) has been generated through a German (BMBF [Bundesministerium fuer Biotechnologische Forschung] network, about 325 mutants) and a European (EUROFAN, standing for European functional analysis network, about 875 mutants) scientific approach, both of which aim to study *S. cerevisiae* genes of unknown function. At present, access to the mutant and plasmid collection is limited to BMBF Network and EUROFAN members. However, in future it is intended to make the collection available to the whole scientific community.

B. Yeast Genetic Analysis

The yeast *S. cerevisiae* provides a simple model for the organization of eukaryotic cells. As a model microorganism it displays several advantages compared with multicellular eukarotic organisms, such as its fast growth (with generation times less than 2 h for haploid cells), its ease of handling and storage, as well as its accessibility to genetic manipulation. Haploid *S. cerevisiae* strains exist in two mating types, *MAT*a and *MAT*α, which are both stable in heterothallic *S. cerevisiae* strains. Mixing the two mating types results in diploid strains (Lindegren and Lindegren, 1943) which are stable and proliferate resulting in a clone of diploid cells. However, specific conditions such as nitrogen limitation can be used to stimulate meiosis in the diploids, resulting in asci with four haploid spores (Fowell, 1969). Such asci can be dissected with a micromanipulator, and the four progeny of one diploid cell can be followed after separation and germination of the respective spores (tetrad analysis). A semi-automatic micromanipulator for tetrad analysis of yeast asci has been developed by C. Singer. The Singer MSM micromanipulator (Singer Instruments, Watched, Somerset, UK) has a transmission of 1:30 000 and a precise step motor to place segregants at defined positions on agar plates (Figure 1). The Singer MSM micromanipulator is more easy to handle and needs less experience than the micromanipulator of De Fonbrune (1949), which is completely manual.

The particular life cycle of *S. cerevisiae* makes it possible to isolate recessive mutations from haploid strains as well as dominant mutations from diploid strains (Figure 2). After tetrad analysis, the recombination frequency between two mutations can be measured easily within the asci and genetic maps can be easily established.

Three types of asci are derived from tetrad analysis with a pair of heterozygous markers (AB/ab): the parental ditype (PD) contains four segregants of which two have the genotype of one parent and the other two have the genotype of the other parental cell ($2 \times$ AB and $2 \times$ ab). The

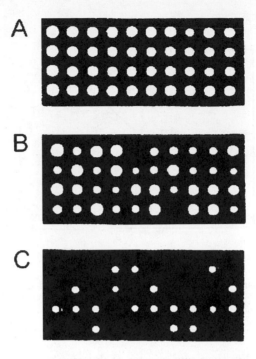

Figure 1. Tetrad analysis of *S. cerevisiae*. Asci were obtained after sporulation of the diploid strain CEN.PK2 and were dissected using the Singer MSM micromanipulator. Growth of segregant colonies was monitored after 2–3 days incubation on YEPD plates at 30°C. (A) Segregants from a homozygous wild-type diploid. (B) Segregants resulting from diploids carrying a heterozygous mutation which reduces growth. Note that small and large colonies segregate 2:2. (C) Segregants resulting from diploids having a heterozygous mutation in an essential gene. Note that only two segregants grow indicating 2:2 segregation of the deletion in the essential gene.

non-parental ditype (NPD) contains four segregants which all carry recombined markers (2 × Ab and 2 × aB). Finally, in the tetrad type (TT) two segregants display the genotype of the parental strains (AB and ab), and two segregants show the recombined genotypes (Ab and aB). In cases where the two genes segregate at random, the ratio of PD:NPD:TT asci is 1:1:4. If two genes are linked to each other, the number of PD is larger than that of the NPD ditypes (PD > NPD), and the number of parental PD is lower than one-quarter of the TT (PD < 1/4 TT). The genetic distance between two linked genes can be calculated according to the equation of Perkins (1949): $x = 100 \, (TT + 6 \, NPD) / 2 \, (PD + NPD + TT)$, where x is the distance between the two genes in centimorgans.

In cases where a centromere-linked gene is used as an additional genetic marker for tetrad analysis, a centromere linkage of any of the two genes of interest can be detected easily by following its recombination frequency with the centromere-linked marker. In cases of centromere linkage, the ratio of PD to NPD shown by comparing the centromere marker and the gene of interest is 1:1, and the amount of PD or NPD is much

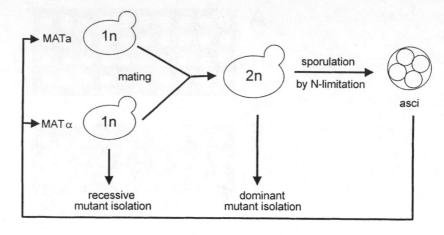

Figure 2. The life cycle of *S. cerevisiae* and the possibilities to isolate and study recessive and dominant mutations.

higher than one-quarter of the number of tetrad types (PD = NPD > 1/4 TT).

From such mapping data, a genetic map of *S. cerevisiae* was established by Mortimer and co-workers (Mortimer and Schild, 1980, 1985; Mortimer *et al.*, 1989). From this map it was concluded that *S. cerevisiae* contains 16 chromosomes, a fact that was later confirmed by pulsed-field gel electrophoresis (Carle and Olson, 1984).

The well-established genetic map of *S. cerevisiae* and its compact genomic organization provided the basis for the complete DNA sequencing of the yeast genome. This worldwide effort was successfully completed in April 1996. DNA sequencing revealed a genome size of 12.5 Mb, encoding approximately 6200 genes which are now accessible for systematic functional analysis. The precise homologous recombination apparatus of *S. cerevisiae* allows one to perform specific gene deletions by selective integration of prototrophic gene markers or dominant resistance genes (see Chapter 5). In addition to the isolation of mutants by classical genetics, this allows the creation of gene-specific mutations by reverse genetics. Previously the genetic map of *S. cerevisiae* was mainly derived from mapping data (Mortimer *et al.*, 1989), but genome sequencing has now established a physical map for gene localization. In general, the DNA sequencing confirmed the high quality of the earlier genetic map (Mortimer *et al.*, 1989) and we now know that, on average, 1 cM corresponds to approximately 3 kB. However, this varies because of the differing recombination frequencies at different regions of the chromosome.

The ability to create well-defined mutations and deletions for each gene (Rothstein, 1983, 1991; Chapter 5) allows one to investigate the biological function of yeast genes with unknown function by searching for phenotypical anormalities of the respective mutants. Such mutants can be collected, a genetic archive of gene deletion mutants can be established, and these mutants can be made available to the scientific community. Here we report such attempts which, at present, cover more than 1000 genes of the S. cerevisiae genome. The major advantage of the present collection compared with previous mutant collections is that all deletions have been created in the isogenic backgrounds of not more than three strains. In addition, the deletion cassettes for each gene are also being collected, which allows one to introduce each deletion into any S. cerevisiae strain of interest. This is a major breakthrough for the functional analysis of genes because the genetic background strongly interferes with the mutant phenotype in many cases.

At least two types of null mutant can be created by reverse genetics. In gene disruption mutants the ORF of a particular gene is interrupted by the insertion of a genetic marker. In general this results in a nonfunctional protein. However, depending on the position of the disruption within the ORF, the residual proteins may display partial function. Furthermore, if there is significant selective pressure against the mutation, revertants will probably arise through selection for outlooping of the disruption marker to restore the original protein. Alternatively gene function can be destroyed via a gene deletion. In this approach, the major part of the target ORF is replaced with a selection marker, thereby deleting part of the ORF. In general, such gene deletions abolish the gene function irreversibly, even if there are selective pressures against such mutations. Only mutations in a second gene (epistatic suppressor mutations) can compensate for such deletions. Such epistatic mutations can easily be detected in tetrad analysis because they result in a 3:1 segregation of the wild-type phenotype in tetratype asci.

◆◆◆◆◆◆ II. EUROSCARF

A. Establishment of EUROSCARF

The EUROSCARF strain collection was established in 1994 after the sequencing of the S. cerevisiae genome had revealed a considerable number of genes with unknown function. Funded by the German government, a consortium of 16 research groups established a scientific network, the aim of which was to delete approximately 325 genes of unknown function from the yeast genome (BMBF Functional Analysis Network). The resulting 325 deletion mutants provided the basis for a systematic screen for possible phenotypes of the respective mutations (K-D. Entian et al., 1998, submitted for publication). Using the strain CEN.PK2, approximately one-third of the deletion mutants first revealed a phenotype during tetrad analysis according to the slower growth or germination of the mutant segregants (Figure 1). In about 20% of cases the deleted genes were essential (Figure 1). For approximately two-thirds of the mutants, phenotypes were discovered some of which were indicative of the function of the corresponding gene (K-D. Entian et al., 1998, submitted for publication).

The deletion mutants were stored centrally and made available to the participating research groups. In 1996 the European Community also decided to support the systematic study of yeast genes of unknown function. The resulting EUROFAN project is a collaboration involving 144 European research groups, which is currently studying approximately 1000 genes of unknown function in the first phase (Oliver, 1996). These genes were deleted from the yeast genome by reverse genetics and the resulting deletion mutants are being made accessible to the participating laboratories. The selection of genes to be targeted was made in collaboration with the BMBF Functional Analysis Network, resulting in a EUROSCARF collection of approximately 1200 deletion mutants. Furthermore, deletion cassettes, with long flanking regions of homology to the target locus, are also being collected so that the corresponding mutation can be introduced easily into other *S. cerevisiae* strains with different genetic backgrounds.

Within the next phase of the EUROFAN project, which started in October 1997, another 2200 *S. cerevisiae* genes will be deleted. These deletions will be part of a worldwide effort with US and Canadian research groups, which aims to establish a collection of deletion mutants for all 6200 *S. cerevisiae* genes. In addition to the currently used deletion strategies, each of these deletion mutants will be tagged by a specific and unique DNA sequence that is not present in the yeast genome. This allows the easy identification of each mutant either by hybridization or by DNA sequencing of the respective polymerase chain reaction (PCR) fragments (see below). Such tagged mutants are very useful for functional analysis in selective screening procedures involving rapid identification of the enriched mutations (see section v).

In the light of current progress in gene deletion strategies it can be assumed that the EUROSCARF collection will cover deletions of all 6200 *S. cerevisiae* genes. This will enable easy access for the scientific community to each *S. cerevisiae* mutant of interest, and in cases where the deletion cassette is also available, such deletions can be easily introduced into any *S. cerevisiae* strain background of interest. This will avoid redundant efforts in the study of genome functions as the basic tools become generally available. Furthermore, the easy access to deletion mutants will also improve the use of *S. cerevisiae* as a model organism because they can be used by scientists studying other eukaryotic organisms if functional similarities become obvious (see Chapters 21 and 22).

B. The EUROSCARF Strain Collection

At present, gene deletions are being made in the three *S. cerevisiae* strains FY1679 (a S288C derivative; B. Dujon, personal communication), W303, and CEN.PK2, each of which have different genetic backgrounds. They are all available with various combinations of the auxotrophic mutations *trp1*, *his3*, *leu2* and *ura3*, which are often used as selectable markers for yeast transformation or for different genetic test systems (see Table 1). The strain S288C was used as the source of DNA for *S. cerevisiae* genome

Table I. Genotypes of strains used for gene deletions

Strain name	Genotype					
CEN.PK2	*MATa*	*ura3-52*	*trp1-289*	*leu2-3,112*	*his3Δ1*	*MAL2-8ᶜ SUC2*
	MATα	*ura3-52*	*trp1-289*	*leu2-3,112*	*his3Δ1*	*MAL2-8ᶜ SUC2*
W303	*MATa*	*ura3-1*	*trp1Δ2*	*leu2-3,112*	*his3-11*	*ade2-1 can1-100*
	MATα	*ura3-1*	*trp1Δ2*	*leu2-3,112*	*his3-11*	*ade2-1 can1-100*
FY1679	*MATa*	*ura3-52*	*trp1Δ63*	*leu2Δ1*	*his3Δ200*	*GAL2*
	MATα	*ura3-52*	*TRP1*	*LEU2*	*HIS3*	*GAL2*

sequencing. Most deletions are being made in FY1679, so that mutants are isogenic to the sequenced strain. Unfortunately, this strain has a poor sporulation frequency, which makes its genetic analysis relatively difficult.

In order to prove that the deletion cassettes are suitable for constructing deletions in genetic backgrounds other than strain FY1679, all deletions are also being introduced into a second strain, which is either W303 or CEN.PK2. W303 has a much better sporulation efficiency than FY1679, which makes the former strain more suitable for genetic analysis. W303 was used in a pilot project for functional analysis which mainly involved the screening of the respective mutants for altered sensitivities to drugs and metabolic inhibitors. The strain CEN.PK2 was developed especially for gene functional analysis within the BMBF Functional Analysis Network (K-D. Entian *et al.*, unpublished). Its sporulation efficiency is as good as that of W303, and its major advantage is its fast growth rate, with doubling times of about 80 min for haploid strains. The available CEN.PK2 derivatives are isogenic and mutations that reduce germination or growth can be detected easily by a clear 2:2 segregation in tetrads in which wild type segregants become visible within 24 h after tetrad dissection in contrast to the mutant segregants.

Deletions are being created either by replacement of the target gene with suitable prototrophic markers such as *LEU2, HIS3, URA3,* and *TRP1* (Rothstein, 1983, 1991), or with a dominant kanamycin resistance marker (see also Chapter 5). For gene replacements with prototrophic markers the respective deletion cassettes were obtained after replacing approximately 60–90% of the target's coding region by the prototrophic marker. This deletion strategy, however, is time consuming and its success generally depends on suitable restriction sites within the target gene.

Based on a PCR deletion strategy (Baudin *et al.*, 1993), an efficient deletion method was recently developed that uses heterologous modules such as the kanamycin resistance gene (*kanMX*) as a selection marker (Wach *et al.*, 1994). In contrast to the prototrophic markers that usually share homologous sequences within the *S. cerevisiae* genome, no sequences homologous to the *kanMX* deletion cassette are present in the *S. cerevisiae* genome. Therefore, short sequences of about 40 bases can be used for recombination to target the *kanMX* module to replace the gene of interest. This allows one to construct a deletion cassette with short flanking recombination sequences by PCR. Even with strains that are not strongly isogenic to the S288C sequencing strain, this deletion strategy works with a more than 90% rate of success (see Chapter 5 for details).

The PCR-mediated (Baudin *et al.*, 1993) deletion strategy was further improved recently by the introduction of directed repeats of *loxP* recombination sites flanking the *kanMX* deletion cassette (Güldner *et al.*, 1996). These 34 bases lie at each end of the kanamycin resistance gene and this deletion cassette also generates stable deletions. However, the kanamycin resistance gene can subsequently be removed from the genome after transformation of the deletion mutant with the Cre-recombinase of bacteriophage P1, which catalyzes a site-specific recombination event between the *loxP* sites. Using this *loxP–kanMX–loxP* deletion strategy allows one to remove the kanamycin resistance gene from the genome, thereby making it possible to perform a second deletion using the same resistance marker. This is a remarkable advantage, which allows the sequential replacement of isogenes, and it is also suitable for the generation of deletions in industrial strains in which removal of the resistance marker is recommended after transformation.

Currently, the strain collection contains about 900 deletion mutants of strain FY1679. Most of the deletions were introduced via replacement of the respective gene by the kanamycin resistance gene as described above (see also Chapter 5). All deletions for non-essential genes are available in both mating types. Deletions within essential genes are collected as heterozygous diploids. In some cases, plasmids, which have the essential gene under the control of a regulatory promotor, are also available for functional studies (see below).

About 700 deletions are available in the CEN.PK2 background, of which about 30% were deleted by conventional gene replacement strategies with prototrophic markers and about 70% were deleted with the kanamycin resistance gene. All deletions are available in both mating types, and deletions of essential genes are collected in heterozygous diploids.

In addition, about 500 deletions are available in the W303 background and, as for the other mutations, they exist in *MATa* and *MATα* backgrounds. Most of the W303 deletion mutants were constructed using the kanamycin resistance marker.

Upon receipt of the strains at EUROSCARF, each strain is streaked out for single colonies on YEPD plates (1% yeast extract, 2% bacto peptone and 4% glucose) and replica-plated onto a routine set of synthetic drop-out plates (each lacking adenine, uracil, leucine, histidine, or tryptophan) and tested for kanamycin resistance. The colony growth is estimated and particular care is taken to monitor any form of papillar growth because such growth is indicative of a heterogenous population or genetic instability of the deletion mutant. If colonies of uniform size are obtained on plates, a single colony is inoculated into 5 ml YEPD liquid medium and grown to stationary phase. For storage, glycerol is

added to a final concentration of 17.5% (w/w) and the strain is stored at
–80°C. The strains are cultivated in three-fold redundancy. If stored cultures have to be replicated, the number of generation times is kept to a
minimum to avoid the accumulation of second-site mutations.

Upon request, the strains are streaked out onto YEDP plates and tested
for uniform growth. For mailing, strains are either soaked onto sterile
filter papers wrapped in sterile aluminium foil, or alternatively, streaked
out onto a YEPD agar slope in a screw-capped vial.

Strains carrying mutations that affect essential genes or genes that
strongly reduce growth on glucose are collected as heterozygous diploids
to minimize any selective pressure against the respective mutation.
Mutants with strongly reduced growth on glucose need special attention
and a tetrad analysis is recommended upon receipt to confirm the 2:2 segregation of the mutation. This reduces the risk that second-site mutations
have accumulated in the diploids which, in most cases, will result in a non
2:2 segregation of the growth phenotype. Upon request, such analyses
can be performed by EUROSCARF and haploid segregants obtained
directly from the tetrad analysis will be mailed.

EUROSCARF strains are also stored in liquid nitrogen as an energy-
independent back-up. However, viability under such storage conditions
is strongly reduced. An alternative method for strain cultivation is used at
the Yeast Genetic Stock Center, Berkeley, where strains are conserved
with evaporated milk at 4°C (see below). This method is also considered
for the EUROFAN collection in future.

C. The Collection of Deletion Cassettes

Although the deletion strategy using short flanking regions (described
above) works for most of the genes for all three strains described here,
problems may arise if the strains are more diverged. To overcome this,
deletion cassettes with long flanking sequences of 0.5 to 1 kb were also
constructed. In general, the large flanking sequences were obtained by
PCR reactions and most of them were checked for PCR-induced errors by
sequencing. At present about 500 deletion cassettes with long flanking
sequences are available in the collection (Table 2).

Upon receipt, the plasmids containing the deletion cassettes are used to
transform *Escherichia coli* strain DH5α (for genotypes see Sambrock *et al.*,
1989). The plasmid is purified from the transformed *E. coli* strain and its
quality is checked after agarose electrophoresis. Plasmids are stored as
ethanol precipitates at –20°C and –80°C.

About 0.5–1 μg of lyophilized plasmid DNA is mailed using screw-
capped vials. The DNA should be used immediately for *E. coli* transformation upon receipt. If requested, larger amounts of freshly purified and
quality controlled plasmids (tested by restriction analysis) can also be
obtained from EUROSCARF. This makes it possible to use the plasmids
directly for gene deletions.

Table 2. Systematic names of genes for which deletions and plasmids are available at EUROSCARF

CHR 11	YBR030w	YBR114w*	YBR201w	YBR285w	YDL088c
YBL001c	YBR032w	YBR121c*	YBR203w	YBR287w	YDL089w
YBL004w	YBR041w	YBR125c*	YBR204c	YBR288c	YDL091c
YBL006c	YBR042c	YBR128c	YBR207w	YBR293w*	YDL100c
YBL009w	YBR043c	YBR129c	YBR209w	YBR294w*	YDL103c
YBL010c	YBR044c	YBR130c	YBR210w	YBR295w*	YDL105w
YBL018c	YBR045c	YBR131w	YBR211c	YBR296c*	YDL109c
YBL019w	YBR047w	YBR133c	YBR214w	YBR88c*	YDL110c
YBL024w	YBR050c*	YBR137w	YBR216c		YDL112w
YBL025w	YBR051w*	YBR138c	YBR217w	**CHR III**	YDL113c
YBL028c	YBR052c*	YBR142w	YBR220c	YCR008w*	YDL114w
YBL029w	YBR053c*	YBR147w*	YBR222c		YDL115c
YBL031w	YBR054w*	YBR148w*	YBR223c	**CHR IV**	YDL116w
YBL032w	YBR056w*	YBR150c*	YBR225w	YDL003w	YDL117w
YBL036c	YBR057c*	YBR151w*	YBR227c	YDL005c	YDL118w
YBL037w	YBR058c*	YBR152w*	YBR228w	YDL008w	YDL119c
YBL042c	YBR059c*	YBR155w*	YBR229c	YDL009c	YDL120w
YBL043w	YBR061c*	YBR156c*	YBR231c	YDL010w	YDL121c
YBL046w	YBR062c*	YBR157c*	YBR233w	YDL012c	YDL124w
YBL047c	YBR063c*	YBR158w*	YBR238c*	YDL015c	YDL125c
YBL048w	YBR065c	YBR159w*	YBR239c*	YDL018c	YDL128w
YBL049w	YBR066c	YBR161w	YBR240c*	YDL019c	YDL129w
YBL051c	YBR071w	YBR162c	YBR242w*	YDL021w	YDL131w
YBL052c	YBR073w	YBR163w*	YBR244w*	YDL024c	YDL132w
YBL054w	YBR074w	YBR164c*	YBR245c*	YDL025c	YDL133w
YBL055c	YBR075w	YBR165w*	YBR246w*	YDL027c	YDL136w
YBL056w	YBR076w	YBR167c*	YBR247c*	YDL048c	YDL138w*
YBL057c	YBR078w	YBR168w*	YBR250w*	YDL053c	YDL139c
YBL059w	YBR079c	YBR169c*	YBR254c	YDL054c	YDL142c
YBL064c	YBR086c*	YBR170c*	YBR255w	YDL057w	YDL144c
YBL066c	YBR087w*	YBR171w*	YBR257w	YDL059c	YDL146w
YBL067c	YBR091c*	YBR175w	YBR258c	YDL060w	YDL147w
YBL071c	YBR094w*	YBR176w*	YBR259w	YDL063c	YDL148c
YBR002c	YBR095c*	YBR179c*	YBR260c	YDL065c	YDL149w
YBR004c	YBR096w*	YBR180w	YBR264c	YDL070w	YDL157c
YBR005w	YBR097c*	YBR182c	YBR266c	YDL072c	YDL158c
YBR006w	YBR098w*	YBR183w	YBR270c	YDL073w	YDL161w
YBR007c	YBR101c*	YBR184w	YBR271w	YDL074c	YDL166c
YBR008c	YBR102c*	YBR185c	YBR272c	YDL076c	YDL167c
YBR014c	YBR103w*	YBR186w	YBR273c	YDL077c	YDL171c
YBR016w	YBR104w*	YBR187w	YBR274w	YDL080c	YDL172c
YBR017c*	YBR105c*	YBR188c	YBR276c	YDL082w	YDL173w
YBR022w	YBR106w*	YBR193c	YBR280c	YDL084w	YDL175c
YBR025c	YBR107c*	YBR194w	YBR281c	YDL085w	YDL176w
YBR027c	YBR108w*	YBR197c	YBR283c	YDL086w	YDL177c
YBR028c	YBR111c*	YBR198c	YBR284w	YDL087c	YDL179w

YDL180w	YDR056c	YGL134w	YGR110w	YGR269w	YJL050w
YDL182w	YDR057w	YGL136c	YGR111w	YGR272c	YJL051w
YDL183c	YDR058c	YGL138c	YGR112w	YGR273c	YJL054w
YDL186w	YDR060w	YGL139w	YGR113w	YGR275w	YJL055w
YDL189w	YDR061w	YGL140c	YGR117c	YGR276c	YJL056c
YDL193w	YDR063w	YGL141w	YGR120c	YGR280c	YJL057c
YDL199c	YDR065w	YGL142c	YGR136w	YGR284c	YJL058c
YDL201w	YDR066c	YGL144c	YGR142w		YJL059w
YDL202w	YDR067c	YGL146c	YGR145w	**CHR IX**	YJL060w*
YDL203c	YDR068w	YGL159w	YGR178c	YIL001w*	YJL062w
YDL204w	YDR071c	YGL161c	YGR179c	YIL003w*	YJL064w
YDL206w	YDR072c	YGL179c	YGR187c	YIL005w*	YJL065c
YDL207w	YDR073w	YGL180w	YGR189c	YIR001c*	YJL066c
YDL209c	YDR078c	YGL183c	YGR194c	YIR002c*	YJL068c
YDL211c	YDR080w	YGL184c	YGR195w	YIR003w*	YJL069c
YDL213c	YDR082w	YGL185c	YGR196c	YIR004w*	YJL070c
YDL214c	YDR083w	YGL186c	YGR198w	YIR005w*	YJL071w
YDL216c	YDR084c	YGL194c	YGR200c	YIR007w*	YJL072c
YDL217c	YDR087c	YGL196w	YGR205w	YIR009w*	YJL073w*
YDL218w		YGL197w	YGR210c	YIR010w*	YJL074c*
YDL219w	**CHR VII**	YGL201c	YGR211w	YIR012w*	YJL076w*
YDL222c	YGL001c	YGL202w	YGR212w		YJL077c*
YDL223c	YGL002w	YGL227w	YGR216c	**CHR X**	YJL078c*
YDL224c	YGL005c	YGL228w	YGR221c	YJL004c	YJL079c*
YDL225w	YGL012w	YGL229c	YGR223c	YJL006c	YJL082w*
YDL231c	YGL047w	YGL231c	YGR224w	YJL008c	YJL083w*
YDL233w	YGL050w	YGL232w	YGR225w	YJL010c	YJL084c*
YDL234c	YGL051w	YGL236c*	YGR226c	YJL011c	YJL089w*
YDL235c	YGL053w	YGL240w	YGR231c	YJL012c	YJL090c*
YDL237w	YGL061c	YGL241w	YGR232w	YJL013c	YJL091c*
YDL238c	YGL064c	YGL242c	YGR235c	YJL016w	YJL093c*
YDL239c	YGL078c	YGL243w	YGR237c	YJL017w	YJL094c
YDL243c	YGL085w	YGL244w	YGR239c	YJL018w	YJL094c*
YDR013w	YGL094c	YGL245w	YGR241c	YJL019w	YJL096w*
YDR014w	YGL096w	YGL246c	YGR243w	YJL020c	YJL097w*
YDR015c	YGL098w	YGL249w	YGR244c	YJL027c	YJL098w*
YDR018c	YGL099w	YGR001c	YGR245c	YJL029c	YJL100w
YDR020c	YGL100w	YGR016w	YGR247w	YJL033w	YJL105w
YDR021w	YGL101w	YGR042w*	YGR248w	YJL035c	YJL107c
YDR022c	YGL105w	YGR043c*	YGR250c	YJL036w*	YJL108c
YDR026c	YGL114w	YGR045c*	YGR252w	YJL037w	YJL112w
YDR027c	YGL120c	YGR046w*	YGR255c	YJL038c	YJL118w
YDR029w	YGL121c	YGR048w*	YGR256w	YJL039c	YJL122w
YDR030c	YGL124c	YGR054w*	YGR260w	YJL043w*	YJL123c
YDR032c	YGL125w	YGR055w*	YGR262c	YJL045w	YJL124c
YDR033w	YGL128c	YGR056w*	YGR263c	YJL046w	YJL125c
YDR036c	YGL129c	YGR057c*	YGR266w	YJL047c	YJL126w
YDR051c	YGL131c	YGR060w*	YGR267c	YJL048c	YJL131c
YDR055w	YGL133w	YGR106c	YGR268c	YJL049w	YJL132w

Table 2. continued

YJL134w	YJR019c	YJR120w*	YLL011w*	YLR018c	YLR091w*
YJL135w	YJR022w	YJR122w*	YLL012w*	YLR019w	YLR092w*
YJL137c	YJR024c	YJR126c*	YLL013c*	YLR020c	YLR093c*
YJL144w	YJR025c	YJR127c*	YLL014w*	YLR021w	YLR094c*
YJL145w	YJR030c	YJR129c*	YLL015w*	YLR022c	YLR095c*
YJL146w	YJR031c	YJR130c*	YLL022c	YLR023c	YLR097c*
YJL147c	YJR032w	YJR132w*	YLL023c	YLR024c	YLR098c*
YJL148w	YJR033c	YJR133w*	YLL025w	YLR028c	YLR099c*
YJL149w	YJR036c	YJR134c*	YLL027w	YLR030w	YLR100w*
YJL159w*	YJR039w	YJR136c*	YLL028w	YLR031w	YLR102c
YJL160c*	YJR041c	YJR138w*	YLL029w	YLR033w	YLR103c
YJL161	YJR043c	YJR139c*	YLL030c	YLR034c	YLR104w
YJL162c*	YJR044c	YJR140c*	YLL031c	YLR035c	YLR106c
YJL163c*	YJR046w	YJR141w*	YLL032c	YLR036c	YLR107w
YJL168c*	YJR053w	YJR142w*	YLL033w	YLR040c	YLR108c
YJL169w	YJR054w	YJR144w*	YLL034c	YLR042c	YLR114c
YJL170c*	YJR056c	YJR155w	YLL035w	YLR045c	YLR115w
YJL171c*	YJR059w		YLL038c	YLR046c	YLR116w
YJL178c*	YJR061w	**CHR XI**	YLL040c	YLR047c	YLR117c
YJL181w*	YJR065c	YKL012w*	YLL042c	YLR049c	YLR118c
YJL183w*	YJR067c	YKL013c*	YLL044w	YLR050c	YLR119w
YJL185c*	YJR070c	YKL017c*	YLL048c	YLR051c	YLR121c
YJL186w*	YJR072c	YKL018w*	YLL049w	YLR052w	YLR122c
YJL192c*	YJR074w	YKL125w*	YLL051c	YLR053c	YLR123c
YJL193w	YJR075w	YKL128c*	YLL052c	YLR054c	YLR124w
YJL195c	YJR080c	YKL129c*	YLL053c	YLR057w	YLR125w
YJL196c	YJR092w*	YKL130c*	YLL054c	YLR059c	YLR126c
YJL198w	YJR096w*	YKL132c*	YLL055w	YLR062c	YLR127c
YJL199c	YJR097w*	YKL133c*	YLL056c	YLR063w	YLR128w
YJL201w	YJR098c*	YKL134c*	YLL057c	YLR064w	YLR129w
YJL204c	YJR099w*	YKL135c*	YLL058w	YLR065c	YLR130c
YJL206c	YJR100c*	YKL137w*	YLL059c	YLR066w	YLR132c
YJL207c	YJR101w*	YKL144c*	YLL060c	YLR068w*	YLR135w
YJL213w	YJR102c*	YKL146w*	YLL061w	YLR070c	YLR136c
YJL217w	YJR105w*	/147c*	YLL062c	YLR072w	YLR137w
YJL222w	YJR106w*	YKL175w*	YLL063c	YLR073c	YLR138w
YJL225c	YJR107w*	YKL177w*	YLR001c*	YLR074c	YLR141w
YJR001w	YJR108w*	YKL185w*	YLR007w	YLR076c*	YLR143w
YJR002w	YJR110w*	YKL188c*	YLR008c	YLR077w	YLR144c
YJR003c	YJR111c*		YLR009w	YLR080w	YLR145w
YJR008w	YJR112w*	**CHR XII**	YLR010c	YLR081w*	YLR146c
YJR011c	YJR113c*	YLL002w*	YLR011w	YLR082c	YLR147c
YJR012c	YJR115w*	YLL003w*	YLR012c	YLR084c*	YLR149c
YJR013w	YJR116w*	YLL004w*	YLR013w	YLR085c*	YLR151c
YJR014w	YJR117w*	YLL005c*	YLR015w	YLR086w*	YLR152c
YJR015w	YJR118c*	YLL007c*	YLR016c	YLR087c*	YLR153c
YJR018w	YJR119c*	YLL010c*	YLR017w	YLR089c*	YLR154c

CHR XIV
YNL006w
YNL008c
YNL010w
YNL011c
YNL014w
YNL018c
YNL019c
YNL020c
YNL021w
YNL022c
YNL023c
YNL024c*
YNL026w
YNL027w
YNL032w
YNL033w
YNL034w
YNL035c
YNL036w*
YNL038w*
YNL040w
YNL041c*
YNL042w*
YNL044w
YNL045w
YNL046w
YNL048w
YNL050c
YNL051w
YNL054w
YNL056w
YNL058c
YNL059c
YNL063w
YNL065w
YNL066w
YNL072w
YNL075w
YNL078w
YNL080c
YNL081c
YNL083w
YNL086w
YNL087w
YNL091w
YNL092w
YNL094w
YNL095c

YNL097c
YNL099c
YNL100w
YNL101w
YNL106c
YNL107w
YNL108c
YNL110c
YNL115c
YNL116w
YNL119w
YNL123w
YNL124w
YNL125c
YNL126w
YNL127w
YNL128w
YNL129w
YNL132w
YNL133c
YNL134c
YNL136w
YNL148c
YNL150w
YNL155w
YNL158w
YNL159c
YNL164c
YNL165w
YNL166c
YNL168c
YNL172w*
YNL173c*
YNL175c
YNL175c*
YNL176c*
YNL177c
YNL180c*
YNL182c*
YNL186c*
YNL187w*
YNL191w*
YNL193w*
YNL194c*
YNL195c*
YNL196c
YNL200c
YNL206c
YNL207w

YNL208w
YNL211c
YNL212w
YNL213c
YNL214w
YNL215w
YNL217w
YNL218w
YNL223w
YNL224c
YNL225c
YNL227c
YNL230c
YNL231c
YNL232w
YNL233w
YNL234w
YNL240c
YNL242w
YNL245c*
YNL246w*
YNL249c*
YNL251c*
YNL252c*
YNL253w*
YNL254c*
YNL255c*
YNL256w*
YNL258c*
YNL260c*
YNL263c*
YNL264c*
YNL265c*
YNL270c
YNL273w
YNL274c
YNL275w
YNL275w*
YNL278w
YNL279w
YNL281w
YNL282w
YNL283c
YNL284c
YNL285w
YNL286w
YNL288w
YNL292w
YNL293w

YNL294c
YNL295w
YNL297c
YNL299w
YNL300w
YNL303w
YNL304w
YNL305c
YNL306w
YNL308c
YNL309w
YNL310c
YNL311c
YNL313c
YNL317w
YNL320w
YNL321w
YNL323w
YNL325c
YNL326c
YNL327w
YNL328c
YNL331c
YNL332w
YNL336w
YNR004w
YNR007c
YNR008w
YNR009w
YNR013c
YNR014w
YNR015w
YNR018w
YNR019w
YNR020c
YNR021w
YNR022c*
YNR023w
YNR024w*
YNR027w*
YNR028w
YNR029c*
YNR030w
YNR031c
YNR038w
YNR039c
YNR043w
YNR047w
YNR048w

YNR049c
YNR051c
YNR053c
YNR056c*
YNR057c*
YNR058w*
YNR059w
YNR061c*
YNR062c*
YNR063w*
YNR064c*
YNR068c
YNR069c
YNR070w
YNR071c
YNR073c
YNR074c
YNR075w

CHR XV
YOL003c*
YOL007c*
YOL008w*
YOL009c*
YOL010w*
YOL011w*
YOL012c*
YOL018c
YOL021c*
YOL022c
YOL025w
YOL028c
YOL029c
YOL030w*
YOL031c
YOL032w
YOL034w*
YOL036w
YOL053w*
YOL054w*
YOL055c*
YOL056w*
YOL057w*
YOL059w*
YOL060c*
YOL061w*
YOL062c*
YOL072w
YOL083w

YOL087c
YOL088c
YOL089c
YOL091w
YOL092w
YOL093w
YOL095c
YOL098c
YOL100w
YOL101c
YOL102c
YOL104c
YOL105c
YOL107w
YOL111c
YOL112w
YOL113w
YOL114c
YOL115w
YOL117w
YOL118c
YOL119c
YOL124c
YOL125w
YOL130w
YOL132w
YOL135c
YOL137w
YOL138c
YOL141w
YOL142w
YOL144w
YOL149w
YOL151w
YOL152w
YOL154w
YOL155c
YOR007c
YOR083w*
YOR085w*
YOR086c
YOR086c*
YOR087w*
/088w*
YOR095c*
YOR097c*
YOR100c*
YOR109w
YOR109w*

Table 2. continued

YOR110w*	YOR137c*	YOR172w	YOR272w	**CHR XVI**	YPL244c*
YOR111w*	YOR138c*	YOR180c*	YOR273c	YPL173w*	YPL245w*
YOR112w*	YOR141c*	YOR191w*	YOR275c	YPL180w*	YPL246c*
YOR114w*	YOR145c	YOR192c*	YOR279c	YPL181w*	YPL247c*
YOR115c*	YOR152c	YOR193w*	YOR301w	YPL183c*	YPL252c*
YOR117w*	YOR154w	YOR195w*	YOR306c	YPL184c*	YPL253c*
YOR118w*	YOR155c	YOR197w*	YOR311c	YPL186c*	YPL254w*
YOR119c*	YOR161c	YOR205c*	YOR315w	YPL187w*	
YOR129c*	YOR162c	YOR206w*	YOR319w	YPL235w*	
YOR130c*	YOR164c	YOR267c	YOR320c	YPL236c*	
YOR131c*	YOR165w	YOR268c	YOR322c	YPL239w*	
YOR134w*	YOR166c	YOR271c	YOR324c	YPL242c*	

*Strains marked with an asterisk have been constructed in the BMBF functional analysis network and are freely available (see Section IIE).

D. The Collection of Cognate and Regulated Clones

Most of the genes that were deleted are also available on centromeric plasmids as complete genes with their native promoter and terminator sequences. These can be used for *trans*-complementation. This is especially important after a phenotype has been observed for the corresponding deletion mutant, and the plasmid is used to confirm that the phenotype is due to this deletion rather than some second-site mutation.

For the study of essential genes, the corresponding ORFs also will be fused to regulated promoters such as *GAL1* (Mumberg *et al.*, 1994). The regulated expression makes it possible to conditionally switch off the expression of the essential gene under investigation (by altering the carbon source in the case of the *GAL1* promoter), thereby making it possible to investigate the manner of cell death upon loss of the essential function. If such plasmids become available they will be also collected by EUROSCARF.

The handling, storage and mailing of plasmids with cognate and regulated clones is the same as described for the deletion cassettes (section IIC).

E. Access to the Strain and Plasmid Collection

The use of the strain and plasmid collections in the beginning was restricted to the participating groups of the BMBF Functional Analysis Network and EUROFAN. The 325 deletion mutants belonging to the BMBF Functional Analysis Network became available to the public at the end of 1997. The EUROFAN strains and plasmids will probably remain restricted until the EUROFAN II project ends in 1999. However, earlier requests can be considered for all deletion mutants and plasmids, if the research group that constructed the deletion agrees to its distribution.

Requests for strains and plasmids should be sent by e-mail (EUROSCARF@em.uni-frankfurt.de) or using the following mailing address: EUROSCARF, Institute for Microbiology, Marie-Curie-Str. 9, D-60439 Frankfurt, Germany; Fax: +49 - 69 / 798 29527.

Please note that personal requests to the authors of this chapter cannot be answered. The handling fees are 30 US$ for each strain, and 50 US$ for each plasmid. In cases where a courier service is requested for sending the strains and plasmids, these costs have to be covered by the recipient.

◆◆◆◆◆◆ III. OTHER YEAST MUTANT AND PLASMID COLLECTIONS

A. The Yeast Genetic Stock Center

The Yeast Genetic Stock Center (YGSC) was started by Bob Mortimer as early as 1960 at the University of California at Berkeley. In 1972 it became an official collection funded by the US National Science Foundation, and Rebecca Contopoulou became its curator. This collection was of great value for any scientist working with yeast because mutants became readily available, which greatly improved the progress in yeast research. The YGSC collection is still operating today, and approximately 1200 strains are available, covering approximately 450 genes. The major strength of these strains is that they are multiply marked. Furthermore, YGSC also offers a good collection of strains that are valuable specifically for their use in yeast-specific genetic techniques. This includes strains for gene mapping, teaching of yeast genetics, transformation, allelism tests, estimation of recombination frequencies, and tests for aneuploidy. Additionally, some mitochondrial mutations are also collected and recently, some wild *S. cerevisiae* isolates of natural fermenting strains were also added. Note that a large collection of wild yeast isolates is held by the National Collection of Yeast Cultures (http://www.ifrn.bbsrc. ac.uk/NCYC/). This collection is not discussed further in this chapter, which focuses mainly on collections of yeast mutants.

Most of the mutants in the YGSC collection result from point mutations, and only in a few cases have the DNA mutations been identified within the respective genes. Furthermore, the mutations were made in different genetic backgrounds, which sometimes makes their use difficult. However, many mutations were backcrossed by the YGSC so that the genetic background became more isogenic and most mutations are available in both mating types. Today, with the excellent tools of reverse genetics, in some cases it is more convenient to introduce the respective mutation into an isogenic background by reverse genetics than to use classical genetics to clean up the genetic background. However, many of the YGSC mutant alleles are still very useful for functional studies of the corresponding gene. The specific mutation can be identified and its behavior compared with the wild-type allele. In particular, nonsense and missense mutations may help to reveal the molecular function of the corresponding protein. Although the YGSC and EUROSCARF collections reveal some overlaps, there are still many interesting mutations in the YGSC collection which will not become readily available by reverse genetics. This includes temperature-sensitive strains and strains suitable for the analysis of genetic instabilities such as gene conversion and recombination events.

YGSC strains are preserved using the evaporated milk method at 4°C, and, as a back-up, in glycerol cultures at –80°C. The milk preservation method is described in the YGSC catalog, which is now available in its 8th edition. The catalog is easily accessible using http://zenith.berkeley.edu/~jontib/YGSC/ or can be ordered from YGSC (using the addresses below). A fee of 15 US$ is charged per strain for non-profit organizations, whereas 60 US$ is charged for industrial use.

Requests for strains should be sent by using e-mail (ygsc305@violet.berkeley.edu) or using the following mailing address: Yeast Genetic Stock Center, Department of Molecular and Cell Biology, University of California, 305 Donner Laboratory, Berkeley, CA 94720-3206, USA; Fax: +1 - 510/642-8589.

The YGSC will accept any strains of *S. cerevisiae* which carry well-defined mutations that are of general interest to investigators. The deposition of strains should be accompanied by a full description of the strains including information on methods for assaying specific markers and using the strains.

B. The ATCC Plasmid Collection

The American Type Culture Collection (ATTC) provides several plasmids useful for molecular biological research with *S. cerevisiae*. The collection holds a set of clones of membrane protein genes as well as some *S. cerevisiae* genomic and cDNA libraries. Furthermore, vectors used for *S. cerevisiae* transformations and *S. cerevisiae* recipient strains are available.

ATCC also offers a set of clones constructed by Olson and co-workers (Olson *et al.*, 1986; Link and Olson, 1991; Riles *et al.*, 1993) for the physical mapping of *S. cerevisiae* genes, and some of the cosmids that were used for the sequencing of the yeast genome.

The ATCC catalog and prices for strains and plasmids are accessible at the World Wide Web site http://www.atcc.org/. Requests for strains and plasmids should be sent by using e-mail (sales@atcc.org) or using the following mailing address: American Type Culture Collection, 12301 Parklawn Drive, Rockville, MD 20852, USA; Fax: +1 - 301/816-4361.

◆◆◆◆◆◆ IV. CONCLUSIONS AND FUTURE DEVELOPMENTS

A. Aspects for the Analysis of Deletion Mutants

For the isolation of mutants, two basic principles can be followed. First, a large number of yeast mutants can be screened individually for a specific phenotype. An example of this screening approach is the replica-plating method (Lederberg and Lederberg, 1952). Second, growth conditions can be used that favor the growth of, and thereby enrich for, a particular set of mutants. The use of 5-fluoroorotic acid to select for uracil auxotrophic mutants is an example of this (Boeke *et al.*, 1984). In general, mutant isolation using the screening approach is more labor-intensive than the selective approach. However, sometimes selective systems are not sufficiently sensitive and many mutants are missed, especially those with reduced growth rates. In the era of whole genome analysis, these two approaches are also applicable for the analysis of deletion mutants. Screening analyses (mutant scanning) test each deletion mutant individually for a large number of possible phenotypes (Entian *et al.*, 1998; submitted for publication), whereas selective growth conditions can be used to diminish mutants with a particular phenotype from a population of transposon-mutagenized mutants (genetic footprinting; Smith *et al.*, 1995, 1996; Chapter 7). Mutant scanning is a random approach which allows one to establish an extensive database of phenotypes, whereas mutant selection is non-random and only addresses a particular phenotype. The major advantage of mutant scanning is that phenotypes are also discovered that would not have been tested for in the mutant selection system. This is extremely valuable because it helps to reduce misinterpretations of mutant phenotypes, for example, where a particular phenotype is the indirect consequence of a different primary phenotype. In many cases, mutant scanning leads to contradictory phenotypes that show that the mutation is not well understood (Entian *et al.*, submitted for publication). The major disadvantages of mutant scanning are that it is labor intensive and it accumulates results whose immediate interpretation is very difficult. The major strength of mutant selection is its easy handling and its phenotype-directed strategy. However, the identification of the selected mutations can be labor intensive.

B. A Bar-Coded Mutant Collection

Recently, methods for the identification of deletion mutants were greatly improved by the introduction of a unique 20-base sequence tag for each yeast deletion mutant (Shoemaker *et al.*, 1996). Approximately 9000 tags were designed that have similar melting temperatures for nucleic acid hybridization, display no secondary structure, show no similarity to each other, and are not present in the yeast genome. Hence, each of the 6200 possible yeast deletions can be assigned a particular tag (molecular bar-coding) which can be easily identified by PCR or hybridization (see Chapter 8 for a description of hybrid assay technology).

In addition to its advantages for phenotypic analysis, the bar-coding of deletion mutants also will improve the quality of a collection of deletion mutants. Because the genetic markers are very similar for all deletion mutants, no simple tests are currently available to distinguish a particular deletion. Therefore, in contrast to previous mutant collections, there is a much higher risk that mix-ups will occur. The bar-coding of the deletions will facilitate the evaluation of deletions by oligonucleotide hybridization and thus provide an efficient method for quality control of the collection.

Starting in 1997, a worldwide collaboration of yeast laboratories intends to delete all 6200 genes of the yeast genome using the bar-coding system. These mutants also will be collected at EUROSCARF so that, in the near future, bar-coded mutants will be available to the scientific community. If possible, the corresponding deletion cassettes also will be collected so that bar-coding deletions can be introduced easily into alternative *S. cerevisiae* genetic backgrounds.

C. Future Organization of the Mutant Collection and Services

The systematic deletion of all yeast genes also allows one to collect the deletions (and probably the deletion cassettes) in accordance to their location on the physical map of the *S. cerevisiae* genome. Additionally, mutations within certain aspects of cellular biology (e.g. metabolism, cell cycle, nuclear proteins, protein secretion, cell structure, etc.) can be grouped together and made accessible as mutant sets.

In many cases the genetic background of the yeast strain strongly influences the phenotype observed, and hence a strain collection can never be comprehensive for all scientific questions. Therefore, EUROSCARF also provides routine services such as inbreeding of strains as well as gene deletion and expression analysis.

A comprehensive catalog of EUROSCARF strains, plasmids and services will be made available on the World Wide Web using http://www.rz.uni-frankfurt.de/FB/fb16/mikro/euroscarf for entry.

Acknowledgements

The authors would like to thank Drs R. Mortimer and R. Contopoulou for their valuable information on the Yeast Genetic Stock Center and Drs A. Brown and A. Plummer for their helpful suggestions during the preparation of the manuscript. EUROSCARF is funded by the German Minister for Biotechnology (BMBF) and the European Union (EUROFAN project). It also received support from the Yeast Industrial Platform (YIP).

References

Baudin, A., Ozier-Kalogeropoulos, D., Denouel, A., Lacroute, F. and Cullin, C. (1993). A simple and efficient method for direct gene deletion in *Saccharomyces cerevisiae*. *Nucl. Acids Res.* **21**, 3329–3330.

Boeke, J. D., Lacroute, F. and Fink, G. R. (1984). A positive selection for mutants lacking orotidine-5′-phosphate decarboxylase activity in yeast: 5-fluoro-orotic acid resistance. *Mol. Gen. Genet.* **197**, 345–346.

Carle, G. F. and Olson, M. (1984). Separation of chromosomal DNA molecules from yeast by orthogonal-field-alternation gel electrophoresis. *Nucl. Acids Res.* **12**, 5647–5664.

Fonbrune, de, P. (1949). *Technique de Micromanipulation*. Masson, Paris.

Fowell, R. R. (1969). In *The Yeasts* (A. H. Rose and J. S. Harrison, eds), vol I, pp. 303–383. Academic Press, London.

Güldner, U., Heck, S., Fiedler, T., Beinhauer, J. and Hegemann, J. H. (1996). A new efficient gene disruption cassette for repeated use in budding yeast. *Nucl. Acids Res.* **24**, 2519–2524.

Lederberg, J. and Lederberg, E. M. (1952). Replica plating and indirect selection of bacterial mutants. *J. Bacteriol.* **63**, 399–406.

Lindegren, C. C. and Lindegren, G. (1943). A new method of hybridizing yeast. *Proc. Natl Acad. Sci. USA* **29**, 306–308.

Link, A. J. and Olson, M. V. (1991). Physical map of the *Saccharomyces cerevisiae* genome at 110-kilobase resolution. *Genetics* **127**, 681–698.

Mortimer, R. K. and Schild, D. (1980). Genetic map of *Saccharomyces cerevisiae*. *Microbiol. Rev.* **44**, 519–571.

Mortimer, R. K. and Schild, D. (1985). Genetic map of *Saccharomyces cerevisiae*, edition 9. *Microbiol. Rev.* **49**, 181–212.

Mortimer, R. K., Schild, D., Contopoulou, C. R. and Kans, J. A. (1989). Genetic map of *Saccharomyces cerevisiae*, edition 10. *Yeast* **5**, 321–403.

Mumberg, D., Müller, R. and Funk, M. (1994). Regulatable promoters of *Saccharomyces cerevisiae*: comparison of transcriptional activity and their use for heterologous expression. *Nucl. Acids Res.* **22**, 5767–5768.

Oliver, S. (1996). A network approach to the systematic analysis of yeast gene function. *Trends Genet.* **12**, 241–242.

Olson, M., Dutchik, J. E., Graham, M. Y., Brodeur, G. M., Helms, C. Frank, M., MacCollin, M., *et al.* (1986). Random-clone strategy for genomic restriction mapping in yeast. *Proc. Natl Acad. Sci. USA* **83**, 7826–7830.

Perkins, D. D. (1949). Biochemical mutants in the sput fungus *Ustilago maydis*. *Genetics* **34**, 607–626.

Riles, L., Durchik, J. E., Baktha, A., McCauley, B. K., Thayer, E. C., Leckie, M. P., Braden, V. V., *et al.* (1993). Physical maps of the six smallest chromosomes of *Saccharomyces cerevisiae* at a resolution of 2.6 kilobase pairs. *Genetics* **134**, 81–150.

Rothstein, R. (1983). One-step gene disruption in yeast. *Methods Enzymol.* **101**, 202–211.

Rothstein, R. (1991). Targeting, disruption, replacement and allele rescue: integrative DNA transformation in yeast. *Methods Enzymol.* **194**, 282–301.

Sambrock, J, Fritsch, E. F. and Maniatis, T. (1989). *Molecular Cloning: A Laboratory Manual*, 2nd edn., Cold Spring Harbor Laboratory, New York.

Shoemaker, D. D., Lashkari, D. A., Morris, D., Mittmann, M. and Davis, R. W. (1996). Quantitative phenotypic analysis of yeast deletion mutants using a highly parallel molecular bar-coding strategy. *Nature Genet.* **14**, 450–456.

Smith, V., Botstein, D. and Brown, P. O. (1995). Genetic footprinting: a genomic strategy for determining a gene's function given its sequence. *Proc. Natl Acad. Sci. USA* **92**, 767–773.

Smith, V., Chow, K. N., Lashkari, D., Botstein, D. and Brown, P. O. (1996). Functional analysis of the genes of yeast chromosome V by genetic footprinting. *Science* **274**, 2069–2074.

Wach, A., Brachat, A., Pohlmann, R. and Philippsen, P. (1994). New heterologous modules for classical PCR-based gene disruptions in *Saccharomyces cerevisiae*. *Yeast* **10**, 1793–1808.

24 Yeast Gene Analysis in the Next Millennium

Alistair J. P. Brown[1] and Mick F. Tuite[2]

[1]*Department of Molecular and Cell Biology, Institute of Medical Sciences, University of Aberdeen, Aberdeen and* [2]*Research School of Biosciences, University of Kent, Canterbury, Kent, UK*

◆◆

CONTENTS

List of Abbreviations

GFP	Green fluorescent protein
HAT	Hybrid array technology
NMR	Nuclear magnetic resonance
ORF	Open reading frame
PCR	Polymerase chain reaction
SAGE	Serial analysis of gene expression

◆◆◆◆◆◆ I. INTRODUCTION

The end of this millennium has been an exciting time for yeast molecular geneticists. The development of a comprehensive array of molecular tools, which complement the genetic, cellular and physiological dissection of *Saccharomyces cerevisiae*, has greatly facilitated the isolation and characterization of genes from yeast. Perhaps more importantly, they have opened the way to the precise manipulation of the yeast genome, allowing the phenotypic effects of well-defined mutations, even single base changes, to be analyzed (gene manipulation first, phenotypic analysis second). Furthermore, these methods have allowed the rapid characterization of mutations that confer an interesting phenotype upon the yeast cell, for example by gap-repair (genetic screens first, gene analysis second). In this way, the ever more sophisticated yeast molecular genetic toolbox has provided an important axis around which has developed our understanding of the biology of this important organism.

In parallel with these exciting developments, we have seen the initiation, development and completion of the yeast genome sequencing project – a remarkable collaborative achievement (Goffeau *et al.*, 1996, 1997; Mewes *et al.*, 1997). This global project was significant in many ways. First it led to the first public release of a eukaryotic genome sequence. While expanding our knowledge base greatly, the genome sequence also strikingly revealed our lack of understanding of yeast biology. Currently about 43% of the 6000 genes in the yeast genome are classified as "functionally characterized" (Mewes *et al.*, 1997). The large number of "orphan" genes with no discernible function illustrates starkly how classical genetic and molecular approaches have, over the past decades, failed to reveal these genes in yeast or, for that matter, any organism (Dujon, 1996; Oliver, 1996a). Fresh experimental approaches have therefore been required to elucidate their function, with one of the first steps being to generate a large object-oriented database of gene function by the European Functional Analysis Network (EUROFAN; Oliver, 1996b). Herein lies a second significant achievement of the genome sequencing effort: the forging of major international alliances in biology through worldwide collaboration. Such genome-targeted alliances provide an important framework for the elucidation of the function of orphan genes.

The dramatic advances in yeast gene analysis over the past decade have had three major consequences. Firstly, *S. cerevisiae* has become one of the most attractive model organisms for academic studies of fundamental aspects of eukaryotic biology such as the cell cycle, gene regulation and genome maintenance. Secondly, yeast is being used increasingly as a tool for the isolation and analysis of genes from higher eukaryotic organisms. Thirdly, *S. cerevisiae* remains a useful host for heterologous production of some commercially valuable proteins, and biotechnology and pharmaceutical companies are now seeking to generate novel recombinant yeast strains for use in high throughput screens for bioactive molecules. For example, expressing heterologous receptors at the yeast cell surface that are coupled to endogenous signal transduction pathways allows one to screen for novel agonists/antagonists that bind the target receptor.

Is the yeast molecular genetic toolbox now complete? What further additions are necessary or possible as we enter the next millennium? In this chapter we look to the future and speculate about some areas in which we anticipate major advances will be made.

◆◆◆◆◆◆ II. NEW TECHNOLOGIES

The genome sequence databases are already large, but they will expand with alacrity in the next decade as sequencing technologies and their automation continue to be enhanced. In addition, most of the experimental approaches described in this chapter will generate massive amounts of new data. This is particularly so for functional data about individual genes which will accumulate alongside comparative genome sequence information. It is important, therefore, that in parallel with advances in bench science, user-friendly software packages are developed to facilitate data collation and the remote analysis of these data by individual researchers around the globe. Many significant advances in the bench science will depend upon advanced facilities for efficient and interactive data analysis.

The continued advances in current technologies will not only improve the efficiency of data generation, but will also open the way to new experimental approaches. For example, advances in mass spectrometry will not only facilitate further characterization of the yeast proteome and lipidome (Kerwin *et al.*, 1996; Shevchenko *et al.*, 1996), but might also allow the characterization of complex and possibly heterogeneous macromolecules such as the mannoproteins, complex carbohydrates and other components of the yeast cell wall. The characterization of phenotypic changes in the metabilome (i.e. the profile of metabolites and their concentrations in the yeast cell) will provide a useful diagnostic tool for the elucidation of gene function (see Chapter 17). Most of the current spectroscopic methods allow detailed analysis of metabolite concentrations *in vitro* in cell-free extracts, and metabolic fluxes can be studied *in vivo* using NMR (Brindle, 1988). However, current limitations in NMR technologies mean that even these *in vivo* studies must be performed on cells under non-physiological conditions. The ability to perform such studies on chemostat cultures in which yeast cells are growing in steady-state (de Jong-Gubbels *et al.*, 1995) would represent a considerable step forward. This depends, however, upon significant advances in NMR technology.

As in other branches of science, the drive towards miniaturization will undoubtedly bring significant benefits as we enter the next millennium. In the context of yeast gene analysis, this is best exemplified by hybrid array technology (HAT), which involves hybridization to high-density arrays of oligonucleotides or DNA probes immobilized on silicon or glass chips (Chee *et al.*, 1996; Schena *et al.*, 1996). The development of such technology represents a major advance on current hybridization methods because it provides the means to analyze the whole yeast genome on a 1.6 cm^2 chip (Winzeler, 1997), thereby increasing the speed and efficiency with which global questions can be addressed. The impact of this technology upon studies of yeast genetics and gene expression is reviewed below and elsewhere in this volume (see Chapter 8).

The miniaturization of other forms of molecular sensing will also influence yeast genome research in the next millennium by providing rapid assays for other types of molecule. For example, it should become possible to analyse the yeast proteome on a chip carrying a high density array of recombinant antibody fragments against each yeast protein. Alternatively, specialized sensor chips might assay combinations of small metabolites, thereby producing a metabolic profile of the yeast cell. Clearly such advances would revolutionize phenotypic studies of yeast mutants.

◆◆◆◆◆◆ III. MORE GENOME SEQUENCES

If, in principle, we know the nucleotide sequence of every gene in yeast, what challenges now face us with regard to our understanding of *S. cerevisiae* biology? The first point is that we have yet to complete the yeast gene catalog. Current studies are focusing largely on open reading frames (ORFs) of more than 100 codons, and hence some short but functional genes are being ignored. This issue has been highlighted by Velculescu and co-workers (1995, 1997) who, using serial analysis of gene expression (SAGE), have shown that some short ORFs are transcribed and presumably represent functional genes. So having characterized the yeast genome, one new challenge is to identify the complete gene set.

The next millennium will continue to bring rapid expansion in genome sequence information for other yeast species of academic and/or medical importance. There are, after all, well over 500 species. Genome sequencing of *Candida albicans,* an asexual, pathogenic yeast, started in earnest in 1997 with the data already accumulating in both the public domain and commercial databases. Many (but not all) *C. albicans* genes sequenced to date display a high degree of similarity to *S. cerevisiae* genes (60–75% amino acid sequence identity). Comparison of the *S. cerevisiae* and *C. albicans* genomes will answer global questions about the degree of synteny between the genomes of these relatively closely related species, and about genome evolution in the presence or absence of a sexual cycle (see Chapter 21).

The genome sequence of the fission yeast *Schizosaccharomyces pombe* will be completed by the year 2000 (see http://www.sanger.ac.uk/Projects/S_pombe/). Because *S. pombe* is distant in evolutionary terms from *S. cerevisiae* and *C. albicans,* a comparison of the genome sequences of these three yeast species will prove enormously useful. For example, these yeasts occupy different environmental niches, and therefore each genome is expected to carry a functional gene set peculiar to their respective lifestyles. The identification of such genes is likely to promote more accurate dissection of the relationship between genome structure, gene function and lifestyle. Undoubtedly, pharmaceutical companies will also exploit this kind of information in their drive to develop new antifungal therapies. In contrast, the comparison of genes that are shared by these yeasts will provide invaluable information on the minimal set of conserved functions that are required for the growth of a unicellular eukaryote. The determination of further fungal genome sequences and the accumulation of genome sequences from multicellular organisms can only enhance and extend these types of analyses.

Much of the current genome sequencing effort is being made by commercial enterprises whose data are not freely available. For example, the *C. albicans* genome is being sequenced simultaneously both in the public and private sectors. While irritating to academics in the short term, this is unlikely to prove a long-term problem, because the commercial value of genome sequence data is likely to have a limited lifespan as academic projects release analogous data into the public domain.

◆◆◆◆◆◆ IV. INTERACTIVE MATRICES

The completion of the *S. cerevisiae* genome sequence was only the start of the efforts to define fully the genetic make-up of this species – a fact starkly illustrated by the large group of orphan genes in *S. cerevisiae* (Dujon, 1996). Clearly a major task is to characterize the function of each orphan gene and complete the object-oriented database of gene function pursued by EUROFAN (Oliver, 1996a,b). Many of the chapters in this volume describe molecular, genetic and biochemical approaches that will prove useful in pursuing this goal. However, a further important consideration is that the functions of many genes will only be completely understood once their interactions with other genes or gene products have been established. Efforts must now focus on generating matrices or networks of genes and gene products which reflect their genetic and biochemical interactions in response to environmental or physiological triggers.

Standard experimental approaches focus on individual genes or pathways. For example, workers interested in cell signaling have naturally focused, at least initially, upon the identification of individual components of a particular signaling pathway. However, as our understanding of one particular pathway has increased, interactions with other signaling pathways have become apparent. For example, not only does the mating pathway of *S. cerevisiae* interact with the cell division cycle and the control of cell polarity, but it also has links, through common signaling components, to the control of pseudohyphal development, and this pathway in turn responds to nutrient signaling (Kron and Gow, 1995). Alternatively, cell wall biosynthesis is regulated by numerous signaling pathways involved in controlling cell growth, division, development and polarity. These examples illustrate how specific biochemical pathways can have complex regulatory interconnections. As we enter the next millennium therefore, we need to develop experimental strategies that build interactive matrices of gene function.

A. Genetic Interactions

Interactions between two (or more) genes can be revealed by smart screens, which highlight phenotypes associated with combinations of mutations (see Chapter 16). How can such screens be improved in the future? Two advances will combine to make this possible. The first advance is the development of HAT (Chee *et al.*, 1996; Schena *et al.*, 1996), which can be applied to the rapid identification of large numbers of yeast mutants within mixed populations. This is where the second advance comes in. Such screens will be dependent upon the generation of new libraries of *S. cerevisiae* mutants some of which are already under construction (Shoemaker *et al.*, 1996). For example, a complete set of yeast mutants, in which each non-essential gene has been disrupted in an isogenic haploid background, will soon be available. Knock-outs of essential functions will also be available in a heterozygous form in diploid strains. Each knock-out will be individually "bar-coded" with a unique hybridization tag, thereby allowing rapid identification, in a single hybridization, of null mutants that have been lost from a population (e.g. because of defective growth under the conditions of interest) or of mutants that have grown under the conditions of interest (Shoemaker *et al.*, 1996). How can such libraries be used to analyse genetic interactions? In principle, libraries of bar-coded mutants could be transformed with a plasmid that overexpresses a specific ORF or that carries the dominant negative allele of a target ORF. Transformants could then be screened, by HAT, to detect null mutations that are synthetically lethal in combination with the plasmid. Alternatively, libraries of bar-coded mutants could be transformed to introduce a specific lethal mutation, and survivors could be screened rapidly using the bar codes to identify null mutations that suppress the lethal defect. Many such phenotypes will be conditional, depending upon the growth conditions, the imposition of stress, or the provision of osmotic stabilizers, for example. Significant advances will be made in our elucidation of genetic interactions through the imaginative exploitation of bar-coded yeast mutants in novel screens.

A further approach to elucidating genetic interactions would be possible if some way could be found to combine the bar-coded mutants with the transposon mutagenesis approach of Smith and co-workers (1996; Chapter 7, this volume). They have described how to identify rapidly genes that are required for growth under a particular set of growth conditions, essentially by determining those genes within which insertion of the yeast transposon Ty1 cannot be tolerated. Using a polymerase chain reaction (PCR)-based strategy, the positions of Ty1 insertion events can be mapped on the chromosome, thereby revealing genes that lack any insertions and which are presumably essential for growth under the conditions tested. If Ty1 mutagenesis could be combined with the bar-coding strategy, finding some method of identifying which genes are essential in specific bar-coded null mutant backgrounds, this would allow the rapid generation of a matrix of synthetic-lethal mutations.

B. Interactions Operating at the Level of Gene Expression

HAT has other applications that will reveal further genetic interactions in yeast. Global pictures of the transcriptome (i.e. the complete mRNA population within the yeast cell grown under a particular set of conditions) can be gained by hybridizing cDNA probes, prepared from cellular poly(A)-containing mRNA, against high density arrays representing the complete yeast genome (Winzeler, 1997). Analogous approaches such as SAGE (Velculescu *et al.*, 1995, 1997) are likely to be overtaken by HAT because of the potential speed and efficiency of the latter approach. In principle, it will become possible to identify specific changes in the yeast transcriptome in response to changes in growth conditions, or to specific mutations. Any genetic interactions that are mediated via changes in mRNA abundance will be identifiable using this powerful technology.

Global changes in gene expression can also be studied at the level of the proteome (i.e. the profile of soluble proteins within the yeast cell under a particular set of conditions; Shevchenko *et al.*, 1996; Payne and Garrels, 1997). In the near future, most yeast proteins that can be detected by standard two-dimensional electrophoretic methods will have been identified. While some proteins will not be present on this map because they are insoluble and/or subject to a high degree of covalent modification (e.g. cell wall mannoproteins), it will be possible readily to identify changes in a large subset of yeast proteins in response to genetic or physiological manipulations. It will also facilitate the identification of changes in post-translational modifications to specific proteins. This is particularly important because many genetic or biochemical interactions are manifested through alterations in, for example, the phosphorylation status of particular proteins (Chapter 14).

C. Networks of Physical Interactions

In addition to its applications in mutant screens and transcriptome analysis, HAT can contribute to the analysis of interactions in other ways. Two-hybrid screens are proving a powerful approach to identify physical interactions between specific proteins (Phizicky and Fields, 1995; Chapter 15). These screens could be accelerated by (a) designing screens based on auxotrophic markers such that only those cells carrying putative positives can grow to form colonies on selective media, (b) harvesting colonies from these plates and PCR-amplifying the population of inserts responsible for the two-hybrid interactions, and (c) hybridizing the labeled mixture of PCR products to yeast genome arrays on chips. In this way it will be possible to identify genes that encode proteins that interact physically with the bait of interest within hours of harvesting the clones. Analogous advances could be made to three-hybrid screens that have been developed recently to detect protein:RNA interactions (SenGupta *et al.*, 1996). Hence, the analysis of protein:protein and protein:RNA interactions will be much faster in the next millennium, thereby increasing the feasibility of generating large matrices that map these types of physical interaction within the yeast cell (Fromont-Racine *et al.*, 1997) and providing tangible links between different areas of metabolism.

While two-hybrid interactions provide an important tool to define physical interactions within a complex, they can also reveal interactions between complexes, thereby highlighting links between areas of metabolism. For example, through their studies of components involved in nuclear RNA metabolism, the European Network, TAPIR (Two-hybrid Analysis of Protein Involved in RNA metabolism) is revealing interesting connections between early and late events in pre-mRNA splicing, mRNA metabolism and export from the nucleus (Fromont-Racine *et al.*, 1997).

Crosslinking experiments have provided an extremely powerful route towards the characterization of interactions *in vitro*. Procedures have been developed for the introduction of large molecules into the yeast cell (Oliviera *et al.*, 1993) and it is not inconceivable that further advances in the *in vivo* analysis of macromolecular complexes could be made through the introduction into the yeast cell of tagged protein or RNA molecules carrying crossreactive side-groups. This might provide a useful approach for the purification of crosslinked complexes for which function (or inhibitory properties) have been demonstrated experimentally *in vivo*.

◆◆◆◆◆◆ V. PROTEIN LOCALIZATION

Protein localization has proved a useful tool in the determination of protein function, where the use of green fluorescent protein (GFP) fusions has extended the experimental armory beyond the use of immunofluorescence and *lacZ* fusions (Niedenthal *et al.*, 1997). In parallel with the emergence of GFP as a localization tag, advances in confocal microscopy have improved resolution and sensitivity thereby allowing detection of nonabundant proteins. Detection of GFP fluorescence does not require a cofactor, and this has made it possible to follow kinetic changes in the intracellular localization of specific GFP fusion proteins over time: for example, during progression through the cell cycle. How can such elegant methods be extended in the future? First, the sensitivity of the procedures could be increased by exploiting codon-optimized and fluorescence-enhanced versions of GFP (Cormack *et al.*, 1997), thereby facilitating kinetic analyses of low-abundance proteins. In addition, new GFP molecules with altered spectral properties have been generated to make red and blue GFPs (Mitra *et al.*, 1996). Therefore, it is now feasible to follow the location of combinations of specific fusion proteins within the same cell. This is likely to prove a powerful tool in the analysis of proteins whose putative functions probably involve even temporary co-localization within the same subcellular compartment. Such an approach might prove extremely useful in several ways: for example, in the analysis of vesicle recycling between compartments in the secretory pathway (Presley *et al.*, 1997), or for the characterization of spatial and temporal expression of different members of a gene family within the yeast cell (such as the 20 putative hexose transporters).

◆◆◆◆◆◆ VI. ANALYSIS OF SUBTLE INTERACTIONS

The disruption of most yeast genes exerts subtle, rather than major phenotypes upon the yeast cell (Oliver, 1996a,b) and the detection of such phenotypes represents a major challenge to yeast molecular geneticists (Dujon, 1996). Could this issue be addressed by constructing sets of sensitive plasmid-borne reporter genes that generate color or fluorescence in response to a slight activation of a specific stress-responsive pathway, increased respiratory activity, activation of protein kinase A, loss of cell polarity, entry to G_0, or nitrogen starvation, for example? If different reporters were used (green, red or blue fluorescent proteins, for example), combinations of reporter constructs could then be transformed into the bar-coded yeast mutants (see above). Mutants that activate specific combinations of responses could be isolated by a fluorescent-activated cell sorter and characterized rapidly by HAT. This type of approach might provide a useful means of detecting subtle changes in phenotype in the future.

◆◆◆◆◆◆ VII. IN CONCLUSION

This chapter has focused on potential advances in our approaches to further understand the biology of S. cerevisiae. However, as our understanding expands, our ability to exploit yeast as a tool in the analysis of genes from other organisms will also be enhanced. For example, as we learn more about MAP kinase pathways and cell cycle regulation in S. cerevisiae, we will be better able to subvert these regulatory circuits by replacing specific components with their homologs from human, nematode or insect cells. In the foreseeable future it will be feasible to move entire signaling pathways from a mammalian or plant cell into yeast. Having created adapter molecules that link yeast inputs and outputs to the heterologous regulatory pathway, signaling through the pathway could be reported by activation of a colorometric or fluorescent signal, for example. This represents a mere extension of the principle of complementing individual yeast mutations with a heterologous homolog, an approach that has led to the isolation and analysis of several interesting human gene sequences in yeast (e.g. Lee and Nurse, 1987; Dever et al., 1993; Wang and Reed, 1993; Guzder et al., 1995; Shpakovski et al., 1995; Tugendreich et al., 1995). Replacing functional yeast genes with heterologous genes will prove particularly useful in commercial screens for chemical agonists or antagonists of the corresponding gene products.

Not surprisingly this chapter has taken a relatively global and somewhat speculative look at the future of yeast molecular genetics. Clearly, workers in this field will have to evaluate enormous amounts of data, a problem not encountered by previous generations of yeast molecular geneticists. This is not to say that all of the next generation of yeast molecular geneticists will be required to work on global problems, thereby stimulating, yet further, the growth of the databases. Quite the contrary. Having assimilated the relevant information, there will be no substitute for specialists performing careful and rigorous analyses of specific aspects of yeast biology. Nevertheless, given the complex networks of interactions that exist in the yeast cell, these specialists will have to be aware of relevant observations made in apparently unrelated areas of yeast biology, and they will have to integrate their findings into the global picture.

References

Brindle, K. M. (1988). ^{31}P NMR magnetization-transfer measurements of flux between inorganic phosphate and adenosine 5′-triphosphate in yeast cells genetically modified to overproduce phosphoglycerate kinase. *Biochemistry* **27**, 6187–6196.

Chee, M., Yang, R., Hubbell, E., Berno, A., Huang, X. C., Stern, D., *et al.* (1996). Accessing genetic information with high density DNA arrays. *Science* **274**, 610–614.

Cormack, B., Bertram, G., Egerton, M., Gow, N. A. R., Falkow, S. and Brown, A. J. P. (1997). Yeast enhanced green fluorescent protein (yEGFP): a reporter of gene expression in *Candida albicans*. *Microbiology* **143**, 303–311.

de Jong-Gubbels, P., Vanrolleghem, P., Heijnen, S., van Dijken, J. P. and Pronk, J. T. (1995). Regulation of carbon metabolism in chemostat cultures of *Saccharomyces cerevisiae* grown on mixtures of glucose and ethanol. *Yeast* **11**, 407–418.

Dever, T. E., Chen, J. J., Barber, G. N., Cigan, A. M., Feng, L., Donahue, T. F., *et al.* (1993). Mammalian eukaryotic initiation factor 2 alpha kinases functionally substitute for *GCN2* protein kinase in the *GCN4* translational control mechanism of yeast. *Proc. Natl Acad. Sci. USA* **90**, 4616–4620.

Dujon, B. (1996). The yeast genome project: what did we learn? *Trends Genet.* **12**, 263–270.

Fromont-Racine, M., Rain, J. C. and Legrain, P. (1997). Towards a functional analysis of the yeast genome through exhaustive two-hybrid screens. *Nature Genetics* **16**, 277–282.

Goffeau, A., Barrell, B. G., Bussey, H., Davis, R. W., Dujon, B., Feldmann, E. J., *et al.* (1996). Life with 6000 genes. *Science* **274**, 546, 563–567.

Goffeau, A., Aert, R., Agostini-Carbone, M. L., Ahmed, A., Aigle, M., Alberghina, L., Albermann, K. *et al.* (1997). The Yeast Genome Directory. *Nature* **387**, 5.

Guzder, S. N., Sung, P., Prakash, S. and Prakash, L. (1995). Lethality in yeast of trichthiodystrophy (TTD) mutations in the human xeroderma pigmentosum group D gene. Implications for transcriptional defect in TTD. *J. Biol. Chem.* **270**, 17660–17663.

Kerwin, J. L., Wiens, A. M. and Ericsson, L. H. (1996). Identification of fatty acids by electrospray mass spectrometry and tandem mass spectrometry. *J. Mass Spec.* **31**, 184–192.

Kron, S. J. and Gow, N. A. R. (1995). Budding yeast morphogenesis: signalling, cytoskeleton and cell cycle. *Curr. Opin. Cell Biol.* **7**, 845–855.

Lee, M. G. and Nurse, P. (1987). Complementation used to clone a human homologue of the fission yeast cell cycle control gene *cdc2*. *Nature* **327**, 31–35.

Mewes, H. W., Albermann, K., Bahr, M., Frishman, D., Gleissner, A., Hani, J., *et al.* (1997). Overview of the yeast genome. *Nature* **387**, 7–65.

Mitra, R. D., Silva, C. M. and Youvan, D. C. (1996). Fluorescence resonance energy transfer between blue-emitting and red-shifted excitation derivatives of the green fluorescent protein. *Gene* **173**, 13–17.

Niedenthal, R. K., Riles, L., Johnston, M. and Hegemann, J. H. (1997). Green fluorescent protein as a marker for gene expression and subcellular localization in budding yeast. *Yeast* **12**, 773–786.

Oliver, S. G. (1996a). From DNA sequence to biological function. *Nature* **379**, 597–600.

Oliver, S. G. (1996b). A network approach to the systematic analysis of yeast gene function. *Trends Genet.* **12**, 241–242.

Oliviera, C. C., Goossen, B., Zanchin, N. I. T., McCarthy, J. E. G., Hentze, M. and Stripcke, R. (1993). Translational repression by the human iron-regulatory factor (IRF) in *Saccharomyces cerevisiae*. *Nucleic Acids Res.* **21**, 5316–5322.

Payne, W. E. and Garrels, J. I. (1997). Yeast protein database (YPD): a database for the complete proteome of *Saccharomyces cerevisiae*. *Nucleic Acids Res.* **25**, 57–62.

Phizicky, E. M. and Fields, S. (1995). Protein–protein interactions: methods for detection and analysis. *Microbiol. Rev.* **59**, 94–123.

Presley, J. F., Cole, N. B., Schroer, T. A., Hirschberg, K., Zaal, K. J. M. and Lippincott-Schwarz, J. (1997). ER-to-Golgi transport visualised in living cells. *Nature* **389**, 81–85.

Schena, M., Shalon, D., Heller, R., Chai, A., Brown, P. O. and Davis, R. W. (1996). Parallel human genome analysis: microarray-based expression monitoring of 1000 genes. *Proc. Natl Acad. Sci. USA* **93**, 10614–10619.

SenGupta, D. J., Zhang, B., Kraemer, B., Pochart, P., Fields, S. and Wickens, M. (1996). A three-hybrid system to detect RNA–protein interactions *in vivo*. *Proc. Natl Acad. Sci. USA* **93**, 8496–8501.

Shevchenko, A., Jensen, O. N., Podtelejnikov, A. V., Sagliocco, F., Wilm, M., Vorm, O., *et al.* (1996). Linking genome and proteome by mass spectrometry: large-scale identification of yeast proteins from two dimensional gels. *Proc. Natl Acad. Sci. USA* **93**, 14440–14445.

Shoemaker, D. D., Lashkari, D. A., Morris, D., Mittmann, M. and Davis, R. (1996). Quantitative phenotypic analysis of yeast deletion mutants using a highly parallel molecular bar-coding strategy. *Nature Genetics* **14**, 450–456.

Shpakovski, G. V., Acker, J., Wintzerith, M., Lacroix, J. F., Thuriaux, P. and Vigneron, M. (1995). Four subunits that are shared by the three classes of RNA polymerase are functionally interchangeable between *Homo sapiens* and *Saccharomyces cerevisiae*. *Mol. Cell. Biol.* **15**, 4702–4710.

Smith, V., Chou, K. N., Lashkari, D., Botstein, D. and Brown, P. O. (1996). Functional analysis of the genes of yeast chromosome V by genetic footprinting. *Science* **274**, 2069–2074.

Tugendreich, S., Tomkiel, J., Earnshaw, W. and Hieter, P. (1995). CDC27Hs colocalizes with CDC16Hs to the centrosome and mitotic spindle and is essential for the metaphase to anaphase transition. *Cell* **81**, 261–268.

Velculescu, V. E., Zhang, L., Vogelstein, B. and Kinzler, K. W. (1995). Serial analysis of gene expression. *Science* **270**, 484–487.

Velculescu, V. E., Zhang, L., Zhou, W., Vogelstein, J., Basrai, M. A., Bassett, D. E., *et al.* (1997). Characterisation of the yeast transcriptome. *Cell* **88**, 243–251.

Wang, M. M. and Reed, R. R. (1993). Molecular cloning of the olfactory neuronal transcription factor Olf-1 by genetic selection in yeast. *Nature* **364**, 121–126.

Winzeler, E. (1997). Functional genomics of *Saccharomyces cerevisiae*. *Am. Soc. Microbiol. News* **63**, 312–317.

Appendix I: The Ribosomal Protein Genes of *Saccharomyces cerevisiae*

Rudi J. Planta[1], Ian Stansfield[2] and Mick F. Tuite[3]
[1]*Department of Biochemistry and Molecular Biology, Vrije Universiteit, Amsterdam, The Netherlands,*
[2]*Department of Molecular and Cellular Biology, Institute of Medical Sciences, University of Aberdeen, Aberdeen, UK, and* [3]*Department of Biosciences, University of Kent, Canterbury, Kent, UK*

◆◆

Using the criteria discussed in Chapter 19, a complete inventory of the ribosomal proteins of *Saccharomyces cerevisiae* has been completed and is shown in Table 1.

In numbering the various yeast ribosomal proteins, we have followed as close as possible the nomenclature proposed by Wool *et al.* (1995) for the rat ribosomal proteins. We have deviated from the rat numbering system in 14 cases to avoid the designations with an "a" and to remove the gaps in the rat system. There is no equivalent of the rat ribosomal protein L28 in yeast, and therefore this protein might be mammalian specific. Furthermore, in the rat ribosomal protein nomenclature there is no L2, but on the other hand an L7 and an L7a exist. Because it has been demonstrated that rat L8 is the functional equivalent of *Escherichia coli* L2, it seems logical to designate L8 and its yeast equivalent L2, and to name L7a as the new L8. Other changes are based on similar considerations. We have designated the products of the *NAB1* genes as S0, because the rat equivalent Sa has not conclusively been identified as a genuine ribosomal protein, and might rather be a kind of a translation factor.

The proposed nomenclature for yeast ribosomal proteins (Table 1) agrees with the rat ribosomal protein nomenclature for 64 out of 79 rat ribosomal proteins. The great majority of the yeast ribosomal proteins have new names, at variance with any one of the three major nomenclatures for yeast ribosomal proteins proposed by Kruiswijk and Planta (1974), Warner and Gorenstein (1978) and Otaka and Osawa (1981), respectively. Furthermore, the yeast ribosomal protein genes are scattered over the entire yeast genome. Of the 78 cytoplasmic ribosomal protein genes in yeast, 59 are duplicated, and of the 137 gene copies 99 contain an intron, and three contain even two introns.

METHODS IN MICROBIOLOGY, VOLUME 26
ISBN 0–12–521526–6

Table 1. Nomenclature for the ribosomal proteins of Saccharomyces cerevisiae

Name of rp	Previous designations				Homologs	Genome location	CAI	Intron (?)	Length (aa)
S0A B	NAB1A NAB1B				rat Sa	YGR214w YLR048w	0.67 0.72	+ +	252 252
S1A B	rp10A rp10B	PLC1 PLC2			rat S3a	YLR441c YML063w	0.70 0.77	– –	255 255
S2	S4	rp12	YS5	SUP44	rat S2; Ec S5	YGL123w	0.80	–	254
S3	S3	rp13	YS3		mammalian S3	YNL178w	0.80	–	240
S4A B	S7A S7B	rp5	YS6		rat, Hs S4	YJR145c YHR203c	0.69 0.71	+ +	261 261
S5	S2	rp14	YS8		mammalian S5	YJR123w	0.75	–	225
S6A B	S10A S10B	rp9	YS4		rat, mouse, Hs S6	YPL090c YBR181c	0.84 0.85	+ +	236 236
S7A B		rp30			Hs S7; Xen. S8	YOR096w YNL096c	0.75 0.60	+ +	190 190
S8A B	S14A S14B	rp19	YS9		mammalian S8	YBL072c YER102w	0.75 0.72	+ (leader) + (leader)	200 200
S9A B	S13	rp21	YS11	SUP46	rat S9; Ec S4	YPL081w YBR189w	0.51 0.81	+ +	197 195
S10A B					rat S10	YOR293w YMR230w	0.84 0.80	+ +	105 105
S11A B	S18A S18B	rp41A rp41B	YS12		rat, Hs S11; Ec S17	YDR025w YBR048w	0.70 0.73	+ +	156 156
S12	S12				rat S12	YOR369c	0.84	–	143

S13	S27a	YS15		rat S13	YDR064w	0.78	+	151
S14A	rp59A		CRY1	mammalian S14; Ec S11	YCR031c	0.76	+	137
B	rp59B		CRY2		YJL191w	0.59	+	138
S15	rp52	S21		rat S15; Ec S19	YOL040c	0.77	–	142
S16A	rp61R			rat S16	YMR143w	0.65	+	143
B					YDL083c	0.76	+	143
S17A	rp51A			rat S17	YML024w	0.81	+	136
B	rp51B				YDR447c	0.76	+	136
S18A				rat S18; Ec S13	YDR450w	0.78	+	146
B					YML026c	0.73	+	146
S19A	rp55A	YS16A		rat S19	YOL121c	0.79	+	144
B	rp55B	YS16B			YNL302c	0.76	+	144
S20			URP2	rat, Hs S20; Xen. S22; Ec S10	YHL015w	0.83	–	121
S21A				rat S21	YKR057w	0.60	+	87
B					YJL136c	0.60	+	87
S22A	rp50	YS22		rat S15a	YJL190c	0.81	–	130
B					YLR367w	0.52	++ (leader)	130
S23A	rp37	YS14		rat S23; Ec S12	YGR118w	0.73	+	145
B					YPR132w	0.79	+	145
S24A				rat S24	YER074w	0.82	+	135
B					YIL069c	0.76	+	135
S25A	rp45	YS23		rat S25	YGR027c	0.72	+ (leader)	108
B					YLR333c	0.61	+ (leader)	108

Table I. continued

Name of rp	Previous designations	Homologs	Genome location	CAI	Intron (?)	Length (aa)
S26A	S26A	rat S26	YGL189c	0.78	+ (leader)	119
B	S26B		YER131w	0.71	+ (leader)	119
S27A	rp61 S27A YS20	mammalian S27	YKL156w	0.36	+	82
B	S27B		YHR021c	0.66	+	82
S28A	S33A YS27	mammalian S28	YOR167c	0.53	–	67
B	S33B		YLR264w	0.56	–	67
S29A	S36A YS29	rat S29	YLR388w	0.65	–	56
B	S36B		YDL061c	0.76	–	56
S30A	S30A	mammalian S30	YLR287c	0.75	+	63
B	S30B		YOR182c	0.75	+	63
S31	S37 YS24 UBl3	rat S27a	YLR167w	0.81	–	152
P0	A0	rat, Hs P0; Ec L10e	YLR340w	0.79	–	312
P1α	YP1α A1	rat, Hs P1; Ec L12eIIA	YDL081c	0.71	–	106
β	L44' YP1β Ax	rat, Hs P1; Ec L12eIIB	YDL130w	0.71	+	106
P2α	YP2α A2	Hs P2; Ec L12eIB	YOL039w	0.80	–	106
β	YP2β	rat, Hs P2; Ec L12eIA	YDR382w	0.76	–	106
L1A	SSM1A	rat L10a;	YPL220w	0.82	–	217
B	SSM1B	eu, arch.bact. L1	YGL135w	0.83	–	217
L2A	rp8 L5A YL6	rat L8; Ec L2	YFR031c	0.77	+	254
B	L5B		YIL018w	0.76	+	254

466

		rp	YL			ORF			
L3	L3	rp1	YL1	TCM1	rat L3	YOR063w	0.83	–	387
L4A	L2A	rp2	YL2		rat L4; Xen., Dros. L1	YBR031w	0.80	–	362
B	L2B					YDR012w	0.81	–	362
L5	L1a		YL3		rat L5	YPL131w	0.83	–	297
L6A	L17A	rp18	YL16		rat, Hs L6	YML073c	0.67	+	176
B	L17B					YLR448w	0.63	+	176
L7A	L6A	rp11	YL8		rat L7; Ec L30	YGL076c	0.76	++	244
B	L6B					YPL198w	0.72	++	244
L8A	L4A	rp6	YL5		rat, mouse, Hs L7a	YHL033c	0.84	–	256
B	L4B					YLL045c	0.85	–	256
L9A	L8A	rp24	YL11		rat L9; Ec L6	YGL147c	0.77	–	191
B	L8B					YNL067w	0.78	–	191
L10	L10			GRC5	rat L10	YLR075w	0.83	–	221
L11A	L16A	rp39A	YL22		rat L11; Ec L5	YPR102c	0.78	–	174
B	L16B	rp39B				YGR085c	0.73	–	174
L12A	L15A		YL23		rat L12(a); Ec L11	YEL054c	0.61	–	165
B	L15B					YDR418w	0.77	–	165
L13A					rat L13	YDL082w	0.65	+	199
B						YMR142c	0.74	+	199
L14A					mammalian L14	YKL006w	0.68	+	138
B						YHL001w	0.68	+	138
L15A	L13A	rp15R	YL10		rat L15	YLR029c	0.78	–	204
B	L13B					YMR121c	0.44	–	204
L16A	L21A	rp22	YL15		rat L13a	YIL133c	0.61	+	199
B	L21B	rp23				YNL069c	0.72	+	198

Table I. continued

Name of rp	Previous designations	Homologs	Genome location	CAI	Intron (?)	Length (aa)
L17A	L20A · YL17	rat, Hs L17; Ec L22	YKL180w	0.81	+	184
B	L20B		YJL177w	0.68	+	184
L18A	L18A · rp28A	rat L18	YOL120c	0.81	+	186
B	L18B · rp28B		YNL301c	0.68	+	186
L19A	L19A · rp15L · YL14	rat L19	YBR084c	0.69	+	189
B	L19B		YBL027w	0.71	+	189
L20A	L18A	rat L18a	YMR242c	0.60	–	180
B	L18B		YOR312c	0.70	+	174
L21A	L21A · URP1	rat L21	YBR191w	0.69	+	160
B	L21B		YPL079w	0.73	+	160
L22A	L22A · L1c · rp4 · YL31	rat L22	YLR061w	0.86	+	121
B	L22B		YFL034c	0.29	+	122
L23A	L23A · L17aA · YL32	rat L23; Ec L14	YBL087c	0.62	+	137
B	L23B · L17aB		YER117w	0.65	+	137
L24A	L24A · L30A · rp29 · YL21	rat L24	YGL031c	0.76	+ (leader)	155
B	L24B · L30B		YGR148c	0.76	+ (leader)	155
L25	L25 · rp16L · YL25	rat L23a; Ec L23	YOL127w	0.75	+	142
L26A	L26A · L33A · YL33	rat L26	YLR344w	0.63	+	127
B	L26B · L33B		YGR034w	0.68	+	129
L27A	L27A	mammalian L27	YHR010w	0.74	+	136
B	L27B		YDR471w	0.52	+	136
L28	L28 · rp44 · CYH2 · YL24	rat, mouse L27a	YGL103w	0.71	+	149
L29	L29 · YL43	rat L29	YFR032c	0.68	+ (leader)	59

Name	rp	YL		gene	ORF	CAI		aa	Homolog
L30	rp73	YL38	L32		YGL030w	0.86	+	105	rat, mouse L30
L31A		YL28	L34A		YDL075w	0.74	+	113	rat L31
B			L34B		YLR406c	0.60	+	113	
L32					YBL092w	0.82	+ (leader)	130	rat, mammalian L32
L33A	rp47	YL37	L37A		YPL143w	0.75	+	107	rat L35a
B			L37B		YOR234c	0.73	+	107	
L34A					YER056c	0.76	+	121	rat L34
B					YIL052c	0.78	+	121	
L35A				SOS1	YDL191w	0.80	+	120	rat L35
B				SOS2	YDL136w	0.76	+	120	
L36A		YL39	L39		YMR194w	0.62	+	100	rat L36
B					YPL249c	0.62	+	100	
L37A		YL35	L43		YLR185w	0.70	+	88	rat L37
B					DR500c	0.71	+	88	
L38					YLR325c	0.78	−	78	rat L38
L39		YL40	L46	PUB2	YJL189w	0.92	+	51	rat L39
L40A				UBI1	YIL148w	0.74	+	128	rat L40
B				UBI2	YKR094c	0.74	+	128	
L41A		YL41	L47A		YDL184c	0.42	−	25	Hs L41
B			L47B		YDL133c	0.42	−	25	
L42A		YL27	L41A		YNL162w	0.63	+	116	rat, Hs L36a
B			L41B		YHR141c	0.80	+	106	
L43A					YPR043w	0.87	+	92	human L37a
B					YJR094w	0.57	+	92	

rp, ribosomal protein; CAI, codon adaptation index; aa, amino acid; Ec, *Escherichia coli*; Hs, *Homo sapiens*; Xen., *Xenopus laevis*; eu, eubacteria; arch. bact., archaebacteria; Dros., *Drosophila melanogaster*.

Acknowledgements

The help of Dr W. H. Mager in the preparation of Table 1 is gratefully acknowledged. A full account of the data on which this table is based will be published elsewhere (R. J. Planta and W. H. Mager, manuscript in preparation). We thank Drs J.-P. G. Ballesta, J. R. Warner and W. Suzuki for valuable comments during the draft of Table 1.

References

Kruiswijk, T. and Planta, R. J. (1974). Analysis of the protein composition of yeast ribosomal subunits by two-dimensional polyacrylamide gel electrophoresis. *Mol. Biol. Rep.* **1**, 409–415.

Otaka, E. and Osawa, S. (1981). Yeast ribosomal proteins: V. Correlation of several nomenclatures and proposal of a standard nomenclature. *Mol. Gen. Genet.* **181**, 176–182.

Warner, J. A. and Gorenstein, C. G. (1978). The ribosomal proteins of *Saccharomyces cerevisiae*. In *Methods in Cell Biology* (D. M. Prescott, ed.), vol. 20, pp. 45–60. Academic Press, New York.

Wool, I. G., Chan, Y.-L. and Glück, A. (1995). Structure and evolution of mammalian ribosomal proteins. *Biochem. Cell Biol.* **73**, 933–947.

Appendix II: Yeast Growth

Alistair J. P. Brown[1] and Mick F. Tuite[2]

[1]*Department of Molecular and Cell Biology, Institute of Medical Sciences, University of Aberdeen, Aberdeen, UK, and* [2]*Research School of Biosciences, University of Kent, Canterbury, Kent, UK*

◆◆◆

The aim of this appendix is to provide some useful tips on the growth and storage of strains of *Saccharomyces cerevisiae*. Most of these tips will be obvious to those with experience in the field, but we hope they will be helpful to those who are about to handle this organism for the first time. We refer readers to an excellent article by Fred Sherman (1991) on "Getting started with Yeast" and to the volume by Guthrie and Fink (1991), which provides useful practical details of many specific yeast methods.

Table I. Rich Media For Yeast Growth[a]

Medium	Constituents	Use
YPD (YEPD)	1% Bacto-yeast extract 2% Bacto-peptone 2% glucose (dextrose)[b,c] [2% Bacto-agar]	For general growth
YPAD	1% Bacto-yeast extract 2% Bacto-peptone 2% glucose (dextrose)[c] 100 mg l[-1] adenine hemisulfate [2% Bacto-agar]	For yeast transformations[d], DNA preparation, and slants[e]
2× YPAD	2-fold concentrated YPAD	For yeast transformations[d]
YPDG	1% Bacto-yeast extract 2% Bacto-peptone 3% glycerol 0.1% glucose (dextrose) [2% Bacto-agar]	For differentiating *petites* from *grandes*

[a]Unless stated otherwise, all media can be sterilized by autoclaving.
[b]Alternative fermentative C-sources (e.g. fructose or maltose) are used at a concentration of 2% (w/v). Alternative non-fermentative C-sources (e.g. glycerol, ethanol, lactate or pyruvate) are used at a concentration of 3% (w/v).
[c]Glucose will caramelize if autoclaved excessively. If necessary, concentrated glucose solutions (20%) can be filter sterilized or autoclaved separately.
[d]Gietz and Woods, Chapter 4.
[e]Sherman (1991).

Recipes for commonly used rich and defined growth media are provided in Tables 1 and 2, respectively. Sporulation media are described in Table 3, and details on the preparation of growth supplements are provided in Table 4.

Table 2. Defined media for yeast growth[a]

Medium	Constituents	Use
SD (GYNB)	0.67% Bacto-yeast nitrogen base (without amino acids)[b] 2% glucose (dextrose)[c,d] [2% Bacto-agar] Each appropriate supplement[e]	For growth under defined or selective conditions
Drop out medium	SD (GYNB) lacking one specific growth supplement	
SC[f,g]	935 ml Mix 1 10 ml Mix 2 50 ml 40% glucose 5 ml 1M Na$_2$HPO$_4$ (pH 7)	For growth under defined or selective conditions. For yeast transformations[h]
	SC Mix 1 (935 ml of 1× stock): 6.7 g Bacto-yeast nitrogen base (without amino acids)[b] 10 mg adenine 40 mg uracil 50 mg tyrosine [30 g Bacto-agar]	
	SC Mix 2 (100 ml of 100× stock)[i]: 200 mg arginine 200 mg histidine 600 mg isoleucine 600 mg leucine 400 mg lysine 100 mg methionine 600 mg phenylalanine 1000 mg threonine 400 mg tryptophan	

[a]Unless stated otherwise, all media can be sterilized by autoclaving.
[b]For details of ingredients, see Sherman (1991).
[c]Alternative fermentative C-sources (e.g. fructose or maltose) are used at a concentration of 2% (w/v). Alternative non-fermentative C-sources (e.g. glycerol, ethanol, lactate or pyruvate) are used at a concentration of 3% (w/v).
[d]Glucose will caramelize if autoclaved excessively. If necessary, concentrated glucose solutions (20% (w/v)) can be filter sterilized or autoclaved separately.
[e]See Table 4.
[f]Strathern et al. (1994).
[g]For "SC–Ura", omit the uracil from Mix 1. For "SC–Leu", omit the leucine from Mix 2. For "SC–His", omit the histidine from Mix 2.
[h]Gietz and Woods, Chapter 4.
[i]Filter sterilize and store in aliquots at 4°C.

Table 3. Yeast sporulation media[a]

Medium	Constituents	Comments
Presporulation medium	0.8% Bacto-yeast extract 0.3% Bacto-peptone 10% glucose (dextrose) [2% Bacto-agar]	Grow strains on this medium for 1–2 days then transfer to sporulation medium
Sporulation medium[b]	1% potassium acetate 0.1% Bacto-yeast extract 0.05% glucose (dextrose) [2% Bacto-agar]	After several divisions, strains sporulate in 3–5 days
Minimal sporulation medium[b]	1% potassium acetate [2% Bacto-agar]	Strains sporulate in 3–5 days

[a]Sherman (1991).
[b]Adding the appropriate growth supplements for auxotrophic diploids usually increases the sporulation frequency (Sherman, 1991).

Table 4. Growth supplements for *Saccharomyces cerevisiae*

Supplement[a]	Stock concentration (g/100 ml)	Final concentration (μg/ml)	Dilution
Adenine hemisulfate[b,d]	0.2	20	100×
Uracil[c,d]	0.2	20	100×
L-Aspartic acid[e, d]	1.0	100	100×
L-Glutamic acid[d]	1.0	100	100×
L-Isoleucine	0.3	30	100×
L-Leucine	0.3	30	100×
L-Lysine-HCl	0.3	30	100×
L-Phenylalanine[d]	0.5	50	100×
L-Serine	4.0	400	100×
L-Threonine[e,d]	2.0	200	100×
L-Tyrosine	0.3	30	100×
L-Valine	1.5	150	100×
L-Arginine-HCl	1.0	20	500×
L-Histidine-HCl	1.0	20	500×
L-Methionine	1.0	20	500×
L-Tryptophan	1.0	20	500×

[a]Use HCl salts of amino acids where possible. All supplements can be autoclaved except aspartic acid and threonine, which must be added to media after autoclaving.
[b]Dissolve adenine hemisulfate in dilute HCl.
[c]Dissolve uracil in dilute NaOH.
[d]Store at room temperature.
[e]Add to media after autoclaving (Sherman, 1991).

A. Testing Genotypes

Having obtained a yeast strain from a culture collection or another laboratory, the first step is to confirm the genotype of the strain. The full genotype should also be checked when strains are revived after long-term storage. To check specific markers, the strain first should be streaked onto a suitable solid growth medium (usually YPD, depending upon the genotype; Table 1) to grow single colonies. Auxotrophic requirements are then checked by replica-plating cells from individual colonies onto defined media lacking individual supplements (drop-out media; Table 2) and drug resistance markers are tested by plating onto suitable drug-containing media.

Replica plating is best performed by suspending cells from a single colony in 200 µl of sterile H_2O or saline (0.9% w/v NaCl), and spotting 10 µl of this suspension onto each plate. This ensures that approximately equal numbers of cells are placed on each plate thereby facilitating comparison of growth between plates. Plates are generally grown at 28–30°C for 2 days (not overnight!). Some mutants require longer incubation periods or different incubation temperatures (see below).

Conditional phenotypes must also be checked, for example by monitoring growth of temperature-sensitive strains at their permissive and non-permissive temperatures. This can also be done by replica-plating. Most temperature-sensitive strains are tested for their ability to grow at 25°C and 36°C, but care should be taken with incubation temperatures because permissive and non-permissive temperatures vary amongst different temperature-sensitive strains. Check the published procedures carefully for handling individual strains.

Note that reversions and suppressor mutations arise at significant frequencies for some types of mutant, particularly where the original mutant has a negative impact on growth or survival under storage conditions. In such cases the markers should be checked regularly (i.e. before and after each experiment). Also, note that respiratory-deficient (petite) mutants arise at high frequencies (usually 0.1–1.0%) in most strains. Such mutants form small (*petite*) colonies on YPDG (Table 1), whereas respiratory-competent strains form large (*grande*) colonies on this medium.

B. Storing Strains

Once the markers have been checked, the strain should be stored in your local strain collection. There are various ways of storing yeast strains, but the most common procedure is to add sterile glycerol to a final concentration of 15% (v/v), mix thoroughly, and store at –80°C in sterile screw-cap 2 ml vials. Cells may then be removed quickly by scraping off a small amount of ice from the surface of unthawed stocks and using this cell sample to inoculate cultures. Cell stocks kept at –80°C should not be thawed and refrozen. Frequently used strains should be stored on plates at 4°C (to minimize disturbance of frozen stocks), and these strains should be restreaked at regular intervals (usually monthly) to reduce the accumulation of spontaneous mutations. In our experience yeast strains stored in glycerol at –80°C will remain viable for 10–20 years although some strains do die much quicker. Alternative methods of strain storage are described by Entian and Kötter in Chapter 23.

C. Growing Strains

Most yeast strains should be grown in liquid medium at 30°C and at 200 r.p.m. to ensure good aeration. The most obvious exceptions to this are temperature-sensitive strains, which require alternative incubation temperatures. Aeration is inefficient if more than 50 ml of liquid medium is used in a 250 ml conical flask, and so as a general guide, culture volumes should always be less than one-fifth of the flask's capacity. Alternatively, baffled flasks can be used to increase aeration efficiency.

Robust yeast strains have doubling times of about 2 h in rich media (Table 1) and 2.5–3 h in defined media (Table 2). However, different strains grow at different rates, especially those carrying deleterious mutations. Therefore, it is often useful to measure the growth rates of unfamiliar strains under the conditions of interest before embarking on an extensive program of experiments.

Growth rates also depend on the carbon source: strains grow most efficiently on fermentable carbon sources such as glucose or fructose, and less efficiently on other carbon sources such as galactose, raffinose, glycerol, pyruvate and ethanol. Even with aeration, most yeast strains ferment glucose to ethanol, and then assimilate the ethanol once the glucose is exhausted from the medium. Therefore, yeast cultures often display diauxy, and the physiological characteristics of yeast cells change during the various stages of growth in batch cultures.

Cell physiology exerts significant effects upon a wide range of phenomena (and *vice versa*), and therefore most workers perform their experiments on mid-exponential phase yeast cultures. Starter cultures are set up by inoculating 2–5 ml of liquid medium with cells from a single colony, and growing at 30°C at 200 r.p.m. with aeration for about 2–4 days (depending upon the strain and the growth medium). Starter cultures that have been stored for longer than 1–2 weeks at 4°C should not be used because cell viability decreases significantly during a prolonged stationary phase. Fresh starter cultures are then used to inoculate experimental cultures (2–100 μl into 50 ml), which are then grown overnight at 30°C at 200 r.p.m. with aeration to mid-exponential phase ($OD_{600} = \sim 0.5$). This is sufficient for many kinds of experiment, although better-defined starter cultures are required for experiments that demand minimal physiological variation between replicate cultures.

D. Monitoring Growth

Most frequently, growth rates are measured simply by monitoring the OD_{600} of liquid cultures. It should be noted, however, that the relationship between OD_{600} and cell concentration is not linear over the whole growth range (lag, exponential, transition and stationary phases), and that cell aggregation or significant changes in cell size or morphology disturb this relationship. Alternatively, growth can be monitored by measuring cell concentration (by plating, or microscopy using a hemocytometer), by measuring the wet or dry weight of the biomass, or by measuring the protein, RNA or DNA content of the culture.

Measuring the OD_{600} of cultures does not differentiate between viable and dead cells. The proportion of viable cells in a culture can be determined most conveniently using dyes such as propidium iodide (see Chapter 20). Where necessary, the total cell concentration (estimated by microscopy) can be compared with the viable cell concentration (estimated by counting colonies on YPD).

E. Plasmid-containing Strains

DNA transformation is mutagenic, and transformants frequently display clonal variation. For this reason, assays should be performed on several individual transformants. Therefore, several individual transformants should be picked from transformation plates, and these should be purified by restreaking for single colonies on selective medium, their markers checked, and then the strains stored as described in section II above.

Many ARS (YRp) or 2 μm-based (YEp) plasmids are unstable and plasmid-free cells can be generated at significant rates. Therefore, strains bearing such plasmids should be maintained and grown under selective pressure (i.e. on defined media lacking the relevant supplement). Some experiments need to be performed on cells grown in rich media. In such cases, starter cultures of plasmid-bearing strains should be prepared in selective media, then inoculated into rich media and grown overnight. At the time of harvesting for analysis, the proportion of plasmid-containing cells should be measured by comparing the concentration of colony-forming units on defined media with or without the appropriate supplement.

YIp (yeast integrating) plasmids are generally more stable than YRp or YEp plasmids. As a result, it is safe to grow most YIp-bearing strains in rich media, although it is good practice to store them under selective conditions. The stability of YIp-bearing strains depends on the nature of the integration event. For example, if the YIp generates a tandem repeat upon integration into the genome, this can revert to wild-type by an intramolecular homologous recombination event. Hence markers should be checked regularly and preferably during each experiment.

Some plasmids are deleterious to the host cell. Clearly, selection pressures will operate against cells that contain such plasmids (e.g. those that impose a significant metabolic load on the cell or that encode a toxic product). In such cases, cultures will accumulate plasmid-free cells and cells with rearranged plasmids, and clonal variation in the YRp and YEp plasmid copy number will arise. This problem can be minimized by routinely analysing *fresh* transformants.

References

Guthrie, C. and Fink, G. R. (eds) (1991). *Guide to Yeast Genetics and Molecular Biology: Methods in Enzymology, vol. 194.* Academic Press, San Diego.

Sherman, F. (1991). Getting started with Yeast. In *Guide to Yeast Genetics and Molecular Biology: Methods in Enzymology, vol. 194* (C. Guthrie and G. R. Fink, eds), pp. 3–21. Academic Press, San Diego.

Strathern, J. N., Mastrangelo, M., Rinkel, L. A. and Garfinkel, D. J. (1994). Ty insertional mutagenesis. In *Molecular Genetics of Yeast: A Practical Approach* (J. R. Johnston, ed.), pp. 111–119. Oxford University Press, Oxford.

Appendix II

References

Appendix III: Useful World Wide Web Addresses for Yeast Researchers

Alistair J. P. Brown

Department of Molecular and Cell Biology, Institute of Medical Sciences, University of Aberdeen, Aberdeen, UK

◆◆◆

This appendix lists World Wide Web (WWW) sites that we and others have found useful. The list comes with a health warning: it was up to date at the time of preparation (August 1997), but WWW sites come and go with great alacrity. Those with suggestions for further useful, yeast-related WWW sites are welcome to e-mail the author with this information for inclusion in new editions of this appendix (al.brown@abdn.ac.uk).

WWW address	Comments	Reference
Strain and vector collections		
http://www.rz.uni-frankfurt.de/FB/fb16/mikro/euroscarf	EUROFAN strain and plasmid collection	Entian & Kötter, Chapter 23
http://zenith.berkeley.edu/~jontib/YGSC/	Yeast Genetic Stock Center at Berkeley	Entian & Kötter, Chapter 23
http://www.atcc.org/	ATCC (American Type Culture Collection) yeast strain collection	Entian & Kötter, Chapter 23
http://www.ifrn.bbsrc.ac.uk/NCYC/	NCYC (National Collection of Yeast Cultures) yeast strain collection	Entian & Kötter, Chapter 23
http://www.atcg.com/vectordb/	List of some vectors for yeast research	
Dedicated molecular biology servers		
http://www.ncbi.nlm.nih.gov/	NCBI (National Center for Biotechnology Information) home page, USA	Moore, Chapter 22
http://www.ncbi.nlm.nih.gov/Web/Genbank/index.html	Entry to GenBank, USA	
http://www.mips.biochem.mpg.de/	MIPS (Martinsried Institute for Protein Sequences) home page, Germany	Mewes et al., Chapter 3
http://www.embl-heidelberg.de/Services/index.html	EMBL WWW services, Switzerland	
http://expasy.hcuge.ch/expasy-top.html	ExPASy WWW molecular biology server (includes entry to SWISS-PROT; see below), Switzerland	
http://gserv1.dl.ac.uk/SEQNET/home.html	Entry to Seqnet; Daresbury Laboratory, UK	
http://www.sanger.ac.uk/	Sanger Centre home page, UK	
http://www.sanger.ac.uk/srs/srsc	Network browser for databanks in molecular biology, UK	Mewes et al., Chapter 3
http://www.nig.ac.jp/	National Institute of Genetics home page, Japan	

Yeast genome sites

URL	Description	Reference
http://speedy.mips.biochem.mpg.de/mips/yeast/	MIPS yeast genome page	*Mewes et al.*, Chapter 3
ftp://mips.embnet.org/yeast/tables/mips_orfs_table.ascii	MIPS table of yeast gene names (old and new format)	*Mewes et al.*, Chapter 3
http://genome-www.stanford.edu/Saccharomyces/	SGD (*Saccharomyces* Genome Database) home page	*Mewes et al.*, Chapter 3
http://genome-www.stanford.edu/cgi-bin/dbrun/SacchDB?find+locus	SGD yeast genome page; Gene Name Registry	Tuite, Appendix IV
http://www.ncbi.nlm.nih.gov/Entrez/	Entry to NCBI browser	Moore, Chapter 22
http://www.expasy.ch/sprot/sp-docu.html	SWISS-PROT list of documents including *S. cerevisiae, S. pombe* and *C. albicans* genome entries	Walsh and Barrell (1996)
http://www.sanger.ac.uk/yeast/home.html	Sanger Centre yeast genome home page	
http://sequence-www.stanford.edu/group/yeast_deletion_project/deletion.html	Stanford yeast deletion project	
http://embl-heidelberg.de/~genequiz/yeast.html	EMBL GeneQuiz resource for 3D structural information related to the yeast genome	*Mewes et al.*, Chapter 3
http://acer.gen.tcd.ie/~khwolfe/yeast/nova/index.html	Wolfe's year gene duplications	Oliver, Chapter 1

Yeast protein databases

URL	Description	Reference
http://quest7.proteome.com/YPDhome.html	YPD (yeast protein database) home page	Molina *et al.*, Chapter 20
http://quest7.proteome.com/graphs.html	YPD theoretical 2D gels for subsets of yeast proteins	
http://expasy.hcuge.ch/sprot/prosite.html	PROSITE, a database of biologically significant protein sites, patterns and profiles	Molina *et al.*, Chapter 20

WWW address	Comments	Reference
http://www.ibgc.u-bordeaux2.fr/YPM	University of Bordeaux II 2D gel database of yeast proteins	Moore, Chapter 22; Boguski et al. (1993) Boguski (1995)
http://genecrunch.sgi.ch/	Functional prediction of yeast proteins	
http://siva.chsl.org/#2dgel	Software for construction analysis of 2D gel protein databases	
Yeast protocols		
http://www.umanitoba.ca/faculties/medicine/human_genetics/gietz/Trafo.html	Gietz yeast transformation home page	
http://www.fhcrc.org/~gottschling/homepage.html	Gottschling yeast protocols home page	
Genome cross-referencing		
http://www.ncbi.nlm.nih.gov/dbEST/index.html	Entry to dbEST (database of Expressed Sequence Tags), USA	Moore, Chapter 22
http://www.ncbi.nlm.nih.gov/XREFdb/	NIH (National Intitutes of Health) Organismal Cross-Referencing Databases	Moore, Chapter 22
http://www.ncbi.nlm.nih.gov/Bassett/Yeast/	Database of yeast sequences homologous to known human disease genes	Moore, Chapter 22
http://www.ncbi.nlm.nih.gov/Omim/	Genetic disorders associated with a particular area of the human genome	Moore, Chapter 22
http://www.ncbi.nlm.nih.gov/UniGene/index.html	Unigene database of overlapping ESTs	Moore, Chapter 22
http://www-bio.llnl.gov/bbrp/image/idist_add.html	List of distributors of EST cDNAs	Moore, Chapter 22
http://www.ncbi.nlm.nih.gov/BLAST/	NCBI site for BLAST homology searches	Moore, Chapter 22

General molecular biology tools

http://www.fmi.ch/biology/research_tools.html — Pedro's research tools – a constantly updated list of all molecular biology tools available on the Web

http://www.dna.affrc.go.jp/~nakamura/codon.html — Searchable index of codon usage tables for 5440 organisms

Metabolic control analysis (MCA)

http://gepasi.dbs.aber.ac.uk/metab/mca_home.htm — Mendez's MCA tutorials and leads into MCA software — Teusink *et al.*, Chapter 17

http://www.bi.umist.ac.uk/courses/2IRM/MCA/default.htm — Butler's MCA course — Teusink *et al.*, Chapter 17

http://ir2lcb.cnrs-mrs.fr/lcbpage/athel/mcafaq.html — Cornish-Bowden's FAQ site — Teusink *et al.*, Chapter 17

http://www.bio.net:80/hypermail/BTK-MCA/ — MCA newsgroup — Teusink *et al.*, Chapter 17

Candida albicans

http://alces.med.umn.edu:80/Candida.html — General information on *Candida albicans* — Magee, Chapter 21

http://alces.med.umn.edu:80/bin/genelist?seqs — *Candida albicans* genome sequence information

Newsgroups and e-mail addresses

http://www.bio.net/ — General newsgroup

http://www.bio.net/hypermail/YEAST/
news:bionet.molbio.yeast — Newsgroup for yeast molecular biology

http://www.panix.com/~candida/ — General *Candida* links academic/clinical/personal pages

gopher://gopher.gdb.org:70/11/biol-search/yeast-email — Facility for searching e-mail addresses of yeast researchers

2D, two dimensional; 3D, three dimensional.

References

Boguski, M. S. (1995). The turning point in genome research. *Trends Biochem. Sci.* **20**, 295–296.

Boguski, M. S., Lowe, T. M. J. and Tolstoshev, C. M. (1993). dbEST – database for "expressed sequence tags". *Nature Genetics* **4**, 332–333.

Walsh, S. and Barrell, B. (1996). The *Saccharomyces cerevisiae* genome on the World Wide Web. *Trends Genet.* **12**, 276–277.

Appendix IV: Nomenclature for Yeast Genes and Proteins

Mick F. Tuite

Department of Biosciences, University of Kent, Canterbury, Kent, UK

◆◆

There is a universally accepted nomenclature for *Saccharomyces cerevisiae* genes and their gene products and gene names are often used as a guide to a phenotype or to the identity of the gene product, or even its function. As the number of *S. cerevisiae* genes of known function approaches 3000, the challenge to derive meaningful gene (and protein) names increases. Nevertheless, it is imperative that, once a researcher has an indication of the function of a gene or its product, the gene be assigned an appropriate name that conforms to the standard nomenclature. In this short article the rules for naming yeast genes and their gene products are reviewed and the steps one must take to register a gene name outlined.

A. Chromosomal Genes

Figure 1 outlines the basic rules for naming a gene of *S. cerevisiae*. Where possible the three letter name should give an indication of the function of the gene or the major phenotype associated with a defect in that gene. For example:

- *PDI1:* encodes the enzyme **p**rotein **d**isulfide **i**somerase;
- *ADE2:* encodes an enzyme involved in **ade**nine biosynthesis;
- *SUP35:* encodes a **sup**pressor of nonsense mutations.

Gene designations that give no clues to the function of the gene or its product should be avoided, e.g. *FUN* (**fun**ction **un**known) or *YFG* (**y**our **f**avorite **g**ene).

To avoid confusion between the wild-type gene and a dominant mutant allele of that gene, the wild-type gene should be designated with the superscript "+" symbol (i.e. *YFG1⁺* is the wild-type gene, *YFG1* is a dominant mutant allele). In practice, if one is discussing the wild-type gene without the need to refer to a dominant allele, the superscript "+" symbol is usually dropped.

Allele numbers (e.g. *yfg1-1, yfg1-2*) are used to differentiate between genetic lesions that give rise to the same (in this case) recessive genetic defect although different alleles for a given gene may give a different spectrum of phenotypes.

An assigned gene symbol should only have one associated description. Thus, for a given three-letter symbol, all genes designated with that

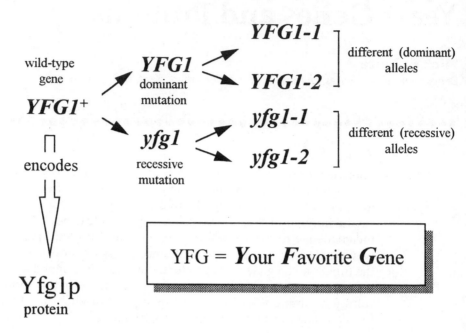

Figure 1. The nomenclature for genes of *Saccharomyces cerevisiae*, their mutant alleles and encoded polypeptide. *YFG1* is a fictitious gene (**Y**our **F**avorite **G**ene).

symbol should have a related phenotype, gene product or gene function. For example, although the *ADE1* and *ADE2* genes encode different enzymes, these enzymes are involved in adenine biosynthesis and mutations in these genes cause the same phenotype: in this case a defect in adenine biosynthesis.

The ability to create a specific, large-scale disruption of a given gene sequence by recombinant DNA technology has led to the establishment of two other types of allele designation:

- *yfg1-Δ1*: denotes a partial or complete deletion of the *YFG1* gene;
- *yfg1::URA3*: denotes the insertion of (in this case) a functional *URA3* gene into the *YFG1* gene thereby generating a non-functional, recessive allele of the latter.

There are of course several accepted exceptions to the above rules although these are relatively small in number. The reader is referred to Sherman (1991) for a discussion of these exceptions.

Where an open reading frame (ORF) has been identified by the yeast genome project, but for which no function has yet been assigned, the normal practice is to retain the standard ORF designation. Such designations have the following format, e.g. YBR086c: Y = yeast; B = chromosome number (A = chromosome I, B = chromosome II, etc.); R = right arm (L = left arm) 086 = designated ORF number; c = which DNA strand the ORF is on (c = the Crick strand, w = the Watson strand). In identifying such ORFs, those encoding proteins less than 100 amino acids in length are usually discarded unless there is an indication of amino acid identity to a known protein.

B. Phenotypes

Where a mutation in a given gene gives a phenotype that is consistent with the designated gene name, then the phenotypes can be denoted by cognate symbols in Roman type together with either a superscript + or –. For example, a mutation in the *ADE2* gene leads to a requirement for adenine to be exogenously provided owing to a defect in the adenine biosynthetic pathway, i.e. adenine auxotrophy. The respective phenotypes are designated thus: Ade$^+$, for the wild-type prototrophic strain; and Ade$^-$, for the mutant auxotrophic strain.

C. Extrachromosomal Gene Nomenclature

There are several genetic elements in yeast that are not inherited according to Mendel's rules. These include mutations in the mitochondrial genome which lead to a defect in respiration and thus an inability to use non-fermentable carbon sources (Grivell, 1994). In addition, there are several small cytoplasmic genomes which have a very limited coding capacity and which may or may not have an effect on phenotype. These include the 2μ plasmid (Williamson, 1991) and the various viral RNA-like genomes associated with the killer phenotype (Wickner, 1996). The recent discovery of non-nucleic acid-based genetic determinants, namely the [*PSI*] and [*URE3*] prions (Tuite and Lindquist, 1996) also requires a nomenclature to distinguish them from cytoplasmic genes.

The recommended nomenclature for "non-Mendelian" genes is to enclose the gene designations in square brackets; a summary is given in Table 1.

D. Protein Names

Where the precise function of an encoded gene product is known then the standard enzyme/protein name can be used. However, where an encoded protein has no known function or cannot be assigned a specific enzymatic/structural activity, then the protein nomenclature shown in Figure 1 should be used. For example, while the product of the *GCN4* gene is an extremely well-characterized DNA-binding protein that regulates the transcription of a large number of amino acid biosynthetic genes, it is usually referred to simply as Gcn4p.

Table I. The nomenclature for non-Mendelian genetic elements in *Saccharomyces cerevisiae*

Genetic element	Nomenclature[a]	Phenotype
Mitochondrial DNA	[*rho*⁺]	Wild-type
	[*rho*⁻]/[*rho*°]	Defective respiration
Double-stranded RNA	[*KIL*-k₁]	Wild-type, secretes killer toxin
	[*KIL*-o]	Sensitive to killer toxin
2 μm-circle plasmid	[*cir*⁺]	Contains plasmid
	[*cir*°]	No plasmid (no change in phenotype)
Sup35p prion protein	[*PSI*⁺]	Prion-form of Sup35p, low efficiency translation termination
	[*psi*⁻]	No prion, normal translation termination (wild-type?)
Ure2p prion protein	[*URE3*]	Prion-form of Ure2p, altered regulation of N₂ metabolism
	[*ure3*]	No prion, normal regulation of N₂ metabolism (wild-type?)

[a] Use of o indicates loss of the corresponding genome, e.g. [*rho*°] mutants have no mitochondrial genome, [*KIL*-o] mutants do not contain the M double-stranded RNA genome.

E. Reserving a Gene Name

With the ever decreasing number of informative three letter names for *S. cerevisiae* genes it is important that as soon as a function is assigned to a gene, that the chosen gene name is reserved. In order to do this, and to avoid the problem of gene-naming conflicts, all new gene names should be registered with the *S. cerevisiae* genome database (SGD) at the University of Stanford, California, USA. The SGD maintains a complete list of all *S. cerevisiae* gene names in the so-called "Gene Name Registry" and have taken on the important role of registering new names and ensuring that the gene names assigned are unique. They also resolve any issues of conflict over gene nomenclature.

The onus is on the individual researcher to identify an appropriate unique gene name and then to contact the SGD *prior* to publication of the name. The SGD ensures that the gene name has not been reserved by another researcher and then registers the name for up to 6 months. If, at the end of 6 months, the researcher fails to provide evidence for continuing work on the gene in question, the gene name will then become freely available for other researchers to reserve and adopt.

All communications relating to reserving a gene name should be sent by electronic mail to yeast-curator@genome.standford.edu. Further information on how to reserve a gene name with the SGD can be found on the SGD web site (see Appendix III).

References

Grivell, L. A. (1984). In *Genetic Maps* (S. J. O'Brien, ed.), pp. 234–240. Cold Spring Harbor Laboratory, Cold Spring Harbor, New York.

Sherman, F. (1991). Getting started with yeast. *Methods Enzymol.* **194**, 3–21.

Tuite, M. F. and Lindquist, S. L. (1996). Maintenance and inheritance of yeast prions. *Trends Genet.* **12**, 1467–1471.

Wickner, R. B. (1996). Double-stranded RNA viruses of *Saccharomyces cerevisiae*. *Microbiol. Rev.* **60**, 250–265.

Williamson, D. H. (1991). Nucleus: chromosomes and plasmids. In *The Yeasts, Vol. 4: Yeast Organelles* (A. E. Wheals, A. H. Rose and J. S. Harrison, eds), pp. 433–488. Academic Press, London.

Appendix IV

Index

Page numbers in *italic* refer to illustrations and tables.

Multicopy suppressors, in genetic screening, 277–280
Mutagenesis, insertion mutations, 162
  PCR mutagenesis, 91, *92*
  see also Transposon mutagenesis; Mini–Mu transposon; Ty mutagenesis; Deletion mutants; Gene targeting
Mutant proteins
  in lipid modification testing, 245
  non-phosphorylatable, in function testing, 243–244
  see also Deletion mutants; Yeast mutant/plasmid collections
MYO5 gene, identification, *MYO3* inactivation in, 285
Myristoylation, 244–245

National Collection of Yeast Cultures (NCYC), 445
NCBI, in human homolog searching, 419
*Nif*S gene, 4
NMT1 gene, in myristoylation, 244
Nomenclature
  genes
    chromosomal, 485–487, *486*
    extrachromosomal, 487, *488*
    name registration, 488
  phenotypes, 487
  proteins, *486*, 487
    ribosomal, 463, *464–469*
Non-allelic non-complementation, 285–287
Nonsense error assays, 365–367, *366*
Northern analysis, 121–123, *124–125*
  protocols, *122, 123*
Novel genes, 3–5, 36–40
Nrf1, 423

Olf1, 423
OMIM, 419
One hybrid assay, for DNA-binding proteins, 421–423, *422*
Open reading frames, 36–40
  see also Bioinformatics of yeast genome
Orphan genes *see* Novel genes
Optical density for growth monitoring, 475–476
Orthologs, human, of yeast genes, 417–418
Osmolarity, and MAP kinase signaling pathways, *342*, 345
Osmosensing pathway, 389–390
  Sln1p-Ssk1p, 382–383
Osmostress, 338
Osmotic sensitivity assay, 389
Oxidative stress, genes, 340–341

Poly(A) binding protein (PABP), 353
PACT2 plasmid, activation domain vector, 260, *261*

PAGE, in protein synthesis study, 355–356
Palmitoylation, 245
Paromomycin, sensitivity mutations, 363
PAS2 plasmid, DNA-binding domain vector, 258, *259*
PCK1 promoter, for *Candida albicans*, 411
PCR
  competition experiment quantitation, 316
  in deletion mutant generation, 5
  in genetic footprinting, 112–113, *113*
    in data interpretation, 114–115
  for mutagenesis, 91, *92*
  reverse transcriptase, for mRNA quantitation, 120
  vectorette, 175–177, *175, 176*
  see also Gene targeting, PCR-based
PEG *see* Polyethylene glycol
PEG202 plasmid, DNA-binding domain vector, 258, *259*
Peptides, synthetic, as antigens, 203–204
Perchloric acid quenching, 318, 319
PFA6a plasmids
  heterologous modules, 75–79, *76–77*
  pFA6a-*kanMX4*, 68, *69*, 80
PGAD424 plasmid, activation domain vector, 260, *261*
PGBT9 plasmid, DNA-binding domain vector, 258, *259*
PGI1 deletion, 279
PGK1/lacZ assay, for nonsense error assay, 365–366, *366*
PH domains, 379
Phenotypes
  analysis, quantitative, 6–7
  genetic screening, 282
  nomenclature, 487
Pheromones
  resistance, Ty mutagenesis analysis, 105
  response pathway assays, 386
Phosphatases
  in phosphorylation, 240–241
  protein, and phosphorylation state, 243
Phosphoaminoacid analysis, 241–242
Phosphofructokinase (PFK)
  in metabolic control analysis *see* Metabolic control analysis
Phosphorylation of intracellular proteins, 240–244
  determining, 240–241
  and protein function, 243–244
  residues phosphorylated, 241–242
  sites, 242
Photolithography in hybridization array technology, 132
PJG4–5 plasmid, activation domain vector, 260, *261*
Pkc1 mutants, 382
Plasmids

Two-hybrid protein interaction assay (*cont.*)
  see also One hybrid assay; Three hybrid
    assay
Ty mutagenesis, 101–117
  example uses, 105–106
  in genetic footprinting analysis, 110–116
    data interpretation, 114–116
    mutagenesis induction, 110–111, *111*
    oligonucleotide design, 113–114
    PCR analysis, 112–113, *113*
    rationale, 110
    selections, 112
  insertional
    his3-AI/ade2–AI indicator genes, 102–105
    marked pGTy elements, 105
  problems with, 109–110
  technique development, 101–102
  in yeast artificial chromosome
      fragmentation, 106–108
    protocols, *106, 107*
    strategy, 106
  see also Transposon mutagenesis; Mini-Mu
      transposon
Ty transposons, 183

Ubiquitin, protein destabilizing fusion system,
    86, 87
Ubiquitination, 247–248
Ultrasonication of DNA, in activation-domain
    library construction, 262
Unfolded protein response (UPR), 341
Unigene database, in human homolog
    searching, 419
Upstream activating sequences (UAS), 256
URA3 gene, in counterselection genetic
    screening, 271
'URA blaster' for *Candida albicans* gene
    disruption, 408–410, *409*

Vectorette PCR, 175–177
  protocol, *176*
  strategy, *175*

Web addresses *see* World wide web addresses
Western blotting
  quantitative, for antibody specificity, 205,
    206–207
  in subcellular protein localization, 209
Whole chromosome analysis, 16–31
  in *Candida albicans*, 397–400, *398, 399*
  end marking, 23–24, *24*
  fluorescent in situ hybridization, 22–23
  immunofluorescence, 22
  manipulations, 27–28

mapping
  genetic, 25, 27
  physical, 25, *26*
  pulsed-field gel, 17–22
    aneupoloid detection, 20–21, *21*
    principles, 17–18
    size changes, 22
    size variation assessment, 19
    species identification, 19–20, *20*
    techniques, 18–19
  related genomes, comparative anatomy,
    28–29
  tools
    genetic, 17
    physical, 16
World wide web addresses, 480–483
  Candida albicans, 405
  for human homolog searching, 418–420, 421
  yeast genome, 47–8

XREFdb, 48

Yap1p factor, in stress responses, 340–341
YCL017c ORF, in yeast, 4
YCpSUPEX1 expression vector, *146*
YCpSUPEX expression vector, 150
Yeast artificial chromosomes (YAC), 16
  double strand break mapping, 25, *26*
  Ty1–mediated fragmentation, 106–108
Yeast extract peptone dextrose medium (YPD),
    471
Yeast Genetic Stock Center (YGSC), 445–446
Yeast genome
  bioinformatics *see* Bioinformatics of yeast
      genome
  functional analysis *see* Functional analysis,
      yeast genome
Yeast mutant/plasmid collections, 431–432
  American Type Tissue Collection, 446
  bar-coded mutant collection, 447–448
  cognate/regulated clone collection, 444
  EUROSCARF, 435–445
    access, 444–445
    deletion cassette collection, 439, *440–444*
    establishment, 435–436
    strain collection, 436–439
  future organization, 447
  Yeast Genetic Stock Center, 445–446
Yeast protein database (YPD), 48, 379

ZDS1/ZDS2, isolation in cellular processes,
    279–80
Zinc finger proteins, Msn2p/Msn4p, and
    stress response, 340